This book presents the first comprehensive account of the properties of plasma loops, the fundamental structural elements of the solar corona. Plasma loops occur in many sizes, and range in temperature from tens of thousands to millions of degrees. They not only define the structure of individual active regions but connect different active regions – even across the solar equator. Loops also play an integral and decisive role in the enormous solar explosions called flares. Over recent years a wealth of space and ground-based observations of loops has been obtained in various widely spaced regions of the electromagnetic spectrum. In this book the authors have selected the best observational material from the literature on which to base a detailed account of the properties of flare and non-flare loops. The book also explores the larger implications of the loop structures in our understanding of solar and stellar coronae. The text is enhanced by a large number of illustrations and unique and beautiful photographs obtained from the ground and from space.

This book will be of interest to postgraduate researchers in solar and stellar physics, space scientists and plasma physicists.

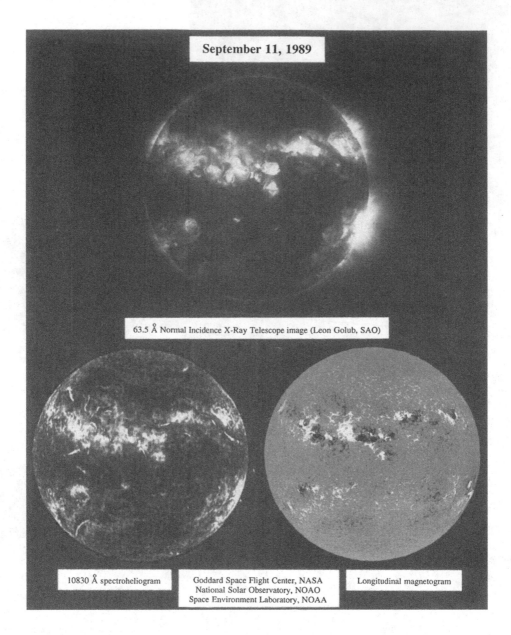

September 11, 1989

63.5 Å Normal Incidence X-Ray Telescope image (Leon Golub, SAO)

10830 Å spectroheliogram

Goddard Space Flight Center, NASA
National Solar Observatory, NOAO
Space Environment Laboratory, NOAA

Longitudinal magnetogram

Cambridge astrophysics series

Editors: R.F. Carswell, D.N.C. Lin *and* J.E. Pringle

Plasma loops in the solar corona

In this series

PLASMA LOOPS IN THE SOLAR CORONA

R.J. BRAY,
L.E. CRAM,
C.J. DURRANT
and R.E. LOUGHHEAD

The right of the
University of Cambridge
to print and sell
all manner of books
was granted by
Henry VIII in 1534.
The University has printed
and published continuously
since 1584.

CAMBRIDGE UNIVERSITY PRESS
Cambridge
New York   Port Chester
Melbourne   Sydney

CAMBRIDGE UNIVERSITY PRESS
Cambridge, New York, Melbourne, Madrid, Cape Town, Singapore, São Paulo

Cambridge University Press
The Edinburgh Building, Cambridge CB2 2RU, UK

Published in the United States of America by Cambridge University Press, New York

www.cambridge.org
Information on this title: www.cambridge.org/9780521351072

First published 1991
This digitally printed first paperback version 2005

A catalogue record for this publication is available from the British Library

Library of Congress Cataloguing in Publication data

Plasma loops in the solar corona/R. J. Bray . . . [et al.].
p.   cm. – (Cambridge astrophysics series)
Includes bibliographical references and index.
ISBN 0-521-35107-3 (hardcover)
1. Sun – Corona.   2. Sun – Loop prominences.   I. Bray, R. J.
II. Series.
QB529.P57   1991
523.7′5 – dc20 90-41704 CIP

ISBN-13 978-0-521-35107-2 hardback
ISBN-10 0-521-35107-3 hardback

ISBN-13 978-0-521-02223-1 paperback
ISBN-10 0-521-02223-1 paperback

Contents

Frontispiece courtesy of Dr L. Golub (SAO) and Dr J. Harvey (NOAO) using data obtained by SAO, NOAO/Kitt Peak, NASA/GSFC, NOAA/SEL and IBM Research.

Preface

The discovery that a significant part of the energy emission from the solar corona is concentrated along well-defined curved paths – called *loops* – represents a major advance in our understanding of the Sun. Such plasma loops are the basic structural elements of the corona, particularly in and over active regions. Moreover, they play a decisive role in the origin and physics of solar flares. Our new insight is due largely to the wealth of space observations of the Sun obtained, in particular, from the manned satellite Skylab (1973–4) and the unmanned satellites Solar Maximum Mission and Hinotori which followed. Ground-based observations in the visible and microwave regions of the electromagnetic spectrum have also played a vital role. The literature on coronal plasma loops is vast and includes not only hundreds of research papers but also the proceedings of numerous symposia and workshops. This book presents for the first time a comprehensive, unified and well-illustrated account of the properties of coronal loops based on the best space and ground-based observations currently available (Chapters 2–4). A magnetohydrodynamic analysis of the stability and dynamics of loops is presented in Chapter 5, while the final chapter (Chapter 6) explores the wider implications of the loop regime on our understanding of both the solar corona and stellar coronae.

The history of the observation of coronal loops (Chapter 1) is marked by a number of significant events over a period ranging from the earliest description of a solar prominence in 1239 to the launch of Skylab in 1973. This latter date may reasonably be taken as the start of the modern era of coronal loop observations. Well-defined loop structures were first seen as prominences at total eclipses of the Sun. Clear descriptions of loop prominences based on non-eclipse observations were given in the 1870s by A. Secchi and C.A. Young. By the 1950s, thanks to the widespread distribution of birefringent filters tuned to the Hα line, extensive observations of 'cool' loops were available. By 1967 the distinction between (cool) flare and non-flare loops was well recognized. However, to observe '*hot*' loops (temperature $> 10^6$ K) in visual region lines such as $\lambda 530.3$ (nm) of Fe XIV and $\lambda 637.4$ of Fe X required the development of the coronagraph. In fact its inventor, B. Lyot, was the first observer to obtain good limb photographs in these lines. In 1971 R.B. Dunn at Sacramento Peak Observatory remarked that many coronal scenes in $\lambda 530.3$ consist exclusively of loops and 'arches' – a conclusion soon to be thoroughly verified by observations from Skylab.

To observe the corona on the disk required space vehicles. The rocket programs were highly successful leading, for example, to the discovery in 1968 of X-ray loops connecting different active regions. By 1974 G.E. Brueckner and J.-D.F. Bartoe were able to conclude from rocket observations in the extreme ultraviolet (EUV) that *all coronal EUV emission is concentrated either in loops or in their footpoints*. On the other hand, observations from the pre-Skylab solar satellites OSO 1 to OSO 7 (1962–71) usually did not have adequate spatial resolution to reveal the loop structure of active regions and flares. By contrast, EUV and soft X-ray observations obtained from Skylab have provided much of our knowledge – described in this book – of the properties of non-flare coronal loops. Skylab was the first manned solar satellite observatory and the first to make extensive use of photography in space observations of the Sun.

Cool non-flare loops (Chapter 2) range in temperature from $\sim 1 \times 10^6$ K down to $\sim 20\,000$ K, the lower part of the range being characteristic of observations made in the Hα and Lα lines of H I. The application of tunable narrow-band birefringent filters for the Hα line has greatly expanded our quantitative knowledge of the properties of cool loops, including their physical conditions and their morphological and dynamical properties. Observations at the Culgoora Solar Observatory by two of the present authors and their colleagues achieved a spatial resolution better than $1''$ of arc, which is significantly superior to the resolution so far achieved from space in the EUV and X-ray spectral regions and from the ground in the microwave region. Moreover, disk observations in Hα have given us the most complete account to date of the physical conditions in a single loop, flare or non-flare, but so far only for one specimen. On the other hand, the full three-dimensional loop structure of active regions only becomes apparent when we scrutinize the Sun in EUV lines hotter than those of He II. Our knowledge – which is extensive – comes mainly from Skylab spectroheliograms and spatially-resolved spectrograms obtained in such lines. The footpoints of both Hα and EUV loops are found to be located on sunspots or on a single sunspot and an area of opposite magnetic polarity. The material motions in a cool loop may be strongly supersonic, with non-gravitational forces operating on the loop material. Regardless of temperature all cool loops have similar properties, although the diameter (thickness) increases with temperature.

Hot non-flare loops (Chapter 3) range in temperature from $> 1 \times 10^6$ K to $\sim 3 \times 10^6$ K. They are observed in appropriately hot EUV lines, soft X-rays (line + continuum) and in the microwave region of the radio spectrum. Hot loops are thicker, longer, higher and longer-lived than cool loops, but their other morphological and physical properties (with the exception of temperature) are similar. Hot and cool loops are not co-spatial although they may lie close together. Hot EUV emission is confined largely to active regions and to the loops emanating from them. The enhanced EUV emission of a loop indicates that its density is higher than that of its surroundings. Frequently, hot EUV loops link two or more active regions in a huge magnetic complex, even across the solar equator. Individual EUV loops join regions of opposite magnetic polarity, but observations of higher resolution are needed to resolve central portions of active regions into (EUV) loops.

In soft X-rays, active regions are seen to be made up of loops arching between areas of opposite magnetic polarity. Several classes of soft X-ray loops can be distinguished, including so-called 'interconnecting' loops which join different active regions and play a role in the evolution of complexes of activity. These loops terminate in areas of enhanced magnetic field near the edges of active regions, *not* in sunspots. So-called 'quiet region' loops are located over filament cavities and are rooted in the chromospheric network. They can extend up to heights of 500 000 km, i.e. to over 70 % of the solar radius. X-ray observations of loops are of great importance: they give information about loops *joining* active regions and they give insight into how the associated magnetic fields emerge, evolve, fragment and disperse. Radio loops are believed to be the radio counterparts of X-ray loops. First observed in 1980, they also offer a unique capability since they provide a potential means of observing the strength and direction of the vector magnetic field. Radio emission from hot loops is due to thermal bremsstrahlung or gyroresonance radiation, or both. They are best mapped at a wavelength of 20 cm (frequency = 1.5 GHz): in fact, at this wavelength emission from the entire loop is routinely detected. The spatial resolution so far achieved with such observations is 3–13″.

Skylab was never intended as a *flare* mission since its launch in 1973 took place during the declining phase of the rather weak solar cycle which peaked in 1968–9 (cycle 20). Nevertheless, flare observations from Skylab have made a considerable contribution to our understanding of the role of loops in flares. The Solar Maximum Mission satellite (SMM) (launched in 1980), on the other hand, has been described as the most powerful package ever flown for studying solar flares. SMM, together with the Japanese satellite Hinotori (launched in 1981) operated during times of high solar activity and observed all phases of numerous flares. Together they have furnished a wealth of EUV and hard and soft X-ray data which have laid the foundations for a systematic study of flare loops. The results of all this work are described in Chapter 4.

Important information about flare loops has come from no less than five regions of the spectrum, viz., visible, EUV, soft and hard X-ray, and microwave. Properly coordinated and simultaneous multi-wavelength observations are of course much less common than observations obtained in only one or two wavelength regions. Furthermore, in any given wavelength region it often happens that a particular set of observations fails to reveal emission from the *whole* of a flare loop or loop system. For example, hard X-ray images showing emission extending over an entire loop are rare. Nevertheless, consistent interpretations of such observations are usually possible and these are invariably based on the hypothetical involvement of a loop or loops. The primary energy release in a flare is believed to occur at or near the top of a loop and modern interpretations of the flare event now enable us to distinguish between the primary phenomenon and the complex secondary and tertiary phenomena characteristic of flares.

Disk and limb observations in the Hα line have provided detailed knowledge of the properties of *cool* flare loops ($T \simeq 20\,000$ K). But three factors combine to make the EUV line observations of greater importance: (a) the EUV flare lines encompass almost the full range of temperatures encountered in flares, $20\,000$ K-2×10^7 K; (b) the spatial resolution of the best EUV spectroheliograms is much higher than that attained by any X-ray observations up to the present; and (c) a more complete picture can be obtained by combining images taken simultaneously in lines spanning a wide temperature range. Furthermore, EUV observations have provided information on flare loops during all four phases of a flare: pre-flare, rise, maximum and decay. However, to probe the ultra-hot regions of a flare, $T > 2 \times 10^7$ K, one must turn to X-ray observations where, in fact, those at the limb have provided clear evidence of the loop structure of X-ray flares. One Skylab astronaut (E.G. Gibson) remarked that 'one or more well defined loops were the only structures of flare intensity [in soft X-rays] observed during the rise phase and near flare maximum'. Hard X-ray emission from flares also involves a loop mechanism, the frequent location of hard X-ray bursts at the footpoints of an (assumed) loop being attributed to thick target bremsstrahlung from fast electrons precipitating downwards along the legs of a magnetic flux loop into the dense chromospheric layers.

Finally, observations of flares in the microwave region of the radio spectrum enable us closely to approach, in space and time, the primary flare event – the injection of energetic electrons into a magnetic flux loop at or near its apex. The microwave emission is due to gyrosynchrotron radiation from highly energetic electrons trapped near the top of a loop. The energy band of the non-thermal electrons believed to be responsible is 100–1000 keV; they are considerably more energetic than those thought responsible for the hard X-ray emission. Observations with the Very Large Array in New Mexico and other radio interferometers have given us a reasonably clear picture of the properties of microwave bursts and of their relationship to the loops in which they occur.

Simultaneous observations of flare loops at different wavelengths have provided a considerable body of reliable information about various significant spatial and temporal correlations. Some of these are remarkably close. Pairs of wavelengths for which useful data exists include Hα and Fe XIV λ530.3, EUV and Hα, hard and soft X-rays, Hα and hard X-rays, Hα and microwave, EUV and hard X-rays, and hard X-rays and microwave. Our observational account of flare loops in Chapter 4 concludes with a summary of their morphological, dynamical and physical properties and a comparison with the corresponding numerical data for non-flare loops.

In Chapter 5 we turn to a consideration of the physics of coronal loops from a theoretical standpoint. Until 1978, models of the corona and the transition region had utilized almost exclusively plane-parallel geometry, the physical properties depending only upon height above the photosphere. The physical processes operating at these levels were well known at this stage. Heat is predominantly carried out of the hot corona by conduction down the temperature gradient into the lower atmosphere, where the cooler and denser material is able to radiate the heat away. In the corona, the thermal conductivity is very high and the temperature gradients are correspondingly small. But the thermal conductivity drops rapidly with temperature so that an ever-steepening temperature gradient is required in the transition region in order to maintain the necessary conductive heat flux. An extended, almost isothermal, corona and a thin transition region of high temperature gradient result. The only major problem was the source of the heat that caused the coronal plasma to rise to temperatures in excess of 10^6 K in the first place.

In 1978, two sets of authors – R. Rosner, W.H. Tucker & G.S. Vaiana and I.J.D. Craig, A.N. McClymont & J.H. Underwood – analysed how these physical processes would operate in a loop geometry. Assuming that the plasma loops delineate magnetic flux loops, the strength of the field will ensure that there is generally little exchange of mass, momentum and energy between neighbouring flux loops in active regions. They therefore modelled an isolated loop in which the physical properties varied along the axis. Models of closed loops are constrained by the fact that they must be matched to conditions in the lower atmosphere at *both footpoints*. This constraint leads to relationships, known as scaling laws, between the global parameters describing the loops. Scaling laws allow the gross characteristics of static and dynamic hot loops to be understood in quite simple terms. However, these loop models fail to account for the observed dynamic behaviour of transition region material and for the existence of large-scale loops with temperatures in the range 10^5–10^6 K.

The dynamical properties of coronal loops have always been a major concern. They play a crucial role in ensuring the thermal stability of hot loops. A static loop is thermally unstable but it was soon recognized that the dynamical coupling between the coronal loop and its chromospheric footpoints allows the loop to fill and drain as the changing energy budget requires. Transverse motions of loops allow mechanical disturbances to propagate throughout the magnetized system. These disturbances assume the form of waves associated with fluctuating electric currents if the time scales are short enough, but appear more like quasi-steady direct currents if the time scales are long. The extreme forms provide the basis for rival theories of coronal heating, a subject which still remains unresolved. Whether transverse motions of loops may also lead on occasion to a significant exchange of mass and energy between neighbouring loops is also unresolved at present.

Finally, in Chapter 6 we turn to the wider aspects of the loop model of the coronae of the Sun and stars, recognizing that the behaviour of coronal loops and loop systems is determined ultimately by processes occurring in the invisible solar and stellar interiors. Chief among these are the generation of magnetic fields by regenerative dynamo action, the formation of time-dependent, inhomogeneous 'ropy' magnetic topology by turbulent magneto-convection, and the extrusion of magnetic flux into the visible atmosphere by magnetic buoyancy and other phenomena. The bewildering complexity of the observed coronal response to these processes has been documented in previous chapters. Here, we review the theory of these processes and of the coronal response to them. We conclude that, despite a great deal of imaginative and thorough work, many uncertainties remain regarding the fundamental mechanisms involved in the evolution of the magnetic topology of the corona, and in the generation and conversion of mass and energy flow in coronal structures. It is emphasized that further progress must recognize that coronal loops cannot be understood in isolation: investigation of the global electrodynamic coupling promises to elucidate some of the more pressing and difficult problems.

The fruitfulness of the synergy between solar and stellar physics lies in the fact that the Sun lets us study phenomena in detail, while stars offer an opportunity to investigate a wider range of physical conditions. Over the past decade or so, the successful flights of several EUV and X-ray satellites combined with increasing sensitivity and flexibility of ground-based radio and optical instrumentation have provided a fascinating glimpse of the coronae of many stars. Inferences based on this work and on studies of the Sun lead to the conclusion that plasma loops may be a ubiquitous constituent of stellar coronae. Moreover, in some stars these loops occur on a much grander scale than the loops seen in the solar corona. The existence of different levels of coronal activity among stars is evidently related to such fundamental stellar properties as the mass, age and angular momentum. There are excellent prospects for improving our understanding of the behaviour of plasma loops in the coronae of the Sun and the stars, and in other astrophysical contexts, by exploring these relations further.

The broad scope of this volume has necessitated a significant division of labour between the authors. Chapters 1 to 4 were principally the responsibility of Bray and Loughhead; Chapter 5 of Durrant and Chapter 6 of Cram.

The question of physical units is a difficult one in a volume covering both observational astronomy and theoretical astrophysics. The cgs system is used in most reports pertaining to astronomical observations. On the other hand, the SI system has become more common in theoretical astrophysics. Thus we have chosen to use cgs units in Chapters 1–4, which are principally observational, and SI in Chapters 5 and 6, both of which contain a good deal of astrophysical electromagnetic theory. The following conversions should be noted:

SI–cgs Conversions

$1 \, cm = 10^{-2} \, m$	$1 \, m = 10^2 \, cm$
$1 \, cm^2 = 10^{-4} \, m^2$	$1 \, m^2 = 10^4 \, cm^2$
$1 \, erg = 10^{-7} \, J$	$1 \, J = 10^7 \, erg$
$1 \, dyne = 10^{-5} \, N$	$1 \, N = 10^5 \, dyne$
$1 \, dyne \, cm^{-2} = 10^{-1} \, Pa$	$1 \, Pa = 10 \, dyne \, cm^{-2}$
$1 \, erg \, s^{-1} = 10^{-7} \, W$	$1 \, W = 10^7 \, erg \, s^{-1}$
$1 \, statamp = 3.33565 \times 10^{-10} \, A$	$1 \, A = 2.99792 \times 10^9 \, statamp$
$1 \, statvolt = 299.792 \, V$	$1 \, V = 3.33565 \times 10^{-3} \, statvolt$
$1 \, Mx = 10^{-8} \, Wb$	$1 \, Wb = 10^8 \, Mx$
$1 \, G = 10^{-4} \, T$	$1 \, T = 10^4 \, G$

R.J.B., L.E.C., C.J.D., R.E.L.

Acknowledgements

We wish to thank the Librarian of the National Measurement Laboratory, Ms Robin Shelley-Jones, and her staff for their whole-hearted assistance. We also thank Mr R.M. Rattle for his skillful preparation of the figures, many of which are based on photographs obtained at the CSIRO Solar Observatory, Culgoora.

Overseas colleagues, with characteristic generosity, have supplied a number of significant photographs, for which we are grateful to the following persons and organizations:

> J.M. Beckers and R.N. Smartt, National Solar Observatory (Fig. 4.2)
>
> V. Bumba and J. Kleczek, Ondrejov Observatory (Fig. 2.1)
>
> C.-C. Cheng, U.S. Naval Research Laboratory (Figs. 4.4, 4.5, 4.7, 4.10)
>
> K.P. Dere, U.S. Naval Research Laboratory (Fig. 3.3)
>
> R.F. Howard, National Solar Observatory, and Z. Svestka, Laboratory for Space Research (Fig. 3.9)
>
> G. Hurford, California Institute of Technology (Fig. 4.18)
>
> A.S. Krieger, Radiation Science Inc. (Figs. 3.11, 3.12)
>
> M. McCabe, Institute for Astronomy, Hawaii (Fig. 4.1)
>
> A. Ray, Tata Institute of Fundamental Research (Fig. 3.13)
>
> N.R. Sheeley, U.S. Naval Research Laboratory (Figs. 2.10, 2.11, 2.15, 3.2, 3.4, 3.5)
>
> T. Velusamy, Radio Astronomy Centre, Ootacamund (Fig. 4.17)
>
> K.G. Widing, U.S. Naval Research Laboratory (Figs. 4.6, 4.20).
>
> Dr J.W. Harvey provided, on behalf of NSF/NOAO, NASA/GSFC and NOAA/SEL, the superb magnetogram displayed as Fig. 6.2.

For permission to reproduce copyright material we are indebted to

Harvard University Press, D. Reidel Publishing Co.; the Editors of *Annual Review of Astronomy and Astrophysics*, *Astronomy and Astrophysics*, *Astrophysical Journal*, *Monthly Notices of the Royal Astronomical Society*, *The Observatory*, *Proceedings of the Astronomical Society of Australia*, *Science*, and *Solar Physics*; and the following individuals: T.R. Ayres, P.V. Foukal, T. Gold, B. Haisch, P. Hoyng, A.S. Krieger, M.R. Kundu, R. Pallavicini, V. Pizzo, R. Rosner, F. Walter and G. Withbroe.

For observational and instrumental support at Culgoora extending over many years we wish to thank – amongst others – Mr G.J. Blows, Mr E.J. Tappere and Mr J.G. Winter.

This work was supported in part by the Australian Research Council and the Science Foundation for Physics within the University of Sydney.

Finally, we thank the CSIRO Executive for supporting a research program in optical solar astronomy from the mid-1950s to the mid-1980s.

R.J. BRAY
R.E. LOUGHHEAD
Commonwealth Scientific and Industrial Research Organization, National Measurement Laboratory, Sydney, Australia

L.E. CRAM
University of Sydney, School of Physics (Astrophysics), Sydney, Australia

C.J. DURRANT
University of Sydney, Department of Applied Mathematics, Sydney, Australia

Other books by the authors

Sunspots
R.J. Bray & R.E. Loughhead (London: Chapman & Hall, 1964: republished by Dover Publications, Inc., 1979)

The Solar Granulation
R.J. Bray & R.E. Loughhead (London: Chapman & Hall, 1967)

The Solar Chromosphere
R.J. Bray & R.E. Loughhead (London: Chapman & Hall, 1974)

The Solar Granulation
2nd ed., R.J. Bray, R.E. Loughhead & C.J. Durrant (Cambridge: Cambridge University Press, 1984)

The Physics of Sunspots
L.E. Cram & J.H. Thomas (eds.) (Sunspot: Sacramento Peak Observatory, 1981)

FGK and T Tauri Stars
L.E. Cram & L.V. Kuhi (Washington: NASA; Paris: CNRS, 1989

Illustrated Glossary for Solar and Solar-Terrestrial Physics
A. Bruzek & C.J. Durrant (eds.) (Dordrecht: D. Reidel Publishing Co., 1977)

The Atmosphere of the Sun
C.J. Durrant (London: Adam Hilger, 1987)

Abbreviations

AC Aerospace Corporation
ACRIM active cavity radiometer irradiance monitor (SMM)
AOSO Advanced Orbiting Solar Observatory
AS&E American Science and Engineering
BCS bent crystal spectrometer (SMM)
BMR bipolar magnetic region
CD column density
CNRS Centre National de la Recherche Scientifique
C/P coronagraph/polarimeter (SMM)
CSIRO Commonwealth Scientific and Industrial Research Organization
EUV extreme ultraviolet
FCS flat crystal spectrometer (SMM)
FLM hard X-ray monitor on Hinotori
GRE gamma-ray experiment (SMM)
GSFC Goddard Space Flight Center
HCO Harvard College Observatory
HRTS High Resolution Telescope and Spectrograph (NRL rocket)
HXIS hard X-ray imaging spectrometer (SMM)
HXM hard X-ray monitor on Hinotori
HXR hard X-ray
HXRBS hard X-ray burst spectrometer (OSO 5, OSO 8 and SMM)
ISEE-3 International Sun Earth Explorer 3
LDE long-decay enhancement
MSFC Marshall Space Flight Center
NASA National Aeronautics and Space Administration
NOAA National Oceanic and Atmospheric Administration
NRAO National Radio Astronomy Observatory
NRL Naval Research Laboratory
OAO Orbiting Astronomical Observatory

OSO   Orbiting Solar Observatory
OVRO   Owens Valley Radio Observatory
PVO   Pioneer Venus Orbiter
SGR   solar gamma-ray detector (Hinotori)
SMM   Solar Maximum Mission
SOX   soft X-ray crystal spectrometer on Hinotori
SXRP   soft X-ray polychromator (SMM)
SXT   hard X-ray imaging telescope on Hinotori
UV   ultraviolet
UVSP   ultraviolet spectrometer and polarimeter (SMM)
VLA   Very Large Array
VLBI   very long baseline interferometry
WSRT   Westerbork Synthesis Radio Telescope
X-REA   X-ray event analyser
XRP   (soft) X-ray polychromater (SMM)

1

Historical introduction

1.1 Eclipse observations of prominences: Middle Ages to 1868

Historically, visual observations of prominences were the first to reveal the existence of well-defined loop structures arching upwards from the surface of the Sun high into the overlying corona. Regular visual observations of prominences obtained during total eclipses of the Sun date from 1842, but sporadic reports go back to the Middle Ages. Most of the early descriptions of the eclipses refer only to the corona but there are specific references to prominences in medieval Russian chronicles. For example, in the first volume of his well-known book *Le Soleil* (1875, p. 330) the Italian astronomer Father Angelo Secchi (1818–78) cited the description of a prominence observed at the eclipse of 1239. According to Secchi, this was the most ancient eclipse for which a detailed description of the corona was then extant. Early eclipses were extensively described by the historian R. Grant in his *History of Physical Astronomy* (1852). Grant concluded, pessimistically but realistically, that 'Down to the beginning of the eighteenth century, the accounts respecting total eclipses of the sun, contain very few remarks which are of advantage in forming the basis of any physical enquiry'.

By contrast, the arrival of the eighteenth century saw a growth in the spirit of exact enquiry in many branches of science. The eclipse of 1733 was noteworthy for the zeal with which a number of Lutheran pastors in the rural districts of Sweden responded to a call by the Royal Swedish Academy to observe the various phenomena. Their observations were collected together and methodically arranged by the Swedish astronomer (and inventor of the centigrade temperature scale) A. Celsius (1701–44). They were regarded by Grant as forming the most complete description of a total eclipse observed prior to that of 1842. Observations of three or four small prominences seen during the eclipse were communicated by the astronomer B. Vassenius to the Royal Society of London. He described them as reddish

1

clouds entirely detached from the lunar limb and floating – as he supposed – in the lunar atmosphere. His prominence drawings are reproduced in Secchi's book (*loc. cit.*, Fig. 131).

The eclipse of 1842 marked the beginning of the serious study of the phenomena observable during a total eclipse of the Sun and, in particular, signalled the start of regular (eclipse) observations of prominences. This eclipse, whose track passed through Austria, northern Italy, and the southern provinces of France, Germany and Russia, was noteworthy for the number of experienced astronomers who sought to observe it: these included Airy, Arago, Baily, Littrow and Struve. The accounts by the various observers (quoted by Grant) clearly demonstrated that a prominence had to be regarded as a true solar phenomenon. One astronomer, the Italian Santini, even adduced a connection with *sunspot groups*, having noticed an extensive chain of sunspots near the western limb prior to the eclipse, not far removed from prominences observed during the eclipse itself.

Nevertheless, it was not until after the eclipses of 1850 (Honolulu) and 1851 (Norway and Sweden) and the discussions and comparisons of observations which followed that astronomers generally became satisfied that prominences were real solar phenomena. During the latter eclipse, a professional photographer named Berkowski succeeded in obtaining a daguerrotype of a few prominences, together with the inner corona. However, the first systematic application of photography took place during the total eclipse of 1860 (Spain) as a result of the independent efforts of Secchi and the well-known British amateur astronomer Warren de la Rue (1815–89), working at two sites separated by some 400 km. In the first volume of his book (*loc. cit.*, pp. 377 *et seq.*) Secchi described at length the prominence photographs obtained by himself and de la Rue and reproduced engravings of several of these photographs. The same volume contains numerous drawings of prominences by various observers obtained at earlier eclipses. Mostly these drawings show the prominences as mere blobs, with little or no structure. However, the drawings based on Secchi's own photographs show more detail: one of them depicts what appears to be one-half of a single loop prominence ('K' in Secchi's Fig. 134) – called by Secchi a 'horn'.

The eclipse of 1868 (India and Malacca) must be counted as a watershed in the history of solar physics since it marked (a) the first employment of spectroscopy, and (b) the discovery of the 'open-slit' method for observing prominences and the chromosphere, obviating the need for an eclipse. Credit for this discovery belongs to P.J. Janssen (1824–1907) (who attended

Fig. 1.1. Loop system observed by Young in the Hα line using the open-slit method on 5 October 1871. In his book *The Sun* (1895) Young described such loops as '. . . jets of liquid fire, rising and falling in graceful parabolas'.

the eclipse) and J.N. Lockyer (1836–1920) (who did not). The circumstances leading to this crucial advance in technique have been described by two of the present authors in their book *The Solar Chromosphere* (1974) and need not be repeated here. Following this advance, it was no longer necessary to wait for the fleeting and comparatively rare moments of a total eclipse to study the properties of prominences. The new technique was soon put into use at a number of observatories, prominent among its practitioners being Secchi himself and his Italian colleagues, together with Lockyer, Young, Spörer, Zöllner, Rayet and Wolf. As a result of the application of the open-slit method, together with visual spectroscopy using a narrow slit in the normal way, the 1870s witnessed a great increase in knowledge of prominences in general and of prominence loops in particular.

1.2 Visual and spectroscopic observations of prominence loops: Secchi and Young

Clear recognition of the existence of loop prominences emerged during the 1870s from the observations of a number of workers, most notably C.A. Young (1834–1908) of Princeton University and, of course, Secchi. These two assiduous observers left clear and detailed descriptions of their observations and conclusions. It was quickly realized that prominences could be classified into two main types, *quiescent* and *active*, the latter being called *eruptive* by Secchi. Sometimes the first type were called 'cloud' and the second 'flame' prominences, but all observers agreed with a division into two main classes, a distinction that has been maintained to the present day. Secchi went on to sub-divide the two main classes into a number of sub-classes, but Young pointed out that it was not always easy to distinguish between these.

The active prominences were described by Secchi as being bright and showing rapid changes. Moreover, their spectra revealed metallic lines in addition to the hydrogen and helium lines exhibited by quiescent prominences. Finally, they appeared in the immediate neighbourhood of sunspots, sometimes appearing to proceed directly from the umbra itself.

Both Secchi and Young used the word 'jet' to describe the sub-class of active prominences that we now call loop prominences. In his book *The Sun* (1895) Young described them as '...(presenting) most exactly the appearance of jets of liquid fire, rising and falling in graceful parabolas'. Secchi used the term 'parabolic arcs' as well as 'jets' to indicate loops. However, neither author actually established the true shapes of the loops, a task which has been successfully accomplished only in recent times (see Section 2.3). We can surmise that they *assumed* the shapes were parabolic since they both supposed that material was ejected from the surface and was then subjected to purely gravitational forces.

Figure 1.1 shows a drawing by Young of a well-developed loop system located near a sunspot adjacent to the solar limb and observed on 5 October 1871. This drawing gives a reasonable representation of Hα loops similar to those seen on modern photographs, but no information is given by Young that would allow us to decide whether the loops he observed belonged to a flare or a non-flare system. Other reasonably accurate representations of loop systems dating back to the same period are to be found in the second volume of Secchi's book *Le Soleil* (1877: see Plates D8, D9, E3 and G10).

Young found that for the most part eruptive prominences attained altitudes of not more than 32000–48000 km, but occasionally they rose much higher. Secchi gave a characteristic limit for 'metallic' prominences of 44000–51000 km. These figures are similar to those given in Tables 4.16 and 2.8 for flare and non-flare loops observed in the Hα line, based on modern measurements.

An important discovery resulting from visual spectroscopic observations was the dynamical nature of the prominence gases. Secchi reproduced a number of drawings showing distorted and shifted lines of hydrogen observed by Young and Lockyer. He correctly interpreted these as velocity shifts arising from the Doppler effect, which had been discovered some 30 years earlier. On the magnitude of the line-of-sight velocities, Secchi remarked that 'The largest displacements that we have seen and measured correspond to a velocity of about 30 to 90 kilometres (per second)'. On the other hand Young remarked that the velocity 'often' exceeded 160 km s^{-1} and 'sometimes, though very rarely' reached 320 km s^{-1}. The velocities found by Secchi and Young are, in fact, typical of those found in loops and flare surges respectively.

With the exception of the height and velocity measurements mentioned above, most of the new work was largely qualitative. The wavelengths of the various lines observed in the spectra of quiescent and active prominences were measured and, using laboratory data, the lines were identified with lines of known terrestrial elements. Diagnosis of the physical conditions in prominences lay far in the future, however, owing to the absence of theories of the atom and its spectral line emission.

1.3 Early observations of cool loops with the spectroheliograph, spectrohelioscope and birefringent filter

In the 1890s and following decades progress in understanding the nature of flare and non-flare loops was advanced by the invention successively of the spectroheliograph, spectrohelioscope, and birefringent filter. The observations were obtained, firstly, in the H or K lines of singly-ionized calcium (Ca II) and later, after red-sensitive emulsions became available (~1908), in the Hα line of hydrogen. Loops observed in these strong chromospheric lines belong to the category of so-called 'cool' loops. The distinction is based on the temperature of the loop plasma: loops are described as 'cool' or 'hot' depending on whether their temperature is less than or greater than about 1 000 000 K (Section 2.1).

The first successful spectroheliograph was constructed in 1892 by the American astronomer G.E. Hale (1868–1938), who later became the founder and first director of Mt Wilson Observatory, California. This instrument was attached to the 30-cm refractor of the Kenwood Observatory near Chicago, a private observatory financed by Hale's father. His next instrument, a much more powerful one, was the famous Rumford spectroheliograph, completed in 1899 and attached to the large 100-cm refractor of the Yerkes Observatory. Following Hale's move to Mt Wilson, these were followed in 1905 by a '5-foot' (1.5-m) spectroheliograph and later by a 13-foot (4-m) instrument. At Meudon near Paris a large spectroheliograph was completed in 1909 by H.A. Deslandres (1853–1948), whose first spectroheliograph dated back to 1893.

Hale and his colleague F. Ellerman (1869–1940) described their first Mt Wilson instrument in a paper published in the *Astrophysical Journal* in

1906. The modern reader of this paper may be surprised at the high standard of engineering design and construction incorporated into solar instrumentation at this epoch. By the 1920s there were a number of spectroheliographs distributed around the world, a good illustrated account of which is to be found in a review article by Abetti (1929) in the *Handbuch der Astrophysik*. In addition, the book *The Solar Chromosphere* (1974) by two of the present authors contains a short account of the early history of the instrument.

In an important paper published in 1908, P. Fox of the Yerkes Observatory discussed limb observations of loops obtained with the Rumford spectroheliograph in the H or K lines and, in particular, the association of loops with structures seen on disk spectroheliograms. Fig. 1.2 is a composite of a photograph of a loop system beyond the limb taken by Fox on 14 August 1907 and a disk spectroheliogram taken the same day. The similarity between this system and that observed by Young on 5 October 1871 (our Fig. 1.1) was noted by Fox. Fox concluded that the bases or footpoints of the loops coincided with especially bright points in the Ca II flocculi (faculae) present on the disk spectroheliograms. These he called 'eruptions'; from his description of one such event there is no doubt that they were what we now call *flares*. Fig. 1.2 therefore represents an early photograph of a typical cool flare loop system, modern observations of which are described in Chapter 4.

A dedicated observer of prominences of all types was the astronomer E. Pettit of Mt Wilson Observatory. Using H or K line observations obtained at Yerkes and Mt Wilson (4-m spectroheliograph) dating back to the 1920s, he discussed in a paper published in 1936 motions in prominences belonging to what he called the 'sun-spot' type. He concluded that some prominences of this type (class 3*b*) '...rise and fall like a fountain in closed loops within the spot-area...and are distinctive of sun-spots alone...'. The apparent heights of the highest and lowest loops were found to be 216000 and 27000 km respectively. In 1938 motions in loops – apparently of the same class – were studied by McMath & Pettit using cinematographic observations obtained at the McMath–Hulbert Observatory. They found that apparent velocities of 100 km s^{-1} were 'quite common' and that the motion was downward from the top. This type of motion is characteristic of loops associated with flares (see Section 4.3.2), suggesting that it was this category of cool loops that they studied.

Hale pointed out in 1930 that the spectrohelioscope is better adapted than the spectroheliograph to the study of rapidly moving structures, particularly on the solar disk. He noted with satisfaction the score of

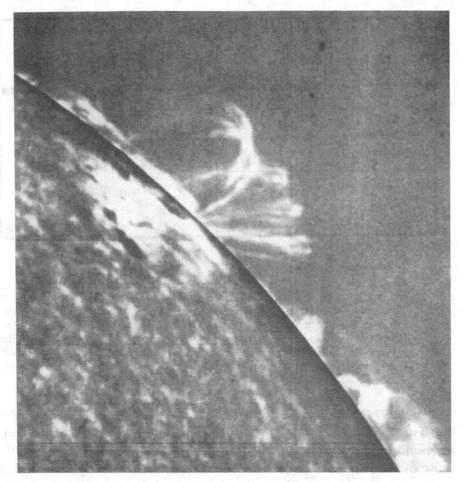

Fig. 1.2. Composite limb and disk photograph obtained by Fox in a line of Ca II with the Rumford spectroheliograph on 14 August 1907. This early photograph gives a good representation of a typical (cool) flare loop system observed at the limb.

spectrohelioscopes already constructed or ordered following publication of the design of his own instrument the previous year! In particular, by means of the line-shifter, the spectrohelioscope can render visible the *entire extent* of high velocity structures such as loops. Furthermore, it readily gives a qualitative – but not quantitative (see Section 2.3.4) – measure of the magnitude of the line-of-sight velocity at various points along a loop. During the 1930s and early 1940s, accordingly, a number of workers applied the new instrument to observe loops seen on the disk (called 'dark flocculi' at the start of the 1930s) in the Hα line, including Hale himself, M. Waldmeier in Switzerland, and M.A. Ellison in England.

A paper published by Ellison in 1944 is of considerable importance and

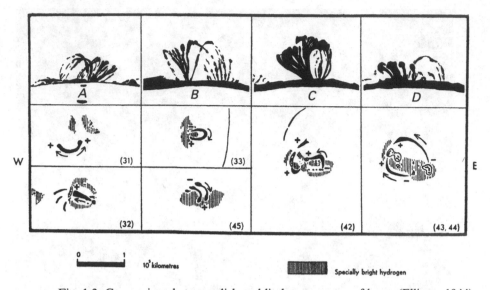

Fig. 1.3. Comparison between disk and limb appearance of loops (Ellison, 1944). The lower six drawings represent Ellison's own disk observations made with a spectrohelioscope at Sherborne, Dorset, England. The upper four were taken from Fox (the same as our Fig. 1.2), Evershed (Kodaikanal), Mt Wilson, and Mt Wilson respectively. The limb system of Fox represents a flare loop system; it is possible that some of the other limb drawings chosen by Ellison also represent such systems.

several of his results are incorporated into our account of cool active region loops in the following chapter. Ellison's chief aim was to correlate the disk and limb appearances of Pettit's class 3 prominences and, in particular, those of class 3*b*, the 'single and double arches...frequently seen with the spectrohelioscope connecting adjacent spots of the same group'. To this end he made use of his own spectrohelioscope observations made at Sherborne in England, together with earlier limb observations of Fox (Yerkes), Evershed (Kodaikanal) and of the Mt Wilson observers.

Some of Ellison's own disk observations are illustrated in the six drawings in the lower part of Fig. 1.3. No. 31 shows a loop with matter descending down both sides with no visible spot at either footpoint, while No. 32 shows a loop with gas flowing from the following to the leading spot. No. 33 illustrates a loop rising from a small spot and descending to a point south of the spot. Double arches are depicted in Nos. 43, 44 and 45, while No. 42 shows material rising from the following spot and descending into a spot in the middle of the group. The disk observations of Ellison evidently show fewer loops at any one time than the limb drawings at the top of Fig. 1.3: although partly an observational effect, this may be due in part to the circumstance that some of the limb systems that he chose to reproduce were loop systems associated with flares, which tend to have more loops than non-flare systems (cf. Section 4.3.1).

In a short paper published in 1933 the French solar physicist B. Lyot (1897–1952) proposed a way of building a new type of optical filter utilizing the phenomenon of birefringence. In the years ahead this device was destined to revolutionize observations of loops and other prominences both at the limb and on the disk, supplanting for this and other applications the spectroheliograph and spectrohelioscope. The first concrete realization of such a filter was due to the Swedish astronomer Y. Öhman, who in 1937 constructed a simple Hα filter having a bandwidth of 5 nm. Such a bandwidth was far too large for serious solar observations but in the following year Lyot constructed a temperature-controlled filter with a bandwidth of only 0.3 nm.

In 1939 Lyot installed his filter at the Pic-du-Midi Observatory and used it to photograph the fine structure in loops and other prominences. The exposure times were only 0.2 or 0.3 s and the exposures were made either automatically at the rate of two per minute or manually, by visually choosing moments of atmospheric steadiness. The photographs were taken on 35-mm film which, on viewing, could be speeded up some 600 times in order to reveal motions. Fig. 38 of Lyot's 1939 paper shows a fine loop system photographed in this way, the quality being such that extensive fine detail is visible in individual loops.

Lyot's films of the late 1930s were taken with a 0.3-nm filter, while in the early 1940s he employed an improved filter having the narrower bandwidth of 0.15 nm. These bandwidths are quite narrow enough for limb observations of loops and other prominences but are too broad for the effective study of such features on the disk. However, by the late 1940s the Paris firm Optique et Précision de Levallois had undertaken with Lyot's aid the commercial production of birefringent filters having a bandwidth of only 0.065 nm. By the 1950s these and similar commercial filters had become widely distributed around the world and were extensively used to obtain cinematographic observations of all types of solar phenomena, both on the disk and at the limb; the best spatial resolution achieved was usually no better than 2–3″ of arc.

These developments stimulated the construction by CSIRO (Australia) of the first really narrow band birefringent filter (Steel, Smartt & Giovanelli, 1961). It had a bandwidth of 0.0125 nm and was fully tunable over a range of ±1.6 nm centred on Hα. In the later 1960s the filter was fitted with a computer-controlled tuning system and installed in a 30-cm refractor at the CSIRO Solar Observatory, Culgoora, Australia, by two of the present authors and their colleagues. In addition to other types of observations, the CSIRO filter was used to obtain high-resolution, spectrally-pure disk photographs of active region and flare-associated loops at seven wavelengths nearly simultaneously in Hα. With the aid of a seeing monitor, a spatial resolution better than 1″ of arc was achieved on occasion. The filter was finally withdrawn from service in 1982. Observations of cool loops obtained with this and other birefringent filters around the world are described in Chapters 2 and 4.

Referring to limb observations of cool loops (mainly in Hα) carried out over a number of years by McMath and Pettit, Waldmeier, Ellison and others, Tandberg-Hanssen remarked in 1967: 'It is possible that what these authors have discussed is not a uniquely defined active object. We seem to have to deal with two different types of loops, namely (1) loop prominences that are intimately connected with sunspots, not only with the active region associated with sunspots, which we shall refer to as sunspot loops, and (2) flare loops that seem to grow out of flares; these loops will also occur near sunspots because their triggering or generating flares occur near sunspots...'. Out of this perception has grown the important distinction made today between (cool) active region loops (non-flare) and flare loops – a distinction which is reflected in the way the subject matter of the present book is sub-divided. Nevertheless, it sometimes happens that limb observations by themselves fail to provide adequate information to establish without ambiguity to which category a particular cool or hot loop system properly belongs; in this respect disk observations are superior.

1.4 Coronagraph observations of hot loops in the green and red coronal lines: Lyot and Dunn

The observation of hot loops in the visual region of the spectrum required two major instrumental developments, both of which were due to Lyot: (a) the construction of a suitable filter for the brightest of the coronal emission lines, namely λ530.3 of Fe XIV and λ637.4 of Fe X; and (b) the development of a specialized telescope at a high altitude site able to detect the faint coronal detail without the benefit of a total eclipse – i.e. a *coronagraph*. The high ionization potentials of these lines are indicative of temperatures of the order of 1×10^6–2×10^6 K (cf. Table 3.4); loops observed in these lines therefore fall into the category of hot loops.

The invention of the birefringent filter by Lyot and Öhman has been described in the previous section. The green line had been independently discovered by Young and Harkness at the eclipse of 1869, but its true coronal origin and its correct wavelength (530.3 nm) were not established until the eclipse of 1898 (Lockyer, Fowler, Evershed, and Campbell). The red line (637.4 nm) was discovered at the eclipse of 1914. Lyot obtained the first good out-of-eclipse photographs of the corona in these lines in 1939, the images showing much more detail than typical integrated, white-light photographs. In 1940–1 Lyot improved his earlier filter by replacing the polaroids by 'spath', i.e. calcite. The new filter, mounted on a coronagraph at the Pic-du-Midi Observatory, was then employed to obtain much improved three-wavelength observations of loops and other structures, the third wavelength being Hα.

Lyot's first (somewhat primitive) coronagraph was set up at the Pic-du-Midi Observatory in 1930. On 8 August 1930 it was used to obtain the first spectrogram of the corona outside of an eclipse, which included, of course, the green line – the brightest line in the visual emission spectrum. A more sophisticated coronagraph, equipped with a grating spectrograph, was set up in 1935; it is described in the George Darwin lecture delivered by Lyot in 1939. Using a curved slit, Lyot obtained excellent detailed spectrograms of the inner corona and discovered the line $\lambda 569.4$, not previously observed at eclipses. This line, due to Ca XV, is characteristic of particularly energetic coronal events and, when present, may appear along only a portion of a loop (Section 3.2.3).

Describing a photograph taken in the green line in 1941, Lyot remarked: '... la couronne a une structure très complexe, elle apparaît formée d'arches enchevêtrées, de jets fins et de petits nuages brillants'. (Translated, this reads: '... the corona has a very complex structure, it appears to be formed of entangled arches, fine jets, and small bright clouds'.) Furthermore, he noticed that the arches (loops) appeared to be sharper and more regular in the red line, a fact confirmed by more recent observers (Section 3.2.3). Figure 1.4 shows a beautiful pair of loop photographs taken in the red line on 2 September 1941 and published by Lyot in 1944. Note the relatively small changes in the structures over the one hour interval which separated the exposures. The clarity and contrast of the loops in Fig. 1.4 is comparable to that obtained with modern equipment – a tribute to the excellence of Lyot's coronagraph and filter and of the Pic-du-Midi site.

Following the work of Lyot, a systematic programme of coronal photography was started in 1955 at Sacramento Peak Observatory after the American solar physicist and instrumentation specialist R.B. Dunn joined the staff. Dunn modified a small (15-cm) telescope by replacing the objective by one of coronal quality and adding a dust tube and a 0.2-nm birefringent filter for the green and red coronal lines of his own manufacture. In the years that followed the Sacramento Peak workers

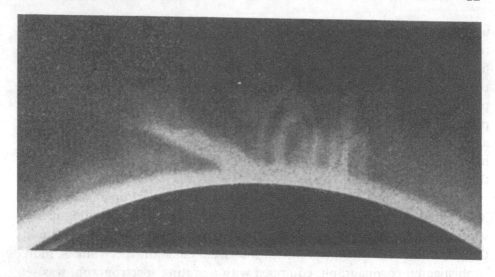

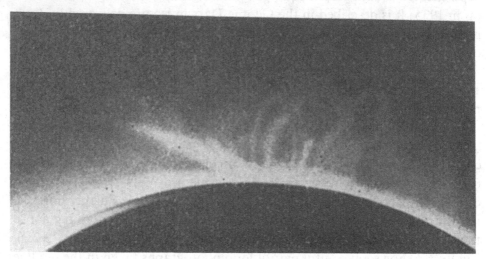

Fig. 1.4. Hot loop system photographed in the red coronal line λ637.4 at the Pic-du-Midi Observatory on 2 September 1941 (Lyot, 1944). The loops show little change over the 1-hour interval separating the photographs. These excellent photographs may be compared with the photograph of a hot flare loop system taken in the green coronal line λ530.3 at Haleakala on 6 March 1970 and reproduced in Fig. 4.1.

obtained extensive films of the corona in the λ530.3 line. The main results of this work and, in particular, the properties of coronal loops – a term used at Sacramento Peak since 1956 – were summarized by Dunn in a paper published in 1971 (cf. Section 3.2.2).

From his scrutiny of these films Dunn reached the general conclusion that 'Some coronal scenes look "open" . . . and some look to be all loops and arches or "closed" . . . The differences are presumed to be due to the magnetic field structure. The coronal stuctures are related to the magnetic field, and we say that they "map" the magnetic field. This is misleading since the force-free calculations show that the field should uniformly permeate the entire area and should not lie only in loops. It would seem to be more accurate to say that the corona defines particular flux tubes . . . This is best illustrated by the coronal loop system . . . The arches and loops may be very basic coronal structures since many scenes appear to contain nothing else . . .'. These remarks were made in 1971 and demonstrate that the fundamental loop structure of the active corona was recognized on the basis of films taken in the visible region from the ground prior to the launch of the Skylab satellite (May, 1973). This satellite, whose genesis is described in Section 1.7, yielded a wealth of information on the loop structure of the corona as observed in the extreme ultraviolet (EUV) and X-ray regions of the spectrum; it inaugurated the modern era of space observations of the Sun.

1.5 Start of the space era: rocket observations of the Sun in the EUV and X-ray regions of the spectrum

Despite the achievements of the ground-based observers described in the previous section, observations in the visible coronal lines offered little scope for further progress in elucidating the stucture of the active corona. In particular, the overwhelming brightness of the photosphere at these wavelengths make it impossible to observe the corona against the solar *disk*. Even at the limb heroic efforts had been needed – culminating in the coronagraph – to suppress the scattered light from the photosphere sufficiently to make the coronal structure visible.

On the other hand, when the high temperature of the corona became generally appreciated during the 1940s, it must have occurred to many people that observations of the EUV and X-ray Sun could be expected to reveal the structure of the low corona on the disk as well as at the limb. Below a wavelength of 150 nm, which marks the start of the EUV region, the contribution of the photospheric layers virtually vanishes and the radiation comes from overlying material at chromospheric and coronal temperatures. Moreover, this region of the spectrum displays a wealth of strong resonance lines emitted by many of the abundant ions of various elements at various stages of ionization which are formed in the temperature range $\sim 10^4$–10^7 K. The soft X-ray region below 10 nm is dominated by the presence of emission lines of very highly ionized stages of a number of elements, superimposed on a weak continuum background of coronal origin.

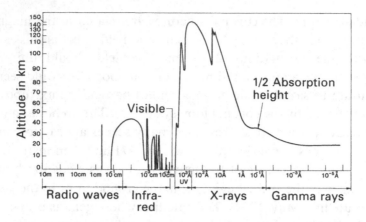

Fig. 1.5. Absorption by the Earth's atmosphere of solar radiation over the wavelength range from metre waves (radio) to gamma rays. The envelope indicates the height at which the remaining overlying atmosphere absorbs just *one-half* of the incident radiation. It is evident that to observe at all wavelengths in the EUV and X-ray regions of the spectrum, a rocket or other space vehicle must be located at a height of at least 150 km (Noyes, 1982).

The key to further progress evidently lay in (a) the construction of space vehicles – i.e. rockets and satellites – to operate at heights where atmospheric absorption of solar EUV and X-rays is negligible; and (b) the development of pointed and guided instruments capable of obtaining spectrally- and spatially-resolved observations. This formidable, world-wide undertaking (which, of course, still continues) has involved the skills of literally thousands of technicians, engineers, scientists and astronauts, and the expenditure of hundreds of millions of dollars. Although solar observation from space has been dominated by the U.S.A., the availability of experienced rocket engineers from Germany immediately following World War 2, together with competition from the U.S.S.R., has played a vital role in the development of the large launch vehicles required by the U.S. program. Other countries which have been prominent in developing solar space vehicles and instrumentation, and in analysing the results obtained, include Great Britain, France, Italy and Japan. Our aim in the present and following sections is to identify the significant landmarks, paying particular attention to advances which gradually revealed the ubiquity of the loop phenomenon.

Absorption by the Earth's atmosphere of solar radiation over the wavelength range metre waves to gamma rays is graphically shown in Fig. 1.5. The envelope indicates the height at which the overlying atmosphere absorbs just *one-half* the incident radiation. It can be seen from this curve that to observe the Sun in the region of the EUV extending from 10 to

100 nm, the space vehicle must be located at a height of at least 150 km; the principal absorbers are O_3 (near ultraviolet), O_2, O^+, N_2 and N. The first ultraviolet spectrogram of the Sun was obtained on 10 October 1946 at a height of 90 km from a captured German V-2 rocket. This spectrogram, taken by a team from the U.S. Naval Research Laboratory, recorded the spectrum down to a wavelength of 210 nm. By September 1952, the shortest continuum wavelength so far photographed had been pushed down to 190 nm (Aerobee rocket) and in the same year the Lα emission line of H I at 121.6 nm was photographed.

The next few years saw substantial improvements in rocket techniques. These included greater trajectory heights and the development of biaxial Sun followers, together with more sophisticated payloads – e.g. spectrographs. These improvements led to the securing of nearly monochromatic (but partially overlapping) images of the Sun in various EUV lines. In 1959 the first good-quality image was obtained in Lα by the NRL group, who followed this achievement by obtaining images in other lines in 1963. Improved images were obtained in 1966, the wavelengths now covering the range 15–70 nm with a (claimed) spatial resolution of 10″ of arc. In 1969 Purcell and Tousey obtained the first EUV spectroheliograms of a solar flare with even higher resolution.

By 1974 Brueckner and Bartoe of NRL were able to claim a resolution of approximately 4″ for spectroheliograms taken in various EUV lines of He I, He II, O IV, O V and Ne VII, as well as for broad-band EUV images. These rocket photographs are comparable in quality to photographs obtained during the manned Skylab missions of 1973 and 1974. In fact, the detail shown on the broad-band images was such that Brueckner and Bartoe were able to conclude that '...all coronal emission in active regions is concentrated either in loops or in their footprints...'. Furthermore, they were able to discern loops connecting different active regions: such interconnecting EUV loops had been detected the preceding year by Tousey and other NRL workers on high quality Skylab spectroheliograms.

X-rays from the Sun in the soft X-ray range 1–10 nm are absorbed in the E-region of the ionosphere at a height of approximately 100–150 km (cf. Fig. 1.5). Solar X-rays were first detected on 5 August 1948 from a V-2 rocket using beryllium windows and photographic recording. X-ray emission from flares was first observed in 1956 and the first direct evidence for the association of enhanced X-ray emission with active regions was obtained during the total eclipse of 12 October 1958 by H. Friedman and his colleagues at NRL. On 19 April 1960 the same group obtained the first crude X-ray photograph of the Sun, using a pinhole camera on an Aerobee rocket. Rotational precession of the rocket during the 5-min exposure was not compensated by the biaxial pointing system, so that each point of the image was drawn out into a 160° arc; this historic photograph is reproduced in Fig. 1.6.

By 1966 the best X-ray photographs to date had been obtained by a Goddard Space Flight Center (GSFC) team using a grazing-incidence telescope on an Aerobee rocket, together with filters passing the bands 0.3–1.1 nm, 0.8–2.0 nm, and the two bands 0.3–1.3 and 4.4–6.0 nm. These showed relatively featureless images of the active regions present against a dark background. Improved spatial resolution, however, soon made it possible to detect individual structures. Studying an importance 1N flare photographed on 8 June 1968, G.S. Vaiana of American Science and Engineering (AS&E) claimed the presence of details in the X-ray image only 2″ of arc in size!

The earliest X-ray photographs of the Sun had been obtained with pinhole cameras, and the pre-1968 observations carried out by the AS&E group were limited to a resolution of approximately 1′. However, by the beginning of 1968 several rocket flights conducted independently by GSFC and AS&E had established the superiority of grazing-incidence X-ray telescopes. In 1973 Vaiana and his colleagues published eight X-ray photographs obtained from various rocket flights over the period 1963–73 which dramatically demonstrated the improvement in resolution over the decade immediately preceding Skylab. For the best of the AS&E observations a resolution of the order of several arc seconds was claimed and, indeed, a flight made on 8 March 1973 yielded observations of active region loops with a definition superior to that of photographs subsequently obtained from Skylab (Section 3.4.2).

The improvement in resolution during this period led to the discovery of X-ray loops interconnecting different active regions. In an important paper published in 1968 the AS&E team pointed out that '. . . The X-ray images near the limb . . . show a looping and interconnecting of active regions which are not found in the Hα images; one sees X-ray structures extending 100 000 km or more above the Hα plage . . . at the limb there are loops and structures similar to those seen in white-light photographs of the corona'. These remarks were based on photographs obtained during the rocket flight of 8 June 1968 referred to above. Later flights revealed more details of the loop structure and by 1973 Vaina and his colleagues were referring to the coronal X-ray loops surrounding a developed active region and to loops joining areas of opposite (photospheric) magnetic polarity.

In summary, the major achievements of the rocket programme over the

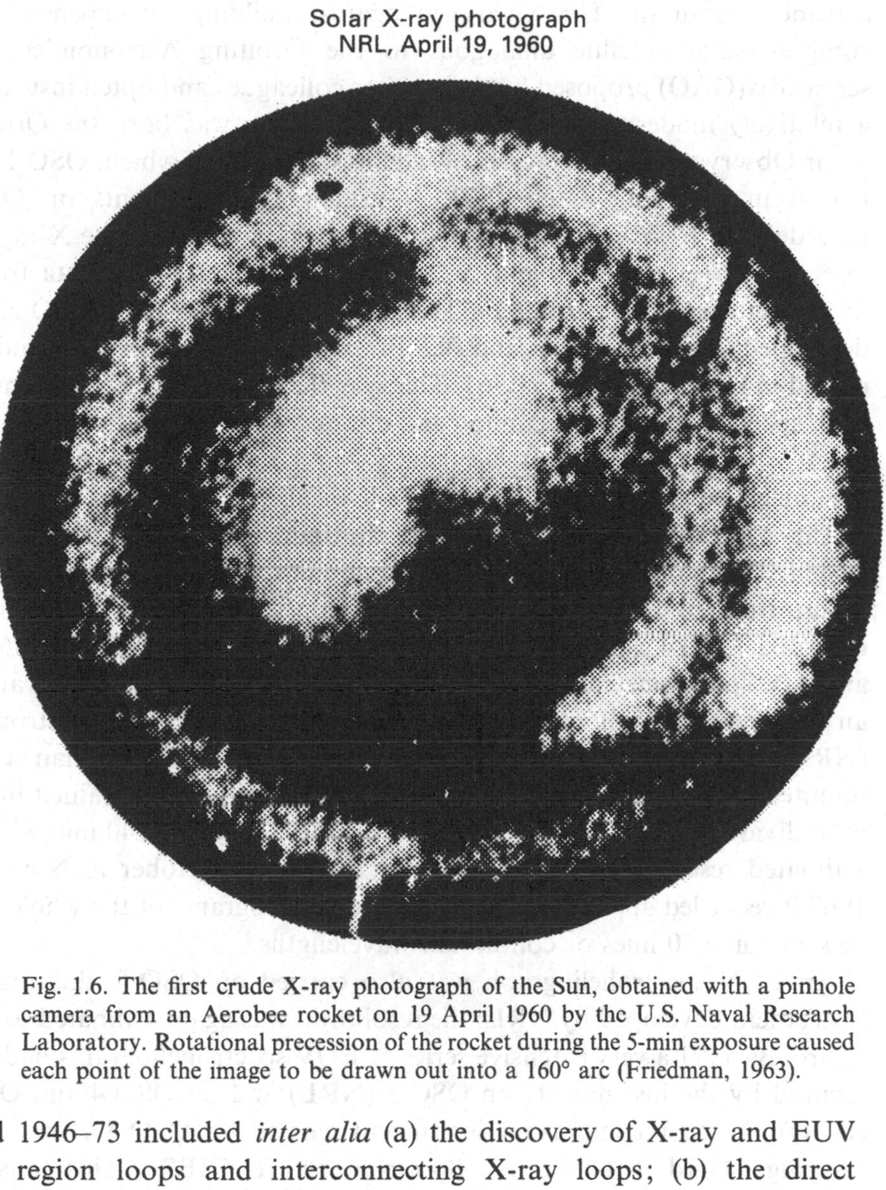

Solar X-ray photograph
NRL, April 19, 1960

Fig. 1.6. The first crude X-ray photograph of the Sun, obtained with a pinhole camera from an Aerobee rocket on 19 April 1960 by the U.S. Naval Research Laboratory. Rotational precession of the rocket during the 5-min exposure caused each point of the image to be drawn out into a 160° arc (Friedman, 1963).

period 1946–73 included *inter alia* (a) the discovery of X-ray and EUV active region loops and interconnecting X-ray loops; (b) the direct detection of X-ray emission from flares; and (c) the development of advanced instrumentation which was subsequently to form the basis for equipment employed in manned and unmanned solar satellite observatories.

1.6   Pre-Skylab solar satellites: OSO 1 to OSO 7

In the northern autumn of 1958 the U.S. National Aeronautics and Space Administration (NASA) was founded. In the following year the solar astronomers in the U.S.A. argued against building an expensive and complex solar satellite analogous to the Orbiting Astronomical Observatory (OAO) proposed by their stellar colleagues and opted instead for a relatively modest solar satellite program. Thus was born the Orbiting Solar Observatory (OSO) series of satellites, the first of which, OSO 1, was placed in orbit in March 1962. Non-imaging instruments on OSO 1 recorded the EUV solar flux over the range 17–40 nm and the X-ray flux integrated over the two bands 0.2–0.8 nm and 0.1–1.1 nm. Data from 5 flares were recorded during 1962. The second of the series, OSO 2, was developed jointly by GSFC and Ball Bros. Research Corporation and was placed in orbit in February 1965. This was the first of the series to employ both spectral and raster scanning and obtained the first spectroheliograms from orbit – recorded photoelectrically – in Lα and in the He II line at 30.4 nm. The spatial resolution was ∼ 1′ of arc.

The pace quickened in 1967 with the launch of OSO 3 and OSO 4. Together with other satellites, both carried out thousands of hours of solar monitoring in the X-ray band 0.001–10 nm, many X-ray bursts being recorded. OSO 4 carried instruments for both spectral and raster scanning: an EUV spectrometer/spectroheliograph (Harvard College Observatory), an X-ray spectroheliometer (AS&E), and a Bragg crystal spectrometer (NRL). The X-ray images had a spatial resolution no better than several minutes of arc. On the other hand, the HCO instrument obtained images in individual EUV emission lines over the range 28–140 nm with an estimated resolution of 1′ of arc; over the period October to November 1967 it recorded approximately 4000 spectroheliograms of the whole solar disk in some 50 lines or continuum wavelengths.

An EUV spectroheliograph was also carried on OSO 5, this time an instrument developed by NRL; the resolution was again estimated to be 1′ of arc. By 1971 a very extensive series of EUV spectroheliograms had been secured by the instruments on OSO 5 (NRL) and on OSO 4 and OSO 6 (HCO). A second important instrument carried on OSO 5 was a (non-imaging) Hard X-ray Burst Spectrometer (HXRBS). This was the forerunner to similar highly productive instruments subsequently carried on two post-Skylab satellites, OSO 8 and the Solar Maximum Mission (SMM).

OSO 6 was launched in 1969 and carried a Harvard College Observatory experiment similar to that on OSO 4. However, the spatial resolution had improved to 35″ and EUV spectroheliograms could now be obtained both of the full disk and of selected localized areas 7′ × 7′ in extent. The instrument could obtain images at any wavelength between 28 nm and

138 nm, together with spectrograms over the same range; the spectral resolution was approximately 0.3 nm. The 288 kg spacecraft carried a number of other experiments.

Launched in September 1971, OSO 7 was the last member of the OSO series prior to Skylab. Both at X-ray and EUV wavelengths it represented a significant advance over OSO 6. The GSFC experiment consisted of a pair of X-ray spectroheliographs covering the ranges 0.17–0.80 nm and 0.80–1.59 nm and a (co-aligned) EUV spectroheliograph covering the range 12–40 nm. The spectral resolution in the EUV was 0.08 nm, while the spatial resolution of all three instruments was effectively 20″. A given solar event could be observed simultaneously at four selected wavelengths within the region 0.18–0.40 nm; a number of flares were observed from beginning to end.

The spatial resolution achieved by the pre-Skylab members of the OSO series (i.e., excluding OSO 8, launched in 1975) was not usually adequate to clearly reveal the loop structure of active regions and flares. In fact, the resolution achieved by the rockets mentioned in the previous section was superior both in the EUV and X-ray regions of the spectrum to that achieved by OSOs 1–7. Indeed, as we have seen, it was the rocket programme that was responsible for the discovery of X-ray loops.

The important contributions of the OSO program were twofold. Firstly, it contributed to the development of specialized instruments for observing spectrally- and spatially-resolved features on the Sun in the EUV and X-ray regions of the spectrum – a contribution it shared with the rocket programme. This instrumental work helped to lay the foundations for the equipment and observational procedures subsequently employed on Skylab and the satellites which followed. (Soviet satellites of the Cosmos series also contributed to the study of X-ray flares during the OSO period.) Secondly, by producing literally thousands of hours of recordings of spectra and spectroheliograms of the quiet and active Sun, including flares during all their phases, the OSO program helped to stimulate the development of the diagnostic techniques needed to interpret EUV and X-ray spectral observations of coronal structures.

A historical treatment of diagnostic techniques lies beyond the scope of the present chapter but, where necessary, the techniques themselves are outlined in the chapters that follow.

We shall conclude our account of pre-Skylab solar satellites by briefly mentioning several purely monitoring (i.e., non-imaging) satellites. The most important of these were members of the Solrad series, developed by the Naval Research Laboratory. In fact, Solrad 1, launched in 1960, was the first successful X-ray and EUV satellite. It performed real-time (non-recording) monitoring in the X-ray band 0.2–0.8 nm and in Lα (121.6 nm), detecting more than 100 X-ray events during its 6-month life span. A later member of the Solrad series, Solrad 9, was launched in 1970. Other X-ray monitoring satellites during the pre-Skylab period included Ariel 1 (U.K.), Electrons 2 and 4 (U.S.S.R.), and Orbiting Geophysical Observatories 1 and 3 (U.S.A.). Collectively, these satellites – together with OSOs 3 and 4 – carried out thousands of hours of solar X-ray monitoring. Soft X-rays define the time of flare maximum and hard X-rays characterize the impulsive stage of a flare; non-imaging X-ray flux data obtained during flares therefore provide important complementary data.

1.7   The advent of Skylab

Skylab was placed in orbit by a Saturn V rocket on 14 May 1973 and operated until 8 February 1974. Between these dates three manned flights took place: 25 May–22 June 1973, 28 July–25 September 1973, and 16 November 1973–8 February 1974. Skylab represented the first manned solar satellite observatory and the first use of photography in long-term space observations of the Sun.

The genesis of Skylab was described in the year of its launch by L. Goldberg of Kitt Peak Observatory during the course of the Russell memorial lecture delivered that year. In 1962 U.S. astronomers had recommended to NASA the development of a spacecraft more advanced than members of the OSO series – a so-called Advanced Orbiting Solar Observatory. AOSO was to have a pointing stability of 1″ of arc and a spatial resolution for short exposures of 0″.1, but became a casualty of the first large reduction in NASA's budget in December 1965. Eventually it was decided to use hardware left over from the manned lunar program (Apollo), including the Lunar Excursion Module and the third stage of a Saturn V rocket, to construct a *manned* solar observatory – Skylab. During launch a nearly disastrous accident occurred – the ripping-off of one of the two solar power panels and the failure to deploy of the other. The necessary repairs were carried out by the first team of astronauts, equipped with special tools for the job. The mission was thereby saved, although with reduced electrical power capacity.

Six major instruments were carried on Skylab: (a) a white-light coronagraph (High Altitude Observatory, 370–700 nm); (b) two grazing-incidence X-ray telescopes (AS&E, 0.35–6.0 nm; Marshall Space Flight Center, 0.3–5.3 nm); (c) an EUV spectrometer/spectroheliometer (HCO, 28–135 nm); (d) an EUV spectroheliograph (NRL, 17–63 nm); and (e) a

UV spectrograph (NRL, 97–394 nm). All of these instruments employed photographic recording except the Harvard spectrometer/spectroheliometer. The spatial resolution claimed for the various instruments represented a considerable advance over that previously obtained from solar satellites, ranging from $\sim 2''$ for the two X-ray telescopes, through $5''$ for the Harvard spectroheliometer, to $\sim 8''$ for the coronagraph. (The best of these figures may be over-optimistic, however; cf. Section 3.4.1.) The last two instruments, together with one of the X-ray telescopes, could be operated in an unmanned mode, i.e. between manned flights.

The launch of Skylab initiated a new era in the observation of the Sun from space. Within a very short time a wealth of new information became available about the detailed structure of the corona and, in particular, about the behaviour and properties of coronal loops. Even at the time of writing (1989), most of our knowledge of non-flare coronal loops stems back to Skylab observations in the EUV and soft X-ray regions of the spectrum. Some idea of the enormous productivity of Skylab can be gauged from the fact that between 28 May 1973 and 7 February 1974 the AS&E X-ray telescope alone obtained nearly 32 000 photographs of the Sun!

Skylab was never intended as a flare mission, since its launch took place during the declining phase of the rather weak solar cycle which peaked in 1968–9 (cycle 20). Nevertheless, as we shall see in Chapter 4, flare observations obtained from Skylab have made a considerable contribution to our understanding of the role of loops in flares.

Skylab has been followed by a succession of important (unmanned) solar satellites, most notably OSO 8, launched in 1975 (Section 4.4.1), the Solar Maximum Mission (1980; Section 4.5.3) and the Japanese Hinotori satellite (1981; Section 4.5.3). Together with Skylab, these satellites have furnished the wealth of EUV and X-ray data which has laid the foundations for a systematic study of the properties of flare and non-flare coronal loops. This data, together with vital complementary information from the visible and microwave regions of the spectrum, is described at length in the chapters that follow.

1.8 Chronological summary

1239   Prominence observed at eclipse described in Russian chronicles
1733   Well-observed eclipse in Sweden leads to description of prominences (Vassenius)
1842   Start of regular (eclipse) observations of prominences
1850–1   Prominences now generally recognized as a true solar phenomenon

1860   First systematic application of (eclipse) photography to prominences by Secchi and de la Rue

1868   Discovery by Janssen and Lockyer of open-slit method of observing prominences outside of an eclipse

1871–7   Young and Secchi describe cool Hα loops ('parabolic arcs and jets')

1898   Coronal origin and correct wavelength of λ530.3 line established (Lockyer, Fowler, Evershed and Campbell)

1899   Completion by Hale of Rumford spectroheliograph at Yerkes Observatory

1908   Fox publishes spectroheliogram of a cool flare loop system obtained in a line of Ca II

1914   Discovery of the red coronal line at 637.4 nm

1929   Hale publishes design of spectrohelioscope

1930   Lyot establishes his first coronagraph at the Pic-du-Midi Observatory

1936   Pettit describes properties of sunspot loops observed at Yerkes and Mt Wilson

1940–4   Lyot obtains good observations of hot coronal loops in λ530.3 and λ637.4 and describes complex structure of the monochromatic corona

1944   Disk and limb appearances of cool loops correlated by Ellison using spectrohelioscope observations

1946   Ultraviolet spectrum of Sun photographed from V-2 rocket

1948   Detection of X-rays from Sun

1952   Emission line Lα of H I photographed from rocket

1955   Cinematography of corona started at Sacramento Peak Observatory (Dunn)

1956   Detection of X-rays from flares

1958   NASA founded

1958   Association of X-rays with active regions established

1959   Good-quality rocket photographs of Sun in Lα obtained by NRL

1960   Solrad 1 launched to monitor solar Lα and X-ray flux

1960   Crude (pinhole) X-ray photograph of Sun obtained from Aerobee rocket

1962   OSO 1 placed in orbit

1965   OSO 2 obtains the first EUV spectroheliogram from orbit

1966   NRL claims resolution of 10″ of arc for EUV rocket images

1966   GSFC uses grazing-incidence telescope to obtain good-quality X-ray images

1967–9   Launch of OSOs 3–6
1968   AS&E observe interconnecting active region loops and fine detail in a flare on X-ray images obtained from rocket
1969   OSO 6 obtains EUV spectroheliograms with resolution of 35″
1970   Launch of Solrad 9
1970–1   Computer-controlled 0.0125-nm filter for the Hα line brought into operation by CSIRO at Culgoora
1971   OSO 7 in orbit – resolution achieved now 20″
1971   Fundamental loop structure of the active corona described by Dunn from observations in visible-region coronal lines
1973   Launch of Skylab

References

Abetti, G. (1929). Solar physics. In *Handbuch der Astrophysik*, ed. G. Eberhard, A. Kohlschütter & H. Ludendorff, vol. 4, pp. 57–230. Berlin: Springer.

Bray, R.J. & Loughhead, R.E. (1974). *The Solar Chromosphere*. London: Chapman & Hall.

Bray, R.J. & Winter, J.G. (1970). High-resolution photography of the solar chromosphere. VIII. Computer control of a tunable $\frac{1}{8}$ Å filter. *Solar Physics*, **15**, 309–16.

Brueckner, G.E. & Bartoe, J.-D.F. (1974). The fine structure of the solar atmosphere in the far ultraviolet. *Solar Physics*, **38**, 133–56.

Clerke, A.M. (1887). *A Popular History of Astronomy during the Nineteenth Century*. Edinburgh: A. & C. Black.

Dunn, R.B. (1971). Coronal events observed in 5303 Å. In *Physics of the Solar Corona*, ed. C.J. Macris, pp. 114–29. Dordrecht: Reidel.

Ellison, M.A. (1944). Sunspot prominences – some comparisons between limb and disk appearances. *Monthly Notices of the Royal Astronomical Society*, **104**, 22–32.

Evershed, J. & Evershed, M.A. (1917). Results of prominence observations. *Memoirs of the Kodaikanal Observatory*, vol. I, part II, 55–126.

Fox, P. (1908). The distribution of eruptive prominences on the solar disk. *Astrophysical Journal*, **28**, 253–8.

Friedman, H. (1963). Ultraviolet and X-rays from the Sun. *Annual Review of Astronomy and Astrophysics*, **1**, 59–96.

Goldberg, L. (1967). Ultraviolet and X-rays from the Sun. *Annual Review of Astronomy and Astrophysics*, **5**, 279–324.

Goldberg, L. (1974). Research with solar satellites. *Astrophysical Journal*, **191**, 1–37.

Grant, R. (1852). *History of Physical Astronomy*. London: Baldwin.

Hale, G.E. (1930). The spectrohelioscope and its work. Part II. The motions of the hydrogen flocculi near sun-spots. *Astrophysical Journal*, **71**, 73–101.

Hale, G.E. (1935). The brightness of prominences as shown by the spectrohelioscope. *Monthly Notices of the Royal Astronomical Society*, **95**, 467–8.

Hale, G.E. & Ellerman, F. (1903). The Rumford spectroheliograph of the

Yerkes Observatory. *Publications of the Yerkes Observatory*, **3**, part 1, pp. 3–25.

Hale, G.E. & Ellerman, F. (1906). The five-foot spectroheliograph of the solar observatory. *Astrophysical Journal*, **23**, 54–63.

Huber, M.C.E. & 6 co-authors (1973). The Harvard experiment on OSO 6: instrumentation, calibration, operation, and description of observations. *Astrophysical Journal*, **183**, 291–312.

Loughhead, R.E. & Tappere, E.J. (1971). High-resolution photography of the solar chromosphere. IX. Limb observations of high spectral purity. *Solar Physics*, **19**, 44–51.

Lyot, B. (1933). Un monochromateur à grand champ utilisant les interférences en lumière polarisée. *Comptes Rendus hebdomadaires des Séances de l'Académie des Sciences*, **197**, 1593–5.

Lyot, B. (1939). A study of the solar corona and prominences without eclipses. *Monthly Notices of the Royal Astronomical Society*, **99**, 580–94.

Lyot, B. (1944). Le filtre monochromatique polarisant et ses applications en physique solaire. *Annales d'Astrophysique*, **7**, 31–79.

McMath, R.R. & Pettit, E. (1938). Prominences studies. *Astrophysical Journal*, **88**, 244–77.

Meadows, A.J. (1970). *Early Solar Physics*. Oxford: Pergamon Press.

Mitchell, S.A. (1929). Eclipses of the Sun. In *Handbuch der Astrophysik*, ed. G. Eberhard, A. Kohlschütter & H. Ludendorff, vol. 4, pp. 231–357. Berlin: Springer.

Neupert, W.M. (1969). X-rays from the Sun. *Annual Review of Astronomy and Astrophysics*, **7**, 121–48.

Neupert, W.M., Thomas, R.J. & Chapman, R.D. (1974). Spatial distribution of soft X-ray and EUV emission associated with a chromospheric flare of importance 1B on August 2, 1972. *Solar Physics*, **34**, 349–75.

Noyes, R.W. (1971). Ultraviolet studies of the solar atmosphere. *Annual Review of Astronomy and Astrophysics*, **9**, 209–36.

Noyes, R.W. (1982). *The Sun, Our Star*. Cambridge: Harvard University Press.

Öhman, Y. (1938). A new monochromator. *Nature*, **141**, 157–8, 291.

Pettit, E. (1936). The motions of prominences of the eruptive and sun-spot types. *Astrophysical Journal*, **84**, 319–45.

Pettit, E. (1943). The properties of solar prominences as related to type. *Astrophysical Journal*, **98**, 6–19.

Secchi, A. (1875–7). *Le Soleil*, 2nd ed. Paris: Gauthier-Villars.

Steel, W.H., Smartt, R.N. & Giovanelli, R.G. (1961). A $\frac{1}{8}$ Å birefringent filter for solar research. *Australian Journal of Physics*, **14**, 201–11.

Tandberg-Hanssen, E. (1967). *Solar Activity*. Waltham: Blaisdell.

Tousey, R. (1953). Solar work at high altitude from rockets. In *The Sun*, ed. G.P. Kuiper, pp. 658–76.

Tousey, R. (1967). Highlights of twenty years of optical space research. *Applied Optics*, **6**, 2044–70.

Tousey, R. (1971). Survey of new solar results. In *New Techniques in Space Astronomy*, ed. F. Labuhn & R. Lüst, I.A.U. Symposium No. 41, pp. 233–50. Dordrecht: Reidel.

Tousey, R. (1972). High angular resolution observations from rockets: solar XUV observations. In *Space Research XII*, ed. S.A. Bowhill, L.D. Jaffe & M.J. Rycroft, vol. 2, pp. 1719–37. Berlin: Akademie-Verlag.

Tousey, R. (1977). Apollo Telescope Mount of Skylab: an overview. *Applied Optics*, **16**, 825–36.

Vaiana, G.S., Krieger, A.S. & Timothy, A.F. (1973). Identification and analysis of structures on the corona from X-ray photography. *Solar Physics*, **32**, 81–116.

Vaiana, G.S., Reidy, W.P., Zehnpfennig, T., VanSpeybroeck, L. & Giaconni, R. (1968). X-ray structures of the Sun during the importance 1N flare of 8 June 1968. *Science*, **161**, 564–7.

Vidal-Madjar, A. & 5 co-authors (1987). Galactic ultraviolet astronomy. *Reports on Progress in Physics*, **50**, 65–113.

Waldmeier, M. (1937). Chromosphärische Vorgänge bei der Enstehung von Sonnenflecken. *Zeitschrift für Astrophysik*, **14**, 91–103.

Withbroe, G.L. & Gurman, J.B. (1973). Models of the chromospheric-coronal transition layer and lower corona derived from extreme-ultraviolet observations. *Astrophysical Journal*, **183**, 279–89.

Young, C.A. (1895). *The Sun*. London: Kegan Paul.

2

Cool loops: observed properties

2.1 Introduction

Coronal loops are a phenomenon of active regions (Chapter 1) and there is growing evidence that they are in fact the dominant structures in the higher levels (inner corona) of the Sun's atmosphere. Our knowledge of loops has greatly expanded in recent years as a result of space observations in the far ultraviolet and X-ray regions of the spectrum. However, the success of the space work should not be allowed to obscure the fact that a considerable amount of quantitative information on the morphological, dynamical, and physical properties of coronal loops has been derived from ground-based observations in the visible and near-visible regions. In fact, observations at these wavelengths have achieved significantly higher spatial resolution (better than $1''$ of arc) than almost all of the space observations so far obtained. Our aim in this and the following chapter is to bring together *all* the available data and thus present an integrated and consistent picture of the properties of non-flare coronal loops.

Observations show that coronal loops, depending on their temperature, can be divided into two distinct categories. The properties of the two types differ radically. Loops formed at temperatures in excess of $\sim 1 \times 10^6$ K are conventionally referred to as 'hot' loops, while those formed at lower temperatures are termed 'cool' loops. It is convenient to deal with the two types separately, cool loops in the present chapter and hot loops in Chapter 3.

The range of temperature covered by the category of cool loop extends from $\sim 1 \times 10^6$ K down to ~ 20000 K; the lower figure represents a temperature characteristic of loops observed in Hα. Taken as a whole, the data described in this chapter show that, with the exception of temperature, all cool loops appear to have similar properties and, in fact, can be regarded as manifestations of the same basic physical structure.

Over the years changes have occurred in the nomenclature of the subject.

Under the classical scheme of prominence designation introduced by Menzel & Evans (1953) loop prominences were classified as AS1, the 'A' indicating that the material originates from above, the 'S' an association with sunspots and the '1' the shape in the form of a loop. However, we now know that two important qualifications are required. Firstly, in some cases material is unmistakeably observed to originate from below, rising from one footpoint to the loop apex and descending the other leg (Section 2.3.4). Secondly, it is now known that there are two distinct sub-classes of loop prominences (Tandberg-Hanssen, 1977), namely flare and non-flare (Section 2.3.2). Flare loops have an integral and intimate relationship with major flares and are described in Chapter 4. Various other names for non-flare loops have been used in the literature, such as 'active region prominence' (or loop) and 'sunspot loop'. However, in recent years the term 'active region loop' has come into general use and is used in this book.

We begin with (cool) loops observed at the limb in Hα and other visible region lines, with particular attention to observations of velocities and accelerations (Section 2.2.2) and magnetic fields (Section 2.2.3). Disk observations in Hα not only give further information on velocities and accelerations (Section 2.3.4) but also provide considerable data on such properties as location of footpoints, lifetime and evolution, diameter, height, and other geometrical parameters (Sections 2.3.2 and 2.3.3).

The solar extreme ultraviolet (EUV) spectrum spans the range 150 to ~10 nm and is dominated by resonance emission lines of H I (Lyman series) and of various ionization stages of He, C, N, O, Si, S, Ne, Mg and Fe. Lines commonly used to obtain EUV observations of cool loops are listed in Table 2.4, together with their 'formation temperatures' (Section 2.4.1). The general characteristics of EUV loop systems in active regions and their change with temperature are described in Section 2.4.2. In Section 2.4.3 we turn to the properties of individual loops, including the question of the spatial relationship between EUV loops of different temperature.

Physical conditions in cool loops are next discussed. In the visible region it is again convenient to separate disk and limb observations. Disk observations (in Hα) have given us the most complete account to date of the physical conditions in a single loop and are therefore discussed first (Section 2.5.2). Limb observations offer considerable potential for inferring physical conditions from observations not only in Hα but also in other visible and near-visible lines and continua (Section 2.5.3).

In the EUV region the available data are less complete and it is practicable to deal with the disk and limb observations together. In Section 2.5.4 we give a short introduction to the subject of EUV diagnostics,

Fig. 2.1. Fine example of an Hα active region loop system (Bumba & Kleczek, 1961). Most or all of the loops are anchored to sunspots whose locations (shown in the drawing) were established with the aid of auxiliary data.

tailored to the particular circumstances applying to coronal loops. The physical conditions in cool loops derived from EUV observations are discussed in Section 2.5.5.

For the convenience of the reader, the chapter ends with tables summarizing all the available quantitative data concerning the morphological, dynamical and physical properties of cool loops (Tables 2.8 and 2.9).

2.2   Limb observations in Hα and other visible region lines
2.2.1   *Introduction*

Figure 2.1 shows a particularly fine example of an active region loop system beyond the limb (Bumba & Kleczek, 1961). These authors observed this or a similar system at no less than five consecutive east and west limb passages. Using auxiliary data, they established that the individual loops were anchored to underlying sunspots, although this particular aspect of active region loop behaviour is more easily studied from disk observations (Section 2.3). Other photographs of active region loops have been published by, for example, Lategan & Jarrett (1982, Fig.

2(*a*)), showing a single neat loop, and Foukal (1978, Fig. 2), showing the characteristic condensations or 'knots' often present in Hα loop systems.

The number of loops in a single system may range from just one up to perhaps 10 or so; an upper limit is hard to establish from limb observations owing to the confusion caused by overlapping when several loops are present. Apparent, i.e. projected, heights measured by various authors range from 26400 km (Lategan & Jarrett, 1982), through 50000 km (Tandberg-Hanssen, 1977; Makhmudov, Nicolsky & Zhugzhda, 1980), to 57000 km (Kleczek, 1963). However, true heights cannot be determined from limb observations unless the loop geometry is known. Limb observations do not seem to have yielded any systematic determinations of lifetimes, but Tandberg-Hanssen quotes a typical value of ~ 15 min.

In the following sections we shall consider two important types of limb observation in detail, namely those which have yielded measurements of loop velocities and accelerations (Section 2.2.2) and magnetic fields (Section 2.2.3). Once again, limb observations are handicapped by the circumstance that in general it is impossible to establish the loop geometry and hence to convert line-of-sight measurements of the velocity and magnetic fields to true values. A second handicap is that it is not always possible to establish with certainty whether the feature in question is a flare loop or a non-flare loop. However, we have endeavoured to satisfy ourselves on the basis of the published evidence that the loops whose properties are described in this chapter fall into the non-flare category.

2.2.2 *Velocities and accelerations*

It has long been known that there are three types of motion associated with active region loops (de Jager, 1959, p. 237): (a) a flow down both legs, starting at the top of the loop; (b) a flow up one leg and down the other; and (c) a mainly horizontal back-and-forth motion of the whole loop ('oscillation'). Examples of (a) have been given by Kleczek (1963) and Lategan & Jarrett (1982) and of (c) by Vrsnak (1984), while Martin (1973) has illustrated a loop which started with upward flow in one leg and downward in the other but exhibited after 1 hr a downward flow in both legs. Bodily upward motions of the whole loop have also been observed (Kleczek, 1963; Vrsnak, 1984).

Two methods are commonly used to determine the velocity and acceleration of the material in loops observed at the limb. The first relies on a straightforward determination of the Doppler shift of a suitable line, usually Hα. This gives the line-of-sight velocity, which can be converted to the true velocity along the loop if a plausible geometry is assumed. The

Table 2.1. *Flow velocities in active region loops observed at the limb*

Reference	Line-of-sight velocity (km s^{-1})	'True' velocity km s^{-1})	Method
Kleczek (1963)	—	87–524	'knots'
	—	0–175	'knots'
Tandberg-Hanssen (1977)	—	∼30	quoted value
Foukal (1978)	∼25	≳45	Doppler shifts
	—	40–60	'knots'
Makhmudov *et al.* (1980)	—	22–150	'knots'
Lategan & Jarrett (1982)	−20–+26	—	Doppler shifts

second method measures the projected motions in the plane of the sky of the condensations or 'knots' which are often prominent features of loop photographs (Kleczek, 1963; Foukal, 1978; Makhmudov, Nikolsky & Zhugzhda, 1980). To derive true velocities requires, once again, knowledge of the geometry and, in addition, the assumption that the apparent motions of the knots represent the genuine motion of loop material and not merely changing conditions of excitation. Table 2.1 summarizes the line-of-sight and 'true' loop velocities that have been obtained by these techniques.

With the exception of the single very high figure of Kleczek (1963), Table 2.1 gives a consistent picture: the majority of the inferred true velocities lie in the range 20–150 km s^{-1}. This range is consistent with values derived from measurements of loops observed on the *disk* (Section 2.3.4).

A number of authors have sought to measure the acceleration of loop material and, in particular, to compare the inferred values with the gravitational acceleration, $g_\odot = 0.27$ km s^{-2}. Unfortunately, however, the results may be subject to considerable error since the derivation of the acceleration compounds any errors in the measured velocities or displacements. Makhmudov, Nikolsky & Zhugzhda (1980) found that downward-moving knots were decelerated by a force counteracting the gravitational force; the magnitude of the deceleration sometimes exceeded $g_\odot$. Positive accelerations were always less than $g_\odot$ but negative accelerations varied between wider limits. On the other hand, Foukal (1978, 1981) and Lategan & Jarrett (1982) found that their observed velocities were close to those expected for free fall along assumed field lines from the observed heights. The velocity of free fall from height h, $(2g_\odot h)^{\frac{1}{2}}$, is for example 73 km s^{-1} for $h = 10\,000$ km and 150 km s^{-1} for $h = 40\,000$ km.

Table 2.2. *Magnetic fields in active region loops observed at the limb*

Reference	Line-of-sight field component $B_\parallel$ (G)	Total field B (G)	Measuring aperture	Method
Tandberg-Hanssen (1974)	20–100	—	—	Zeeman effect (quoted value)
Tandberg-Hanssen & Malville (1974)	7–47	—	$10'' \times 10''$	Zeeman effect
Bommier *et al.* (1981)	—	45	$5''$	Hanle effect
Athay *et al.* (1983)	—	4–46	$10'' \times 4''$ $10'' \times 7''$	Hanle effect
Vrsnak (1984)	—	45	—	Inferred from period of oscillation

Although further measurements of acceleration are needed, both limb and disk (Section 2.3) observations imply that, on occasion, accelerating and decelerating forces other than gravity operate on the material in loops.

2.2.3 *Magnetic fields*

One of the most difficult properties of prominences to measure is their magnetic field; nevertheless, a number of workers have succeeded in producing concordant results using the Zeeman or Hanle effects. Athay *et al.* (1983) measured the linear polarization in two resolved components of the He I D_3 line and obtained complete Stokes profiles for thirteen prominences, mostly relatively quiescent ones. They used the Hanle effect to interpret the results, obtaining the vector magnetic field at a number of locations. One of the prominences ('prominence N') showed two arches extending down from the main body to the chromosphere. In this prominence, the total field B ranged from 4 to 46 G; where the two loops met the main body the field was 6 G. Probably owing to the large size of the measuring aperture (cf. Table 2.2), the field was not measured in the two loops themselves.

The theory underlying the application of the Hanle effect to the determination of vector magnetic fields in prominences has been outlined by Bommier, Leroy & Sahal-Bréchot (1981). They applied the method to a loop prominence system observed at the Pic-du-Midi Observatory, employing a measuring aperture of diameter 5″. The field was assumed to lie in the direction of the loop. The authors note that results obtained with the Hanle method are sensitive to the assumed geometry of the prominence system, since the polarization depends on the scattering angle. They remark that B is larger in active as compared to quiescent prominences.

The Zeeman method normally employs either Hα or Hβ and has been more widely used than the Hanle method even though it yields only the longitudinal component of the field, $B_{\parallel}$. Tandberg-Hanssen & Malville (1974) measured $B_{\parallel}$ in a number of active prominences at the limb, finding an upper limit of 150–200 G. A system of loops gave longitudinal fields varying from 7 G to 47 G (measuring aperture, $10″ \times 10″$). Tandberg-Hanssen (1974, p. 43) states that a typical value for $B_{\parallel}$ in loops is 20–100 G, and that at their maximum height they show stronger fields than any other type of prominence at the same height. Furthermore, he remarks that J.W. Harvey has shown the orientation of loops coincides well with that of magnetic field lines computed from potential theory (Tandberg-Hanssen, 1974, p. 75 and Fig. III.8).

A general review of methods and results of prominence magnetic field measurements has been published by Leroy (1979). Earlier measurements have also been summarized by Tandberg-Hanssen & Malville (1974) and by Athay *et al.* (1983). Finally, we should mention that a novel method of estimating the total field B in a loop has been introduced by Vrsnak (1984). He measured the bodily oscillations (mainly horizontal) of an Hα loop in the plane of the sky, finding a period of 8 min. Assuming that the motion was controlled by the magnetic field, he was able to show that the observed period was consistent with a value of B of 45 G.

The various measurements or estimates of $B_{\parallel}$ and B in active region loops at the limb are collected together in Table 2.2. Although the results are reasonably consistent, they probably represent *lower limits* to the true field along the axis of a loop in view of the rather large measuring apertures used. However, the limb measurements are valuable since (to the authors' knowledge) no *disk* measurements have yet been attempted. Nor have any measurements so far been attempted in the EUV spectral region.

2.3   Disk observations in Hα
2.3.1   *Introduction*
On the disk an active region loop appears in the Hα line as a thin, curved, dark feature linking a sunspot with another spot or area of opposite magnetic polarity. Usually, however, the whole of a loop cannot

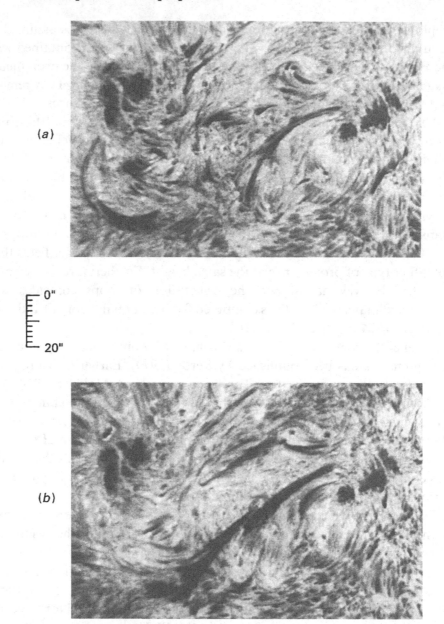

Fig. 2.2 (*a*) and (*b*), for legend see facing page.

be observed at a single wavelength in Hα. Not only is the velocity of the material in the loop relatively high but also the angle between the loop axis and the line of sight to the observer varies markedly from one part of the loop to another. Observations must therefore be made at a number of wavelengths on both sides of line centre extending out to, say, Hα±0.1 nm or Hα±0.2 nm. This practice has been followed by the authors and their colleagues at Culgoora since 1970 (Bray & Loughhead, 1974) and is

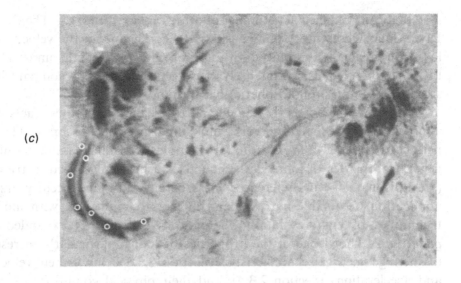

(c)

0"

20"

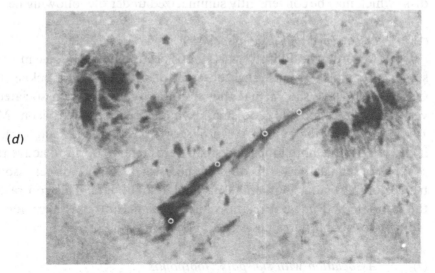

(d)

Fig. 2.2. Active region loop of 6 June 1980 photographed at (*a*) Hα+0.06 nm and
(*b*) Hα−0.06 nm (Bray & Loughhead, 1983). For filtergrams at Hα±0.1 nm, see
Figs. 2.2(*c*) and (*d*). A geometrical reconstruction of this loop is shown in Fig. 2.7.
Fig. 2.2, cont. Active region loop of 6 June 1980 photographed at (*c*) Hα+0.1 nm
and (*d*) Hα−0.1 nm (Bray & Loughhead, 1983). The small white circles identify
the points of the loop at which microphotometry was carried out on each of seven
frames of a *Δλ*-sequence – the first step in determining flow velocities and physical
conditions in the loop. For filtergrams at Hα±0.06 nm, see Fig. 2.2(*a*) and (*b*).

systematically pursued at a number of other observatories, e.g. The Ottawa River Solar Observatory (Gaizauskas, 1979). The higher the velocity of a loop the greater its visibility; with appropriate wavelength tuning it can then be observed as an intensely dark object against a comparatively structureless (quasi-photospheric) background.

Well before the advent of tunable narrow band filters, these facts were appreciated by visual observers employing spectrohelioscopes. The experienced observer M.A. Ellison, for example, determined a number of the general properties of loops with the aid of a manually-tuned instrument (see Ellison, 1944, and references cited therein). In more recent years the application of tunable narrow band filters in conjunction with modern techniques of high resolution photography has greatly expanded our *quantitative* knowledge of the properties of loops, particularly in respect of their geometrical characteristics (Section 2.3.3 below), their velocities and accelerations (Section 2.3.4), and their physical conditions (Section 2.5.2).

2.3.2 *General properties*

Before proceeding to discuss the topics listed above we shall first outline the general properties of Hα active region loops observed on the disk, which may be conveniently summarized under the following headings:

(1) *Occurrence and flare association*

According to Ellison (1944) loops occur only during the most active stages of complex groups. Disk observations made at the Lockheed Solar Observatory have shown that they are almost invariably associated with very small flares, possibly too small to have been seen by Ellison (Martin, 1973). However, the authors' own observations, made at Culgoora, make it clear that while both loops and small flares tend to occur in active regions at times of high activity, this does not necessarily imply a direct association between a particular loop and a particular small flare. This contrasts with the intimate and integral relationship between Hα (post-) flare loops and major two-ribbon flares described in Section 4.3.

(2) *Association with sunspots: footpoints*

Hα loops terminate in or near sunspots at one or both ends (Ellison, 1944; Tandberg-Hanssen, 1974, p. 5). The actual location of the footpoints with respect to the associated sunspot or spots is shown by high resolution photographs in the authors' collection. Figure 2.2 gives an example of a point of termination just inside the penumbra/photosphere boundary,

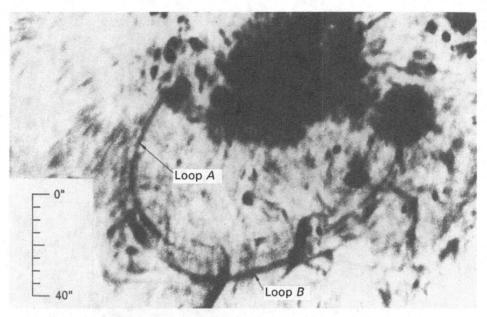

Fig. 2.3. Composite photograph showing two active region loops connecting two sunspots of opposite magnetic polarity (Chen & Loughhead, 1983). The photograph represents the superposition of filtergrams obtained 0.06, 0.15 and 0.25 nm to the red of Hα line centre. Geometrical reconstructions of these loops are shown in Fig. 2.6.

while Figs. 2.3 and 2.4 give examples where the apparent point of termination lies on the umbra/penumbra boundary. In these latter cases, however, the true footpoints may be located well within the umbra (Bray & Loughhead, 1985). Photographs of somewhat lower resolution showing a dark loop crossing the penumbra of a spot and apparently pouring material into the umbra have been published by Martin (1973, Figs. 8 and 9).

(3)   *Direction of flow*

      Two types of flow are observed: (a) a unidirectional flow along the axis of a loop, in other words an ascent in one leg and a descent in the other (Ellison, 1944; Tandberg-Hanssen, 1977; Bray & Loughhead, 1983); and (b) a downflow from the top towards both footpoints (Martin, 1973; Chen & Loughhead, 1983). Category (a) is illustrated in Fig. 2.2; according to Ellison, in this type of flow the direction of flow is independent of the magnetic polarities of the spot(s) involved. Figure 2.3 shows a good example of flow down both legs of each loop in a two loop system. Martin (1973) has remarked that flow of type (a) is a commonly observed characteristic of loops in new and complex active regions.

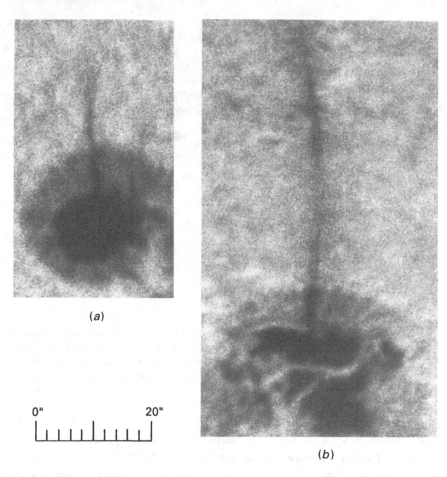

(a)

0"   20"

(b)

Fig. 2.4. Thin loops apparently terminating on the umbra/penumbra boundaries of two sunspots (Bray & Loughhead, 1986). (*a*) Loop of 7 June 1982, photographed at Hα+0.25 nm; (*b*) loop of 6 June 1980, photographed at Hα+0.1 nm.

(4)   *Number of loops in a system*

According to Ellison (1944) single or double loops are quite common, but complex formations of loops are rare. The composite photograph of Fig. 2.3, taken from Chen & Loughhead (1983), shows the entire extent of two prominent dark loops, but these authors remark that there were other less prominent loops present which could not be traced in their entirety. By contrast, (post-) flare loop systems commonly contain a large number of loops in the form of an arcade (Section 4.3.1).

(5)   *Lifetime and evolution*

Ellison (1944) found that on occasion the motion along a loop may continue for up to several hours. This is roughly the same interval over

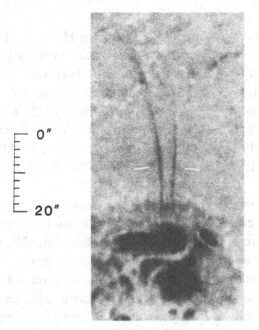

0"

20"

Fig. 2.5. Portions of two very thin (sub-arc-sec) loops photographed at Hα+0.1 nm (Bray & Loughhead, 1985). The white lines indicate the locations in the loops at which microphotometry was carried out in order to determine the true widths (see text). Note that the left-hand loop shows a perceptible increase in width further out from the sunspot.

which loop *systems* are observed to persist. For example, the loops of Fig. 2.3 belonged to a system which lasted for at least 5–6 hours (Chen & Loughhead, 1983). Similarly, Bray & Loughhead (1985) observed a system of numerous very thin loops which, as a whole, lasted for at least $3\frac{1}{2}$ hours. On the other hand, the two individual members shown in Fig. 2.5 persisted only for a few minutes.

For prominences in general, fine structures in the form of threads and knots are known to last for periods of the order of minutes (Engvold, 1979). Substantial changes can occur in similar intervals in any given large and well-developed dark loop observed on the disk. Thus, referring to the loop of Fig. 2.2, Bray & Loughhead (1983) reported substantial changes in visibility (contrast) and morphology in various parts of the loop during intervals of only 2–6 min. Furthermore, they noticed that the distribution of apparent Doppler shifts along this particular loop changed significantly over a period of the order of 15 min (Loughhead & Bray, 1984). Tandberg-Hanssen (1977) gives 15 min as a representative figure for the lifetime of a (single) loop, while Martin (1973, Fig. 8) has illustrated evolutionary changes in a long-lived loop over a period of ~1 hr.

(6)   *Structure*

In common with other prominences (Tandberg-Hanssen, 1974), active region loops often show considerable evidence of structure, including fine detail down to the limit of resolution of existing observations, i.e. $<1''$ of arc (Bray & Loughhead, 1985). This is apparent in the off-band filtergrams of Figs. 2.2–2.5 and in Figs. 8 and 9 of Martin (1973). Clearly the presence of even finer structure, below the resolution limit of present observations, cannot be ruled out (Loughhead, Bray & Wang, 1985); the question of filamentation in EUV loops is discussed in Section 2.5.5.

2.3.3   *Diameter, height and other geometrical parameters*

The diameter of the cross-section of a loop (i.e. its thickness) is typically only a few sec of arc and may be much smaller (Loughhead & Bray, 1984). Provided the cross-section is circular, measurement of loop thickness can be carried out on the (projected) image of a loop recorded on a high resolution filtergram, correction for geometry being unnecessary. The thickness of the loop of Fig. 2.2 was measured at seven locations by Loughhead, Bray & Wang (1985, Table 1), who obtained values ranging from 930 km to 2100 km (mean = 1600 km).

A special study of the two very thin loops of Fig. 2.5 was undertaken by Bray & Loughhead (1985). These loops were so thin that an accurate deconvolution procedure had to be applied to the measurements, taking proper account of the combined instrumental profile of the (30-cm) telescope and the Earth's atmosphere. They concluded that the true widths of these loops, measured at a height of several thousand kilometres, lay within the fairly narrow range $0''\!.6$ to $0''\!.8$ (440 to 580 km). Furthermore, assuming the magnetic field in the loop at the chromospheric level, B_{chrom}, to be 300 G and at the *photospheric* level, B_{spot}, to be 1000 G, they used a constancy of flux argument to deduce that at the photospheric level – where the loop entered the spot – the loop thickness was $d_{\mathrm{spot}} \lesssim 0''\!.44$ (320 km). However, if instead we take a lower value for B_{chrom} more consistent with the figures in Table 2.2, namely ~ 50 G, we obtain $d_{\mathrm{spot}} \lesssim 0''\!.18$ (130 km).

An interesting (possible) confirmation of such low figures for the loop thickness at the photospheric level comes from a detailed scrutiny of the system shown in Fig. 2.3. Examining photographs taken in the far red wing of Hα, Chen & Loughhead (1983) found that the two prominent loops shown in Fig. 2.3 shared a common footpoint on the right-hand side with a *third* loop, the (apparent) combined thickness of the composite loop just before entering the sunspot being about $1''$. Assuming that each of the three

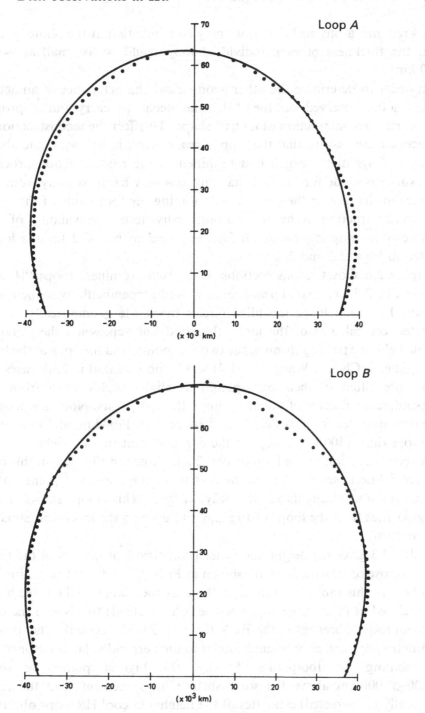

Fig. 2.6. Geometrical reconstructions of loops *A* and *B* of Fig. 2.3, showing the loops in their own planes (Loughhead, Chen & Wang, 1984). The filled circles mark the central axes of the loops. The solid lines show the (tilted) magnetic dipole field lines passing through the footpoints and the apex of each loop (see text).

loops remains a physical entity, we may conclude that at the photospheric level the thickness of each individual loop could be as small as $\sim 0\overset{.}{''}3$ (220 km).

In order to determine the other geometrical characteristics of an active region loop observed on the disk, one needs to carry out a proper geometrical reconstruction of its true shape. To effect the reconstruction it is necessary to assume that the loop is *symmetrical* in its own plane about an axis of symmetry which is determined in the reconstruction process. Allowance is made for the fact that this axis may be tilted away from the normal to the line on the solar surface joining the footpoints of the loop. The reconstruction technique automatically tests the validity of the symmetry assumption which, in fact, is found to be valid for the loops shown in Figs. 2.2 and 2.3.

Figure 2.6 shows reconstructions of the two prominent loops ('*A*' and '*B*') of Fig. 2.3, i.e. their appearance as viewed perpendicularly to their own planes. For each loop, the filled circles mark the geometrically reconstructed central axis of the loop; the solid line represents the (unique) dipole field line passing through the two footpoints and the apex of the loop (Loughhead, Chen & Wang, 1984). It will be noticed that in both cases the loops are tilted in their own planes at slight angles away from the perpendicular bisector of the line joining the footpoints. Note that loop *A* nowhere deviates from its field line by more than 1700 km, and loop *B* by no more than 1400 km, except at the depression at the top right.

A reconstruction of the loop of Fig. 2.2 is shown in Fig. 2.7. In this case the solid line is simply the line of best fit to the measured points, after reconstruction (Loughhead & Bray, 1984). This loop shows more irregularities than the loops of Fig. 2.6, but even so the loop is reasonably symmetrical.

Table 2.3 gives the height and other geometrical properties of the three loops whose reconstructions are shown in Figs. 2.6 and 2.7. Examining the numbers in this table, we notice, firstly, that the planes of the three loops are inclined at fairly large angles to the solar vertical; this is also the case for cool loops observed in the EUV (Section 2.4.3). Secondly, the axes of symmetry are tilted at only small angles to the perpendicular bisector of the line joining the footpoints. Finally, the highest points lie some 40 000–50 000 km above the solar surface, in agreement with the (geometrically uncorrected) estimates of the heights of cool Hα loops observed at the limb (cf. Section 2.2.1). These heights are comparable to those found for cool EUV loops (Section 2.4.3).

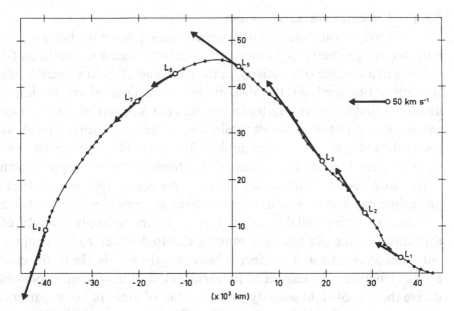

Fig. 2.7. Geometrical reconstruction of the loop of Fig. 2.2, showing the loop in
its own plane (Loughhead & Bray, 1984). The open circles $L_1 \ldots L_8$ represent the
transformed positions of the points of measurement of the line-of-sight velocity.
The arrows at these points represent the magnitudes and directions of the true
(axial) flow velocity V_0, which is supersonic at all locations (see text).

Table 2.3. *Loop heights and other geometrical properties*

Quantity	Loop of 6 June 1980	Loop A 7 June 1982	Loop B 7 June 1982
Azimuthal angle of loop plane, α (deg)	18.0	3.7	3.7
Inclination of loop plane to vertical, β (deg)	30.0	40.7	41.3
Tilt of axis of symmetry from normal, σ (deg)	3.9	6.0	−6.2
Separation of footpoints (km)	85 800	70 500	73 600
Distance between highest point and line joining footpoints (km)	46 000	63 700	71 100
Height of highest point above solar surface (km)	39 800	48 300	53 400

Note: for the sign conventions applying to the various angles, see Loughhead,
Chen & Wang (1984).

2.3.4 Velocities and accelerations

To determine the true flow velocity along the central axis of a (disk) loop whose geometry is known, one needs to measure the line-of-sight velocity at a number of locations. In the past, line-of-sight velocities of dark structures observed on the disk in Hα were determined by implicitly assuming that the wavelength of maximum (dark) contrast was the same as the wavelength of the Doppler shifted line – an easy measurement with a spectrohelioscope: see, for example, Ellison (1944). The desired line-of-sight velocity V_z could then be calculated from the usual Doppler formula.

In more recent years it has become generally appreciated that this procedure can lead to serious error (cf. Bray & Loughhead, 1983, Table II), and that to obtain reliable values it is necessary to apply a *model* of the formation of the Hα line in a moving elevated structure. In conjunction with measurements at a number of wavelengths across Hα of the contrast of the structure with respect to its background, it may then be possible to derive the line-of-sight velocity and a number of other physical parameters. The so-called 'cloud' model provides the best practical basis currently available for this purpose. Given measurements at a number of wavelengths across Hα of the contrast of the structure with respect to its background, one can use the cloud model to obtain values of the source function S, the maximum optical thickness along the line of sight t_0, the line broadening parameter $\Delta\lambda_0$ and the Doppler shift of the line absorption coefficient $\Delta\lambda_1$ ($= \lambda V_z/c$).

The application of the cloud model to the loop illustrated in Fig. 2.2 is described in Section 2.5.2. Values of V_z obtained from this analysis range from 7 km s^{-1} upward at the point of measurement nearest to the right-hand footpoint to 78 km s^{-1} downward at the point on the (inner) loop nearest to the left-hand footpoint (Bray & Loughhead, 1983). Our interest here, however, is in the true flow velocities along the axis of the loop, which can readily be derived by virtue of our knowledge of the geometry of this loop (Loughhead & Bray, 1984). These are shown in the form of vectors in Fig. 2.7, the arrows representing the magnitude and direction of the true flow velocity $\mathbf{V}_0$ at the seven points of measurement. The flow is unidirectional from one footpoint to the other.

The magnitude of the flow speed V_0 is given in Table 2.5, Section 2.5.2. It ranges from 34 km s^{-1} downward at the point L_6 to 125 km s^{-1} upward at L_3. At all locations the flow speed is supersonic: values of the Mach number M are given in the last row of Table 2.5. The maximum Mach number is $M = 6.0$ at the point L_3. On the other hand, all the measured velocities are smaller than the free-fall velocity from the highest point

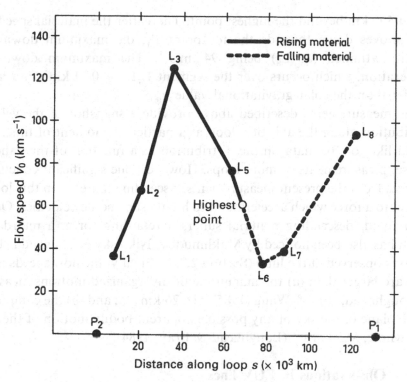

Fig. 2.8. Flow speed V_0 along the central axis of the loop of Fig. 2.7 as a function of s, the distance along the loop measured from the footpoint P_2 (Loughhead & Bray, 1984). Rising material is shown by a solid line, falling material by a dashed line. As before, $L_1 \ldots L_8$ represent the points of measurement. Note that between L_2 and L_3 the speed of the rising material almost doubles, despite the retarding effect of gravity. On the downward leg the material speeds up as it approaches L_8 and the footpoint P_1 beyond. However, it does not attain the velocity of free-fall (~ 150 km s^{-1}) from the highest point above the surface (39 800 km).

of the loop to the solar surface, which for a gravitational acceleration of $g_\odot = 0.27$ km s^{-2} is ~ 150 km s^{-1}.

It is of interest to display the flow speed V_0 as a function of s, the distance along the central axis of the loop measured from one of its footpoints. This is shown in Fig. 2.8. We see that the loop material rises from the footpoint P_2 with an increasing upward velocity despite the retarding effect of gravity. The maximum acceleration takes place over the segment between L_2 and L_3, the average value over this segment being 0.38 km s^{-2} – a value numerically greater than the gravitational acceleration $g_\odot$. The maximum upward velocity is attained at L_3, roughly halfway up the ascending branch of the loop. Between this point and the highest point above the solar surface (which is marked on the graph), the material slows down. A further deceleration occurs until L_6 is reached, a distance along the loop of

$\Delta s \simeq 9500$ km beyond the highest point. Thereafter the material speeds up as it moves down towards the footpoint P_1, the maximum downward velocity attained (at L_8) being 94 km s^{-1}. The maximum downward acceleration, which occurs over the segment $L_7 L_8$, is 0.11 km s^{-2}, a figure smaller than the solar gravitational value $g_\odot$.

The measurements described above provide a snapshot of the velocity distribution along the axis of a loop at a particular moment of time. One would like to have data on the distribution as a function of time and, of course, measurements of more loops. However, one significant conclusion emerges from the present measurements: ascending material in this loop is subject to a force which accelerates it to highly supersonic velocities. On the other hand, descending material suffers a retarding force, a retardation which has also been noticed by Makhmudov, Nikolsky & Zhugzhda (1980) in loops observed at the limb (Section 2.2.2). Finally, the flow speeds in the loop are larger than (a) the magnitude of disorganized motions measured by Loughhead, Bray & Wang (1985), viz. 26 km s^{-1}; and (b) the component in the plane of the sky of any possible apparent bodily motion of the loop as a whole, <5 km s^{-1} (Loughhead & Bray, 1984).

2.4　Observations in EUV lines
2.4.1　*Introduction*

Below 150 nm the contribution of the photospheric layers to the solar spectrum vanishes and the radiation comes from overlying material at chromospheric or coronal temperatures. When this wavelength is reached, the character of the emission has changed from that of the familiar dark-line spectrum in the visible and near ultraviolet to that of a bright-line spectrum (cf. Fig. 2.9). The far ultraviolet extends down to about 10 nm, below which it is customary to categorize the emission as soft X-ray radiation (Section 3.4.1). Some authors refer to the region from 150 to 30 nm as the extreme ultraviolet (EUV) and the region from 30 to ~ 10 nm as the XUV. This distinction is, however, made on instrumental rather than physical grounds (Svestka, 1976, p. 161) and in this book we shall use the term EUV to refer to the whole spectral region from 150 to ~ 10 nm.

The solar EUV spectrum (Fig. 2.9) is dominated by emission from resonance lines of H I (Lyman series), He I, and He II, of intermediate stages of ionization of C, N, O, Si and S, and of highly-ionized stages of Si, Ne, Mg and Fe. Other distinctive features are the Lyman continuum at $\lambda < 91.2$ nm, and the He I and He II continua at $\lambda < 50.4$ nm and $\lambda < 22.8$ nm respectively. Under the conditions of formation normally assumed to apply, the intensity of any given line is a sensitive function of

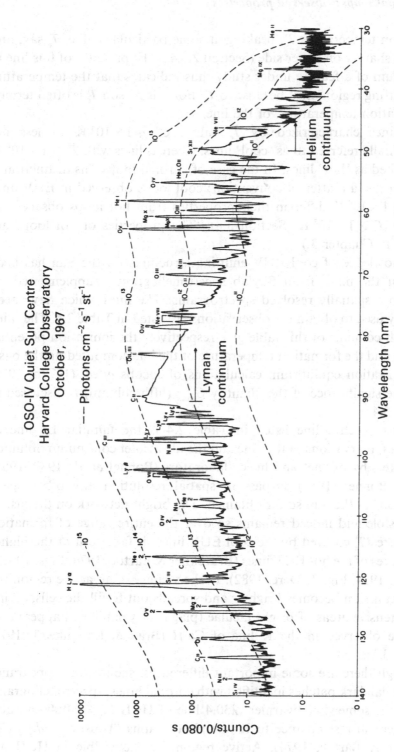

Fig. 2.9. Prominent EUV emission lines visible on a spectral scan at the centre of the solar disk obtained with the spectrometer on Orbiting Solar Observatory IV. Note the Lyman continuum of H I at λ91.2 nm and the He I continuum at λ50.4 nm. (By courtesy of Harvard College Observatory.)

the electron temperature T_e, peaking at some particular value, $\bar{T}_e$ say, and falling off sharply on either side (Section 2.5.4). The presence of this line in the spectrum of a feature under study thus indicates that the temperature in the emitting region must be close to $\bar{T}_e$. For this reason $\bar{T}_e$ is often termed the 'formation temperature' of the line.

EUV lines characterized by $\bar{T}_e$ values of $\sim 1 \times 10^6$ K or less are conventionally referred to as 'cool' lines, whereas lines with $\bar{T}_e > 1 \times 10^6$ K are described as 'hot' lines. In the case of coronal loops this distinction is more than just a matter of convention: cool loops observed in EUV lines with $T_e \ll 1 \times 10^6$ K differ in their properties from hot loops observed in lines with $T_e \gtrsim 1 \times 10^6$ K (Section 3.8). (The properties of hot loops are described in Chapter 3.)

Our knowledge of cool EUV emitting structures on the Sun has been derived in the main from Skylab spectroheliograms, supplemented by results from spatially resolved spectrograms. The lines which have been commonly used to obtain the observations are listed in Table 2.4. The third and fourth columns of this table give respectively the ionization potential of the ion and the formation temperature of the line computed on the basis of the ionization equilibrium calculations of Jacobs *et al.* (1977b, 1978, 1979). The significance of the quantity G_{max} (fifth column) is discussed in Section 2.5.4.

The first spectral line listed in Table 2.4 is the familiar Lα line of hydrogen. Observations in this line and the ultraviolet continuum obtained from rockets by Bonnet and his collaborators (Bonnet *et al.*, 1980, 1982; Foing & Bonnet, 1984) surpass in spatial resolution any other space observations of the Sun so far obtained. The bright network on the disk is clearly visible and indeed remains so over the entire range of formation temperatures $\bar{T}_e$ covered by the cool EUV lines; it fades out at the higher temperatures of the hot EUV lines (Brueckner & Bartoe, 1974, Figs. 10–12; Goldberg, 1974, Fig. 7; Dere, 1982). In the vicinity of an active region the network emission becomes brighter and spreads out to fill the cells, giving rise to extensive areas of bright faculae (plage) very similar in appearance to faculae observed in the K line of Ca II (Bray & Loughhead, 1974, Section 3.3.2).

Although there are some important differences, such as the appearance of occasional dark patches identified with coronal holes, spectroheliograms taken in the somewhat 'warmer' $\lambda 30.4$ line of He II ($\bar{T}_e = 80\,000$ K) also show a general resemblance to K-line photographs (Tousey *et al.*, 1973; Brueckner & Bartoe, 1974). Active region faculae visible in He II are essentially horizontal structures whose intensity, except over sunspots,

Table 2.4. *EUV lines used to obtain observations of cool loops*

Ion	Wavelength (nm)	Ionization potential (eV)	$\bar{T}_e$ (K)	G_{max}
H I	*121.57*	13.7	$\sim 2 \times 10^4$	—
He II	*30.38*	54.4	$\sim 8 \times 10^4$	—
C II	*133.45*	24.4	4.3×10^4	2.6×10^{-4}
	133.57			2.7×10^{-4}
C III	*97.70*	47.9	7.9×10^4	4.3×10^{-4}
C IV	*154.82*	64.5	1.0×10^5	4.3×10^{-4}
N IV	*148.65*	77.5	1.5×10^5	7.6×10^{-4}
N V	*123.88*	97.9	1.8×10^5	3.7×10^{-4}
O VI	*55.41*	77.4	1.8×10^5	3.9×10^{-4}
O V	*62.97*	113.9	2.5×10^5	3.7×10^{-4}
	121.84			5.7×10^{-4}
	137.13			6.0×10^{-4}
O VI	*103.20*	138.1	3.2×10^5	2.7×10^{-4}
Si III	*120.65*	33.5	6.3×10^4	4.1×10^{-4}
	129.67			4.6×10^{-4}
Si VIII	*31.98*	303.1	7.9×10^5	3.1×10^{-4}
Ne V	41.62	126.3	3.2×10^5	3.3×10^{-4}
Ne VI	*40.33*	157.9	4.0×10^5	3.2×10^{-4}
Ne VII	*46.52*	207.2	5.0×10^5	2.3×10^{-4}
Mg VI	*40.33*	186.5	4.0×10^5	4.1×10^{-4}
Mg VII	*43.49*	224.9	6.3×10^5	3.7×10^{-4}
Mg VIII	*43.67*	266.0	7.9×10^5	3.2×10^{-4}
Mg IX	*36.81*	327.9	1.0×10^6	1.7×10^{-4}

(a) Wavelengths in the second column referring to resonance lines (Section 2.5.4) are given in italics;
(b) For each line $\bar{T}_e$ is the temperature at which function $G(T_e)$ defined by Eqn. (2.15) attains its maximum value, G_{max} (Section 2.5.4);
(c) For more complete atomic data see Wiese, Smith & Glennon (1966) and Wiese, Smith & Miles (1969).

shows a strong correlation with the strength of the underlying photospheric magnetic field (Cheng, Smith & Tandberg-Hanssen, 1980; Dere, 1982).

Recent rocket filtergrams with a (claimed) resolution of 1″ of arc have provided strong evidence for the presence of active region loops in Lα. On the *disk* (Bonnet & Tsiropoula, 1982; Haisch et al., 1986) the Lα loops are not impressive: they reveal themselves as diffuse bright and dark arcs against a complex background of enhanced emission. At the limb, on the other hand, the Lα loops are seen with the clarity of loops observed at the limb in the Hα line (see Tsiropoula *et al.*, 1986, Figs. 3(*a*) and 8(*a*)). Their diffuse appearance on the disk has been attributed to their small optical thickness. Earlier discussions of the presence of loops on disk spectroheliograms obtained in Lα and other cool lines are to be found in papers by Foukal (1978), Bonnet *et al.* (1980), Bonnet *et al.* (1982) and Cheng *et al.* (1980).

However, it is not until we turn to lines warmer than the He II line that the full three-dimensional structure of active regions becomes apparent. Then an active region is seen to consist of a complex system of bright loops arching between areas of opposite magnetic polarity. The identification of these structures as *loops* in the geometrical sense is confirmed by photographs of active regions near the limb. The individual loops are oriented in many different directions and frequently overlap one another. In fact, the clarity and sharpness with which a loop is seen depend both on the temperature of the line used in the observations and, of course, on the spatial resolution achieved.

It is interesting to note that the spatial resolution shown by the Skylab spectroheliograms improved significantly between the first and final missions (Sheeley & Golub, 1979); this is reflected in the quality of the published observations. Even with the best resolution available – dubiously claimed by some authors to be as high as 2″ of arc – there is no guarantee that all the vital detail has been resolved. Nevertheless, a considerable amount of information is now available about the properties of cool EUV loops, which is summarized below.

2.4.2 *Active region loop systems: general characteristics*

Figure 2.10 shows a bipolar active region on the disk photographed during the final Skylab mission in the two cool EUV lines Ne VII $\lambda 46.5$ and Mg IX $\lambda 36.8$ as well as the distribution of the underlying photospheric magnetic field (Sheeley, 1980). The first line has a formation temperature of $\bar{T}_e = 5 \times 10^5$ K and is representative of lines formed in the middle of the cool EUV range, while the second ($\bar{T}_e = 1 \times 10^6$ K) is representative of lines formed near the upper limit of the range. It is evident from Fig. 2.10 that the emission from the active region is centred on two bright areas overlying the regions of opposite magnetic polarity in the photosphere; not surprisingly, the outlines of the bright areas follow the contours of the underlying Hα faculae (Cheng, Smith & Tandberg-Hanssen, 1980; Schmahl *et al.*, 1982). With the best available resolution the central bright areas are resolved into aggregates of small, very bright kernels, some of which are noticeably elongated.

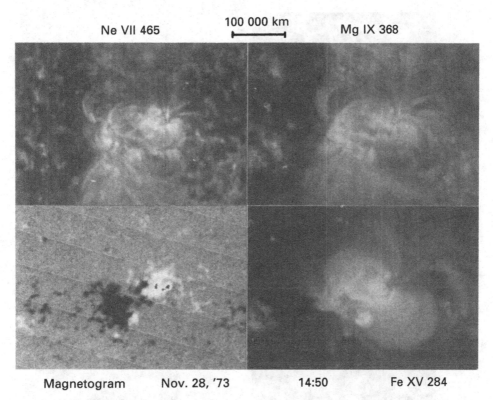

Ne VII 465   100 000 km   Mg IX 368

Magnetogram   Nov. 28, '73   14:50   Fe XV 284

Fig. 2.10. Bipolar active region on the disk observed almost simultaneously in the two cool EUV lines Ne VII $\lambda46.5$ nm and Mg IX $\lambda36.8$ nm, as well as in the hot EUV line Fe XV $\lambda28.4$ nm (Sheeley, 1980). White and black patches on the Kitt Peak magnetogram delineate areas of positive and negative polarity respectively in the underlying photospheric magnetic field. The elongated bright features radiating outwards from the central areas of bright (facular) emission represent the lower ends of coronal loops.

Just outside the central bright areas in Fig. 2.10 a number of bright loop-like features can be seen pointing more or less radially outwards. Limb observations demonstrate that these features represent portions of coronal loops, which may extend up to heights of tens of thousands of kilometres.

There is clearly a close correspondence between the loop systems observed in the Ne VII and Mg IX lines and, in fact, according to Sheeley (1980), the individual features coincide to within ± 2000 km (Section 2.4.3). On the other hand, the individual loops seen in Mg IX are decidedly more diffuse and longer (i.e. emitting along a greater part of their length) than their counterparts in Ne VII; these differences are characteristic of loops formed in these two temperature regimes (Cheng *et al.*, 1980; Sheeley, 1980).

The completeness with which particular loops are outlined by emission

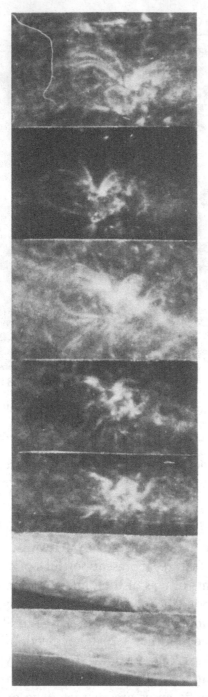
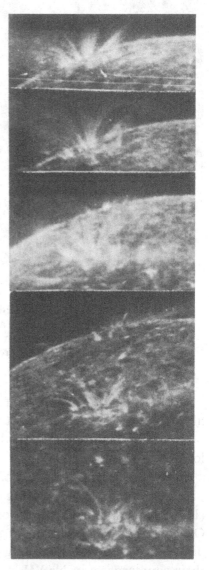

Fig. 2.11. Change in the appearance of an active region observed in the cool EUV line Ne VII $\lambda46.5$ nm during its passage across the disk (Tousey *et al.*, 1973). Near the central meridian some of the loops visible are almost complete. Near the limb complete loops are not visible; instead, the active region has a decidedly spike-like appearance.

Fig. 2.12. McMath active region 12628 observed at the west limb in the cool EUV line O VI λ103.2 nm (Levine, 1976). Despite the obscuration at low heights caused by overlapping, the region consists basically of a number of separate loops lying in planes inclined at widely differing angles to the solar vertical. Individual loops may reach heights of tens of thousands of kilometres (see text).

along their length varies markedly with time. As a result active regions observed in Ne VII and lines of similar temperature often show a more spike- or plume-like structure than a loop-like one. Of course such an appearance does not mean that loops are absent but implies that only their lower ends are bright at the time of observation. This point is well illustrated in Fig. 2.11, which shows how the appearance of an active region in Ne VII λ46.5 changed as it crossed the disk from the east to the west limb (Tousey *et al.*, 1973). Near the central meridian some almost complete loops, tens of thousands of kilometres in height, can be seen connecting the areas of opposite magnetic polarity. Near the limb, on the other hand, complete loops are not visible and the active region has a decidedly spike-like appearance. Note that the overall configuration of the loop system appears to change rather slowly, on a time scale of two or three days.

Foukal (1976) has studied the cool EUV emission over 22 large sunspots and found that the emission is often brighter there than anywhere else in the active region (see also Brueckner & Bartoe, 1974; Sheeley *et al.*, 1975; Dere, 1982). However, the intensity and distribution of the radiation above the spots changed markedly with time and, as a consequence, a large umbra can remain invisible in the cool EUV for as long as several days. Previously Foukal (1975) had recorded two cases where the loops associated with the penumbra of a preceding sunspot were the brightest and most prominent ones present.

A good illustration of the three-dimensional structure of active regions seen in the EUV is provided by Fig. 2.12, which is a photograph of McMath region 12628 at the west limb taken in the line O VI λ103.2 ($\bar{T}_e = 3.2 \times 10^5$ K). Despite the obscuration at low heights caused by the overlapping of different features, it is evident that the region is composed basically of a large number of separate loops lying in planes inclined at widely varying angles to the solar vertical. Some loops are nearly vertical, whereas others are almost horizontal. According to Levine (1976, cf. Figs. 5 and 6) the number of vertical as opposed to horizontal loops increased during the passage of the region across the disk.

Note that the number of loops visible beyond the limb in Fig. 2.12 appears to be much greater than that seen in active regions on the disk in lines of comparable temperature (compare Figs. 2.10 and 2.15). This implies that on the disk only the brightest of the loops actually present show up in emission (Athay, Gurman & Henze, 1983).

When only the lower portions of the bright loops emanating from a bright area above a sunspot near the limb are visible, their divergence with height may give rise to the appearance of a conspicuous bright 'plume' over the spot. This is well illustrated by Foukal (1976, Fig. 3).

Finally, mention should be made of the work of Schmahl *et al.* (1982), who sought to investigate the relationship between EUV loops and dark Hα filaments – i.e. prominences on the disk. In several cases cool EUV loops were visible over unstable filaments which subsequently disappeared, the loops being inclined at quite small angles to the axis of the filament. No loops were observed over stable filaments except at, and beyond, their ends.

2.4.3 *Properties of individual loops*

Because of the generally inadequate spatial resolution of the available EUV observations and the paucity of systematic analyses of this data in the literature, relatively little information is available about the morphological and dynamical properties of individual loops. What is known may be conveniently summarized under the following headings:

(1)   *Shape, size and thickness*

In common with all other solar features, cool EUV loops are always observed in projection against the plane of the solar disk or of the

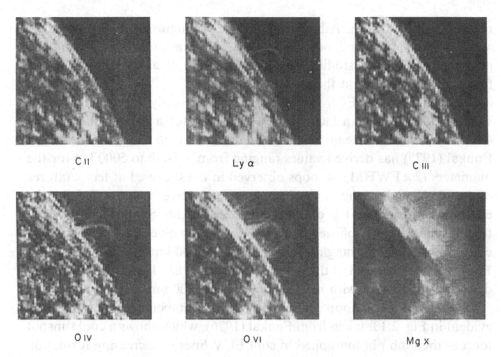

Fig. 2.13. Sunspot loop prominence at the limb observed in cool EUV lines of increasing formation temperature ranging from C II λ133.5 nm to O VI λ103.2 nm (Foukal, 1976). Note the significant increase in loop width with temperature. The last image shows the great enhancement in the active region emission around the loop in the hot EUV line Mg X λ62.5 nm.

sky beyond. Consequently, in the absence of geometrical reconstructions of individual loops along the lines of those accomplished for Hα loops (Section 2.3.3), relatively little can be said about the true shapes and sizes of cool EUV loops. Nevertheless, limb observations such as that in Fig. 2.12 do show that most loops are essentially *planar*. The inclination of the planes containing the loops may vary from nearly vertical to nearly horizontal. Moreover, there is some indication that the preferred inclination of a loop may become more vertical as the region grows older (Section 2.4.2).

Limb observations also yield estimates of the heights attained by the loops: these are typically of the order of tens of thousands of kilometres. In particular, Cheng (1980) has measured the (projected) heights of a well-defined loop prominence in a number of EUV lines and found values ranging from 57000 km in Ne VII to 67000 km in Mg IX. His results indicate an increase in loop height with temperature. Chiuderi (1981) quotes a typical height of 50000 km, while Athay, Gurman & Henze (1983) remark that the heights of cool EUV loops often exceed 60000 km and may reach over 110000 km. Athay *et al.* do not distinguish between ordinary active region loops and post-flare loops, while according to Cheng the prominence that he studied (14 August 1973) 'was probably formed following a flare behind the limb'.

The widths of cool EUV loops seem to increase only slowly with height: Foukal (1976) quotes a factor of less than two for a height difference of 40000 km (see also Cheng, 1980). From measurements of intensity tracings Foukal (1975) has derived values ranging from <2000 to 5000 km for the diameters (2 × FWHM) of loops observed in lines formed at temperatures of 5×10^5 K or less; no corrections appear to have been applied for the effects of instrumental profile and scattered light. Similar photometric tracings have been published by Dere (1982), who gives larger estimates of cool loop diameters, ranging from 6000 to 22000 km (see his Table III). For Lα loops observed at the limb, Tsiropoula *et al.* (1986) have measured diameters – constant with height – of ~ 2000–3500 km.

On the other hand, loop width increases with temperature. This is clearly evident in Fig. 2.13, taken from Foukal (1976), which shows a cool sunspot loop at the limb photographed in cool EUV lines of increasing formation temperature, ranging from C II $\lambda 133.5$ ($\bar{T}_e = 4.3 \times 10^4$ K) to O VI $\lambda 103.2$ ($\bar{T}_e = 3.2 \times 10^5$ K). Fig. 2.10 gives a good illustration of the same effect in the Ne VII and Mg IX lines (see also Sheeley, 1980, Fig. 2). Cheng (1980), however, derives a contrary conclusion from measurements of a loop prominence, finding the width of the loop to be 9″ of arc in Mg IX as opposed to 10″ in Ne VII.

In passing, it may be noted that the faint loop visible on the Skylab Lα photograph in Fig. 2.13 appears to be almost devoid of internal structure. In contrast, a small loop prominence present on a Lα photograph published by Bonnet *et al.* (1980, Fig. 8) shows the very fine, thread-like, internal structure which is revealed when more adequate resolution is achieved (cf. Dere, 1982).

Another parameter, which sometimes arises in studies of loop stability (Chapter 5), is the so-called *aspect ratio*, namely the ratio of the radius of the cross-section of the loop to its length. Cheng, Smith & Tandberg-Hanssen (1980) found one large Ne VII loop to have a radius of about 2″.5 of arc and an apparent length of about 430″, giving an aspect ratio of 0.006. (In default of other information, the length of an observed loop is commonly estimated by assuming the loop to be semicircular.) This is an order of magnitude smaller than that found for hot EUV loops (Section 3.3.3).

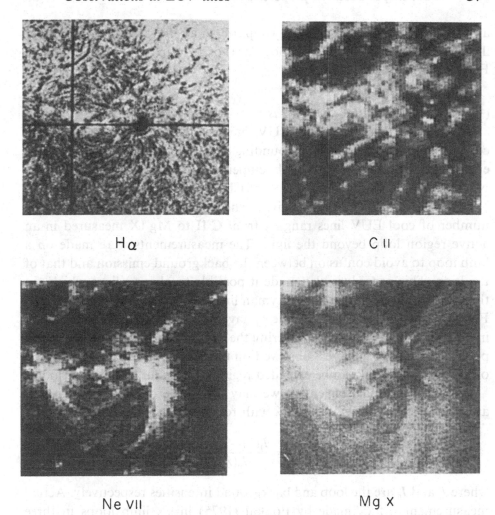

Fig. 2.14. Active region observed near disk centre in Hα, the cool EUV lines C II
λ133.5 nm and Ne VII λ46.5 nm, and the hot EUV line Mg X λ62.5 nm (Foukal,
1976). Over the umbra of the large (preceding) sunspot the intensity of the Ne VII
emission greatly exceeds that of the C II and Mg X emission. A prominent Ne VII
loop can be seen arching between this spot and a group of small followers. (The
scale of the EUV images is 1.1 times greater than that of the Hα photograph.)

(2)   *Location of footpoints*
     The footpoints of cool EUV loops, such as those observed in the
Ne VII and Mg IX lines, are generally located on the peripheries of the two
areas of opposite magnetic polarity in a bipolar active region (Cheng *et al.*,
1980). This tendency is well illustrated in Fig. 2.10. On the other hand,
there is an important class of cool coronal loops which have at least one end
anchored in bright EUV emission over a sunspot (Section 2.4.2). An
example is the Ne VII loop which can be seen in Fig. 2.14 arching between

a large preceding spot and a group of small following spots. Other illustrations of spot-associated loops are to be found in Fig. 2.13 and in Figs. 2–4 of Foukal (1976).

(3)   *Loop intensities and contrast*

Loops observed in cool EUV lines beyond the limb show a strong contrast with respect to their surroundings. The intensity of the background emission is weak but increases with temperature (compare, for example, the Ne VII and Mg IX images in Fig. 2 of Cheng, 1980). Raymond & Foukal (1982) have published a table giving the absolute intensities of a large number of cool EUV lines ranging from C II to Mg IX measured in an active region loop beyond the limb. The measurements were made on a limb loop to avoid confusion between the background emission and that of the loop. This procedure also made it possible to observe lines which, on the disk, would be buried under Lyman lines or the Lyman continuum (cf. Fig. 2.9). Noyes *et al.* (1985, Table 1) have published an extensive table of intensities and identifications covering the range 30.3–134.3 nm. Their data pertain to 'sunspot plumes' observed on the disk with a spatial resolution of 5″, which they take to be extended regions within large magnetic loops.

Following normal convention we may define the contrast of a loop at any given point and wavelength λ with respect to its surroundings as

$$C_1(\lambda) = \frac{I_1(\lambda) - I_0(\lambda)}{I_0(\lambda)},$$

where I_1 and I_0 are the loop and background intensities respectively. Actual measurements of C_1 made by Foukal (1975) in six limb loops in three different active regions yielded values ranging from 1 to 8 in the line O VI $\lambda103.2$ and 2 to 7 in Ne VII $\lambda46.5$. Dere (1982) has published intensity tracings across a number of active region loops observed on the disk but has not given any actual contrast values.

(4)   *Lifetime and evolution*

Virtually no quantitative information has so far been published on the lifetime and evolution of cool EUV loops. However, it is known that loops observed in cool lines may evolve appreciably in just a few hours (Levine & Withbroe, 1977; Cheng *et al.*, 1980; Sheeley, 1980; Athay, Gurman & Henze, 1983). Significant changes can occur on a time scale of half an hour or less and, moreover, such changes are more marked in cooler lines like Ne VII than in warmer lines like Mg IX. This is illustrated in Fig.

2.15, which shows the evolution taking place over 34 min in the Ne VII and Mg IX loops associated with the active region illustrated in Fig. 2.10. Note the prominent loop on the left-hand side which was visible throughout this period in Mg IX but only became fully apparent in Ne VII towards the end. Hence, as Sheeley (1980) remarks, one may have to wait a further 30 min or more to see in Ne VII most of the loops visible on a single Mg IX image, and even then some of the Mg IX loops may fail to appear in Ne VII. Cool loops observed by Habbal, Ronan & Withbroe (1985) in Ne VII showed strong variations over periods of only 5–10 min, whereas hot loops observed in Si XII varied on a time scale (typically) of hours.

It is possible that spot-associated loops may be more stable than other cool EUV loops: Foukal (1976, Figs. 4 and 5) has published photographs of a large spot loop near the limb whose basic form remained relatively unchanged over a period of 27 hr.

(5)   *Velocities*

Attempts to date to measure line-of-sight velocities in cool EUV lines have been hampered by difficulties in instrumental calibration and by uncertainties in determining the reference level for zero velocity (Athay, Gurman & Henze, 1983). Despite this, Athay *et al.* (1982) found evidence of the existence of large-scale velocity patterns above active regions which, around sunspots, correspond to the well-known Evershed inflow observed at chromospheric levels (Bray & Loughhead, 1964, Section 4.4.2). In a subsequent investigation Athay, Gurman & Henze (1983) measured line-of-sight velocities in individual bright loops in a number of active regions on the disk, using observations in the C IV λ154.8 line obtained with the ultraviolet spectrometer and polarimeter on the SMM satellite. The formation temperature of this line is about 1×10^5 K (Table 2.4). The lengths of the loops studied ranged from 22000 to 109000 km and their diameters were less than about 7000 km. Moreover, limb observations made it clear that only the brightest of the C IV loops actually present show up in emission on the disk, and therefore the results refer only to the unusually bright loops in each active region.

The line-of-sight velocities in the C IV loops were found to range from ± 5 to ± 10 km s^{-1} near disk centre and from ± 9 to ± 18 km s^{-1} near the limb, a negative sign denoting a redshift and a positive sign a blueshift. In most cases the flow was directed downwards in both legs of the loop, but there were others showing a unidirectional flow from one end of the loop to the other. This pattern is analogous to that observed in the Hα active region loops discussed in Sections 2.2 and 2.3.

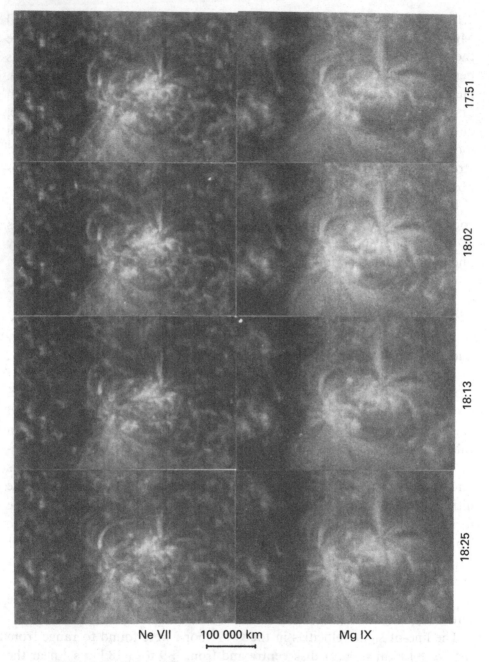

Ne VII 100 000 km Mg IX

17:51

18:02

18:13

18:25

Fig. 2.15. Evolution of the loops associated with the active region of Fig. 2.10. The observations were taken in the cool EUV lines Ne VII λ46.5 nm and Mg IX λ36.8 nm and cover a period of 34 min. During this interval the Ne VII loops underwent more apparent change than the warmer Mg IX loops. The prominent loop on the left-hand side was visible in Mg IX throughout but became fully apparent in Ne VII only towards the end.

With an assumed geometry and using the same C IV line, line-of-sight velocities have also been determined by Kopp *et al.* (1985). Their velocities for three loops (see their Fig. 5) are larger than those of Athay and his co-workers. Attempts to explain the observations by various dynamical loop models (e.g. syphon flow, downflow in both legs) were inconclusive.

As we have seen in Section 2.3.4, only with the aid of a full geometrical reconstruction is it possible to convert measured line-of-sight velocities in a loop to true velocities along the loop axis. Nevertheless, Roussel-Dupré *et al.* (1984) have attempted to determine the velocity along the axis of a small coronal loop spanning a (bipolar) ephemeral region with the aid of EUV spectrograms obtained during the second rocket flight of the U.S. Naval Research Laboratory's high-resolution telescope and spectrograph (HRTS). Using measurements made in the cool C IV λ154.8 line and assuming an arbitrary loop geometry they purport to infer the existence of a supersonic axial velocity of 95 km s^{-1}. However, it would seem from the authors' own description (see their Fig. 4) that the measurements actually refer more to the velocity variation *across* the loop than to the variation along the loop axis.

(6)   *Spatial relationship between loops seen in different lines*

On a casual inspection of EUV spectroheliograms one often gains the impression of seeing the same coronal loop in lines of widely dissimilar formation temperatures. But are such features actually coincident? Foukal (1975) came to the conclusion that they are and, in a series of papers, expounded the hypothesis that a loop consists of a cool core surrounded by a series of concentric sheaths, each one filled with material hotter than that in the adjacent inner sheath. It has since been established, however, that this model certainly does not describe the relationship between cool and hot EUV loops: they are now known not to be coincident (Section 3.7.3). But what of apparently coincident loops visible in cool EUV lines of quite widely different formation temperatures?

Foukal claimed that the emissions of lines formed below 1×10^6 K define cylinders that are coaxial to within 5" of arc. Cheng, Smith & Tandberg-Hanssen (1980) found that loops seen in Ne VII ($\bar{T}_e = 5 \times 10^5$ K) and Mg IX ($\bar{T}_e = 1 \times 10^6$ K) are coincident to within ± 5", while Sheeley (1980) – using observations obtained during the final Skylab mission – claimed coincidence to within ± 3". Dere (1982) also found that loops observed in cool EUV lines are generally coaxial to within the accuracy of the data. By comparison the cross-sectional diameter of such loops may range from

~3″ to 10″ (see above). At present, therefore, the accuracy of the available measurements is inadequate to resolve the question of the possible extent to which loops observed in different cool EUV lines are cospatial.

There is, however, one piece of evidence which cautions against any assumption of close cospatiality along the entire length of a loop. As we have already mentioned above, Cheng (1980) has measured the (projected) heights of a well-defined loop beyond the limb (possibly a post-flare loop) in a number of EUV lines and found that the measured values for the Ne VII and Mg IX lines, for example, differ by some 13″ – an amount exceeding the cross-sectional diameter of 10″!

Cheng *et al.* (1980) have pointed out that the spatial relationship between Ne VII $\lambda46.5$ and Mg IX $\lambda36.8$ loops reflects the variation in the line emissivity with temperature. They place the peak of the emissivity curve of the Ne VII line at 6.3×10^5 K and that of the Mg IX line at 1×10^6 K. Over the temperature range $6 \times 10^5 - 1 \times 10^6$ K there is considerable overlap between the two curves when both lines are in strong emission. Therefore both lines can come from the same place provided the temperature of the material there is between 6×10^5 and 1×10^6 K. On the other hand, when the temperature rises above 1×10^6 K to the point where the emissivity of the Mg IX line drops to 60% of its maximum value the emissivity of the Ne VII line falls by two orders of magnitude below its peak value. This is consistent with the fact that Mg IX loops are found to be more diffuse than their Ne VII counterparts (Section 2.4.2): the Ne VII emission comes from material at a temperature around 6×10^5 K while the Mg IX must come from material at temperatures in the range 6×10^5 K$-\sim 1.5 \times 10^6$ K occupying a greater volume of space. It would therefore appear physically impossible for Ne VII and Mg IX loops to be entirely cospatial.

2.5   Physical conditions in cool loops

2.5.1   *Introduction*

Having reviewed the morphological and dynamical properties of cool loops, we now turn our attention to their physical conditions. As before we shall first discuss the data obtained in the visible region of the spectrum (mainly Hα) and then describe information obtained from EUV observations.

In the visible region it is convenient to separate disk and limb observations. Disk observations (in Hα) are considered first (Section 2.5.2) because – perhaps surprisingly at first sight – they have given us the most complete account currently available of the physical conditions in a single loop. This does not imply, however, that limb observations are of lesser importance. In fact, they offer considerable potential for inferring physical conditions from observations not only in Hα but also in other visible and near-visible lines and continua. A number of well-tried diagnostic techniques are already available, based on work extending over many years on the investigation of the physical conditions in quiescent prominences and the quiet chromosphere (e.g. spicules). However, up to now they have not been widely applied to the (cool) loops which form the subject of the present chapter. Since detailed accounts of the theory underlying these techniques are available in the literature of prominences and the chromosphere, we will refrain from discussing them in detail here.

We next turn to the EUV region of the spectrum where, because the available data is rather sparse, it is practicable to deal with the disk and limb observations together. In contrast to the visible region, the subject of EUV diagnostics is more recent. Therefore, in Section 2.5.4 we give a short introduction to the subject, tailored to the particular circumstances applying to coronal loops. The physical conditions in cool loops derived from EUV observations are summarized in Section 2.5.5.

2.5.2 *Analysis of disk observations in Hα*

In this section we shall first consider the results which have been obtained by application of existing diagnostic techniques and then discuss how future refinements may lead to a more accurate knowledge of the physical conditions in cool loops.

(1) *Application of existing diagnostic techniques*

Loughhead, Bray & Wang (1985) have for the first time presented a fairly complete description of the physical conditions in a cool loop observed on the disk in Hα. Their method requires: (a) a knowledge of t_0 and $\Delta\lambda_0$, the maximum optical thickness along the line of sight through the loop and the line broadening parameter respectively – preferably at several locations; (b) a knowledge of ℓ the *geometrical* path length through the loop at the same locations; and (c) a non-LTE theory of the excitation of hydrogen under conditions realistically simulating those in a structure (such as a loop) elevated above the solar surface and illuminated from below. It is then possible to determine the electron temperature and density T_e and N_e; the particle densities of the first three quantum states of hydrogen N_1, N_2 and N_3; the degree of hydrogen ionization x_H; the gas density ρ and pressure p; the speed of sound a; and – given the axial flow speed along the loop V_0 (Section 2.3.4) – the Mach number M.

The analysis proceeds as follows: assuming uniform conditions along the line of sight through the loop, we have

$$\kappa_0 = t_0/\ell, \tag{2.1}$$

where κ_0 is the maximum value of the line absorption coefficient and ℓ is related to the observed diameter of the loop's cross-section (assumed circular) by the formula

$$\ell = d/\sin\varepsilon, \tag{2.2}$$

ε being the angle between the loop's central axis and the line of sight to the observer. Assuming pure Doppler broadening, we have in addition

$$\kappa_0 = \frac{\pi^{\frac{1}{2}}e^2}{m_e c}\frac{f\lambda^2}{c}\frac{N_2}{\Delta\lambda_0}, \tag{2.3}$$

(cf. Giovannelli, 1967b; Gray, 1976, p. 249), where $f = 0.640742$ is the oscillator strength for $H\alpha$ and m_e is the electron mass. Inserting numerical values we obtain from Eqns. (2.1) and (2.3) the following expression for N_2 in terms of measurable quantities:

$$N_2 = 7.255 \times 10^8 t_0 \frac{\Delta\lambda_0}{\ell}\ \text{cm}^{-3} \tag{2.4}$$

($\Delta\lambda_0$ in nm, ℓ in km).

This equation was applied by Loughhead *et al.* to the active region loop illustrated in Fig. 2.2. For this loop the true geometry was known (Loughhead & Bray, 1984), thus allowing ℓ to be determined at seven locations along the loop from the measured values of d and ε via Eqn. (2.2). Furthermore, the same authors (Bray & Loughhead, 1983) had earlier determined t_0 and $\Delta\lambda_0$ (as well as the source function S and $\Delta\lambda_1$, the Doppler shift of the line absorption coefficient) by applying the so-called 'cloud' model to their measurements of the contrast of the loop with respect to the background at a number of wavelengths in the $H\alpha$ line. The first four rows of Table 2.5 give their values of d, ℓ, t_0 and $\Delta\lambda_0$. Inserting the appropriate figures in Eqn. (2.4) Loughhead *et al.* obtained figures for N_2 at each of the seven points of measurement (Table 2.5).

The best practical basis for proceeding from the 'observed' values of N_2 to figures for T_e and N_e are calculations carried out by Giovanelli (1967a) for the excitation of hydrogen over a range of typical conditions. Adopting a four-level representation of the hydrogen atom (three bound states + a continuum), Giovanelli solved the equations of statistical equilibrium. He modified the net radiative bracket to take account of the effects of self-

Table 2.5. *Physical conditions in an active region loop*

Quantity	Mean value	Range
Loop thickness d (km)	1600	930–2100
Path length through loop ℓ (km)	2300	1100–4200
Maximum optical thickness along the line of sight t_0	0.92	0.70–1.15
Line broadening parameter $\Delta\lambda_0$ (nm)	0.070	0.054–0.092
Temperature T_e (K)	21000	15000–25000
Electron density N_e (cm^{-3})	5.6×10^{10}	3.0×10^{10}–7.6×10^{10}
Hydrogen particle densities (cm^{-3}):		
N_1	1.6×10^8	1.0×10^8–3.7×10^8
N_2	2.5×10^4	0.9×10^4–4.8×10^4
N_3	2.9×10^2	0.7×10^2–5.0×10^2
Degree of hydrogen ionization x_H	0.996	0.988–0.999
Gas density ρ (g cm^{-3})	1.3×10^{-13}	0.7×10^{-13}–1.7×10^{-13}
Gas pressure p (dyne cm^{-2})	0.36	0.14–0.53
Speed of sound a (km s^{-1})	21	18–23
Axial flow speed V_0 (km s^{-1})	67	34–125
Mach No. M	3.2	1.6–6.0

Note: for values of the various quantities at each of the seven loop locations, see Loughhead, Bray & Wang (1985, Table 1).

absorption in a uniform stationary slab permeated by the ambient radiation field. He took the slab to be 2000 km thick, but his numerical results are not sensitive to the exact value adopted (Giovanelli, 1967b). Table 2 of Giovanelli (1967a) gives calculated particle densities for the first three quantum states – N_1, N_2 and N_3 – for values of N_e and T_e ranging from 10^{10} to 10^{12} cm^{-3} and 7500 to 50000 K respectively. Another vital piece of information for our purposes is contained in Giovanelli's second paper (1967b), where the surface intensity of the model structure is calculated as a function of N_e and T_e.

The next step carried out by Loughhead *et al.* was to use Giovanelli's Table 2 to find all pairs of values (N_e, T_e) producing the equality $(N_2)_{\text{obs}} = (N_2)_{\text{calc}}$. This was done for each of the seven points of measurement on the loop axis using Giovanelli's results for a nonthermal velocity ξ of 11 km s^{-1}. (The mean value of ξ inferred from the observations under discussion was 26 km s^{-1}, but Giovanelli's figures are not sensitive to the choice of ξ.) The next step was to compare Giovanelli's theoretical curves for the contrast at Hα line centre as a function of N_e and T_e (Giovanelli, 1967b, Fig. 3; Bray & Loughhead, 1974, Fig. 4.11) with the value predicted by the cloud model for material at rest (i.e. $\Delta\lambda_1 = 0$). This again yielded possible pairs of values (N_e, T_e) which, in conjunction with those determined from N_2, finally established the actual values of N_e and T_e for the observed loop. In this way, for each of seven locations along the axis of the loop, unique values of N_e and T_e were derived which correctly accounted for both the observed value of N_2 and the contrast at $H\alpha$ line centre. The mean value and the range of values of T_e and N_e are given in the fifth and sixth rows respectively of Table 2.5.

With N_e and T_e determined, derivation of the other desired physical quantities is straightforward. Particle densities for hydrogen in its first and third quantum states, N_1 and N_3, were derived by interpolation from Table 2 of Giovanelli (1967a); they are given in the seventh and ninth rows of Table 2.5. The total particle density of hydrogen N_H (i.e. neutral + ionized) was then calculated, and the remaining quantities obtained from the formulae

$$x_H = N_e/N_H, \tag{2.5}$$

$$\rho = N_H m_H + 0.0851 N_H \times 3.9708 m_H, \tag{2.6}$$

$$p = k(N_e + 1.0851 N_H) T_e, \tag{2.7}$$

$$a = \left(\frac{\gamma p}{\rho}\right)^{\frac{1}{2}}; \quad \gamma = \tfrac{5}{3}, \tag{2.8}$$

$$M = V_0/a. \tag{2.9}$$

Here, m_H is the mass of the hydrogen atom and 0.0851 is the abundance ratio of helium to hydrogen by numbers of atoms.

The mean value and the range of values for each of the quantities x_H, ρ, p, a, V_0, and M are given in the last six rows of Table 2.5. It may be noted in passing that the hydrogen is almost completely ionized, x_H being greater than 0.99 everywhere except at the point on the loop at which T_e takes its lowest value ($= 15000$ K), where $x_H = 0.988$. An important conclusion is that all the basic physical quantities appear to show a marked variation from point to point of the loop, as illustrated in Fig. 2.16. For example, ρ varies by a factor of 2.5 from its lowest value, at L_7, to its highest, at L_5, while p varies by a factor of 3.7 between the points of minimum (L_6) and maximum (L_1). Similarly, the Mach number M varies from a low of 1.6 at L_1 to a high of 6.0 at L_3.

Particularly striking is the variation in density: there appears to be a distinct compression near the top of the loop and rarefactions in both the ascending and descending legs. The variation in the pressure is even more

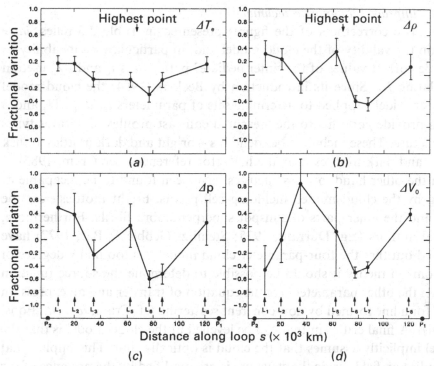

Fig. 2.16. Variation along the loop of Fig. 2.7 of (a) electron temperature T_e, (b) gas density ρ, (c) gas pressure p and (d) axial flow speed V_0 about their respective means: $\Delta T_e = [T_e - (T_e)_{mean}]/(T_e)_{mean}$, etc. The abscissa s is the distance along the loop measured from the ascending footpoint P_2. The points of measurement are labelled $L_1 \ldots L_8$ as before; P_1 is the descending footpoint. Vertical bars indicate the estimated errors of the four quantities at each point of measurement (see Loughhead, Bray & Wang, 1985). Note the wave-like nature of the variations in density and pressure, with an apparent wavelength of some 50000–60000 km, roughly one-half the length of the loop.

marked, although this largely reflects the variation in ρ. If these results are interpreted as evidence for the presence of a *wave* in the loop, then the implied wavelength is some 50000–60000 km, or roughly one-half the length of the loop. On the other hand, Loughhead, Bray & Wang (1985) are at pains to point out that the evidence in Fig. 2.16 for a wave-like variation in ρ and p along the loop is heavily dependent on the correctness of the results at L_5. This happens to be one of the two points at which their values of the source function S (derived by applying the cloud model) differed significantly from values subsequently calculated by Cram (1986) using a more exact treatment (see below). Nevertheless, it may be significant that the variations in all four quantities plotted in Fig. 2.16 are close to zero near the highest point of the loop, suggesting the presence of a node at this location.

(2)   *Improved diagnostic techniques*

The correctness of the figures presented in Table 2.5 relies, *inter alia*, on the validity of the cloud model and, in particular, on its ability to provide correct values of the 'intermediate' parameters t_0 and $\Delta\lambda_0$ used in calculating N_2. Since its introduction by Beckers (1964) the cloud model has been widely applied to determine sets of parameters t_0, $\Delta\lambda_0$, $\Delta\lambda_1$ and S which provide good fits to the measured contrast profiles of various types of structure. These include – beside loops – bright and dark mottles at disk centre and dark mottles near the limb (for references, see Cram, 1985).

On the other hand, other structures have been found to be incapable of fitting by the cloud model, including cell points, bright mottles near the limb and the inner parts of sunspot superpenumbra fibrils. Furthermore, several authors (e.g. Durrant, 1975; Steinitz, Gebbie & Bar, 1977) have pointed out that the four-parameter cloud model has too many degrees of freedom: in theory it should be possible to determine the source function S from the other parameters *via* the equation of transfer and an expression for S for a line formed by non-coherent scattering (cf. Cram, *loc. cit.*, Eqns. (1)–(4)). A final criticism that can be levelled at the cloud model is that the model implicitly assumes that the cloud is optically thin. This implies that the radiation field inside the structure is not modified by the presence of the structure. In an optically thick structure, on the other hand, the presence of the structure itself results in a modification of the radiation field inside the structure. Indeed, the radiation field needs to be computed at every point of the structure if the emergent radiation (or contrast with the surroundings) is to be correctly evaluated. In short, even when parameters can be found which provide an adequate fit to the measured contrast profile, the assumptions embodied in the cloud model may not be valid (Cram, 1975).

Let us now return to the loop whose physical conditions are listed in Table 2.5. For this loop we know not only the true geometry but also the axial flow velocity at several points along its axis (Section 2.3.4). This loop is therefore particularly suited to a proper quantitative calculation of S according to the non-LTE theory of the formation of Hα by non-coherent scattering, since the circumstances of its illumination and the effects of velocities can be calculated in detail. The necessary computations have been carried out by Cram (1986), who found satisfactory agreement with the cloud model values of S determined by Bray & Loughhead (1983) at five out of the seven points considered; at the remaining two there was clear disagreement. No special features of these points (L_3 and L_5 in Figs. 2.7 and 2.16) could be found to explain these discrepancies. Accordingly, Cram concluded that at these points we must question the adequacy both of the application of the cloud model and of the theory of formation of Hα. At the other five points, his calculations support the reliability of the physical conditions listed in Table 2.5.

Areas in which future work might profitably be concentrated (Cram, 1985), in the hope of improving existing diagnostic procedures, include: (a) simultaneous observations in Hα, Hβ, Hγ, the sodium D lines, the K line of Ca II, etc.; (b) investigation of multi-dimensional radiative transfer effects, taking particular account of the geometrical configuration of the specific loop studied; (c) coupling of the radiative and hydrodynamic or magnetohydrodynamic equations; and (d) consideration of the time-dependent ionization and excitation of hydrogen and other elements in loop material undergoing substantial motions (see discussion in Section 2.5.4).

A detailed non-LTE study of the effect of the macroscopic velocity of prominence material on prominence brightness in Hα and other hydrogen lines, and on the hydrogen population levels $N_1, N_2 \ldots N_5$ and on the electron density N_e, has been carried out by Heinzel & Rompolt (1987). They did not apply their theory to specific observations but demonstrated that for velocities of a magnitude that we know to occur in loops (Sections 2.2.2 and 2.3.4), substantial so-called 'Doppler-brightening' or 'Doppler-dimming' effects are to be expected. On the other hand, velocities up to 160–200 km s^{-1} have little effect on the value of N_e.

In connection with (b) the question of *filamentation* once again raises itself, since exact calculations require (ideally) a knowledge of the 'microstructure' of the object under study. Unresolved filamentation leads to an error in ℓ, the path length through the structure; this leads to a proportionate error in N_2 (cf. Eqn. (2.4)). In the analysis leading to the figures in Table 2.5 percentage errors in N_e, ρ and p would then be approximately one-half the percentage errors in N_2 (Loughhead, Bray & Wang, 1985). However, the derived values of T_e remain virtually unaffected.

Finally, one should remain alert to possible new diagnostic methods that are free of any theoretical uncertainties. For example, Bray & Loughhead (1986) have devised a method for determining the optical thickness t_0 of bright or dark Hα loops (a key intermediate parameter, as we have seen) that requires no knowledge of S. Their method is based on the measurement of the change in contrast of a loop as it crosses a sharp change or discontinuity in the underlying background emission. Suitable cases for the application of the method include a loop crossing a sunspot umbra, penumbra, pore, other dark material, a flare, another loop, or the solar

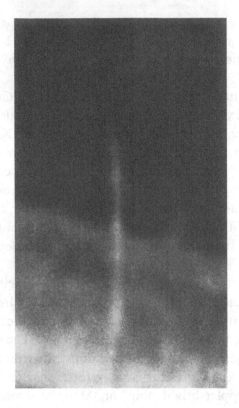

0"   **20"**

Fig. 2.17. Thin loop crossing solar limb, photographed at Hα line centre (Bray & Loughhead, 1986). Measurement of its contrast with respect to the background on either side of the limb discontinuity (see text) yielded its optical thickness, $t_0 = 0.06$.

limb. The last of these is illustrated in Fig. 2.17: here the method yielded the value $t_0 = 0.06$. Similarly, the loop of Fig. 2.4(*a*) gave $t_0 = 0.12$. The method is best suited to loops of small optical thickness and can, in principle, be applied not only to loop observations obtained in Hα and other strong lines in the visible region, but also to those obtained in the EUV and X-ray regions of the spectrum.

2.5.3 *Analysis of limb observations in the visible and near-visible regions*

There are a number of diagnostic methods available for determining N_e and T_e in prominences. Although these methods have mostly been applied to quiescent prominences, they are equally applicable to 'active' prominences (i.e. loops, surges, eruptive prominences, coronal rain)

Table 2.6. *Temperature, electron density and gas pressure in 'active' prominences observed at the limb: visible and near-visible regions*

Reference	T_e (K)	N_e (cm^{-3})	p (dyne cm^{-2})
de Jager (1959)	25 000	2.0×10^{10}	0.14
Orrall (1965)	20 000	1.0×10^{11}	0.58
Hirayama (1979)	7000–$\gtrsim$20 000	3.2×10^9–6.3×10^{12}	—
Foukal (1981)	—	$\leqslant 4 \times 10^9$	—

(Hirayama, 1979); specific applications to flare loops are described in Section 4.8.2. The methods may be summarized as follows:

(1) measurement of the intensity ratios of carefully selected lines of H I, He I and Ca II (Milkey, 1979);

(2) measurement of line widths, yielding the kinetic temperature T_{kin} – which can be equated to T_e – and ξ, the non-thermal velocity (see Section 3.6.2, Eqn. (3.1)). However, notice should be taken of the warning voiced by Tandberg-Hanssen (1974, p. 17): 'The analysis of spectra from active prominences (like loops and surges) leads us to conclude that it is, in general, not possible to find a solution set of T_e and ξ that will satisfy the widths of He lines, as well as H and metal lines';

(3) observations of the confluence of the higher Balmer lines (up to $n \simeq 30$ and beyond) due to the Stark effect. This method depends on determining the quantum number of the last resolvable Balmer line and then applying the well-known Inglis–Teller formula (Tandberg-Hanssen, *loc. cit.*, p. 20);

(4) measurement of the intensity of the electron scattering continuum. This method utilizes Eqn. (3.2) of Section 3.6.2; and

(5) measurement of the intensity of the Balmer continuum. This method does not give N_e and T_e separately, only the quantity $N_e^2 T_e^{-\frac{3}{2}}$ (cf. Bray & Loughhead, 1974, Eqn. (4.15)).

Summaries of values of N_e and T_e in active prominences have been published by de Jager (1959, p. 235) and Hirayama (1979, Table 3 and p. 31). Values for N_e and T_e specifically for loops have been determined by Orrall (1965) and Foukal (1981). All of these values are collected together in Table 2.6; the fourth column gives the gas pressure calculated according to Eqn. (2.7).

When we compare the figures in Table 2.6 with the physical conditions listed in Table 2.5 for the disk loop shown in Fig. 2.2, we find very good agreement for T_e and p. Furthermore, the mean value of N_e for the disk loop lies well within the (rather large) range of values found for active prominences at the limb.

2.5.4 *Diagnostic techniques for EUV lines*

To date almost all analyses of solar EUV line radiation intensities have been based on the simplifying assumption that the lines are formed in regions which are *optically thin* at the wavelengths concerned. In his classic interpretation of the solar ultraviolet emission line spectrum Pottasch (1963, 1964) remarked that the optical depth in most lines is probably only 0.1 or less, although values of the order of 10 or possibly higher might be expected for some lines of the more abundant elements (oxygen, carbon, and helium). Under the latter circumstance the concept of 'effective optical thinness' is sometimes invoked to validate the application of the optically thin approximation (Pottasch, *loc. cit.*; Cheng & Moe, 1977; Dere & Mason, 1981). Actual values of the optical depths of a large number of EUV lines at disk centre have been calculated by Burton *et al.* (1971, Table 5).

In this section we shall first describe the theoretical basis underlying the methods used in the analysis of EUV line intensities and shall then briefly discuss the use of measurements of the intensity ratios of suitably chosen lines to obtain estimates of electron density and temperature. Finally, we shall consider the important question of possible departures from ionization equilibrium in coronal loops.

(1) *Analysis of EUV line intensities: the emission measure*

In the case of an optically thin line the standard expression for the emergent intensity along a given line of sight (see for example, Gray, 1976, p. 126) becomes simply

$$I_\lambda = \frac{1}{4\pi} \int_0^L \varepsilon_\lambda \, \mathrm{d}\ell, \tag{2.10}$$

where L is the physical thickness of the region of formation along the line of sight, ε_λ is the emission per unit volume at wavelength λ and $\mathrm{d}\ell$ is an element of length along the line of sight. On the Sun the flux of the continuous ultraviolet emission is very low and, as a consequence, at the electron densities typical of coronal loops, collisions are frequent enough to ensure that collisional excitation by electrons is the dominant process in raising an ion from the lower state of the transition to the upper excited state (Zirin, 1966, Section 7.1). When the lower state is the ground state of the ion the line is termed a *resonance* line and, in fact, most of the EUV lines identified in the solar spectrum fall into this category (Pottasch, 1964).

If each excitation to the upper state is followed by the emission of a line quantum, the emission per unit volume is

$$\varepsilon_\lambda = \Delta E_{1u} \cdot N_{ion} \cdot C_{1u} N_e, \qquad (2.11)$$

where ΔE_{1u} is the transition energy between the lower state l and the upper state u, C_{1u} the collisional excitation rate per electron and N_{ion} the number density of the ion in its lower state. Apart from atomic constants the quantity C_{1u} is a function only of the electron temperature T_e and is given by the formula

$$C_{1u} = 2.724 \times 10^{15} T_e^{-\frac{1}{2}} (\Delta E_{1u})^{-1} g_{eff} f_{1u} e^{-\Delta E_{1u}/kT_e}, \qquad (2.12)$$

where k is Boltzmann's constant, g_{eff} the effective Gaunt factor and f_{1u} the oscillator strength (Seaton, 1964). In fully ionized solar plasma ($T_e \geqslant 2 \times 10^4$ K) with a hydrogen to helium abundance of 10:1 the number density of hydrogen is approximately $0.8 N_e$. Therefore we can write

$$N_{ion} \equiv N_H \cdot \frac{\Sigma N_{ion}}{N_H} \cdot \frac{N_{ion}}{\Sigma N_{ion}} = 0.8 N_e \cdot A \cdot \frac{N_{ion}}{\Sigma N_{ion}}, \qquad (2.13)$$

where ΣN_{ion} is the total number density of the element concerned (i.e. summed over all stages of ionization) and $A = (\Sigma N_{ion})/N_H$ is its (coronal) abundance relative to hydrogen. Substituting from Eqns. (2.11)–(2.13) in Eqn. (2.10) we obtain for the emergent intensity of the line the expression

$$I_\lambda = 1.73 \times 10^{-16} A \cdot g_{eff} \cdot f_{1u} \int \frac{N_{ion}}{\Sigma N_{ion}} \cdot T_e^{-\frac{1}{2}} e^{-\Delta E_{1u}/kT_e} N_e^2 \, d\ell. \qquad (2.14)$$

The calculation of the fractional ion abundance $N_{ion}/(\Sigma N_{ion})$ requires detailed consideration of the various processes whereby ionization equilibrium is maintained in the gas. While the production of ions is dominated by collisional ionization, the process of recombination is more complicated. Account has to be taken not only of radiative recapture of electrons but also of dielectronic recombination and autoionization effects (Pottasch, 1967; Dere & Mason, 1981). Fortunately, however, it transpires that the quantity $N_{ion}/(\Sigma N_{ion})$ depends only on the electron temperature and hence the quantity

$$G(T_e) = \frac{N_{ion}}{\Sigma N_{ion}} \cdot T_e^{-\frac{1}{2}} e^{-\Delta E_{1u}/kT_e} \qquad (2.15)$$

is itself a function only of T_e. Detailed calculations of $N_{ion}/(\Sigma N_{ion})$ for different ionization stages of various elements have been made by a large number of authors, including Summers (1972, 1974) and Jacobs *et al.* (1977a, b, 1978, 1979). Table 2.4 lists the values of G_{max} for a number of EUV lines commonly used in observing coronal loops, calculated from Eqn. (2.15) using the ionization equilibrium ion abundances published by Jacobs *et al.* (1977b, 1978, 1979). Substitution of the tabulated values of G_{max} in Eqn. (2.14) yields the corresponding emergent line intensities in erg cm^{-2} s^{-1} sr^{-1} per unit wavelength interval.

Under normal circumstances the function $G(T_e)$ rises to a sharp maximum, G_{max} say, at some particular temperature $\bar{T}_e$ and falls off rapidly on either side (cf. Fig. 3.1). As we have already remarked in Section 2.4.1, $\bar{T}_e$ is often loosely described as the 'formation temperature' of the line. Pottasch (1963, 1964) assumed that, to a reasonable approximation, $G(T_e)$ could be equated to zero outside the (narrow) range of temperatures around $\bar{T}_e$ (of width ΔT say) over which $G(T_e)$ exceeds one-third of its maximum value. Inside this range Pottasch replaced $G(T_e)$ by the constant value 0.7 G_{max}, the numerical factor being derived from the normalizing condition

$$\int_{\Delta T} G(T_e) \, dT_e = 0.7 \, G_{max} \cdot \Delta T.$$

Introducing this approximation into Eqn. (2.14) we then obtain

$$I_\lambda = 1.21 \times 10^{-16} \, G_{max} \, A g_{eff} f_{lu} \int_0^L N_e^2 \, d\ell, \qquad (2.16)$$

which provides the essential basis for the analysis of EUV line intensities in the Sun. In general the atomic data needed to apply Eqn. (2.16) to particular lines are not accurately known, but references to the available sources are given in the papers by Pottasch (1967), Dupree (1972), Dere & Mason (1981), Roussel-Dupré *et al.* (1984) and Giampapa *et al.* (1985). Moreover, there remains some doubt as to whether or not the abundances of the elements in the corona differ significantly from those in the photosphere; for a detailed discussion the reader is referred to Withbroe's (1971) review.

It should be noted that in the derivation of Eqn. (2.16) we have implicitly assumed that the upper level of the line transition is a singlet one. When the upper level is multiple the value of the oscillator strength is different for each component of the line, but the values of g_{eff}, G_{max} and $\bar{T}_e$ usually remain the same (Pottasch, 1967). Eqn. (2.16) then applies separately to each component of the multiplet provided the appropriate values of I_λ and f_{lu} are used.

When discussing relatively *broad-band* observations, such as the slitless spectroheliograms from Skylab, we are in effect dealing not with the monochromatic intensity I_λ but with the total line intensity

$$I = \int_0^\infty I_\lambda \, d\lambda.$$

The problem of calculating I has been considered by Withbroe (1970) for the case of a line with a Gaussian Doppler profile formed within a region which is optically thin along the line of sight.

The quantity $\int_0^L N_e^2 \, d\ell$ in Eqn. (2.16) is known as the *emission measure* of the line. Knowledge of the emission measure allows us to estimate the 'mean' electron density $\bar{N}_e$ in a loop by putting

$$\int_0^L N_e^2 \, d\ell = \bar{N}_e^2 \cdot L. \qquad (2.17)$$

The path length L is usually identified with the measured diameter of the loop (assumed circular) at the point of observation, although the possible effect of filamentation may have to be considered (see Section 2.5.5). However, as we have already seen in Section 2.5.2, the true path length through the loop is actually $L/\sin\varepsilon$, where ε is the angle between the axis of the loop and the line of sight to the observer. The correction factor may be considerable but of course can be applied only when the true geometry of the loop has been determined.

An extension or variant of the emission measure, called the *differential emission measure* (DEM), is needed for diagnosing physical conditions in non-isothermal sources. In effect the DEM specifies the emission measure (defined above) over the range T_e to $T_e + dT_e$: good definitions, together with the mathematical methods needed to exploit the DEM technique, have been given by Withbroe (1975), Craig & Brown (1976) and Fludra & Sylwester (1986). Sometimes convergence problems prove a drawback (Pallavicini *et al.*, 1981) but Levine & Withbroe (1977) systematically applied a DEM analysis to observations of an active region loop. However, they concluded that new observations of higher spatial resolution would be needed for a proper test of the theory. In concrete terms, the observations that they used ($\sim 5''$ resolution) could not fully resolve the radius of the loop as recorded in lines formed at different temperatures.

(2)   *Use of line intensity ratios to determine T_e and N_e*
        Provided the abundances of the elements concerned are known it is possible in principle to determine the electron temperature in a coronal

loop from the observed ratio of the intensities of two optically thin lines. To see this we first note that, from Eqn. (2.16),

$$\frac{I_1}{I_2} = \frac{A_1 \cdot g_1 \cdot f_1}{A_2 \cdot g_2 \cdot f_2} \cdot \frac{\int_0^{L_1} G_1(T_e) N_e^2 \, \mathrm{d}\ell}{\int_0^{L_2} G_2(T_e) N_e^2 \, \mathrm{d}\ell}, \tag{2.18}$$

where the subscripts 1 and 2 are used to distinguish between the two lines. Then, if both lines are formed in the same part of the loop and the temperature there is essentially uniform, Eqns. (2.15) and (2.18) give

$$\frac{I_1}{I_2} = \frac{A_1 \cdot g_1 \cdot f_1 [N_{\mathrm{ion}}/(\Sigma N_{\mathrm{ion}})]_1}{A_2 \cdot g_2 \cdot f_2 [N_{\mathrm{ion}}/(\Sigma N_{\mathrm{ion}})]_2} \cdot \mathrm{e}^{-(\Delta E_{1_1 u_1} - \Delta E_{1_2 u_2})/kT_e}, \tag{2.19}$$

thus enabling T_e to be determined. To minimize uncertainties arising from lack of accurate atomic data it may be advantageous to use lines from the same ion rather than two different ions, so that

$$\frac{I_1}{I_2} = \frac{f_1}{f_2} \cdot \mathrm{e}^{-(\Delta E_{1_1 u_1} - \Delta E_{1_2 u_2})/kT_e}. \tag{2.20}$$

If the two lines chosen are resonance lines with a common lower level, it is clear from Eqn. (2.20) that the diagnostic method will lack sensitivity unless the excitation energies of the upper levels are substantially different. This has the experimental disadvantage, however, that both the wavelengths of the lines and their intensities will then be quite different.

Turning now to the question of electron density we see from Eqn. (2.20) that the intensity ratio of two optically thin lines which arise from the same ion and have almost the same excitation energies is nearly independent of temperature. On the other hand, provided one of these lines is connected to a metastable level, the line ratio will be sensitive to changes in the electron density. Good examples of ions possessing suitable line pairs are those of the beryllium sequence such as C III and O V, which have been widely used for inferring electron densities. Here the lowest resonance line, $^1P - ^1S$, is compared with the lowest line of the triplet system $2p^2 \, ^3P - 2s2p$ ^3P (Dupree, 1978, Fig. 6). The upper levels of both transitions are essentially populated by electron collisions from the respective lower levels and hence the ratio of line intensities is proportional to the ratio of the populations of the two lower levels. The density sensitivity arises because the lower level of the triplet system line, the metastable $2s2p \, ^3P$ level, is populated by electron collisions from the ground state which, however, is itself only little affected by the electron density.

Other categories of line ratios are also density sensitive in certain ranges of electron density provided that in each case one of the lines involved is connected to a metastable state. The density range over which a strong density dependence occurs depends on the properties of the ion in question.

The line ratio technique of determining the electron density has recently been applied to coronal loops (Section 2.5.5), but one has to remember that the reliability of the results is affected by the (possibly) limited accuracy of the atomic data available. In addition, of course, the application of the method assumes that the electron density is effectively constant over the region of line formation.

The subject of density diagnostic techniques for optically thin plasmas is an extensive one and for a fuller discussion the interested reader is referred to the articles of Feldman, Doschek & Rosenberg (1977), Dupree (1978), Feldman & Doschek (1978), Dere *et al.* (1979) and Dere & Mason (1981).

(3)   *Possible departures from ionization equilibrium*

Apart from the recent work of Roussel-Dupré *et al.* (1984) described in Section 2.4.3, we have no direct observational knowledge of the true velocities of mass flow along EUV loops. However, measurements of line shifts and line widths over active regions suggest that flow speeds as high as 100 km s^{-1} could be encountered, comparable in fact to the axial velocities found by Loughhead & Bray (1984) in an Hα loop observed on the disk (Section 2.3.4). Consequently, if the temperature of an EUV loop were to vary significantly along its length, an element of fluid might be transported from one point to another more quickly than the ionization balance could adjust to the changed local temperature. This means that departures from ionization equilibrium would then have to be taken into account in analysing EUV line intensities in the loop.

Such departures will occur if the distance travelled by the fluid element, d_i, say, during the mean lifetime of a given ion i is greater than the scale length of the temperature variations along the loop. Since the lifetime of a level is the reciprocal of the sum of the ionization and recombination rates (McWhirter, 1975, p. 213), we have

$$d_i = \frac{V_0}{N_e(S_i + \alpha_i)}, \tag{2.21}$$

where V_0 is the axial flow speed and S_i and α_i are the ionization and recombination coefficients respectively. According to Borrini & Noci (1982) the value of $(S_i + \alpha_i)$ for carbon atoms is of the order of

10^{-11}–10^{-12} cm^3 s^{-1}. Thus, taking a representative loop density of $N_e = 1 \times 10^9$ cm^{-3} (see Table 2.7) and $V_0 = 100$ km s$^{-1} = 1 \times 10^7$ cm s^{-1}, we find that $d_i = 10^4$–10^5 km.

As we shall see in Section 2.5.5, virtually no information is available about the variation of temperature along cool EUV loops, although Cheng (1980) found no 'observable' variation in temperature along the axis of a well-defined loop beyond the limb in any of the spectral lines observed. Most observers have, in fact, simply taken the temperature of the whole loop to be equal to the formation temperature of the line. On the other hand, Loughhead, Bray & Wang (1985) found temperature variations of ~ 5000 K along distances as short as 10000 km in an Hα loop observed on the disk and similar gradients might be expected in cool EUV loops; if so, departures from ionization equilibrium are possible.

Observational evidence for possible departures from ionization equilibrium in cool EUV loops has been adduced from a study of temperature and density sensitive line ratios by Raymond & Foukal (1982) and Doyle *et al.* (1985). The observed loop temperatures tend to be less than the calculated line formation temperatures but, in view of the uncertainties of the method, the apparent differences may not be significant. Accordingly, at present there is no strong evidence for significant departures from ionization equilibrium in cool EUV loops.

2.5.5 *Analysis of observations in EUV lines*

The first attempt to estimate the electron densities in cool EUV loops was made by Foukal (1975). As we have already mentioned in Section 2.4.3, he measured the ratio of the emergent loop intensity I_1 to the intensity I_c of the hot coronal background in the associated active region at a number of points in six different loops observed beyond the limb in the lines O VI $\lambda103.2$ and Ne VII $\lambda46.5$. Then, by expressing I_1 and I_c in terms of the emission measures along the corresponding lines of sight, Foukal obtained the relation

$$\frac{I_1}{I_c} = \frac{[N_e(\text{loop})]^2 L_1 + \beta[N_e(\text{corona})]^2 L_c}{\beta[N_e(\text{corona})]^2 L_c},$$

whence we have

$$\frac{N_e(\text{loop})}{N_e(\text{corona})} = \left[\beta \frac{L_c}{L_1} \left(\frac{I_1}{I_c} - 1 \right) \right]^{\frac{1}{2}}. \tag{2.22}$$

In this equation L_1 and L_c represent the path lengths through the loop and corona respectively, while the factor β takes account of the (reduced) emissivity of the line at the temperature of the hot coronal background (assumed to be 2×10^6 K) as compared to the emissivity at the temperature of the loop. This Foukal took to be the formation temperature of the line ($\bar{T}_e$). Intensity tracings across the loop were used to determine L_l ($= 2 \times$ FWHM), while L_c was arbitrarily put equal to twice the (unforeshortened) width of the active region as seen in the hot $\lambda 62.5$ line of Mg X. When Foukal inserted his estimate of β for each line into Eqn. (2.22) he found that the value of the ratio $N_e(\text{loop})/N_e(\text{corona})$ was less than 2 for the O VI line and less than 1 for the Ne VII line.

Actual values of the electron densities in loops were given in a subsequent paper (Foukal, 1976). Again using the emission measure technique he inferred loop electron densities of 4×10^8, 5×10^8 and 1.4×10^9 cm^{-3} for the three lines C III $\lambda 97.7$, O VI $\lambda 103.2$ and Ne VII $\lambda 46.5$ respectively (cf. Table 2.4). By contrast typical values of the average electron density in active regions derived from recent studies of cool EUV lines are of the order of 10^{10}–10^{12} cm^{-3} (Dere & Mason, 1981, Tables 6.1 and 6.2; Roussel-Dupré *et al.*, 1984). However, a proper comparison is difficult since the active region determinations presumably refer to various admixtures of loop and interloop material.

Cheng (1980) used a somewhat different form of Eqn. (2.14) to measure the electron densities in a well-defined loop prominence beyond the limb in the three cool EUV lines Ne VII $\lambda 46.5$, Mg VIII $\lambda 43.7$ and Mg IX $\lambda 36.8$. He found that the density was almost constant along the loop and did not change with height, thereby ruling out any presumption of hydrostatic equilibrium within the loop. The mean density (i.e. averaged along the length of the loop) varied from 1.4×10^9 cm^{-3} in the Ne VII line to 4.0×10^9 cm^{-3} in the warmer Mg IX line (Table 2.7). Cheng also found that, in any particular line, there was no observable variation in temperature along the loop. Putting this equal to the formation temperature of the line he calculated the gas pressure in the Ne VII loop, $p \simeq 2kN_eT_e$, to be 0.2 dyne cm^{-2} as compared to an estimated coronal background pressure of 0.6 dyne cm^{-2}. The gas pressure in the cool loop was thus about three times smaller than that in the surrounding material.

Subsequently Dere (1982) used the emission measure technique to estimate the electron densities at two heights (base and 19 000 km) in three loops observed in an active region on the disk. (The loops were assumed to rise vertically from the solar surface.) The lines he used were Ne V $\lambda 41.6$, Ne VII $\lambda 46.5$ and Mg VIII $\lambda 43.7$ together with the Mg VI and Ne VI lines at $\lambda 40.3$. Table 2.7 gives the inferred electron densities, averaged over the three loops; the mean values range from 1.7×10^9 to 2.8×10^9 cm^{-3}. Dere

Table 2.7. *Physical conditions in cool EUV loops*

Source	Spectral line	Line (loop) temperature[a] (K)	Electron density (cm^{-3})	Gas pressure[b] (dyne cm^{-2})	Height of measurement (km)
Foukal (1976)	C III λ97.7	7.9×10^4	4×10^8	0.008	
	O VI λ103.2	3.2×10^5	5×10^8	0.04	
	Ne VII λ46.5	5.0×10^5	1.4×10^9	0.19	
Cheng (1980)[d]	Ne VII λ46.5	5.0×10^5	1.5×10^9	0.20	9500
			1.4×10^9	0.19	19200
			1.3×10^9	0.17	28000
			1.3×10^9	0.17	34000
	Mg VIII λ43.7	7.9×10^5	2.3×10^9	0.48	9500
			1.6×10^9	0.34	19200
			1.9×10^9	0.40	28200
	Mg IX λ36.8	1.0×10^6	4.2×10^9	1.11	19200
			3.1×10^9	0.82	28000
			4.6×10^9	1.22	34000
Dere (1982)	Ne V λ41.6	3.2×10^5	1.8×10^9	0.15	0
					19000
	Mg VI + Ne VI	4.0×10^5	2.2×10^9	0.23	0
	λ40.3 (blend)		1.7×10^9	0.18	19000
	Ne VII λ46.5	5.0×10^5	2.8×10^9	0.37	0
			1.7×10^9	0.22	19000
	Mg VIII λ43.7	7.9×10^5	1.7×10^9	0.36	0
			2.0×10^9	0.42	19000
Raymond & Foukal (1982)	O V λ76.0/ O V λ63.0	2.5×10^5	$\sim 1 \times 10^{10}$	~ 0.66	
	Mg VIII λ43.0/ Mg VIII λ43.7	7.9×10^5	$\gtrsim 5 \times 10^8$	$\gtrsim 0.10$	
	O III λ50.8/ O III λ70.2	7.9×10^{4c} (6.3×10^4)			
	O IV λ55.4/ O IV λ78.9	1.6×10^{5c} (1.3×10^5)			
	Ne VII λ89.5/ Ne VII λ46.5	5.6×10^{5c} (2.5×10^5)			
Roussel-Dupré et al. (1984)	C IV λ154.8/ O V λ121.8		3.8×10^{11}		
	N V λ123.9/ O V λ121.8		2.5×10^{11}		
	C IV λ154.8/ N IV λ148.7		2.2×10^{11}		
	Si III λ120.7/ Si III λ129.7	6.3×10^4	$< 1 \times 10^{10}$	< 0.17	
	Si III λ130.1/ Si III λ129.7	6.3×10^4	$> 1 \times 10^{12}$	> 16.70	
	O V λ137.1/ O V λ121.8	2.5×10^5	7.6×10^{10}	5.04	
	C IV λ154.8/ O V λ137.1	(1.4×10^5)			
	N V λ123.9/ O V λ137.1	(1.7×10^5)			
	N V λ123.9/ C IV λ154.8	(1.3×10^5)			

[a] Inferred loop temperatures are placed in parentheses.
[b] Calculated from the formula $p = 1.92kN_e T_e$.
[c] Line formation temperatures are taken from Raymond & Foukal (1982).
[d] According to Cheng (1980) the loop in question 'was probably formed following a flare behind the limb'.

concluded that two of the loops were supported hydrostatically, whereas the third was not.

Given the electron densities and formation temperatures of the lines one can calculate the corresponding gas pressures in the loops. However, Dere found that the pressures so obtained were lower than those implied by the results quoted by Dere & Mason (1981) from line ratio studies based on low spatial resolution observations of active regions. Dere attributed the discrepancy to the effect of filamentation within the loop and, to take account of this, introduced a fill factor α defined by the relation

$$L = \alpha \cdot L_{obs}, \tag{2.23}$$

where L is the effective path length through a loop (assumed cylindrical) and L_{obs} its observed cross-sectional diameter. In the case of the narrowest of the three Ne V loops, pressures consistent with the line ratio results were obtained when α was put equal to 5×10^{-3}, implying an effective path length of only 30 km of $0''.04$ or arc! This is far below the resolution limit of any existing observations. Nevertheless, high-resolution observations in Hα and Lα have in fact revealed the existence of very fine, thread-like structures within individual coronal loops with diameters of $1''$ of arc or less (Sections 2.3.2 and 2.4.3). Aside from the Lα photographs, however, we have as yet no direct knowledge of the internal structure of EUV loops on any scale smaller than about $2''$ at best.

Estimates of both electron densities and temperatures have been made by Raymond & Foukal (1982) using spectra of three active region loops observed beyond the limb. Values of N_e ranging from $\sim 2 \times 10^8$ to $\sim 1 \times 10^{10}$ cm^{-3} were deduced from the ratios of the intensities measured in the pairs of lines O V $\lambda\lambda 76.0, 63.0$ and Mg VIII $\lambda\lambda 43.0, 43.7$. The estimates of T_e derived from the observed intensity ratios of pairs of temperature-sensitive lines of O III, O IV and Ne VII were in all cases somewhat less than the calculated line formation temperatures (see Table 2.7).

More recently, measurements of the intensity ratios of pairs of cool EUV lines have been used by Roussel-Dupré *et al.* (1984) to determine the electron density and temperature in an ephemeral active region on the disk (see their Fig. 2). They proceeded on the plausible yet nonetheless arbitrary assumption that the EUV spectra referred to a single small coronal loop spanning the ephemeral region (cf. Section 2.4.3). Adopting several alternative diagnostic approaches Roussel-Dupré *et al.* inferred values of N_e and T_e at ten closely-spaced points along a line crossing the apex of the loop at an angle of 77° to the loop axis. The values are summarized in Table 2.7 as averages over all the points of measurement. Roussel-Dupré *et al.*

then used Eqn. (2.16) to estimate the effective path lengths through the emitting regions from the integrated intensities of selected lines of C IV, N IV and O V. The resulting values lie between 0.1 and 2 km and are thus even smaller than the figure of 30 km found earlier by Dere (see above).

On the basis of their results Roussel-Dupré *et al.* were led to picture the EUV emission from the loop as originating from very thin sheaths of material less than 1 km thick surrounding an (assumed) cool core. Each sheath is isolated from the others by the magnetic field in the loop and radiates at the temperature giving the maximum abundance of the particular ion involved. The temperatures of the various emitting sheaths range from $\sim 1.6 \times 10^4$ to $\sim 3 \times 10^5$ K. Roussel-Dupré *et al.* also noted that the widths of the lines studied were greater than those expected from their respective formation temperatures. They attributed this to the presence of non-thermal motions within the loop with r.m.s. velocities lying between 10 and 30 km s^{-1}. (Loughhead, Bray & Wang (1985) found a mean non-thermal velocity of 26 km s^{-1} in their study of the physical conditions in the Hα loop shown in Fig. 2.2: see Section 2.5.2.) By comparison, the speed of the directed mass flow along the loop derived from the observed Doppler shifts of the lines on the basis of an assumed loop geometry was 95 km s^{-1} (see Section 2.4.3). The latter figure, however, is very sensitive to any change in the geometrical form attributed to the loop.

The work of Roussel-Dupré *et al.* represents the most ambitious attempt yet made to explore the full range of physical conditions in a cool EUV loop. It must be emphasized, however, that the correctness of the results rests on a number of highly uncertain factors and, in particular, on arbitrary assumptions regarding the loop itself. As the authors themselves admit, the geometry of the loop could be different from that hypothesized, the emission could come not from a single loop but from a number of smaller ones, the assumption of ionization equilibrium might not hold, and so on. Further detailed studies of this kind, and particularly ones involving full-sized coronal loops with known geometry, are clearly required.

High-resolution observations of limb loops obtained in Lα have been used by Tsiropoula *et al.* (1986) to deduce their physical conditions. Using a theory for the formation of Lα applicable to both the optically thick and thin cases, they obtained values for the temperature, neutral hydrogen density and optical thickness. However, in the absence of simultaneous Hα data they were obliged to make certain theoretical assumptions – e.g. hydrostatic equilibrium. Their values for T_e were $\sim 3 \times 10^5$ K, i.e. much higher than the 'formation' temperature T_e for Lα given in Table 2.4.

Table 2.8. *Morphological and dynamical properties of cool loops: summary of data*

Quantity	Value	Wave-length	Disk (D) or Limb (L)	Section
Height, true (km)	40000–53000	Hα	D	2.3.3
Height, projected (km)	26000–57000	Hα	L	2.2.1
	50000–67000	EUV	L	2.4.3
Length (km)	130000	Hα	D[a]	2.3.4[b]
	22000–109000	EUV	D	2.4.3
Separation of footpoints (km)	71000–86000	Hα	D	2.3.3
Diameter (km)				
'normal' loop	1600	Hα	D[a]	2.3.3
	<2000–22000	EUV	D, L	2.4.3
'very thin' loop, chromospheric level	440–580	Hα	D	2.3.3
'very thin' loop, photospheric level	<130–320	Hα	D	2.3.3
Aspect ratio	0.006	EUV	D	2.4.3
Inclination of loop plane to vertical β (deg)	30–41	Hα	D	2.3.3
Tilt of axis of symmetry σ (deg)	4–6	Hα	D	2.3.3
Lifetime				
loop system (hr)	3–6	Hα	D	2.3.2
single loop (min)	≲15	Hα	L	2.3.2
Axial flow speed V_0 (km)	34–125	Hα	D[a]	2.3.4
	20–150	Hα	L	2.2.2
	95	EUV	D	2.4.3
Mach No. M	1.6–6.0	Hα	D[a]	2.3.4
Acceleration $|dV_0/dt|$ (cm s^{-2})	1.1×10^4–3.8×10^4	Hα	D[a]	2.3.4
Velocity of bodily motion (km s^{-1})	<5	Hα	D[a]	2.3.4

[a] Loop of 6 June 1980;   [b] See Fig. 2.8.

2.6 Summary of data: cool loops

To conclude this chapter, we summarize in Tables 2.8 and 2.9 the known quantitative data concerning the morphological, dynamical and physical properties of cool loops. These tables provide a useful bird's-eye view, but for more comprehensive information the reader should refer to

Table 2.9. *Physical conditions in cool loops: summary of data*

Quantity	Value	Wavelength	Disk (D) or Limb (L)	Section
Temperature T_e (K)	21 000	Hα	D[a]	2.5.2
	7000–25 000	Visible/ near visible	L	2.5.3
	$\sim 6 \times 10^4$–1×10^6	EUV	D, L	2.5.5
Electron density N_e	5.6×10^{10}	Hα	D[a]	2.5.2
(cm^{-3})	3×10^9–6×10^{12}	Visible/ near visible	L	2.5.3
	$\sim 4 \times 10^8$–$\sim 3 \times 10^{11}$	EUV	D, L	2.5.5
Gas pressure p	0.36	Hα	D[a]	2.5.2
(dyne cm^{-2})	0.14–0.58	Visible/ near visible	L	2.5.3
	~ 0.1–~ 1.0	EUV	D, L	2.5.5
Gas density ρ	1.3×10^{-13}	Hα	D[a]	2.5.2
(g cm^{-3})				
Degree of hydrogen	0.996	Hα	D[a]	2.5.2
ionization x_H				
Speed of sound a	21	Hα	D[a]	2.5.2
(km s^{-1})				
Non-thermal velocity ξ	26	Hα	D[a]	2.5.2
(km s^{-1})	10–30	EUV	D	2.5.5
Line-of-sight magnetic field $B_\parallel$ (G)	7–100	Visible	L	2.2.3
Total magnetic field B (G)	4–45	Visible	L	2.2.3

[a] Loop of 6 June 1980.

the appropriate sections of the chapter. For convenience, the relevant section numbers are given in the last columns.

From the data presented it is evident that, over the temperature range characterizing cool coronal loops, viz. $\sim 20\,000$ K to $\sim 10^6$ K, both the heights and lengths of the loops are comparable. On the other hand, there is a small but apparently real increase in diameter (thickness) with temperature (Section 2.4.3).

The inferred values of the electron density in cool loops observed in both the visible and EUV regions of the spectrum extend over a wide range of several orders of magnitude. Despite this, the values of the gas pressure listed in Table 2.9 are restricted to a much smaller range, i.e. only a single order of magnitude. This may reflect the circumstance that the stability of a loop, whatever its temperature, depends on the maintenance of approximate pressure equilibrium with the surrounding coronal medium.

At present, more quantitative information is available from the visible than from the EUV region. Nevertheless, taken as a whole, the data in Tables 2.8 and 2.9 show that, with the exception of temperature, all cool loops appear to have similar properties and can be regarded as manifestations of the same basic physical structure.

References

Athay, R.G., Gurman, J.B. & Henze, W. (1983). Fluid motions in the solar chromosphere–corona transition region. III. Active region flows from wide slit Dopplergrams. *Astrophysical Journal*, **269**, 706–14.

Athay, R.G., Gurman, J.B., Henze, W. & Shine, R.A. (1982). Fluid motions in the solar chromosphere–corona transition region. II. Active region flows in C IV from narrow slit Dopplergrams. *Astrophysical Journal*, **261**, 684–99.

Athay, R.G., Querfeld, C.W., Smartt, R.N., Landi Degl'Innocenti, E. & Bommier, V. (1983). Vector magnetic fields in prominences. III. He I D₃ Stokes profile analysis for quiescent and eruptive prominences. *Solar Physics*, **89**, 3–30.

Beckers, J.M. (1964). A study of the fine structures in the solar chromosphere. Thesis, Utrecht; AFCRL–Environmental Research Paper No. 49.

Bommier, V., Leroy, J.L. & Sahal-Bréchot, S. (1981). Determination of the complete vector magnetic field in solar prominences, using the Hanle effect. *Astronomy and Astrophysics*, **100**, 231–40.

Bonnet, R.M., Bruner, E.C., Acton, L.W., Brown, W.A. & Decaudin, M. (1980). High-resolution Lyman-alpha filtergrams of the Sun. *Astrophysical Journal*, **237**, L47–50.

Bonnet, R.M. & Tsiropoula, G. (1982). Density and temperature determination of neutral hydrogen in coronal structures. *Solar Physics*, **75**, 139–43.

Bonnet, R.M. & 5 co-authors (1982). Rocket photographs of fine structure and wave patterns in the solar temperature minimum. *Astronomy and Astrophysics*, **111**, 125–9.

Borrini, G. & Noci, G. (1982). Non-equilibrium ionization in coronal loops. *Solar Physics*, **77**, 153–66.

Bray, R.J. & Loughhead, R.E. (1964). *Sunspots*. London: Chapman & Hall.

Bray, R.J. & Loughhead, R.E. (1974). *The Solar Chromosphere*. London: Chapman & Hall.

Bray, R.J. & Loughhead, R.E. (1983). High-resolution photography of the solar chromosphere. XVI. Hα contrast profiles of active region loops. *Solar Physics*, **85**, 131–40.

Bray, R.J. & Loughhead, R.E. (1985). High-resolution photography of the solar chromosphere. XX. True widths of Hα active region loops. *Astronomy and Astrophysics*, **142**, 199–204.

86

Bray, R.J. & Loughhead, R.E. (1986). High-resolution photography of the solar chromosphere. XXIII. Direct measurement of the optical thickness of Hα solar loops. *Astrophysical Journal*, **301**, 989–91.

Brueckner, G.E. & Bartoe, J.-D.F. (1974). The fine structure of the solar atmosphere in the far ultraviolet. *Solar Physics*, **38**, 133–56.

Bumba, V. & Kleczek, J. (1961). On a sunspot group with an outstanding loop activity. *Observatory*, **81**, 141–3.

Burton, W.M., Jordan, C., Ridgeley, A. & Wilson, R. (1971). The structure of the chromosphere–corona transition region from limb and disk intensities. *Philosophical Transactions of the Royal Society of London A*, **270**, 81–98.

Chen, C.-L. & Loughhead, R.E. (1983). Geometry of Hα active region loops observed on the solar disk. *Proceedings of the Astronomical Society of Australia*, **5**, 204–8.

Cheng, C.-C. (1980). Spatial distribution of XUV emission and density in a loop prominence. *Solar Physics*, **65**, 347–56.

Cheng, C.-C. & Moe, O.K. (1977). Emission measures and structure of the transition region of a sunspot from emission lines in the far ultraviolet. *Solar Physics*, **52**, 327–35.

Cheng, C.-C., Smith, J.B. & Tandberg-Hanssen, E. (1980). Morphology and spatial distribution of XUV and X-ray emissions in an active region observed from Skylab. *Solar Physics*, **67**, 259–65.

Chiuderi, C. (1981). Magnetic heating in the Sun. In *Solar Phenomena in Stars and Stellar Systems*, ed. R.M. Bonnet & A.K. Dupree, pp. 269–88.

Craig, I.J.D. & Brown, J.C. (1976). Fundamental limitations of X-ray spectra as diagnostics of plasma temperature structure. *Astronomy and Astrophysics*, **49**, 239–50.

Cram, L.E. (1975). Interpretation of Hα contrast profiles of chromospheric fine structures. *Solar Physics*, **42**, 53–66.

Cram, L.E. (1985). Interpretation of Hα observations of chromospheric fine structure. In *Chromospheric Diagnostics and Modelling*, ed. B.W. Lites, pp. 53–62. Sunspot: Sacramento Peak Observatory.

Cram, L.E. (1986). Interpretation of Hα contrast profiles of active region loops. *Astrophysical Journal*, **300**, 830–5.

Dere, K.P. (1982). The XUV structure of solar active regions. *Solar Physics*, **75**, 189–203.

Dere, K.P. & Mason, H.E. (1981). Spectroscopic diagnostics of the active region: transition zone and corona. In *Solar Active regions*, ed. F.Q. Orrall, pp. 129–64. Boulder: Colorado Associated University Press.

Dere, K.P., Mason, H.E., Widing, K.G. & Bhatia, A.K. (1979). XUV electron density diagnostics for solar flares. *Astrophysical Journal Supplement Series*, **40**, 341–64.

Doyle, J.G., Raymond, J.C., Noyes, R.W. & Kingston, A.E. (1985). The extreme ultraviolet spectrum of sunspot plumes. II. Spectral diagnostics and implications for cooling. *Astrophysical Journal*, **297**, 816–25.

Dupree, A.K. (1972). Analysis of the extreme-ultraviolet quiet solar spectrum. *Astrophysical Journal*, **178**, 527–41.

Dupree, A.K. (1978). UV and X-ray spectrosocopy in astrophysics. *Advances in Atomic and Molecular Physics*, **14**, 393–431.

Durrant, C.J. (1975). Spectroscopic investigation of the chromosphere. IV. A reassessment of the cloud model. *Solar Physics*, **44**, 41–53.

Ellison, M.A. (1944). Sunspot prominences – some comparisons between limb and disk appearances. *Monthly Notices of the Royal Astronomical Society*, **104**, 22–32.

Engvold, O. (1979). In *Physics of Solar Prominences*, ed. E. Jensen, P. Maltby & F.Q. Orrall, p. 356. Blindern-Oslo: Institute of Theoretical Astrophysics.

Feldman, U. & Doschek, G.A. (1978). The electron density at 10^5 K in different regions of the solar atmosphere derived from an intersystem line of O IV. *Astronomy and Astrophysics*, **65**, 215–22.

Feldman, U., Doschek, G.A. & Rosenberg, F.S. (1977). XUV spectra of the 1973 June 15 solar flare observed from *Skylab*. II. Intersystem and forbidden transitions in transition zone and coronal ions. *Astrophysical Journal*, **215**, 652–65.

Fludra, A. & Sylwester, J. (1986). Comparisons of three methods used for calculation of the differential emission measure. *Solar Physics*, **105**, 323–37.

Foing, B. & Bonnet, R.M. (1984). Characteristic structures of the solar disc observed on rocket UV filtergrams. *Astronomy and Astrophysics*, **136**, 133–41.

Foukal, P. (1975). The temperature structure and pressure balance of magnetic loops in active regions. *Solar Physics*, **43**, 327–36.

Foukal, P.V. (1976). The pressure and energy balance of the cool corona over sunspots. *Astrophysical Journal*, **210**, 575–81.

Foukal, P. (1978). Magnetic loops, downflows, and convection in the solar corona. *Astrophysical Journal*, **223**, 1046–57.

Foukal, P. (1981). Comments on the thermodynamic structure and dynamics of the cool solar corona over sunspots. In *The Physics of Sunspots*, ed. L.E. Cram & J.H. Thomas, pp. 191–209. Sunspot: Sacramento Peak Observatory.

Gaizauskas, V. (1979). Braided structures observed in flare-associated Hα filaments. In *Physics of Solar Prominences*, ed. E. Jensen, P. Maltby & F.Q. Orrall, pp. 272–5. Blindern-Oslo: Institute of Theoretical Astrophysics.

Giampapa, M.S., Golub, L., Peres, G., Serio, S. & Vaiana, G.S. (1985). Closed coronal structures. VI. Far-ultraviolet and X-ray emission from active late-type stars and the applicability of coronal loop models. *Astrophysical Journal*, **289**, 203–12.

Giovanelli, R.G. (1967a). Excitation of hydrogen and Ca II under chromospheric conditions. *Australian Journal of Physics*, **20**, 81–99.

Giovanelli, R.G. (1967b). Structure of the normal chromosphere. In *Solar Physics*, ed. J.N. Xanthakis, pp. 353–81. London: Interscience.

Goldberg, L. (1974). Research with solar satellites. *Astrophysical Journal*, **191**, 1–37.

Gray, D.F. (1976). *The Observation and Analysis of Stellar Photospheres*. New York: Wiley.

Habbal, S.R., Ronan, R. & Withbroe, G.L. (1985). Spatial and temporal variations of solar coronal loops. *Solar Physics*, **98**, 323–40.

Haisch, B.M., Bruner, M.E., Hagyard, M.J. & Bonnet, R.M. (1986). A comparison of photospheric electric current and ultraviolet and X-ray emission in a solar active region. *Astrophysical Journal*, **300**, 428–37.

Heinzel, P. & Rompolt, B. (1987). Hydrogen emission from moving solar prominences. *Solar Physics*, **110**, 171–89.

Hirayama, T. (1979). Prominence spectra and their interpretation. In *Physics of Solar Prominences*, ed. E. Jensen, P. Maltby & F.Q. Orrall, pp. 4–32. Blindern-Oslo: Institute of Theoretical Astrophysics.

Jacobs, V.L., Davis, J., Kepple, P.C. & Blaha, M. (1977a). The influence of autoionization accompanied by excitation on dielectronic recombination and ionization equilibrium. *Astrophysical Journal*, **211**, 605–16.

Jacobs, V.L., Davis, J., Kepple, P.C. & Blaha, M. (1977b). The influence of autoionization accompanied by excitation on the dielectronic recombination and the ionization equilibrium of silicon ions. *Astrophysical Journal*, **215**, 690–9.

Jacobs, V.L., Davis, J., Rogerson, J.E. & Blaha, M. (1978). Ionization equilibrium and radiative energy loss rates for C, N, and O ions in low-density plasmas. *Journal of Quantitative Spectroscopy and Radiative Transfer*, **19**, 591–8.

Jacobs, V.L., Davis, J., Rogerson, J.E. & Blaha, M. (1979). Dielectronic recombination rates, ionization equilibrium, and radiative energy-loss rates for neon, magnesium, and sulfur ions in low-density plasmas. *Astrophysical Journal*, **230**, 627–38.

de Jager, C. (1959). Structure and dynamics of the solar atmosphere. In *Handbuch der Physik*, vol. 52, pp. 80–362. Berlin: Springer.

Kleczek, J. (1963). Rainy loops from October 22, 1956. *Bulletin of the Astronomical Institutes of Czechoslovakia*, **14**, 167–71.

Kopp, R.A., Poletto, G., Noci, G. & Bruner, M. (1985). Analysis of loop flows observed on 27 March, 1980 by the UVSP instrument during the Solar Maximum Mission. *Solar Physics*, **98**, 91–118.

Lategan, A.H. & Jarrett, A.H. (1982). Observed mass motions in limb prominences. *Solar Physics*, **76**, 323–30.

Leroy, J.L. (1979). Polarimetric observations and magnetic field determination in prominences. In *Physics of Solar Prominences*, ed. E. Jensen, P. Maltby & F.Q. Orrall, pp. 56–80. Blindern-Oslo: Institute of Theoretical Astrophysics.

Levine, R.H. (1976). Evidence for opposed currents in active region loops. *Solar Physics*, **46**, 159–70.

Levine, R.H. & Withbroe, G.L. (1977). Physics of an active region loop system. *Solar Physics*, **51**, 83–101.

Loughhead, R.E. & Bray, R.J. (1984). High-resolution photography of the solar chromosphere. XIX. Flow velocities along an active region loop. *Astrophysical Journal*, **283**, 392–7.

Loughhead, R.E., Bray, R.J. & Wang, J.-L. (1985). High-resolution photography of the solar chromosphere. XXI. Determination of the physical conditions in an Hα active-region loop. *Astrophysical Journal*, **294**, 697–701.

Loughhead, R.E., Chen, C.-L. & Wang, J.-L. (1984). High-resolution photography of the solar chromosphere. XVIII. Axial tilt of Hα loops observed on the disk. *Solar Physics*, **92**, 53–65.

Makhmudov, M.M., Nikolsky, G.M. & Zhugzhda, Y.D. (1980). Motions in a loop prominence. *Solar Physics*, **66**, 89–104.

Martin, S.F. (1973). The evolution of prominences and their relationship to active centers. *Solar Physics*, **31**, 3–21.

McWhirter, R.W.P. (1975). The contribution of laboratory measurements to the interpretation of astronomical spectra. In *Atomic and Molecular Processes in Astrophysics*, ed. M.C.E. Huber & H. Nussbaumer, pp. 186–308. Sauverny: Geneva Observatory.

Menzel, D.H. & Evans, J.W. (1953). The behaviour and classification of solar prominences. *Accademia Nazionale dei Lincei, Convegno*, **11**, 119–36.

Milkey, R.W. (1979). Models of thermodynamic properties of prominences. In *Physics of Solar Prominences*, ed. E. Jensen, P. Maltby & F.Q. Orrall, pp. 349–56. Blindern-Oslo: Institute of Theoretical Astrophysics.

Noyes, R.W., Raymond, J.C., Doyle, J.G. & Kingston, A.E. (1985). The extreme ultraviolet spectrum of sunspot plumes. I. Observations. *Astrophysical Journal*, **297**, 805–15.

Orrall, F.Q. (1965). The spectrum of sporadic coronal condensations. In *The Solar Spectrum*, ed. C. de Jager, pp. 308–10. Dordrecht: Reidel.

Pallavicini, R. & 5 co-authors (1981). Closed coronal structures. III. Comparison of static models with X-ray, EUV, and radio observations. *Astrophysical Journal*, **247**, 692–706.

Pottasch, S.R. (1963). The lower solar corona: interpretation of the ultraviolet spectrum. *Astrophysical Journal*, **137**, 945–66.

Pottasch, S.R. (1964). On the interpretation of the solar ultraviolet emission line spectrum. *Space Science Reviews*, **3**, 816–55.

Pottasch, S.R. (1967). The inclusion of dielectronic recombination processes in the interpretation of the solar ultraviolet spectrum. *Bulletin of the Astronomical Institutes of the Netherlands*, **19**, 113–24.

Raymond, J.C. & Foukal, P. (1982). The thermal structure of solar coronal loops and implications for physical models of coronae. *Astrophysical Journal*, **253**, 323–9.

Roussel-Dupré, R., Wrathall, J., Nicholas, K.R., Bartoe, J.D.F. & Brueckner, G.E. (1984). HRTS II EUV observations of a solar ephemeral region. *Astrophysical Journal*, **278**, 428–40.

Schmahl, E.J., Mouradian, Z., Martres, M.J. & Soru-Escaut, I. (1982). EUV arcades: signatures of filament instability. *Solar Physics*, **81**, 91–105.

Seaton, M.J. (1964). The spectrum of the solar corona. *Planetary and Space Science*, **12**, 55–74.

Sheeley, N.R. (1980). Temporal variations of loop structures in the solar atmosphere. *Solar Physics*, **66**, 79–87.

Sheeley, N.R. & 5 co-authors (1975). XUV observations of coronal magnetic fields. *Solar Physics*, **40**, 103–21.

Sheeley, N.R. & Golub, L. (1979). Rapid changes in the fine structure of a coronal 'bright point' and a small coronal 'active region'. *Solar Physics*, **63**, 119–26.

Steinitz, R., Gebbie, K.B. & Bar, V. (1977). The embedded feature model for the interpretation of chromospheric contrast profiles. *Astrophysical Journal*, **213**, 269–77.

Summers, H.P. (1972). The density dependent ionization balance of carbon, oxygen and neon in the solar atmosphere. *Monthly Notices of the Royal Astronomical Society*, **158**, 255–75.

Summers, H.P. (1974). The ionization equilibrium of hydrogen-like to argon-like ions of elements. *Monthly Notices of the Royal Astronomical Society*, **169**, 663–80.

Svestka, Z. (1976). *Solar Flares*. Dordrecht: Reidel.

Tandberg-Hanssen, E. (1974). *Solar Prominences*. Dordrecht: Reidel.

Tandberg-Hanssen, E. (1977). Prominences. In *Illustrated Glossary for Solar and Solar-Terrestrial Physics*, ed. A. Bruzek & C.J. Durrant, pp. 97–109. Dordrecht: Reidel.

Tandberg-Hanssen, E. & Malville, J.M. (1974). Magnetic fields in flares and active prominences. II. The field configuration in some active prominences. *Solar Physics*, **39**, 107–19.

Tousey, R. & 8 co-authors (1973). A preliminary study of the extreme ultraviolet spectroheliograms from Skylab. *Solar Physics*, **33**, 265–80.

Tsiropoula, G., Alissandrakis, C., Bonnet, R.M. & Gouttebroze, P. (1986). Emission of Lyman-α radiation by solar coronal loops. *Astronomy and Astrophysics*, **167**, 35–58.

Vrsnak, B. (1984). The oscillating loop prominence of July 17, 1981. *Solar Physics*, **94**, 289–97.

Wiese, W.L., Smith, M.W. & Glennon, B.M. (1966). *Atomic Transition Probabilities, vol. I*. Washington: National Bureau of Standards.

Wiese, W.L., Smith, M.W. & Miles, B.M. (1969). *Atomic Transition Probabilities, vol. II*. Washington: National Bureau of Standards.

Withbroe, G.L. (1970). Solar XUV limb brightening observations. I. The lithium-like ions. *Solar Physics*, **11**, 42–58.

Withbroe, G.L. (1971). The chemical composition of the photosphere and the corona. In *The Menzel Symposium on Solar Physics, Atomic Spectra, and Gaseous Nebulae*, ed. K.B. Gebbie, pp. 127–48. Boulder: National Bureau of Standards.

Withbroe, G.L. (1975). The analysis of XUV emission lines. *Solar Physics*, **45**, 301–17.

Zirin, H. (1966). *The Solar Atmosphere*. Waltham: Blaisdell.

3

Hot loops: observed properties

3.1 Introduction

As we have seen in the preceding chapter, observations made in cool visible and EUV lines have provided extensive information on the large-scale systems of loops which dominate the structure of the lower corona above active regions. The loops are believed to trace out closed lines of force of the magnetic field which protrude up from beneath the photosphere and expand to fill the whole of the coronal volume above an active region. Hence a picture of the loop systems gives us some insight into the three-dimensional configuration of the magnetic field. But the picture obtained from observations of cool loops is far from complete. For more detail we have to turn to observations of hot loops, filled with material at coronal temperatures of a million degrees or more; these form the subject of the present chapter.

Emission from material at coronal temperatures dominates the EUV and soft X-ray regions of the solar spectrum. The lack of any appreciable photospheric or chromospheric emission at these wavelengths enables the corona to be viewed directly against the disk and, in fact, most of the available information on hot loops has been obtained from such (space) observations. In addition, other important contributions to our knowledge have come from the visible and radio regions of the spectrum. The line and continuum radiation emitted by the corona in the visible region is many orders of magnitude too faint to be detected on the disk against the glare of the underlying photosphere. Nevertheless, it can be observed beyond the limb both during and, with the aid of coronagraphs, outside of a solar eclipse.

The arrangement of the material in this chapter follows and extends that adopted in Chapter 2. We begin by summarizing the information on the morphological properties of hot loops derived from observations made at visible wavelengths, principally in the well-known $\lambda 530.3$ line of Fe XIV (Section 3.2). It was from photographs obtained in this line with ground-based telescopes that the fundamental loop structure of the active corona was in fact first recognized (Section 1.4).

In Section 3.3 we turn to observations made in hot EUV lines; those lines commonly used are listed in Table 3.2, together with their calculated formation temperatures. In the hot EUV well-developed bipolar active regions often display a characteristic 'butterfly' appearance, so described because most of the emission comes from two extended lobes of bright material shaped like the wings of a butterfly. The lobes appear to consist of systems of hot loops joining the two areas of opposite magnetic polarity in the region, although the spatial resolution of existing observations is not always adequate to demonstrate this clearly.

Owing to the limited sensitivity of existing detectors most of our knowledge of X-ray loops has come not from spectral line (i.e. monochromatic) observations but from photographs taken through relatively broad-band filters, transmitting a mixture of line and continuum emission (Section 3.4). In discussing the wealth of information so obtained on the morphological properties of loops within and joining active regions we shall find it convenient to adopt a scheme of classifying X-ray loops according to their location put forward by J.M. Davis & A.S. Krieger in 1982. Outside of active regions X-ray observations reveal the existence of arcades of loops overlying filament cavities; these quiet region X-ray loops are believed to be the counterparts of the coronal loops which have been seen over limb prominences at visible wavelengths. Taken together, the X-ray observations offer valuable insight into how the magnetic fields of active regions emerge, evolve and – as the active regions decay – become fragmented and dispersed over the surface of the Sun.

Like X-rays, radio waves in the centimetric range and above originate in plasma at coronal temperatures and can be observed against the disk. During the past few years the advent of large array-type radio telescopes employing the aperture synthesis technique has made it possible to obtain centimetre and low decimetre observations of the Sun with a spatial resolution comparable to that shown by hot EUV and X-ray images (Section 3.5). The first observation of a radio loop connecting two sunspots of opposite magnetic polarity was reported in 1980 and since then considerable effort has been devoted to the study of these features. Radio observations offer the only means currently available for inferring, in principle, both the strength and direction of the magnetic fields present in hot loops and the associated active regions.

Having completed the description of the morphological properties of hot loops we turn in Section 3.6 to the determination of their physical conditions. The theoretical basis for analysing the measured intensities of emission lines in the hot EUV and X-ray regions of the spectrum is essentially the same as that employed for cool EUV lines (Section 2.5.4). Unfortunately, in the case of soft X-rays available spectroscopic observations of active regions generally lack the spatial resolution needed to resolve individual loops and we have to rely instead on broad-band observations reflecting a mixture of line and continuum radiation. This introduces complications into the analysis, discussed in Section 3.6.4.

In the case of the radio observations a different approach is needed (Section 3.6.5). Here the mechanisms identified as being responsible for the observed thermal emission from radio loops are bremsstrahlung and gyroresonance radiation. The analysis requires a complex modelling procedure which, by itself, is not capable of giving unique, independent results. On the other hand, the radio data can serve as a useful check on results obtained in other regions of the spectrum.

One question of considerable interest is that of the relationship between loops observed in different spectral regions (Section 3.7). However, because of the difficulties of obtaining the necessary simultaneous observations, few systematic comparisons have so far been made. An exception is the relationship between hot and cool loops observed in the EUV. Here detailed studies have established that, while both hot and cool EUV loops may occur close to one another, they are not cospatial. Some rather fragmentary information is also available concerning the complex relationship between X-ray and radio loops (Section 3.7.2).

The chapter ends with tables summarizing all the available quantitative data concerning the morphological and physical properties of hot loops (Tables 3.7 and 3.8). Comparison of these data with the corresponding data for cool loops (Chapter 2, Tables 2.8 and 2.9) indicates that, while hot loops tend to be thicker, longer, higher, and longer-lived than cool loops, their other morphological properties (except temperature) are similar.

3.2 Observations in the visible region
3.2.1 *Introduction*

In the visible region of the spectrum light from the solar corona beyond the limb consists of three components, conventionally denoted by the letters K, F and L. The K component originates from scattering of photospheric light by free electrons (Thomson scattering) and is thus proportional to N_e. The F component is due to scattering by interplanetary dust. Finally, the L component is due to the emission of spectral lines and is termed the *monochromatic* corona; it is visible only in the *inner* corona ($<2R_\odot$) (Koutchmy, 1977).

Table 3.1. *Bright coronal emission lines*

Ion	Wavelength (nm)	Ionization potential (eV)	Average equivalent width (nm)
Fe X	637.4	233	0.5
Fe XIV	530.3	355	2.0
Fe XV	706.0	390	0.08
Ni XV	670.2	422	0.12
Ca XV	569.4	814	0.03

Some of the brightest emission lines in the L corona are listed in Table 3.1, in order of increasing ionization potential; this data is extracted from a more extensive table published by de Jager (1959, Table 16). By far the brightest of the coronal emission lines is λ530.3 of Fe XIV, but even so its equivalent width amounts only to 2.0 nm (Table 3.1). In fact, the *total* line emission amounts to only $\sim 1\%$ of the coronal light (Waldmeier, 1963). To this must be added substantial unwanted light scattered in the Earth's atmosphere and in the telescope itself. Successful observation of coronal loops in visible emission lines therefore demands a good coronagraph at a high-quality site at the greatest possible altitude, together with a narrow-band birefringent or interference filter designed to isolate one or more of the lines listed in Table 3.1, or a spectrograph. Some of the practical considerations governing coronagraph observations have been discussed by Dunn (1971).

Three observatories which have contributed to the loop observations described in the sections that follow are Haleakala (elevation 3120 m), Pic-du-Midi (2860 m), and Sacramento Peak (2760 m). A list of other coronal observatories has been published by Billings (1966, Appendix I) but is now somewhat dated. The nature of the observations is exemplified by those undertaken at the Pic-du-Midi Observatory, where a 20-cm coronagraph has been employed in conjunction with a birefringent filter for simultaneous photography in λ530.3 and λ637.4 and – with special adjustment – λ569.4. A second filter gives the last two of these lines together with λ670.2 of Ni XV (Dollfus, 1971). However, around the world only λ530.3 and λ637.4 have been used extensively (cf. Zirin, 1966). Nevertheless, the line λ569.4 of Ca XV may also be employed when it is bright enough. It is a useful indicator of high excitation since its ionization potential is more than twice that of λ530.3 (Table 3.1).

From an analysis of extensive films of the corona obtained with a 15-cm coronagraph at Sacramento Peak Observatory with a 0.2-nm $\lambda530.3$ filter Dunn (1971) reached the general conclusion that 'Some coronal scenes look "open"...and some look to be all loops and arches or "closed"...The arches and loops may be very basic coronal structures since many scenes appear to contain nothing else'. In fact, the fundamental loop structure of the active corona was recognized from $\lambda530.3$ films some years before the Skylab EUV and X-ray results described in Sections 3.3 and 3.4 below became available.

What, if any, is the distinction between 'arches' and 'loops'? Picat *et al.* (1973) claim that arches exist for days and loops for hours, but Dunn (1971) and Koutchmy (1977) have questioned whether the distinction between arches and loops is worth maintaining. In this book the term loop is used for all loop-like or arch-like phenomena regardless of size or lifetime.

The fact that loops are observed in lines of such high ionization potential as those listed in Table 3.1 is indicative of temperatures of the order of 1×10^6–2×10^6 K; however, discussion of the physical conditions in loops observed in the visible region is reserved for Section 3.6.2.

3.2.2 *Properties of loops observed in $\lambda530.3$*

In the visible region of the spectrum coronal loops have been most widely observed in the Fe XIV $\lambda530.3$ line. For such limb observations it is not always possible to establish whether or not the loops observed are associated with flares. Of necessity, therefore, the following summary of the general properties of active region loops, taken from the descriptions of Kleczek (1963) and Dunn (1971), refers to both flare and non-flare loops:

(1) they occur as *systems* of loops in a single active region, with typical heights of up to 50 000–100 000 km. But the larger loops (called arches by Kleczek and Dunn) may connect two active regions and extend up to heights of 200 000–250 000 km. The systems appear to be rooted in sunspot groups or in plage areas;

(2) the larger loops are generally uniform in intensity along their width and length and have a diameter of 8000–12 000 km (but may be thicker); the smaller loops are less uniform, particularly in their earlier stages, and have a diameter of 3000–8000 km. Picat *et al.* (1973) quote somewhat larger loop diameters, viz. 11 000–22 000 km;

(3) loops are more active when they are small: they tend to grow *in situ*, then fade, and others grow at a higher elevation. The larger loops are very stable;

(4) when seen obliquely, loops appear elliptical, indicating that they are *planar* structures and not spherical shells; and

(5) the smaller loops last for hours (Picat *et al.*) and show only minor disturbance from flare-associated 'waves' in the corona. The larger loops last for days and can suffer disruption from such a wave.

Loops have also been observed away from active regions, associated with quiet region (quiescent) Hα prominences (Fort & Martres, 1974). They appear to be related to the X-ray loops observed over filament cavities (Section 3.4.5). Fine examples of (flare and non-flare) loop systems photographed in λ530.3 at Sacramento Peak Observatory have been reproduced by Dunn (1971); other good illustrations are to be found in de Jager (1959, Fig. 89) and Billings (1966, Fig. 2.11).

3.2.3 *Observations in λλ637.4, 569.4 and other lines*

The Pic-du-Midi observers have made a special study of loops photographed in Fe X λ637.4. Picat *et al.* (1973) found that the loops are clearer and have a higher contrast than in λ530.3; on the other hand, they are somewhat more difficult to photograph since on average, $I_{red} \simeq \frac{1}{6} I_{green}$. Nevertheless, these authors have been able to produce photometric maps (in absolute units) of loops photographed in λ637.4. Fort & Martres (1974) have published drawings illustrating the degree of co-spatiality between (stable) loops observed in Hα, λ637.4, and λ530.3. They conclude that there is little superposition between the λ637.4 and λ530.3 loops. Dollfus (1971) has also concluded that, while there are general similarities, λ637.4 and λ530.3 loops differ in their detailed structure. His observations also show that the structure in Fe X λ637.4 is sharper than that in Fe XIV λ530.3.

A careful spectroscopic study by Tsubaki (1975) based on Sacramento Peak observations has confirmed and extended these conclusions. He analysed spectra obtained at graded heights above the limb in Fe X λ637.4, Fe XI λ789.2, Fe XIV λ530.3, Fe XV λ705.9, and Ni XV λ670.2 and intercompared the intensity fluctuations in a coronal region revealed by the different lines. He found clear differences between lines of high ionization potential (Fe XIV, Fe XV and Ni XV) and those of lower ionization potential (Fe X and Fe XI), whereas consistent similarities were found among lines with similar ionization potentials. A similar result is found for EUV loops (Section 2.4.3).

Observations in Ca XV λ569.4 are fragmentary. When a coronal event is particularly energetic, emission may appear in this line along portion of a loop. According to Dunn (1971) the emission is closely associated with the *tops* of coronal loops.

3.2.4 *Observations in the continuum*

Observations of loops in the white light K corona are far less extensive than those which have been made in the various spectral lines, described above. Non-eclipse photographs by Dollfus (1971, Fig. 7) show a small white light loop in roughly the same location as a more clearly-defined system seen in $\lambda637.4$ and $\lambda530.3$. However, Dollfus finds that in general there is no correlation between the white light and $\lambda530.3$ intensity distributions. A large, diffuse white light loop system overlying a prominence, photographed during an eclipse, has been published by Koutchmy (1977; see also Saito & Hyder, 1968).

Finally, a particularly distinct white light loop system was photographed during the 7 March 1970 eclipse by S.M. Smith and L.M. Weinstein of NASA-Ames. This photograph was obtained with a radially-graded filter and according to Tousey & Koomen (1971), who have reproduced it, shows detail down to $10''$ of arc close to the limb. In fact, these authors remark that it shows better detail close to the limb than any that they had previously seen. Again, however, the white light loops are more diffuse than those typically observed in the various spectral lines. One obvious physical reason for the difference in appearance between white light and spectral line photographs of loops is that the former map, in effect, the electron density N_e while the latter map the ionization temperature of the line in question.

Finally, it is of interest to observe that although white light coronal transients often appear to *look* like loops, the balance of current opinion is that, instead, they are probably expanding three-dimensional shell-like structures with partially depleted interiors.

3.3 Observations in EUV lines

3.3.1 *Introduction*

As we have already remarked in Section 2.4, EUV lines characterized by formation temperatures $\bar{T}_e$ in excess of 1×10^6 K are conventionally referred to as 'hot' lines. In this section we shall describe the properties of coronal loops observed in such lines and, once again, we shall find that our knowledge has come largely from Skylab spectroheliograms. Most observations of hot EUV loops have been made in a relatively small number of high-excitation resonance lines of the ions Mg X, Si XII, Fe XV and Fe XVI, details of which are set out in Table 3.2. The third and fourth columns of this table give respectively the ionization potential of the ion and the formation temperature of the line computed on the basis of the ionization equilibrium calculations of Jacobs *et al.* (1977a, b; 1979). The significance of the quantity G_{max} (fifth column) is discussed in Section 2.5.4.

It can be seen from Table 3.2 that the $\lambda61.0$ ånd $\lambda62.5$ lines of Mg X have

Table 3.2. *EUV lines used to obtain observations of hot loops*

Ion	Wavelength (nm)	Ionization potential (eV)	$\bar{T}_e$ (K)	G_{max}
Mg X	60.99	367.4	1.0×10^6	1.94×10^{-4}
	62.53			1.95×10^{-4}
Si XII	49.93	523.2	2.0×10^6	1.61×10^{-4}
	52.11			1.62×10^{-4}
Fe XV	28.42	456.0	2.0×10^6	5.9×10^{-5}
Fe XVI	33.54	489.5	2.5×10^6	5.1×10^{-5}
	36.08			5.1×10^{-5}

(a) All the lines listed are resonance lines (Section 2.5.4);
(b) For each line $\bar{T}_e$ is the temperature at which the function $G(T_e)$ defined by Eqn. (2.15) attains its maximum value, G_{max} (Section 2.5.4);
(c) For more complete atomic data see Wiese, Smith & Miles (1969) and Fuhr *et al.* (1981).

a formation temperature of about $\bar{T}_e = 1.0 \times 10^6$ K and therefore lie on the borderline between the cool and hot temperature regimes. Nevertheless, the appearance of these lines is normally regarded by observers as indicative of the presence of hot coronal material. However, as Webb (1981, p. 175) remarks, the situation may be a little more complicated. We saw in Section 2.5.4 that the contribution to the observed intensity of a line made by material at a particular electron temperature T_e depends on the value of the function $G(T_e)$ defined by Eqn. (2.15), as well as, of course, on the amount of material at this temperature present. Figure 3.1 shows the variation of $\log_{10} G(T_e)$ with $\log_{10} T_e$ for the Mg X $\lambda\lambda 61.0$ and 62.5 lines, based on the ionization equilibrium calculations of Jacobs *et al.* (1979). The extended tail to the curve on the hot side of the peak contrasts with the relatively sharp cut-off on the cool side and thus justifies the classification of Mg X $\lambda\lambda 61.0$ and 62.5 as (predominantly) hot EUV lines. Notwithstanding, one must be careful, in interpreting observations, not to exclude the possibility of a significant amount of material emitting at cool temperatures between 6.0×10^5 and 1.0×10^6 K. For comparison Fig. 3.1 also shows the corresponding curve calculated for another line of the same element, Mg IX $\lambda 36.8$, which also has a formation temperature of about 1.0×10^6 K but is classified as a cool EUV line (cf. Table 2.4).

In the high temperature regime the EUV emission from the Sun is confined very largely to active regions and the loops emanating from them (see, for example, Tousey *et al.*, 1973, Figs. 3(*a*) and 4(*a*); Sheeley *et al.*, 1975a, Fig. 1). This contrasts strongly with the case of the cool EUV

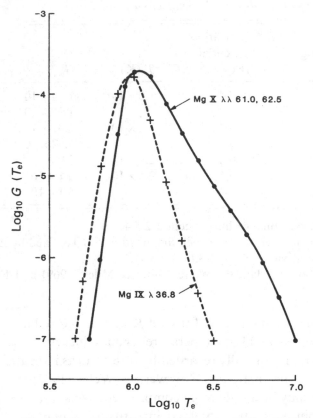

Fig. 3.1. Variation of the quantity $\log_{10} G(T_e)$ defined by Eqn. (2.15) with $\log_{10} T_e$ for the $\lambda\lambda 61.0$ and 62.5 lines of Mg X. Note the extended tail to the curve for Mg X $\lambda 61.0$ on the hot side of the peak; this contrasts with the relatively sharp cut-off on the cool side and justifies the classification of this line as a 'hot' EUV line. Also shown is the corresponding curve for the 'cool' line Mg IX $\lambda 36.8$ (see text).

emission, which comes from all parts of the disk (Section 2.4.1). When photographed in hot EUV lines well-developed bipolar active regions on the disk often present a characteristic 'butterfly' appearance (cf. Levine, 1976), so-called because most of the emission comes from two extended lobes of rather diffuse bright material shaped like the wings of a butterfly. In some cases any detail discernible within the lobes tends to be in the form of blobs or streaks rather than loops. Recognizable loops are readily seen, however, outside the lobes. These may be grouped together to form systems of loops curving around the outside parts of the active region. Where individual hot EUV loops can be clearly distinguished they are found to be broad and irregular and, overall, appear less loop-like than those visible in the cool EUV. Sheeley (1980) remarks that, unlike cool EUV loops, the hot loops are never observed to brighten progressively along their lengths, but appear to brighten and fade *in situ*.

Not all of the loops extending outwards from an active region necessarily return to the same vicinity. Frequently loops arch across the disk for distances of the order of 100 000 km to link two or more active regions into a huge magnetic complex (see, for example, Tousey *et al.*, 1973, Fig. 4(*a*); Sheeley *et al.*, 1975a, Fig. 6; Sakurai & Uchida, 1977, Fig. 3(*b*)). Such linkages are not confined to active regions located in the same hemisphere, i.e. north or south, but may involve regions lying on both sides of the equator (Sheeley *et al.*, 1975a, Fig. 7). Interconnecting loops are generally fainter than those with both ends terminating in or on the outskirts of the same active region. Tousey *et al.* (1973) appear to have been the first to report the existence of hot EUV loops connecting active regions. They remarked that 'this behaviour of the weaker loops was entirely unexpected', overlooking the fact that the presence of hot coronal loops linking active regions was already known to coronal observers working in the visible spectrum (Section 3.2.2); it had also been revealed as early as 1968 by soft X-ray photographs obtained from rockets.

In the following pages we shall first describe the general characteristics of active region loop systems observed in the hot EUV and then discuss in more detail the properties of individual loops.

3.3.2 *Active region loop systems: general characteristics*

Examination of Fig. 2.10 allows us to compare the appearance of a large bipolar active region on the disk in the hot EUV line Fe XV $\lambda 28.4$ ($T_e = 2.0 \times 10^6$ K) with that in the cool Ne VII and Mg IX lines, as well as with the distribution of the underlying photospheric magnetic field (Sheeley, 1980). Most of the Fe XV emission comes from two extended lobes of rather diffuse bright material whose shape gives the active region the characteristic 'butterfly' appearance referred to above. Within the lobes are two smaller patches of distinctly brighter material which, like the remaining parts of the diffuse lobes, seem to show little trace of internal structure. Note that the hot Fe XV emission tends to be relatively weak in those parts of the active region where the cool Ne VII and Mg IX emissions are strong. Nevertheless, there are obvious similarities between the large-scale emission patterns in the three lines and one might suppose that each lobe visible in the Fe XV image is composed of a curved system of faint but largely unresolved loops. As it stands, however, the appearance of this Fe XV photograph could also be explained by supposing that the lobes are clouds of hot gas confined to these particular volumes of space by the closed magnetic field lines joining the areas of opposite polarity within the active region.

At one end of the line dividing the lobes one can distinguish a number

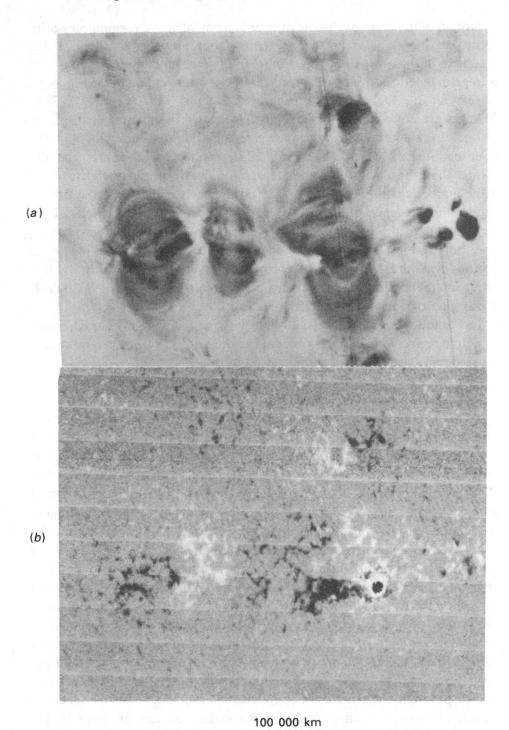

(a)

(b)

100 000 km

Fig. 3.2. (*a*) Photograph of a complex of active regions observed in Fe XV $\lambda 28.4$; (*b*) magnetogram showing the underlying longitudinal photospheric field (Sheeley *et al.*, 1975a). Individual loops are visible within the 'butterfly' lobes of the active regions, as well as loops connecting different active regions. Some loops cross the equator, which runs left to right slightly above the middle of photograph (*a*).

of faint, poorly defined Fe XV loops which appear to radiate outwards like the loops comprising the active region plumes observed in cool EUV lines (Section 2.4.2). The individual Fe XV loops are broad and irregular and appear less loop-like than those visible in cooler lines. Sheeley (1980) remarks that 'it is sometimes easiest to identify individual bright features by concentrating on the dark spaces that separate them'! He attributes the broad and irregular shapes of the bright features partly to the presence of unresolved loops along the line of sight and partly to the obscuration of some of the loops by intervening cool dense material (cf. Section 2.4.1).

More convincing evidence of the presence of individual loops within the butterfly lobes of active regions is provided by Fig. 3.2 (see particularly the upper lobe of the region on the extreme left). It shows a negative print of a complex of interconnected active regions observed on the disk by Sheeley *et al.* (1975a) in Fe XV λ28.4 and demonstrates the close relationship between the structure of the active regions and the large-scale distribution of the underlying photospheric magnetic field. Note the prominent dark areas (seen bright on this negative print) that separate loop systems originating in the same polarity area. According to Bohlin & Sheeley (1978) these dark areas typically have widths of 15″–30″ and lengths of up to 90″, although some are found to be almost circular.

One factor which may contribute to the difficulty of distinguishing individual loops in Figs. 2.10 and 3.2 is the sheer complexity of the structure of a large active region or complex of active regions (cf. Cheng, 1980). Fig. 3.3, taken from Dere (1982), shows negative prints of a less complicated region in Fe XV and in various cool EUV lines. Several quite distinct broad loops can be seen projecting outwards from a compact patch of bright emission. Note that these loops originate near an area of He II facular emission and not over the sunspot marked by an arrow on the magnetogram (see Dere's Fig. 3).

Cheng *et al.* (1980) have studied the structure of a bipolar active region (McMath 12379) on the disk using Skylab spectroheliograms taken in the hot λ33.5 line of Fe XVI and have concluded that the emission originates mainly in a compact system of hot, low-lying loops joining the inner boundaries of the two areas of opposite magnetic polarity (see their Fig. 2). No counterparts of these low-lying loops can be seen in the Fe XV image of Fig. 2.10 and, unfortunately, the scale of the Fe XVI photograph published by Cheng *et al.* (1980, Fig. 1) is much too small to permit an independent assessment. In a separate paper, however, Cheng (1980, Fig. 2) has reproduced Fe XV and Fe XVI spectroheliograms showing four bright streaks linking the opposite polarity areas in a small active region observed on the disk, which he interprets as a system of small, parallel

Mc Math 12651

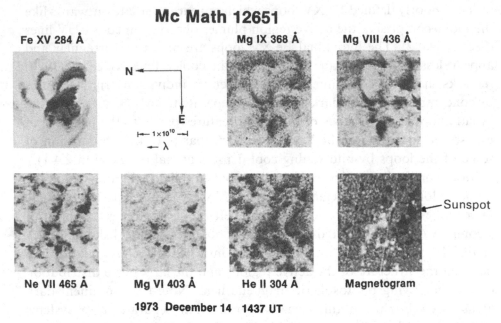

Fig. 3.3. Negative photographs of an active region taken in the hot Fe XV $\lambda 28.4$ line and in various cool EUV lines (Dere, 1982). Several distinct broad loops project outwards from a compact patch of bright emission. Note that they originate near an area of He II facular emission and not over the sunspot marked by the arrow on the magnetogram.

loops. Observations of higher spatial resolution are clearly needed to decide how much of the hot EUV emission from the central areas of active regions does in fact come from hitherto largely unresolved systems of small, low-lying loops.

Figure 3.4 gives a sequence of (negative) Fe XV images of an active region taken at about three-hourly intervals over a period of some 36 hr (Sheeley, 1980). This region displays the typical butterfly appearance described earlier, although in this case the extended emission lobes show considerable internal structure. Nevertheless, only on the outskirts can one distinguish clear loop-like features. Over 36 hr the large-scale emission patterns remained relatively unchanged but, as Sheeley remarks, the individual features present at the end are entirely different from those at the beginning. The large-scale patterns may remain recognizable for several days or more (Sheeley *et al.*, 1975b, Fig. 1; 1980).

The birth, development and decay of small active regions on the disk have been studied by Sheeley *et al.* (1975a) using sequences of Skylab spectroheliograms taken in Fe XV $\lambda 28.4$. Their observations show that the emergence of a new bipolar magnetic region (BMR) in the photosphere is

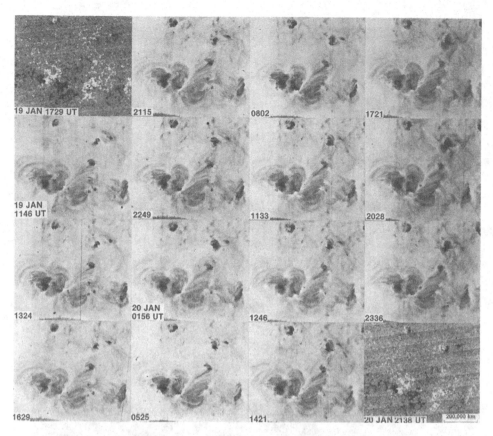

Fig. 3.4. Sequence of (negative) Fe XV λ28.4 images of an active region taken at about three-hourly intervals over a period of some 36 hr (Sheeley, 1980). The region displays a characteristic 'butterfly' appearance. Over 36 hr the large-scale emission patterns remain relatively unchanged, although the individual features present at the end are entirely different from those at the beginning.

accompanied by the appearance of a single bright blob of hot EUV emission over one of the two areas of opposite magnetic polarity. With the passage of time the blob increases in size and develops a characteristic 'plume' (Sheeley *et al.*, 1975a, Figs. 2, 3 and 5) or 'fountain' (*loc. cit.*, Fig. 4). Both the plumes and fountains show evidence of the fine structure in the form of long, somewhat curved (plume) or highly curved (fountain) streaks. These presumably represent loops but in no case do they connect the two areas of opposite polarity in the new BMR itself; rather they appear to join the polarity area over which the bright EUV blob initially developed to neighbouring, pre-existing patches of opposite field outside the new BMR.

At first sight it is hard to reconcile the process of formation of a new active region observed by Sheeley *et al.* with the eventual development of

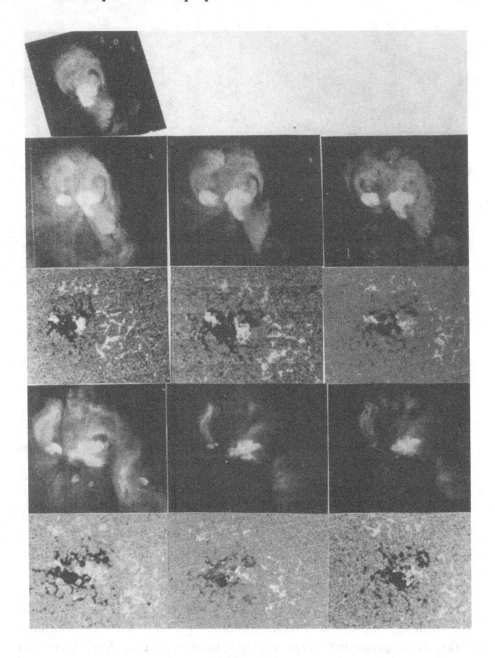

100 000 km

Fig. 3.5. Daily spectroheliograms illustrating the emergence, growth and decay of an active region observed in the hot EUV line Fe XV $\lambda 28.4$ (Sheeley *et al.*, 1975a). A magnetogram taken the same day (except the first) is reproduced under each photograph. For a detailed description see text.

the butterfly structure displayed by well-developed, bipolar active regions (see above). In these active regions the areas of opposite magnetic polarity in the underlying photosphere appear to be linked by a huge three-dimensional system of loops, often largely unresolved, which – when seen in projection on the disk – gives rise to the characteristic butterfly form.

Two points have to be considered. Firstly, most of the emerging regions studied by Sheeley *et al.* occurred at a time of high solar activity within or on the outskirts of regions of pre-existing and, in at least one case, increasing magnetic flux. The appearance of the new regions was thus probably the result of further growth of an existing magnetic complex and, as such, their morphology would have been heavily influenced by the pre-existing field. Whether a similar process of emergence and growth would apply to the development of a new BMR in a hitherto largely field-free region of the disk remains to be answered. Secondly, as Sheeley and his coworkers rightly emphasize, the emergence of a new BMR seems to be accompanied by a process of magnetic reconnection which results in the lines of force of the new field becoming linked to those of the pre-existing field. This suggests that magnetic reconnection may also play an important, but as yet unknown, role in the development of the butterfly form characteristic of large isolated bipolar magnetic regions.

Figure 3.5 gives a sequence of daily Fe XV spectroheliograms illustrating almost the entire process of emergence, growth and decay of an active region (Sheeley *et al.*, 1975a). As far as one can judge from the published magnetograms the region appeared in an area already pervaded by a complex magnetic field of mixed polarities. It began as a single blob of very bright material accompanied by a large region of fainter emission on the lower right. A second concentrated bright blob then appeared to the left (east) of the first, the area of fainter emission at the top became larger, and additional faint emission developed below and to the left of the central bright blobs. After the second day the faint emission below and to the left of the two blobs became weaker; the second bright blob began to decay on the fifth day; and the first bright blob had declined significantly by the sixth day. On the seventh day all that remained of the new active region was a decaying remnant of the first blob, accompanied by some wisps of very faint material at the top.

Figures 3 and 4 of Sheeley *et al.* (1975a) illustrate other instances of the decay and disappearance of active regions on the disk.

At the limb, published photographs of active regions taken in hot EUV lines usually give little indication of loop structure. Figure 3.6, taken from Levine (1976), shows McMath region 12628 near the east limb in the cool

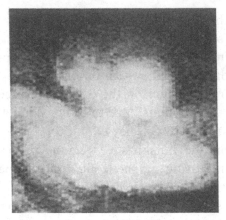

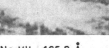

Ne VII 465.2 Å   Si XII 520.7 Å

Fig. 3.6. McMath region 12628 photographed near the east limb in (*a*) the cool Ne VII *λ*46.5 line, and (*b*) the hot Si XII *λ*52.1 line (Levine, 1976). The space occupied by the loops visible in Ne VII appears in Si XII to be filled by an almost amorphous mass of hot material, in which no individual loops can be distinguished.

Ne VII *λ*46.5 line and in the hot Si XII *λ*52.1 line ($\bar{T}_e = 2.0 \times 10^6$ K). Whereas individual loops are clearly distinguishable in Ne VII, none are evident in Si XII. Instead the space occupied by the cool loops appears to be filled by an almost amorphous cloud of hot material, consistent with the similar picture suggested by the Fe XV disk photograph in Fig. 2.10. The same amorphous appearance is shown by a limb photograph of this active region in the somewhat less hot Mg X *λ*62.5 line ($\bar{T}_e = 1.0 \times 10^6$ K) published by Foukal (1975, Fig. 1); see also Sheeley *et al.* (1975a, Fig. 5).

On the other hand, long-exposure Skylab photographs taken in the hot line Fe XV *λ*28.4 and analysed by Stewart, Brueckner & Dere (1986) revealed a system of very high, faint loops above the east limb. The loops apparently connected two active regions and reached heights of ~ 290 000 km – comparable to those reached by soft X-ray loops (cf. Table 3.7).

Habbal, Ronan & Withbroe (1985) have examined the geometry of the hot and cool EUV loops in an active region near the limb. They find that some of the cool (Ne VII) loops appear to extend further out from the limb than the hot (Si XII) loops. This may reflect differences in the heights of the hot and cool loops or could imply that the hot loops tend to lie in planes making greater angles to the solar vertical than those of the cool loops. Clearly a reconstruction of the true geometry of individual loops is required to distinguish between the two possibilities.

In Section 2.4.2 mention has been made of the investigation by Schmahl *et al.* (1982) of the relationship between cool EUV loops and dark H*α* filaments. These authors reached similar conclusions for hot EUV loops as for cool EUV loops.

3.3.3 *Properties of individual loops*

As in the case of cool EUV loops, relatively little solid information is available about the morphological and dynamical properties of hot EUV loops. What is known may be conveniently summarized under the following headings:

(1)   *Shape, size and thickness*

The only information we have about the true shapes and sizes of hot EUV loops comes from the work of Berton & Sakurai (1985). They used the change in perspective resulting from solar rotation to reconstruct two long-lived loops observed in the Fe XV $\lambda 28.4$ line which linked two active regions on the disk (see their Figs. 1 and 2). To within the accuracy of the method each loop was found to lie in a plane, whose inclination to the solar vertical was 25° in one case and 7° in the other. Figure 3.7 shows the true outlines of the central axes of the loops in their own planes. Both are clearly asymmetrical and reach heights of 119 000 and 131 000 km above the solar surface respectively. Berton and Sakurai show that the observed asymmetry is consistent with the assumption of a potential magnetic field distribution within and between the active regions concerned. However, it should be stressed that the reliability of the reconstructions hinges on the validity of the assumption that the basic geometry of the loops remained unchanged for some 31 hr.

Loops associated with a single active region do not rise to the great heights attained by the giant interconnecting loops described above. Indeed, some active region loops extend up to only relatively modest heights. This is borne out by the results of Kundu, Schmahl & Gerassimenko (1980) who, assuming the loop planes to be vertical, have inferred heights ranging from 5000 to 11 000 km for several active region loops observed on the disk in Fe XVI $\lambda 33.5$, Si XIV $\lambda 41.8$ and Mg X $\lambda 62.5$.

These loops were found to have widths varying from 11 000 to 18 000 km. In contrast, Cheng (1980, Table 3) obtained values ranging from 4800 to 6500 km for the widths in Fe XV of four small loops linking the opposite polarity areas in a small active region on the disk.

From measurements of intensity tracings across limb loops observed in Mg X and Si XII Foukal (1975) has derived widths ($2 \times$ FWHM) ranging from 3000 to 12 000 km, thus seeming to indicate that hot EUV loops may be significantly thicker than cool ones (see Section 2.4.3). However, this result may reflect, at least in part, the greater difficulty of distinguishing

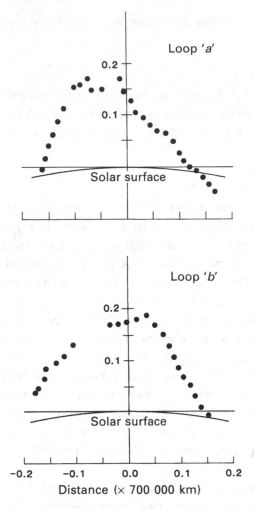

Fig. 3.7. True outlines of the central axes of two long-lived loops in their own planes derived by Berton & Sakurai (1985). The loops connect two active regions and were observed in Fe XV λ28.4. Both are assymetrical and reach heights of 119 000 (loop 'a') and 131 000 km (loop 'b'); their planes are inclined to the (local) solar vertical at angles of 25° and 7° respectively.

individual loops on photographs of active regions at the limb taken in hot EUV lines (compare, for example, the Mg X image in Foukal's Fig. 1 with the simultaneous O IV and O VI pictures).

In passing, mention should be made of the opinion expressed by Dere (1982) that hot EUV loops are spatially resolved when the resolution approaches 2″ of arc but show no evidence of fine structure (filamentation). This is in marked contrast to the situation for cool EUV loops (Sections 2.4.3 and 2.5.5). According to Cheng, Smith & Tandberg-Hanssen (1980) the Fe XVI loops seen in the outer parts of an active region have an aspect ratio (i.e. radius of cross-section divided by length) of about 0.03, which is an order of magnitude greater than that found for cool EUV loops (Section 2.4.3). Cheng (1980, Table 3) found even larger aspect ratios, ranging from 0.09 to 0.18, for four small Fe XV loops observed on the disk.

(2)   *Location of footpoints*
        Examination of Fig. 3.2 shows that individual loops appear to join regions of opposite polarity in the underlying photospheric magnetic field. However, identification of the exact locations of the footpoints is difficult since the emission in hot EUV loops, unlike that in cool loops, tends to fade out towards the footpoints (Sheeley, 1981).

(3)   *Loop intensities*
        Using spectroheliograms taken in the hot EUV lines Fe XV $\lambda 28.4$ and Fe XVI $\lambda 33.5$ Cheng (1980) has measured the absolute intensities at a number of points in four small, parallel loops and in the surrounding background in a small active region on the disk. Further data have been supplied by Raymond & Foukal (1982) in the form of a table giving the absolute intensities of a number of hot EUV lines observed in active region loops beyond the limb.
        Recently Habbal, Ronan & Withbroe (1985) have attempted to study the temporal variation in the emission from loops in an active region near the limb in the hot EUV lines Mg X $\lambda 62.5$ and Si XII $\lambda 52.1$. They plotted the integrated intensity from the whole active region for a period of nearly 20 hr and found relatively little variation. On the other hand, individual hot loops displayed small changes of the order of a few per cent on a time scale of about 30 min. By contrast, the cool EUV loops in the region showed much stronger variations, sometimes appearing or disappearing in 5 to 10 min.
        Retaining the convention of Section 2.4.3 we define the contrast of a loop at any given point and wavelength λ with respect to its surroundings as

$$C_1(\lambda) = \frac{I_1(\lambda) - I_0(\lambda)}{I_0(\lambda)},$$

where I_1 and I_0 are the loop and background intensities respectively. Actual measurements of C_1 made by Foukal (1975) in six limb loops in three different active regions yielded values ranging from 0 to 3 in the hot Mg X $\lambda 62.5$ line and from 0 to 2 in the even hotter Si XII $\lambda 46.5$ line. The upper· limits are considerably smaller than those found for the corresponding loops observed in the cool O VI $\lambda 103.2$ and Ne VII $\lambda 46.5$ lines (Section 2.4.3). This reflects a pronounced rise in the level of background radiation with temperature (cf. Fig. 3.6).

(4)   *Lifetime and evolution*

No measurements have been made of the lifetimes of active region loops observed in hot EUV lines such as Fe XV and Fe XVI, but it is known that they evolve on a time scale of roughly 6 hr (Cheng *et al.*, 1980; Sheeley, 1980). It is possible that loops linking different active regions are longer-lived: six of seven Fe XV loops observed by Berton & Sakurai (1985) persisted for at least 31 hr (compare their Figs. 2(*a*) and 2(*b*)).

3.4   Observations in X-rays
3.4.1   *Introduction*

Below about 10 nm we leave what is conventionally termed the EUV portion of the spectrum (Section 2.4.1) and enter the region of soft X-rays (Donnelly, 1977). In appearance the solar X-ray spectrum resembles the EUV spectrum. It is dominated by the presence of emission lines of highly ionized stages of the elements N, O, Si, S, Ne, Mg and Fe, superimposed on a continuum background. Three processes contribute significantly to the X-ray continuum: thermal bremsstrahlung (free–free emission by electrons), radiative recombinations of electrons and ions (free–bound emission), and two-photon decay of certain metastable states of helium-like and hydrogen-like ions. Their relative contributions depend on the temperature and density of the emitting plasma (Walker, 1972, 1975; Culhane & Acton, 1974). By convention the soft X-ray region is taken to end near 0.1 nm: photons with wavelengths smaller than this have energies exceeding 10 keV and belong to the category of *hard* X-rays. They are observed during and in the aftermath of solar flares (Section 4.5.3). Detailed discussions of the solar X-ray spectrum and its interpretation are to be found in review articles by Walker (1972) and Culhane & Acton (1974).

The region of the solar X-ray spectrum below 2.5 nm is of special interest because it embraces the resonance transitions of many highly ionized ions formed at temperatures truly representative of the conditions in the Sun's inner corona, i.e. $\sim 2 \times 10^6$ K or greater (Walker, 1972; Culhane & Acton, 1974). Despite the difference in physical conditions between hot and cool loops, it is nevertheless valid to use Eqn. (2.15) to compute the formation temperature $\bar{T}_e$ of any specified X-ray line of any of the elements listed above with the aid of the ionization equilibrium calculations of Jacobs *et al.* referenced in Chapter 2.

In 1958 the first direct evidence was obtained of enhanced X-ray emission associated with solar active regions (Kreplin, 1961). Subsequently much effort was devoted to the development of instruments capable of providing the spatial resolution needed to study individual solar features.

In 1968 Vaiana and other workers at the firm of American Science and Engineering (AS&E) succeeded in obtaining rocket observations with a spatial resolution approaching a few sec of arc. These revealed not only the presence of loops above active regions similar to those seen previously on photographs of the corona taken in the visible region of the spectrum (Section 3.2) but also the existence of 'arches' connecting different active regions. The design of the highly productive AS&E telescope carried by Skylab owed much to the experience gained in these earlier rocket experiments (Vaiana *et al.*, 1976, 1977). Between 28 May 1973 and 7 February 1974 the Skylab instrument obtained nearly 32000 X-ray photographs of the Sun, approximately one per cent of which are reproduced in a comprehensive atlas published by Zombeck *et al.* (1978). This atlas also contains a bibliography of no fewer than 192 papers and abstracts written prior to March 1978 in which extensive use had been made of the X-ray data.

The AS&E X-ray telescope was capable of operation in either a direct or spectroscopic mode. In the first mode photographs were taken through a series of filters transmitting in the following wavelength bands: 0.2–1.7 nm; 0.2–1.4 and 1.9–2.2 nm; 0.2–3.2 and 4.4–5.4 nm; 0.2–1.8 and 4.4–4.7 nm; 0.2–1.1 nm; and 0.2–1.4 nm. The observations therefore reflect a mixture of line and continuum radiation. In the second mode an X-ray transmission grating was introduced into the optical path, converting the instrument into a slitless spectrograph whose measured spectral resolution was limited, however, to about 0.014 nm at 0.7 nm in the first order of the grating (Zombeck *et al.*, 1978).

In addition to the AS&E instrument Skylab carried an X-ray telescope developed by the Marshall Space Flight Center and the Aerospace Corporation (Underwood *et al.*, 1977). It too was equipped with broad-band filters. These permitted the detection of somewhat harder (i.e. shorter wavelength) X-rays, making this telescope particularly suited to the study of flares.

A knowledge of the effective spatial resolution is essential to a proper interpretation of X-ray photographs (cf. Vaiana *et al.*, 1977, Sect. 2.1; Davis *et al.*, 1979; van Hoven, 1981, Sect. 4.2.2). Both of the X-ray telescopes on Skylab were of the grazing-incidence type. The principles underlying the optical design of such instruments are well understood (see, for example, Kuang, 1982, and references therein) and great advances have been made in refining the techniques of construction. Nevertheless, imperfections in the alignment, figure and polish of the mirrors significantly degrade the performance of actual telescopes and, in particular, the point spread function may be asymmetric as a result of non-isotropic scattering by surface irregularities (Giacconi *et al.*, 1969). To date solar observations have been obtained with telescopes which concentrate less than half the collected energy into the image, the remainder being scattered over a wide field.

According to Davis & Krieger (1982) the point spread function of the mirrors used in the AS&E telescope on Skylab had a width (FWHM) of 4″ of arc at 0.8 nm, while 50% of the energy in the image of an on-axis point is concentrated within a radius of 48″. By comparison, the grazing-incidence telescope employed by Davis and Krieger to photograph the Sun from a rocket on 8 March 1973 achieved significantly higher spatial resolution, the corresponding mirror figures of merit being 2″ and 31″. Nevertheless, the performance would hardly be adequate to reveal low contrast detail on a scale of 15″ of arc or less. The FWHM of the point spread function of the other X-ray telescope on Skylab is given by Underwood *et al.* (1977) and Maute & Elwert (1981) as approximately 2″.

It is important to remember that in practice the spatial resolution of any solar observation is determined not only by the performance of the telescope but also by the resolving capability of the detector – usually, in the present context, a photographic emulsion. In fact, the resolution of solar X-ray photographs obtained in recent years may have been limited more by the film than by the quality of the X-ray optics (see, for example, Davis *et al.*, 1979, Fig. 6). The literature on the design and performance of X-ray telescopes is voluminous and for an introduction to the subject the interested reader is referred to the review article by Aschenbach (1985).

Soft X-ray photographs of the Sun's disk during the active part of the solar cycle show a wealth of detail. In addition to solar flares, six distinct types of X-ray structure have been identified: coronal holes, X-ray (coronal) bright points (XBP), active regions, active region interconnecting loops, 'quiet region' loops, and filament cavities (see, for example, Vaiana, Krieger & Timothy, 1973, Table II; Zombeck *et al.*, 1978, Fig. 9). Detailed discussions of coronal holes and bright points lie beyond the scope of this book.

As in the EUV (Section 3.3), active regions seen in soft X-rays are believed to be composed basically of bright coronal loops arching between areas of opposite magnetic polarity. However, since the X-ray brightness of an active region may exceed that of 'quiet' parts of the corona by some three orders of magnitude (Vaiana *et al.*, 1976), the actual appearance of the region on a broad-band photograph depends very much on the exposure time (see, for example, Vaiana *et al.*, 1976, Fig. 7). It is also strongly influenced by the spectral bandpass (Zombeck *et al.*, 1978, Fig.

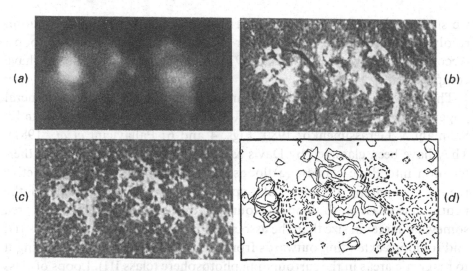

Fig. 3.8. Photographs of a group of active regions taken in (*a*) soft X-rays (0.3–2.3 nm; 4.4–5.6 nm), (*b*) the Hα line, and (*c*) the Ca II K line; map (*d*) shows the underlying longitudinal photospheric magnetic field (Vaiana, Krieger & Timothy, 1973). The active regions overlie areas of strong bipolar magnetic fields. Note that the most intense X-ray emission from each active region comes from the vicinity of the magnetic 'neutral' line.

14). While intermediate density exposures, for example, serve to show details of the loop structures associated with active regions, very short exposures are required to reveal the small cores of intense X-ray emission to be found at the hearts of some of them.

X-ray active regions overlie regions of strong bipolar magnetic fields in the photosphere (see, for example, Vaiana, Krieger & Timothy, 1973, Fig. 8; Poletto *et al.*, 1975, Figs. 1, 2, 8 and 9; Zombeck *et al.*, 1978, Fig. 15; Pallavicini *et al.*, 1979, Figs. 7 and 8; Golub *et al.*, 1982, Fig. 3). The relationship between the X-ray intensity pattern and the distribution of the longitudinal component of the photospheric magnetic field is illustrated in Fig. 3.8, which also shows for comparison photographs of the same group of active regions taken in Hα and the K-line of Ca II (Vaiana, Krieger & Timothy, 1973). It can be seen that the most intense X-ray emission from each active region comes from the vicinity of the magnetic 'neutral' line, a conclusion which seems to apply in general. Moreover, if the field gradient across the neutral line is large, there is frequently a small bright core connecting the two areas of opposite polarity (Krieger, Vaiana & Van Speybroeck, 1971; Vaiana, Krieger & Timothy, 1973). The spectrum of the core is harder than that of the rest of the active region, implying that, if the emission is thermal in origin, the core is hotter. The rocket X-ray photograph in Fig. 3.8 reveals little structure within the cores. However, as we shall see in the next section, observations of higher spatial resolution resolve the core into a small cluster of densely packed, compact loops. According to Howard & Svestka (1977) older active regions do not have compact cores.

The X-ray loops associated with an active region are similar in general appearance to those seen in hot EUV lines (see, for example, Kundu, Schmahl & Gerassimenko, 1980, Figs. 4 and 6; Pallavicini *et al.*, 1981). They have been classified by Davis & Krieger (1982) on the basis of their location into three classes: (a) loops occurring in the core of the active region and joining the areas of opposite polarity on either side of the neutral line (class I); (b) loops surrounding the core and occupying an area somewhat larger in size than the associated Hα or calcium faculae (class II); and (c) loops extending outwards from the active region and connecting it to magnetic areas in the surrounding photosphere (class III). Loops of class III sometimes occur in arcades spanning active region Hα filaments or they may link adjacent active regions, which together form a huge magnetic complex (cf. Section 3.3.1). When interconnecting loops join active regions on opposite sides of the equator they are referred to as 'trans-equatorial' loops.

Many quiet region loops seem to occur in association with the final type of solar X-ray structure listed above, filament cavities. These appear as elongated patches of reduced emission directly overlying quiet region (quiescent) Hα filaments. (In modern usage filaments located within and outside active regions are termed 'active region' and 'quiescent' filaments respectively (Hirayama, 1985). For the latter, however, the term 'quiet region' filament would be more appropriate and is adopted here.) Filament cavities are observed both on the disk (Vaiana, Krieger & Timothy, 1973, Fig. 14; McIntosh *et al.*, 1976, Fig. 4; Webb, Krieger & Rust, 1976, Fig. 2; Golub *et al.*, 1982, Fig. 8) and at the limb (McIntosh *et al.*, 1976, Fig. 5; Serio *et al.*, 1978, Figs. 4 and 5). A filament cavity covers a (projected) area much greater than that of the Hα filament itself and may persist for some time after the filament disappears.

In the following sections we shall first summarize the available information on the general characteristics of active region loop systems observed in X-rays and then discuss in turn the properties of individual active region loops, active region interconnecting loops, and quiet region loops.

3.4.2 *Active region loop systems: general characteristics*

According to Bumba and Howard (1965) the existence of 'complexes of activity' on the Sun was first suggested in a monograph

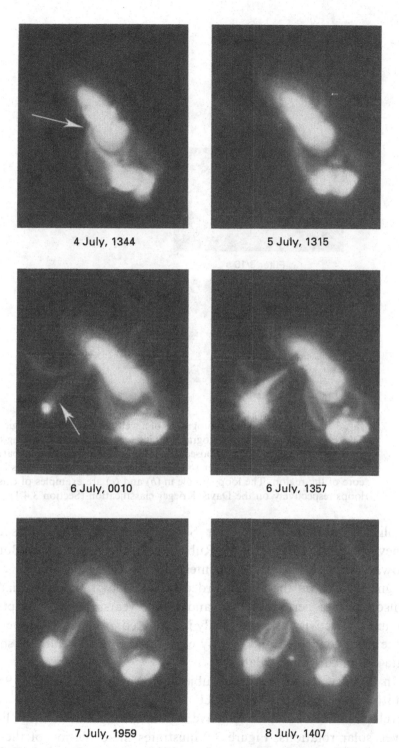

4 July, 1344

5 July, 1315

6 July, 0010

6 July, 1357

7 July, 1959

8 July, 1407

Fig. 3.9. Evolution of a complex of active regions photographed in soft X-rays (0.2–3.2 nm; 4.4–5.4 nm) during its second passage across the disk (Howard & Svestka, 1977). The large region in the centre was the first and basic member. During the lifetime of the complex other regions progressively appeared and became linked to existing ones by interconnecting loops. The latter tend to link an active region to its immediate neighbours rather than to regions further away.

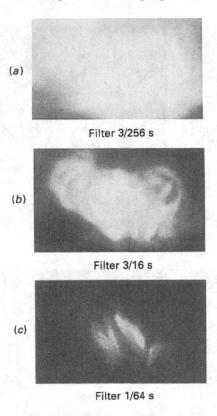

(a)

Filter 3/256 s

(b)

Filter 3/16 s

(c)

Filter 1/64 s

Fig. 3.10. Three soft X-ray photographs of a well-developed active region (Rosner, Tucker & Vaiana, 1978). Photograph (*a*) is an overexposure revealing the overall extent of the region; (*b*) is exposed to show the loops in the outer parts; and (*c*) is taken through a harder filter to reveal the very bright short loops lying at the core of the region. The loops visible in (*b*) and (*c*) are examples of class II and I loops respectively on the Davis–Krieger classification (Section 3.4.1).

published in 1948 by the four Soviet scientists M.S. Eigenson, M.N. Gnevyshev, A.I. Ohl and B.M. Rubashev. In their 1965 paper Bumba and Howard applied this idea to the interpretation of the distribution of 'weak' magnetic fields on the Sun recorded on magnetograms taken daily at Mt Wilson over a period of four and a half years. Their description of a complex of activity as a physically interrelated group of active regions has since been confirmed and greatly elaborated by the wealth of soft X-ray data obtained from Skylab.

In a basic paper on the subject Howard & Svestka (1977) used photographs taken by the AS&E telescope (Section 3.4.1) to study the development of a complex of active regions throughout its entire lifetime of seven solar rotations. Figure 3.9 illustrates the evolution of the complex during its second passage across the disk. The large region in the centre was the first to appear and remained the basic member of the complex. It finally merged with another region on the seventh rotation and decayed behind the limb. During the lifetime of the complex, new regions progressively appeared and became linked to existing ones by interconnecting loops. Note that the interconnecting loops tend to link an active region to its immediate neighbours in the complex rather than to those further away, a behaviour which appears to be a general characteristic of complexes of activity (Chase *et al.*, 1976). Most of the 20 regions ultimately involved in the complex were short-lived, but nine lasted for more than one rotation ($\simeq 27$ days). They were distributed over both the northern and southern hemispheres, although there was a preponderance in the north. The final decay of the complex was accompanied by the development of a coronal hole in the region previously occupied by the complex.

Howard & Svestka (1977) made a detailed study of the origin and behaviour of the interconnecting (and other) loops associated with the complex and reached a number of important conclusions:

(1) interconnecting loops appear when a new active region is born in the neighbourhood of an existing region. The birth of one short interconnecting loop took less than 12 hr;

(2) the emergence of new magnetic flux tends to trigger the appearance of interconnecting loops or to make existing ones brighten;

(3) once magnetic field linkages have been established between active regions these connections tend to last at least as long as the interconnected regions exist as distinct entities. *Individual* interconnecting loops, however, have lifetimes of the order of only days, or even hours;

(4) changes in the shapes of the X-ray loops can sometimes be related directly to variations in the photospheric magnetic field, such as the growth and decay of 'polarity islands', rotation of the sunspot group, or the migration of polarities;

(5) when the complex is fully developed, the appearance of interconnecting loops against the disk is similar to that of the loops associated with individual active regions;

(6) in one case X-ray loops linking different active regions in the complex underwent simultaneous brightenings, apparently unrelated to flares; and

(7) growth or brightening of X-ray loops in the absence of flares is observed quite often in old active regions.

The structure of an individual, well-developed active region observed on

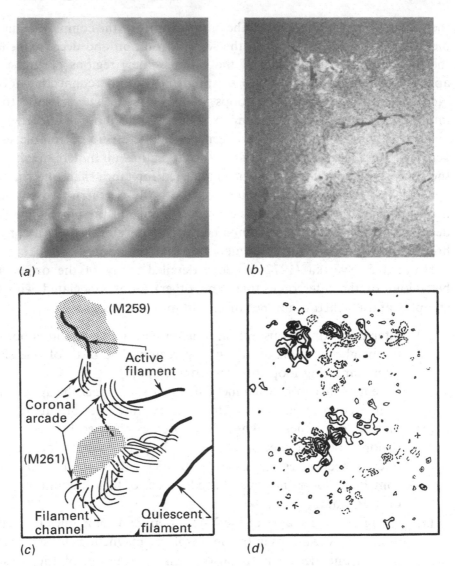

Fig. 3.11. (*a*) X-ray image showing arcades of active region loops over filament cavities; (*b*) simultaneous Hα photograph showing the locations of chromospheric filaments and filament channels; (*c*) schematic drawing depicting the arcade loops above active region Hα filaments and filament channels; and (*d*) map of the underlying longitudinal photospheric magnetic field (Davis & Krieger, 1982). The X-ray photograph (*a*) was obtained from a rocket on 8 March 1973 and shows higher spatial resolution than the X-ray photographs taken subsequently with the AS&E telescope on Skylab (Section 3.4.1).

the disk is illustrated in Fig. 3.10 (Rosner, Tucker & Vaiana, 1978). Photograph (*a*) is overexposed and serves only to indicate the overall extent of the region. On the other hand, photograph (*b*) is exposed to show the loops in the outer parts of the active region surrounding the more intense interior (see also Fig. 1(*b*) of Rosner, Tucker and Vaiana). Finally, photograph (*c*), taken through a somewhat harder (shorter wavelength) filter, brings out the very bright loop-like structures lying at the core (see also Cheng *et al.* 1980, Fig. 2). Note that the most intense emission comes from the two, closely adjacent, small loops just to the right of centre. The loops visible in (*b*) and (*c*) are examples of class II and I active region loops on the Davis–Krieger classification described in Section 3.4.1. A simplified drawing of the class I and II loops associated with another active region on the disk has been published by Pye *et al.* (1978).

By contrast, class III loops extend outwards from the active region and terminate in magnetic areas outside. Some of these loops arch to other active regions and are, of course, interconnecting loops, which are discussed separately in Section 3.4.4. Others, however, do not terminate in another active region. For example, Fig. 3.11 shows four arcades of class III loops which radiate outwards from two neighbouring active regions on the disk but do not terminate in an active region. The photograph was obtained with a rocket-borne, grazing-incidence X-ray telescope on 8 March 1973 and shows significantly higher spatial resolution than that subsequently achieved by the AS&E telescope on Skylab (Davis & Krieger, 1982). Note that the loops in each arcade are separated by a distance roughly equal to their width. The loops emerge from localized areas within the active region which, as best as can be inferred from a magnetogram taken 23 hr later (Fig. 3.11(*d*)), are characterized by high field strengths and gradients but show no preferred polarity. The loops rise to heights of the order of 50 000 km and appear to terminate in localized brightenings in the bright chromospheric network outside the active region.

Figure 3.11(*c*) depicts the relationship between active region loop arcades and active region filaments. Each arcade is located at the end of (not over) a dark Hα filament, effectively forming a 'channel' which represents an extension of the filament itself (see also Serio *et al.*, 1978). In fact, traces of dark Hα material can be seen in Fig. 3.11(*b*) within some of the filament channels. The filaments, and hence presumably the associated filament channels, are believed to delineate neutral lines in the longitudinal photospheric magnetic field (cf. Section 4.2). According to Davis & Krieger (1982) loops are not present above the strongly absorbing (dark) parts of a filament, which are located within X-ray cavities similar to those found above quiet region filaments (Section 3.4.1).

Under the heading of class III loops, mention should also be made of the counterparts in soft X-rays of the large coronal loops observed in the visible coronal lines (Sections 3.2.2 and 3.2.3). McGuire *et al.* (1977, Fig.

3) have analysed Skylab photographs of one such system, which consisted of an arcade of giant loops rising from the southern boundary of an active region complex and terminating near the south pole. The loops reached heights of 500000 km and had footpoints separated by as much as 600000 km. The loop system as such was very stable and remained at the same general location for several solar rotations.

Loop systems associated with individual active regions show both short- and long-term changes (Vaiana *et al.*, 1976). The former affect individual features but leave the overall structure virtually unaltered, whereas the latter involve significant changes in the overall structure and take place over a period of the order of several days (Rosner, Tucker & Vaiana, 1978, Fig. 6). Photographs illustrating the evolution of individual active regions observed on the disk have been published by a number of authors, including Vaiana *et al.* (1976, Figs. 8, 13 and 14), Rosner, Tucker & Vaiana (1978, Fig. 3) and Golub *et al.* (1982, Figs. 1 and 6). Other authors have presented observations showing the loop structure of active regions at the limb (see, for example, Vaiana, Krieger & Timothy, 1973, Fig. 9; Howard & Svestka, 1977, Fig. 8(*d*); Underwood *et al.*, 1977, Figs. 10 and 13). Despite their rather limited spatial resolution these photographs reveal individual loops rising to (projected) heights of the order of 100000 km and more.

Finally, reference should be made to a detailed study by Webb & Zirin (1981) of the relationship between the X-ray loops in an active region and underlying chromospheric features seen in Hα and the K line of Ca II. These authors found that the majority of the brighter X-ray loops occurred within the active region and connected bright facular areas to either the penumbra of a sunspot or to satellite spots. No loops ended in the umbra of a spot.

3.4.3 *Properties of individual active region loops*

As we have seen in Section 3.4.1, the loops associated with an active region have been classified into three classes I, II and III, depending respectively on whether they occur in the core of the active region, surround the core, or extend outwards from the active region to terminate outside. Those class III loops which fall under the category of active region interconnecting loops are discussed in Section 3.4.4. The properties of the remaining class I, II and III loops, in so far as they are known, are described below.

(1)   *Class I loops*

Loops of this class are found in the bright central cores of X-ray active regions (see Fig. 3.10(*c*)). They are small, densely packed features which connect areas of opposite magnetic polarity on either side of the neutral line through the region (Section 3.4.1). According to Davis & Krieger (1982, Table I) the widths of such loops range from 700 to 5000 km and their lengths from 7000 to 20000 km, implying an aspect ratio (loop radius/length) of the order of 0.1. This accords with the estimate of 0.1–0.2 given by Cheng, Smith & Tandberg-Hanssen (1980) for core loops observed in X-rays and the hot EUV line Fe XVI $\lambda 33.5$ (Section 3.3.2). On the other hand, the latter authors give the larger values of 7300–11000 km and 29000–44000 km for the widths and lengths of such loops respectively. The loops they observed showed little change over a period of about 10 hr.

An excellent high resolution digitally enhanced photograph of a soft X-ray loop arcade has been published by Webb *et al.* (1987, Fig. 1(*a*)); the photograph was obtained from a rocket on 13 February 1981. The two largest loops had widths in the range 3500–12000 km and the footpoint separation was ~70000 km. The loops joined areas of opposite polarity.

At present no other data are available on the properties of class I loops.

(2)   *Class II loops*

Likewise, little quantitative information has been published on the properties of the class II loops seen in the outer parts of an active region. At the limb they appear as giant structures rising to heights of 100000 km or more (Section 3.4.2). In fact, Howard & Svestka (1977) observed loops with heights exceeding 200000 km; in one extreme case a loop attained an altitude of 260000 km, or nearly 0.4 times the solar radius. Davis & Krieger (1982) quote a range of 5000–15000 km for the widths of class II active region loops and a range of 10000–100000 km for their lengths. These figures imply an aspect ratio (loop radius/length) of the order of 0.1. On the basis of photometric measurements Gerassimenko, Solodyna & Nolte (1978) give the somewhat larger figure of 20000 km for the widths of three loops observed on the disk.

Finally, mention should be made of the faint, long-lived, loop-like structure observed within an active region by Martens, van den Oord & Hoyng (1985) with the hard X-ray imaging spectrometer (HXIS) carried by the Solar Maximum Mission satellite (SMM). It had an average diameter of 11400 km and a (conjectured) aspect ratio of 0.06. Most of the X-ray emission was in the 3.5–8.0 keV energy band, i.e. at the short wavelength end of the soft X-ray spectrum (cf. Section 3.4.1) and originated in plasma at a temperature of 1×10^7–2×10^7 K, which is more characteristic of flare than normal active region material. At this stage it is not clear that the X-ray source described by Martens, van den Oord and Hoyng is a loop in the strict sense even though they explicitly describe it as such.

(3)   *Class III loops*
        Class III loops extend outwards from an active region and terminate either in another active region or in a surrounding magnetic area. Frequently the loops form bright arcades spanning Hα filaments associated with the active region (cf. Fig. 3.11). Only loops with one end outside active regions are described here, discussion of active region interconnecting loops being reserved for the next section.

        Measurements made by Davis & Krieger (1982) on high-quality rocket photographs indicate that class III loops have widths in the range 10 000 to 30 000 km and lengths in the range 50 000 to 500 000 km, yielding an aspect ratio (loop radius/length) of the order of 0.04. Such loops therefore tend to be wider and longer than the class I and II loops discussed previously. Apart from occasional localized brightenings, individual loops show no evidence of internal structure. However, it should be noted that the presence of low contrast detail within loops on a scale of 15″ of arc (11 000 km) or less would be hard to detect with even the best existing X-ray telescopes (cf. Section 3.4.1). Typically the loops have heights of around 50 000 km (Davis & Krieger, 1982), although they are sometimes seen at the limb to extend up to 100 000 km (Krieger, de Feiter & Vaiana, 1976). These figures agree with the estimates of 60 000–80 000 km given by McIntosh *et al.* (1976).

        As we have already remarked in Section 3.4.2, class III loops appear to be connected to localized areas on the outskirts of an active region where the magnetic field strength and gradients are relatively high. Their other ends are anchored to the bright chromospheric network outlining the boundaries of supergranulation cells. This is very clearly illustrated in Fig. 3.12, taken from Davis & Krieger (1982). Note the localized brightenings in the network which mark the (outer) footpoints of the loops. Davis & Krieger point out that the identification of each loop with a specific network brightening is relatively unambiguous even though the X-ray loops cannot of course be followed right down to the chromosphere.

        The brightness of individual loops seems to vary in a slow and continuous manner throughout their lives, presumably in response to changes in the underlying photospheric magnetic field (Krieger, de Feiter & Vaiana, 1976). More sudden enhancements of the loop brightness are

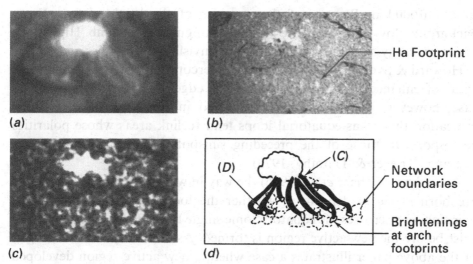

Fig. 3.12. Soft X-ray loops linking the outer parts of an active region to points in the surrounding bright chromospheric network (Davis & Krieger, 1982). (*a*) X-ray image of the active region and its surroundings (see also Fig. 3.11); (*b*) Hα photograph; (*c*) Ca II K line photograph; and (*d*) schematic drawing showing the relationship between the loops and the network. Note the localized brightenings in the network marking the (outer) footpoints of the loops.

observed, but they are generally associated with Hα flares or filament eruptions (Rust & Webb, 1977).

A unique case of brightness *pulsations* at one footpoint of a large loop observed in the band 3.5–5.5 keV has been described by Harrison (1987). The pulsations had a well-defined period of ∼24 min and were observed for 6 hr. Harrison attributed them to either standing or travelling (magnetic) wave packets within the loop.

3.4.4 *Properties of individual active region interconnecting loops*

The first specific investigation of the properties of loops of this type was made by Chase *et al.* (1976). They studied 100 such loops distributed among 94 different active regions and found that they occur only between active regions not too far apart. In fact, most interconnecting loops do not extend for more than 25 heliographic degrees (300 000 km). Loops which cross the solar equator, so-called trans-equatorial loops, are on average about twice as long as those which do not (Chase *et al.*, *loc. cit.*, Fig. 1(*b*)). One bright trans-equatorial loop had a length of 37° or 450 000 km.

In their study of the complex of activity described in Section 3.4.2 Howard & Svestka (1977) found that interconnecting loops generally do not rise as high as loops internal to active regions. Some are quite low-lying (≲25 000 km), although one was observed to extend up to a height of at least 110000 km. Because of the presence of the bright diffuse X-ray background low-lying loops are hard to distinguish at the limb. High loops are more easily seen there but tend to be invisible on the disk.

Howard & Svestka (1977) found that interconnecting loops terminate in areas of enhanced magnetic field near the edges of active regions. In no case, however, were the footpoints rooted in sunspots. There is some indication that trans-equatorial loops tend to link areas whose polarities correspond to those of the preceding sunspots in the two hemispheres (Vaiana, Krieger & Timothy, 1973).

According to Chase *et al.* (1976) the way in which interconnecting loops are born seems to depend on whether the loop is confined to a single hemisphere or crosses the equator. Some single hemisphere loops appear to exist before the new active region is formed. As an example of this, Fig. 2 of the above paper illustrates a case where a new active region developed at the end of a loop extending out from an older active region. On other occasions an active region splits into two and an existing internal loop is transformed into an interconnecting loop. Trans-equatorial loops, on the other hand, appear to develop after a new active region is born.

Individual interconnecting loops typically have lifetimes of the order of one day or less, whereas the loop systems to which they belong may endure for a number of solar rotations (Howard & Svestka, 1977; Rosner, Tucker & Vaiana, 1978, Fig. 6). Howard & Svestka remark that only rarely do loops appear to remain apparently unchanged for several days, and these are generally loops connecting active regions to very old magnetic field remnants.

One aspect of loop behaviour that has received considerable attention is the occurrence of transient brightenings of interconnecting loops (Howard & Svestka, 1977; Nolte *et al.*, 1977; Svestka & Howard, 1979, 1981). As a general rule the visibility of individual interconnecting loops varies in a slow and continuous manner throughout their lives, even though the rather long gaps between successive Skylab observations (6–12 hr) prevent actual light-curves from being obtained (see, for example, Nolte *et al.*, 1977, Fig. 1). However, other loop brightenings are observed which are more transient and conspicuous: they last on average for some 6 to 8 hr and may involve increases in brightness of up to an order of magnitude. Sometimes a loop brightens uniformly along its length, whereas in other cases the top of the loop appears to be the site of the most intense brightening during the initial phase of the event.

Tables I and II of Svestka & Howard (1979) list 25 such events recorded by Skylab. Despite the fact that many, but not all, of them appear to show some indirect association with flares, Svestka and Howard adduce evidence that flares are not the cause of sudden loop brightenings. However, on some occasions the two phenomena do appear to be physically related. It should also be noted that conspicuous brightenings of loops connecting active regions in the same hemisphere can occur within one or two days of the emergence of a new interconnected region, whereas for trans-equatorial loops there is a delay of four to five days. For a detailed description and illustrations of transient loop brightenings the interested reader is referred to Svestka & Howard's (1979) paper.

In a subsequent paper Svestka & Howard (1981) sought to investigate possible dynamical effects associated with loop brightenings. However, with one exception, no evidence was found of any disturbance propagating along an interconnecting loop. In a few cases, however, the brightening was followed by a large-scale twisting of the loop affected.

Finally, mention should be made of the study undertaken by Svestka *et al.* (1977) of the complete life history of one particular interconnecting loop. It belonged to a system of trans-equatorial loops observed by Skylab, in which a conspicuous brightening was observed lasting for about 6 hr. The brightening was accompanied by marked changes in the shapes of the interconnecting loops. Although the brightening could conceivably have been caused by one of the many subflares that occurred during this time, Svestka *et al.* conclude that this was unlikely.

3.4.5 *Properties of individual quiet region loops*

Much of the rather limited information available on quiet region loops pertains to those associated with filament cavities (Section 3.4.1). A good illustration of an arcade of bright loops over a filament channel is given by the Skylab photograph in Fig. 3.13 (Ray & van Hoven, 1982). By studying such an arcade as it crossed the limb McIntosh *et al.* (1976, Fig. 5) demonstrated that the individual loops extend up to great heights. In fact, the heights of the tallest loops observed were about 500 000 km, while their footpoints were separated by distances of 40–45 heliographic degrees (500 000–550 000 km). Serio *et al.* (1978) found that filament cavities are enclosed by a series of loops with heights greater than 50 000 km.

Since the emission from such loops is generally faint, it is not surprising that their tops are not always visible when they are observed on the disk. On the other hand, the lower parts of the loops will then appear brighter as a result of integration of the emission along the line of sight down the legs (cf. Golub *et al.*, 1982, Fig. 8). Serio *et al.* (1978, Fig. 6) have described a case where the tops of the loops above a filament cavity became visible

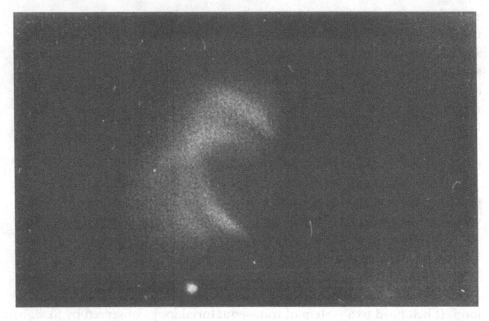

Fig. 3.13. Skylab X-ray photograph showing an arcade of bright, quiet region loops over a filament channel (Ray & van Hoven, 1982). Such loops are generally faint and may reach heights of over 500 000 km, or 0.7 times the solar radius.

several hours after the disappearance of the associated Hα filament. As these authors point out, the X-ray loops seen over filament cavities appear to be related to the large-scale loop systems observed above quiet region (quiescent) Hα prominences in white light and visible coronal lines (see Sections 3.2.2 and 3.2.4).

Observations obtained with the AS&E telescope on Skylab have revealed the frequent occurrence, away from active regions, of transient X-ray enhancements in the lower corona which are associated with the disappearance of Hα filaments (Webb, Krieger & Rust, 1976). Isolated loops or loop arcades are sometimes found to accompany these events, an illustration of which is given in Fig. 3 of the above paper.

Not all neutral lines in the longitudinal magnetic field outside active regions are associated with either existing or recently dissolved filaments. Most of the solar surface is in fact covered with the dispersed remnant fields of long vanished active regions, in which are embedded neutral lines spaced less than 30 heliographic degrees apart (McIntosh *et al.*, 1976). In these regions the X-ray emission is weak and diffuse and appears only on photographs taken through a soft (long wavelength) filter. According to McIntosh *et al.* (1976) loops, when present, appear as isolated structures rather than as the components of long arcades spanning the neutral lines.

On the disk loops over a neutral line are seldom seen but they have been observed beyond the limb as high thin structures by McIntosh *et al.* (1976, Figs. 6 and 7). These loops are somewhat smaller than those over filament cavities, having heights of 200 000–350 000 km and footpoint separations of 20–40 heliographic degrees (250 000–500 000 km).

Since the photospheric magnetic field is concentrated in the bright network outlining the boundaries of the supergranulation cells, one might expect quiet region loops to be rooted in the network. In fact, the high-resolution rocket observations of Davis & Krieger (1982) show two clear examples of loops which do just this (loops 'B' in their Fig. 3). Both loops join different network cells and show localized chromospheric brightenings at their footpoints. Time sequences of observations of this quality would greatly aid the elucidation of the morphology and evolution of both quiet and active region X-ray loops.

In this and the preceding two sections we have described what at first sight might appear to be a bewildering variety of X-ray loops distributed over the solar surface. Closer examination, however, reveals that they all form part of an evolving pattern which begins with the appearance of an active region. When first born, an X-ray active region is relatively small and compact and so must be the loops comprising it. Within hours other loops appear which link the region to other active regions in its vicinity. These inter-region links, but not the individual connecting loops, last at least as long as the regions concerned survive as distinct entities.

As the region grows older, it expands and becomes more diffuse and large peripheral loops appear. Some of these are anchored within the active region at both ends, while others terminate in magnetic areas outside. Of particular interest are the peripheral loops which form arcades spanning channels in the corona which represent (linear) extensions of dark Hα filaments associated with the active region. Eventually the region dissolves but leaves behind the filaments, each with its associated arcade of X-ray loops. During this final stage of their lives the filaments are termed quiet region or quiescent filaments.

After some 100–250 days the quiet region filaments too disappear, but still their associated arcades of loops linger on, spanning the neutral lines in the longitudinal magnetic field over which the filaments had formerly lain. With the further passage of time the loops in the arcade increase in number and become fainter until, ultimately, the magnetic field of the original active region is fragmented and dispersed and no apparent X-ray trace of its presence remains.

3.5 Observations at radio wavelengths

3.5.1 *Introduction*

Thermal radiation from the Sun is detectable over a vast range of the electromagnetic spectrum, from the soft X-ray continuum already described in Section 3.4.1 to metric radio waves. Two factors combine to make the thermal radio radiation from the Sun observed at the Earth's surface peak in the centimetric region. One is the onset of strong tropospheric absorption at wavelengths below about one centimetre, while the other is the intrinsic decrease in the intensity of the radiation at longer wavelengths. The two processes which contribute to the thermal emission from active regions are bremsstrahlung (free–free emission by electrons or 'braking radiation') and gyroresonance radiation by electrons spiralling around magnetic lines of force (Section 3.6.5). Other more exotic processes involving, for example, relativistic particles or very high temperatures (cf. Dulk, 1985) are believed not to be involved in the (non-flare) radio loops which form the subject matter of this section and Section 3.6.5.

Centimetre and low decimetre observations of the Sun with a spatial resolution comparable to that achieved for EUV and X-ray loops are a relatively recent development. This advance has largely been brought about by the construction of large array type radio telescopes employing the method of aperture synthesis. To date, the telescope which has made the most significant contribution is the Very Large Array ('VLA') at Socorro, New Mexico, U.S.A. Other radio telescopes which have achieved high resolution observations of active regions include the Westerbork Synthesis Radio Telescope ('WSRT') in the Netherlands, a 3-element interferometer at the National Radio Astronomy Observatory, U.S.A., a 5-element array at Stanford, U.S.A., and the RATAN-600 in the Northern Caucasus, U.S.S.R.

Until recently an important constraint upon observations obtained with the aperture synthesis technique was the relatively long time required to obtain a radio map of an active region. When this is the case the method is applicable only to relatively stable structures. Typical figures for the period of observation that have been quoted are 12 hr (Kundu & Velusamy, 1980), ~ 5 hr (McConnell & Kundu, 1983), and $\sim 1\frac{3}{4}$ hr (Webb *et al.*, 1983). However, improvements in technique are beginning to allow more rapidly-changing structures to be studied with the same high spatial resolution (cf. Section 4.6). An integral feature of the synthesis method is the need to correct the observed brightness distribution maps for the instrumental pattern and for the effect of any high-brightness sources that may be present during the observing period. This is carried out with the aid of a

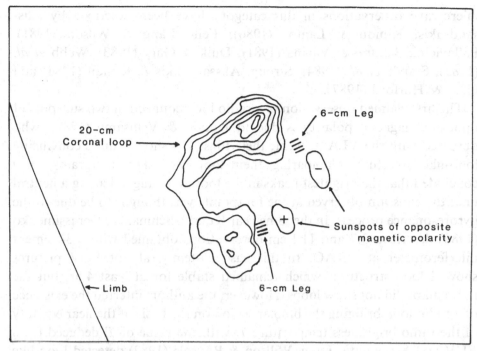

Fig. 3.14. Synthesis map of radio loop observed with the Very Large Array at Socorro, New Mexico at $\lambda = 20$ cm (Kundu & Lang, 1985). The contours trace out equal values of the brightness temperature T_b corresponding to 0.2, 0.4...1.0 times the maximum T_b of 2.0×10^6 K. The 6 cm emission from the legs of the loop is shown schematically. Also shown are the associated sunspots, observed at optical wavelengths.

computer program called 'CLEAN' (Kundu & Velusamy, 1980; Webb *et al.*, 1983).

The technique used by Luo *et al.* (1985) deserves mention: by observing with a small telescope during the course of a solar eclipse they were able to obtain simultaneous observations at four wavelengths in the range 2–21 cm with a spatial resolution (in one direction only) of 20″. Similar multi-wavelength eclipse observations, but with a resolution of only 2″6, have been obtained by Gary & Hurford (1987).

With the various interferometers listed above, somewhat more extensive observations have been obtained of active regions generally than of loops *per se*. These observations are of considerable relevance to loops and have added to our understanding of (a) the dominant emission and absorption mechanisms involved (cf. Section 3.6.5); (b) the complex nature of the magnetic field distribution in the (low) corona above active regions; and (c) the correlation between the distribution of the radio emission and chromospheric features and the photospheric field pattern. Recent microwave observations in this category have been described by Alissandrakis, Kundu & Lantos (1980), Felli, Lang & Wilson (1981), Pallavicini, Sakurai & Vaiana (1981), Dulk & Gary (1983), Webb *et al.* (1983), Staude *et al.* (1984), Strong, Alissandrakis & Kundu (1984) and Gary & Hurford (1987).

The first claimed observation of a radio loop connecting two sunspots of opposite magnetic polarity is due to Kundu & Velusamy (1980), who observed with the VLA at 6 cm; their Figs. 1–4 show a quite convincing loop-like structure. Comparing their loop with Hα filtergrams, they concluded that the brightest peaks in the loop were aligned along a neutral line; the emission observed at the footpoints was thought to be due to the gyroresonance process. In the same year Kundu, Schmahl & Gerassimenko (1980) described 3.7 and 11.1 cm observations obtained with a 3-element interferometer at NRAO. In this case co-temporal hot EUV pictures showed loop structures which remained stable for at least 4 hr, but the radio maps did not show loops. However, the authors inferred the existence of a radio loop bridging the bipolar region on the basis of the near equality of the radio brightness temperature T_b with the value of T_e deduced from EUV and X-ray data. Lang, Willson & Rayrole (1982) detected loop-like structures at 20 cm, also using the VLA (see their Figs. 1A and 1B). They suggested that the loops, which appeared to join the two dominant sunspots and also to connect areas of weaker field, represented 'the radio counterparts of X-ray coronal loops'.

McConnell & Kundu (1983) obtained observations at 6 and 20 cm. At 6 cm they found that the emission was mainly sunspot-associated, but at 20 cm they observed elongated sources connecting regions of opposite magnetic polarity. The geometry was believed to be such that the line of sight lay close to the planes of the sources (loops); the legs of a loop, accordingly, appeared foreshortened. According to Kundu & Lang (1985), the wavelength at which radio loops can be best mapped is in fact 20 cm: at this wavelength 'emission from the entire loop is routinely detected'. Figure 3.14, taken from their paper, shows a VLA synthesis map of a loop observed at this wavelength. Also shown are the associated sunspots and (schematically) the 6 cm emission from the legs of the loop.

A number of workers have sought to determine – with varying degrees of spatial resolution – the association between (a) radio loops or active regions and (b) photospheric magnetograms and pictures obtained in Hα or λ1083 of He I. Working at λ ≃ 20 cm with a resolution of 40″, Dulk & Gary (1983) found good correspondence between their map of circular polarization and a magnetogram (see their Fig. 1); the LH and RH polarizations reflected the extraordinary mode of the (assumed) bremsstrahlung emission. Areas of emission corresponded closely to dark patches on the $\lambda1083$ image. Felli, Lang & Willson (1981) observed an active region at 6 cm with a half-power beam width of $2'' \times 15''$. They also found that the dominant plage-associated radio emission (bremsstrahlung) had a circular polarization which was well correlated with the longitudinal field. A small fraction ($<5\%$) of the observed emission came from small, highly circularly polarized sources associated with sunspots (gyroresonance radiation).

Webb *et al.* (1983) have investigated the same problem and given a lengthy discussion of the various associations observed by them. They concluded that the association with the photospheric field is complex: although about half of the 6 cm radio sources observed were associated with strong photospheric fields, such strong fields did not always produce radio emission at this wavelength. The brightest components were associated with areas of strong field or high field gradients, but not with sunspot umbrae. Some information is also available about the relationship between radio and X-ray loops; this is discussed in Section 3.7.2. Much further work will be required before we can claim to have a proper understanding of the complex associations between radio loops and features observed simultaneously at other wavelengths.

At the present time observations specifically of *loops* are still somewhat sparse; those available are listed in Table 3.3. The wavelengths employed range from 3.7 to 22 cm and the estimated resolution from $3''$ to $13''$ of arc.

3.5.2 *Properties of individual loops*

Quantitative information about the morphological properties of radio loops is meagre. A figure for the length of a loop of 100 000 km has been obtained by Lang, Willson & Rayrole (1982) and Kundu & Lang (1985), while the somewhat smaller figure of 70 000 km has been given by McConnell & Kundu (1983). The width is more difficult to estimate and could be influenced by possible changes in a loop during the rather long observing periods often employed. Figures of 10 000 km and 30 000 km have been quoted by Kundu, Schmahl & Gerassimenko (1980) and McConnell & Kundu (1983) respectively. We have measured radio maps published by Kundu & Velusamy (1980, Fig. 1) and Lang *et al.* (*loc. cit.*, Figs. 1A and 1B), obtaining values of 20 000 km and 16 000 km respectively at the *narrowest* locations: we ignored the bulges and bays that contribute to the characteristic irregular appearance of radio loops. Taking 100 000 km and 15 000 km as a representative length and width respectively, we arrive

Table 3.3. *Microwave observations of loops* (*direct or inferred*)

Reference	Telescope	Wavelength (cm)	Frequency (GHz)	Resolution (sec of arc)
Kundu *et al.* (1980)	3-el. interferometer[a]	3.7, 11	8.1, 2.7	3, 9
Kundu & Velusamy (1980)	VLA[b]	6	5	3.5
Lang & Willson (1982)	WSRT[c]	6	5	—
Lang *et al.* (1982)	VLA	20	1.5	3.7 × 5
McConnell & Kundu (1983)	VLA	20	1.5	13
McConnell & Kundu (1984)	VLA	6	5	3.5
Kundu & Lang (1985)	VLA	6, 20	5, 1.5	—
Willson (1985)	VLA	17–22	1.4–1.7	3 × 4
Webb *et al.* (1987)	VLA	6 20	5 1.5	— 4.7 × 4.7

[a] National Radio Astronomy Observatory; [b] Very Large Array; [c] Westerbork Synthesis Radio Telescope.

at an aspect ratio (Section 2.4.3) of $7500/100\,000 = 0.075$. This value is comparable to those found for active region loops observed in hot EUV lines (Section 3.3.3) and soft X-rays (Section 3.4.3).

No measurements of the heights of radio loops are to be found in the various investigations cited above. However, from model calculations Luo *et al.* (1985, Fig. 4) deduce figures of $\sim 50\,000$ km and $\sim 40\,000$ km for the $s = 3$ and $s = 2$ gyroresonance layers at $\lambda = 21$ cm (cf. Section 3.6.5). These values are consistent with loop heights measured in other regions of the electromagnetic spectrum (cf. Tables 2.8 and 3.7).

Measurements of the lifetimes of radio loops are not available. Clearly they must last for at least several hours to ensure their appearance on synthesis maps typical of those cited in Section 3.5.1.

3.6   Physical conditions in hot loops
3.6.1   *Introduction*

So far in this chapter we have been concerned with what we may broadly categorize as the morphology of hot loops, dealing successively with information derived from observations in the visible region of the spectrum (the well-known coronal lines plus the coronal continuum), hot EUV lines, the soft X-ray region, and the centimetre and low decimetre portion of the radio spectrum. By contrast with cool loops (Chapter 2), however, virtually no data on the *dynamics* of hot loops has so far been obtained.

We now turn our attention to the determination of the physical conditions in hot loops from observations at the various wavelengths. The theoretical basis for analysing the measured intensities of hot coronal lines is essentially the same as that employed for cool EUV lines (Section 2.5.4) irrespective of whether the hot lines are observed in the visible, EUV, or X-ray regions of the spectrum. In each case it is assumed that the lines are optically thin and that the dominant mode of excitation is collisions with electrons. Observations in the visible region are discussed first (Section 3.6.2). Here an upper limit to the electron temperature in a loop can often be established by the simple expedient of fitting a Gaussian profile to the profile of an observed emission line. Furthermore, a direct determination of the electron density in a hot loop can be made by measuring the intensity of the white-light (electron scattering) continuum.

Next we discuss the analysis of hot EUV lines (Section 3.6.3). Such analyses have led to a picture of a hot EUV loop in which its temperature is more or less the same as that of its surroundings but its enhanced emission is due to an increase in its density. When we turn to the soft X-ray region (Section 3.6.4), the complexity of the analysis increases. Due to the limited sensitivity of the detectors presently available, existing spectroscopic observations generally lack the spatial resolution needed to resolve individual X-ray loops. Almost all of our knowledge of the physical conditions in soft X-ray loops has been derived from photographs taken through broad-band filters that integrate over a number of lines and the continuum. Analysis therefore requires an appeal to a *theoretical* model of the coronal X-ray spectrum, whereby the line and continuum contributions are computed, for an assumed density, as functions of temperature. Both the calculations and the measurements are subject to considerable uncertainties.

Finally, the analysis of radio observations of hot loops requires a complex modelling procedure involving no less than eight major steps

(Section 3.6.5)! Not surprisingly, the radio data on its own is not capable of giving unique, independent results. Nevertheless, it may be used to check the physical conditions derived from other regions of the spectrum. Current improvements in observational techniques are likely to make the radio data of growing importance: at present, only radio data are (potentially) capable of giving information about both the strength and direction of the magnetic fields present in hot loops and the surrounding active regions.

In evaluating the results presented in the sections that follow, one must recognize that existing relevant observations are sparse, that in most cases the data are *not* homogeneous, and that existing diagnostic methods are relatively crude, far-reaching assumptions often being required before progress is possible. Despite this, the summary presented in Table 3.8 (Section 3.8) encourages us to the belief that, overall, the existing work has led to a reasonably consistent picture of the physical conditions in hot loops.

3.6.2 *Analysis of observations in the visible region*

The theoretical basis for analysing the measured intensities of visible coronal lines is essentially the same as that employed for EUV lines and described in some detail in Section 2.5.4. Once again it is assumed (a) that the lines are formed in regions which are optically thin at the wavelengths concerned, and (b) that collisional excitation by electrons is the dominant process for raising an ion from the lower state of the transition to the upper excited state. Other accounts of the theory, bearing particularly on the visible coronal lines, have been given by Billings (1966, pp. 122–42), Noci (1971), Fort *et al.* (1973), Tsubaki (1975) and Athay (1976, pp. 314–50).

It is possible to derive an *upper limit* to the electron temperature T_e by the simple expedient of fitting a Gaussian profile to the observed profile of a loop emission line. Subject to certain assumptions (discussed in Bray & Loughhead, 1974, Section 4.3.3), the full width at half maximum intensity (FWHM) is given by the formula

$$\Delta\lambda_{\frac{1}{2}} = 2(\ln 2)^{\frac{1}{2}} \cdot \frac{\lambda}{c} \cdot \left(\frac{2kT_{\text{kin}}}{\mu m_{\text{H}}} + \xi^2\right)^{\frac{1}{2}}$$

$$= 2(\ln 2)^{\frac{1}{2}} \cdot \frac{\lambda}{c} \cdot \left(\frac{2kT_{\text{D}}}{\mu m_{\text{H}}}\right)^{\frac{1}{2}}, \tag{3.1}$$

where T_{kin} is the kinetic temperature of the ion, usually assumed to be a good approximation to T_e, and ξ is the non-thermal velocity of the gas.

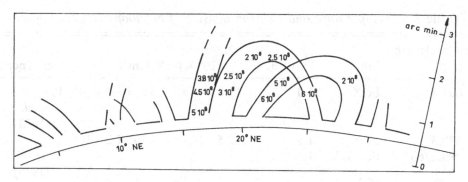

Fig. 3.15. Map of electron density in a loop system observed in Fe X $\lambda637.4$ at the east limb of the Sun by Fort *et al.* (1973). The units for N_e are cm^{-3}. The scale to the right give distance from the (true) limb in minutes of arc.

When ξ is unknown, the line-width method yields the so-called 'Doppler' temperature T_D, which is evidently an upper limit to T_e.

Finally, a direct determination of N_e in loops can be made by measuring the intensity in their white-light continua. The basis of this method is the electron scattering formula

$$\varepsilon(\lambda) = 2\pi F(\lambda)\,\sigma_T\,N_e, \tag{3.2}$$

where $\varepsilon(\lambda)$ is the emission per unit volume of scattered radiation, $F(\lambda)$ is the monochromatic intensity at the top of the photosphere, and $\sigma_T = 8\pi e^4/3m_e^2\,c^4 = 6.65 \times 10^{-25}$ cm^2 is the Thomson scattering coefficient.

Several authors have applied one or more of the techniques described above to determine T_e, T_D and N_e in hot loops, notably Fort *et al.* (1973) and Tsubaki (1975). Fort *et al.* analysed Fe XIV $\lambda530.3$ and Fe X $\lambda637.4$ filtergrams obtained at Pic-du-Midi with an estimated resolution of 5″ of arc, obtaining values of both T_e and N_e. An interesting feature of their paper is a collection of maps showing the distribution of N_e over the loop system recorded on their red and green filtergrams. One of these maps is shown in Fig. 3.15. Within any given loop there is quite a large variation in N_e; nevertheless, these authors believe that the $\lambda637.4$ loops are systematically some five times less dense than the $\lambda530.3$ loops. They estimate their values of N_e to be accurate to no better than a factor of two.

A spectroscopic technique was used by Tsubaki (1975) who, in addition to the above lines, also observed in Fe XI $\lambda789.2$, Fe XV $\lambda705.9$, and Ni XV $\lambda670.2$. He confined himself to determining T_e and T_D at a large number of locations distributed over a prominent loop structure. The N_e values he needed for his analysis he took from Fort *et al.* The results of these investigations are collected in Table 3.4.

Table 3.4. *Temperature and electron density of hot loops: visible region*

Wavelength (nm)	Ion	T_e ($\times 10^6$ K)	T_D ($\times 10^6$ K)	N_e (cm^{-3})	Reference
637.4	Fe X	1.0	1.8	1.5×10^8–2×10^9	Fort *et al.* (1973)
530.3	Fe XIV	1.8	2.5	5×10^8–1.2×10^9	,,
789.2/637.4	Fe XI/X	1.2	—	—	Tsubaki (1975)
705.9/530.3	Fe XV /XIV	2.6	—	—	,,
637.4	Fe X	—	2.13	—	,,
789.2	Fe XI	—	2.15	—	,,
530.3	Fe XIV	—	3.40	—	,,
705.9	Fe XV	—	4.46	—	,,
670.2	Ni XV	—	3.60	—	,,

Considering firstly the temperature, we can draw the following conclusions from Table 3.4: (a) the values of T_e lie roughly in the range 1×10^6–2×10^6 K, the higher ionization states giving the higher values; and (b) T_D is systematically larger than T_e, leading Tsubaki (1975) to the conclusion that the value of the non-thermal velocity ξ is 6–16 km s^{-1}. Secondly, the results of Fort *et al.* (1973, Figs. 3–7) show what appears to be a real variation in the electron density N_e over the loop system studied.

3.6.3 *Analysis of observations in EUV lines*

The diagnostic techniques used to analyse the intensities of hot EUV lines emitted by coronal loops are essentially the same as those employed for cool EUV lines and described in Section 2.5.4. Once again the basic assumption is made that the lines are formed in regions which are optically thin at the wavelengths concerned. There is, however, the complicating factor that the level of background radiation in an active region is much higher for hot EUV lines than for cool lines (Section 3.3.3).

In a pioneering study Foukal (1975) measured the ratio of the emergent loop intensity I_1 to the background intensity I_0 at a number of points in six different loops observed beyond the limb in the two EUV lines Mg X $\lambda 62.5$ and Si XII $\lambda 52.1$. He took these lines to be formed at temperatures of $\bar{T}_e = 1.6 \times 10^6$ K and 2.5×10^6 K respectively, as compared to an assumed background temperature of 2.0×10^6 K. Inserting the measured intensities in Eqn. (2.22) Foukal found that the value of the ratio N_e(loop)/

N_e (corona) was equal to 4 for both lines, implying a loop pressure some 3 to 4 times greater than that in the surrounding material. However, this result depends rather critically on the temperature adopted for the coronal background (2.0×10^6 K) and, as Foukal himself remarks, must be viewed with caution. Nevertheless, unlike their cool counterparts, hot EUV loops appear to show a significant enhancement in density and pressure over their surroundings.

The actual electron density in hot EUV loops was later estimated by Levine & Withbroe (1977) from flux measurements made in the line Mg X $\lambda62.5$. They inferred a value for N_e of 1.8×10^9 cm^{-3} and a corresponding gas pressure of about 0.7 dyne cm^{-2}. These lie within the range of values found for cool EUV loops (cf. Table 2.7). Subsequently, using the emission measure technique Dere (1982) inferred an electron density of 3×10^9 cm^{-3} for Fe XV loops in one active region and 5×10^9 cm^{-3} for those in another region. The corresponding gas pressures were 1.7 and 2.8 dyne cm^{-2}. These values are somewhat smaller than the estimates given by Kundu, Schmahl & Gerassimenko (1980) for several loops observed on the disk in the lines Fe XVI $\lambda33.5$, Si XIV $\lambda41.8$ and Mg X $\lambda62.5$. They derived an electron density of about 2×10^{10} cm^{-3} which, for a temperature of $\sim 3 \times 10^6$ K, gives a gas pressure of $p \simeq 2kN_e T_e = 16.6$ dyne cm^{-2}.

Much lower values of N_e, namely 2–3×10^8 cm^{-3}, have been obtained by Stewart, Brueckner & Dere (1986) for the high, faint, interconnecting loop system described in Section 3.3.2. These values refer to heights in the range 110 000–270 000 km.

Cheng (1980) published Fe XV $\lambda28.4$ and Fe XVI $\lambda33.5$ spectroheliograms showing four bright streaks linking the two opposite polarity areas in a small active region observed on the disk, which he interpreted as a system of small, parallel loops (see Section 3.3.2). No loops had been visible in these locations on the previous day (compare Cheng's Figs. 1 and 2) and, indeed, one of the loops became markedly more prominent over a period of just 7 min due to a rapid enhancement of its density.

Cheng used pairs of spectroheliograms taken in the Fe XV and Fe XVI lines to measure the intensities at a number of points in each loop and in the surrounding background. To analyse the measurements he rewrote the equation for the emergent intensity (see Section 2.5.4, Eqn. (2.16)) in the approximate form

$$I_\lambda = \frac{1}{5\pi} \frac{hc^2}{\lambda^3} AG(T_e) N_e^2 \Delta L, \qquad (3.3)$$

where A is the abundance of iron relative to hydrogen, $G(T_e)$ is (effectively) the function defined by Eqn. (2.15) and ΔL is the path length through the emitting region along the line of sight. Both $G(T_e)$ and N_e are to be regarded as averages over ΔL. Since the spatial distributions of the Fe XV and Fe XVI emissions appear to be the same, the two lines are assumed to be formed at the same temperature. It then follows from Eqn. (3.3) that the observed intensity ratio of the two lines, $I_{28.4}/I_{33.5}$, depends only on the ratio of the values of the function $G(T_e)$ for the two lines; in practice, $I_{28.4}/I_{33.5}$ is found to be a sensitive diagnostic for the temperature (see Cheng's Fig. 3).

Figure 3.16 shows the distribution of temperature in the four loops and in the adjacent parts of the active region derived in this way. It should be emphasized that the temperatures given within the loops are in fact averages over the loop and background material along the line of sight. But since these average values are about the same as the temperatures in the surrounding active region, the actual temperatures in the loops must be about the same as that of the interloop plasma. One can see from Fig. 3.16 that the temperature gradient along the axis of each loop is insignificant (except presumably at the very ends). There is no evidence of an abrupt temperature change between the loop and the surrounding corona.

To determine the electron density Cheng took the observed loop intensity to be the sum of the intrinsic loop intensity and the intensity of the surrounding material along the line of sight. Applying Eqn. (3.3) one then has

$$
\left.
\begin{aligned}
I_{\text{obs}}(\text{loop}) &= \frac{1}{5\pi}\frac{hc^2}{\lambda^3} AG(T_e)\{[N_e(\text{loop})]^2 L_1 + [N_e(\text{interloop})]^2 (L-L_1)\} \\
\text{and} & \\
I_{\text{obs}}(\text{interloop}) &= \frac{1}{5\pi}\frac{hc^2}{\lambda^3} AG(T_e)[N_e(\text{interloop})]^2 L
\end{aligned}
\right\} \quad (3.4)
$$

for the observed loop and interloop intensities respectively. Here L_1 is the path length through the loop and L the total path length along the line of sight. The values of L_1 were obtained directly from the spectroheliograms by assuming the loops to have a circular cross-section. Cheng then used Eqns. (3.4) to determine the three quantities $N_e(\text{loop})$, $N_e(\text{interloop})$ and L by a procedure which effectively assumes a constant intensity across each loop. Although this may not be strictly true, the resulting error is likely to be small.

The electron densities in the four loops and the surrounding active region derived in this way from Fe XV $\lambda 28.4$ spectroheliograms taken at two different times (01.20 and 01.27 UT on 9 June 1973) are summarized in Table 3.5. Also given are the corresponding temperatures, path lengths

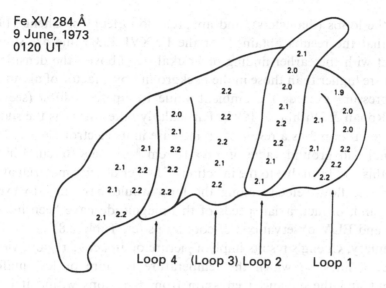

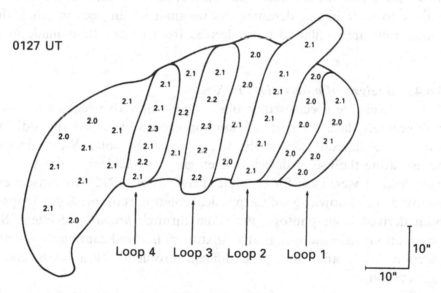

Fig. 3.16. Distribution of temperature (in units of 10^6 K) within a small active region observed in the hot EUV lines Fe XV $\lambda 28.4$ and Fe XVI $\lambda 33.5$ (Cheng, 1980). Four parallel loops link the two opposite polarity areas. There is no evidence of significant temperature gradients along the axes of the loops (except presumably at the very ends) nor of any abrupt temperature change between the loops and their surroundings. The greater brightness of the loops is due to enhancement in the density over their surroundings, not temperature.

through the loops (diameters), and apparent loop lengths. Cheng (1980) remarks that the results obtained for the Fe XVI $\lambda33.5$ line are similar. Consistent with the earlier finding of Foukal (see above), the densities in the loops are higher than those in the background by a factor of about 1.5; the gas pressure exceeds the ambient value by up to $\sim40\%$ (see also Habbal, Ronan & Withbroe, 1985). Particularly noteworthy is the sudden appearance of loop 3 as a result of an increase in its electron density from the original background value of 2.6×10^9 cm^{-3} to 3.6×10^9 cm^{-3} in just 7 min. If this had been due to the injection of higher density material at one footpoint, the flow velocity along the loop would have had to exceed 60 km s^{-1} and, in fact, axial speeds of this magnitude have been inferred from Hα and EUV observations of cool loops (cf. Table 2.8).

In summary, Cheng's results imply a picture of an active region, viewed in hot EUV lines, in which the temperature is more or less uniform throughout and the enhanced emission from the loops within it is due mainly to their higher densities. As we shall see in Section 3.6.4, these conclusions are similar to those derived from observations made in soft X-rays.

3.6.4   *Analysis of observations in X-rays*

Available spectroscopic observations of active regions in soft X-rays generally lack the spatial resolution needed to resolve the individual loops. The difficulty lies in the weakness of the solar X-ray emission, necessitating the use of detectors averaging the incident radiation over a wide field of view (see, for example, Davis *et al.*, 1975). In consequence, almost all our knowledge of the physical conditions in soft X-ray loops has been derived from photographs taken through *broad-band* filters. Such observations reflect a complicated mixture of line and continuum radiation (Section 3.4.1) and thus pose difficult problems of analysis and interpretation.

Owing to its very low density the coronal plasma may be treated as optically thin over the soft X-ray spectrum, even though the optical thickness of some of the prominent lines present would approach unity if the path length were of the order of one solar radius (Haisch & Claflin, 1985). A number of authors have made detailed calculations of the soft X-ray spectrum of an optically thin, uniform plasma of coronal density and composition as a function of temperature (Landini & Fossi, 1970; Tucker & Koren, 1971; Mewe, 1972, 1975; Walker, 1975; Kato, 1976; Mewe & Groenenschild, 1981; Shull, 1981; Gaetz & Salpeter, 1983).

The basis for the calculation of line intensities is essentially the same as that already described for EUV lines in Section 2.5.4. Again it is assumed

Table 3.5. *Physical conditions in loops and interloop regions*[a]: *EUV*

Region	Temperature (K) 01.20 UT	01.27 UT	Electron density (cm⁻³) 01.20 UT	01.27 UT	Gas pressure[b] (dyne cm⁻²) 01.20 UT	01.27 UT	Path length (sec of arc) 01.20 UT	01.27 UT	Loop length (sec of arc)
Loop 1	2.1×10^6	2.0×10^6	2.7×10^9	3.1×10^9	1.6	1.7	9.0	9.0	25
Between loops 1 and 2	2.1×10^6	2.0×10^6	2.4×10^9	2.7×10^9	1.4	1.5	19.0	—	—
Loop 2	2.1×10^6	2.1×10^6	3.6×10^9	3.4×10^9	2.1	2.0	7.0	7.0	40
Between loops 2 and 3	2.2×10^6	2.0×10^6	2.6×10^9	2.7×10^9	1.6	1.5	19.0	—	—
Loop 3	—	2.2×10^6	—	3.6×10^9	—	2.2	—	6.6	35
Between loops 3 and 4	—	2.2×10^6	—	2.7×10^9	—	1.6	—	—	—
Loop 4	2.2×10^6	2.2×10^6	3.7×10^9	3.6×10^9	2.2	2.2	6.6	6.6	32
Beyond loop 4	2.2×10^6	2.1×10^6	2.3×10^9	2.2×10^9	1.4	1.3	8.0	—	—

[a] McMath Active Region No. 12378.
[b] Calculated from the formula $p \simeq 2kN_eT_e$.

that collisional excitation by electrons is the dominant process in raising an ion from the lower state of the transition involved to the upper excited state. The lines are superimposed on a coronal continuum arising from three separate processes: thermal bremsstrahlung, radiative recombination of electrons and ions, and two-photon decay of certain metastable states of helium-like and hydrogen-like ions (Section 3.4.1). For a given density their relative contributions depend on both the wavelength and the temperature of the emitting plasma. Detailed discussions of the methods of calculation and references to the literature are to be found, *inter alia*, in the review articles by Walker (1972, 1975) and Culhane & Acton (1974).

It should be emphasized that both the calculations of the theoretical coronal spectrum and the measurements of solar X-ray intensities are subject to considerable uncertainties. Mewe (1975), for example, estimates the inaccuracies in his computed line fluxes to involve factors of 3 to 5, depending mainly on errors in the collisional excitation cross-sections, the ion concentrations and the element abundances. Moreover, measurements of solar X-ray line intensities, like those of short-wavelength EUV lines, suffer from the difficulty of obtaining absolute laboratory calibrations. In addition, one has to contend with the instability of the windowless detectors used in spectroscopic measurements, a problem often aggravated by the severe change in environmental conditions experienced by the instrument between the final (laboratory) calibration and its arrival in orbit (Walker, 1977).

The way in which a theoretical model of the coronal spectrum is used to infer the temperature of an X-ray emitting structure from broad-band filter observations can be understood by reference to Fig. 3.17. An X-ray telescope of aperture A at the distance D of the Earth images a (projected) area ΔS on the Sun onto a corresponding area Δs in the focal plane such that

$$\frac{\Delta S}{\Delta s} = \frac{D^2}{f^2},$$

where f is the focal length of the telescope. From any point of ΔS the telescope receives radiation over a solid angle

$$\Delta \omega = \frac{A}{D^2},$$

and hence the total energy incident on A from the area ΔS is

$$I_\lambda \cdot \Delta \omega \cdot \Delta S,$$

where I_λ is the emergent intensity of the solar plasma along the line of sight

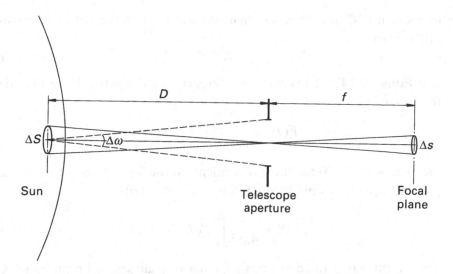

Fig. 3.17. Energy collection by a satellite-borne X-ray telescope (schematic). A (projected) area ΔS on the Sun is imaged onto a corresponding area Δs in the focal plane. D is the Earth–Sun distance, A the telescope aperture, f the focal length, and $\Delta\omega = A/D^2$ is the solid angle subtended by A at a point on ΔS.

per unit solid angle and area at wavelength λ. If $\eta(\lambda)$ is the product of the telescope reflectivity and the filter transmission at wavelength λ, then the flux of energy incident on the focal plane per unit area and unit time is

$$\frac{I_\lambda \cdot \Delta\omega \cdot \Delta S}{\Delta s} \cdot \eta(\lambda) = \frac{A}{f^2} \cdot I_\lambda \cdot \eta(\lambda).$$

Integrating over the passband $\Delta\lambda_i$ of the particular filter i used to make the observations gives a total energy flux incident on the focal plane of

$$W_i = \frac{A}{f^2} \int_{\Delta\lambda_i} I_\lambda \cdot \eta(\lambda) \cdot d\lambda. \tag{3.5}$$

According to Eqn. (2.10) the emergent intensity from an optically thin plasma is

$$I_\lambda = \frac{1}{4\pi} \int_0^L \varepsilon_\lambda(\ell) \cdot d\ell, \tag{3.6}$$

where L is the physical thickness of the emitting region along the line of sight, ε_λ is the emission (line plus continuum) per unit volume, and $d\ell$ is an element of length along the line of sight. Furthermore, all the processes contributing to the emission depend on the electron density N_e and temperature T_e in such a way that one can express the volume emission ε_λ in the form

$$\varepsilon_\lambda = N_e^2 \cdot P(\lambda, T_e) \tag{3.7}$$

(see Eqns. (2.14), (2.15) and the references cited above). If we then define the function

$$F_i(T_e) = \int_{\Delta\lambda_i} P(\lambda, T_e) \cdot \eta_i(\lambda) \cdot d\lambda, \tag{3.8}$$

the expression (3.5) for the flux incident on the focal plane of the telescope per unit area and time through filter i can be written

$$W_i = \frac{A}{4\pi f^2} \int_0^L F_i(T_e) \cdot N_e^2 \cdot d\ell. \tag{3.9}$$

This result has formed the basis for many analyses of broad-band X-ray observations (see, for example, Davis *et al.*, 1975; Landini *et al.*, 1975; Kahler, 1976; Vaiana *et al.*, 1976, 1977; Underwood & McKenzie, 1977; Gerassimenko & Nolte, 1978; Kramer *et al.*, 1978; Maute & Elwert, 1981).

Any appeal to a theoretical coronal spectrum necessarily entails the implicit assumption that the temperature is uniform throughout the plasma. It then follows from Eqn. (3.9) that the ratio of the energy fluxes W_1 and W_2 incident on the focal plane through two different filters $i = 1$ and 2 is given by

$$\frac{W_1}{W_2} = \frac{F_1(T_e)}{F_2(T_e)} \equiv R_{12}(T_e). \tag{3.10}$$

The ratio $R_{12}(T_e)$ is sometimes called the Spectral Hardness Index. It is a function only of the plasma temperature T_e and does not depend on the emission measure of the feature being studied ($\int_0^L N_e^2 \cdot d\ell$). Consequently, one can derive the temperature of the feature by measuring W_1/W_2 and comparing it with values of $R_{12}(T_e)$ calculated from the adopted theoretical coronal spectrum and the values of $\eta_1(\lambda)$ and $\eta_2(\lambda)$ determined by laboratory calibration.

Once the temperature is determined the value of the emission measure can be obtained from an absolute measurement of the energy flux through any one filter. If also the geometrical thickness of the feature along the line of sight can be estimated, a figure can be obtained for the average electron density. This method of deriving the plasma temperature and emission measure from broad-band X-ray observations is referred to as the 'filter ratio technique'.

In general, solar plasmas are found to be very inhomogeneous and may show wide variations in temperature. How reliable therefore is a method for determining temperature and density based on the assumption that the plasma is both homogeneous and isothermal? This question was examined in some depth by Underwood & McKenzie (1977), who came to the pessimistic conclusion that 'in the presence of line-of-sight temperature gradients, the values of "effective temperature" and "emission measure" yielded by this method cannot be directly related to the physical state of the plasma and so are of little utility in the study of coronal processes'! Gerassimenko & Nolte (1978), on the other hand, contended that the filter ratio technique is reasonably valid provided 'appropriately chosen filter pairs' are used to make the observations (but see also Underwood & McKenzie, 1978). Notwithstanding, it is evident that one can never conclusively demonstrate the reliability of results obtained by this technique unless one has independent knowledge of the variation in temperature along the line of sight. In the absence of spectroscopic observations with a spatial resolution adequate to resolve individual loops, the filter ratio method does at least serve to distinguish between the presence of generally hotter or cooler material even though the precise physical significance of the derived loop temperatures is open to question (cf. Dere & Mason, 1981, pp. 137–8).

Although the filter ratio method is simple in principle, its actual application to observations obtained with a grazing-incidence X-ray telescope is an involved and difficult process. To obtain meaningful results it is necessary to deconvolve the brightness measurements to remove – as far as possible – the smearing effect of the point spread function of the telescope employed. In the case of the AS&E telescope on Skylab, for example, 50% of the energy in the image of an on-axis point lies outside a radius of 48″ of arc, whereas the widths of active region loops are typically of the order of only 10″–20″ (Sections 3.4.1 and 3.4.3). Consequently, it is not surprising that relatively few direct determinations have so far been made of the temperatures and electron densities in individual coronal loops.

The first such measurements were those of Stewart & Vorphal (1977), who found temperatures of 2×10^6–3×10^6 K and electron densities of 0.8×10^9–2.4×10^9 cm^{-3} for the brightest loops seen in a complex of active regions observed on the disk over the period 1–6 September 1973. They do not state whether these were active region loops or active region interconnecting loops. The true heights of the loops were unknown but, for an assumed height of $\sim 35\,000$ km, Stewart and Vorpahl's results imply that the density in a loop is several times greater than that in the background.

Subsequently Pye *et al.* (1978) derived values of T_e and N_e for nine individual loops observed from Skylab in McMath region 12628 (see their Table 2 and Fig. 9(a)). Their inferred temperatures lie between 2.6×10^6 and 3.2×10^6 K, while the electron densities range from 1×10^9 to 6×10^9 cm^{-3}. Gerassimenko, Solodyna & Nolte (1978) found temperatures ranging from 2.2×10^6 to 2.5×10^6 K and electron densities from 1.8×10^9 to 2.5×10^9 cm^{-3} for three other loops. It should be remarked, however, that these particular structures actually appear somewhat more blob-like than loop-like on the published illustration (see Fig. 1 of their paper). The emission from the loops remained constant to within about two percent over the $\frac{1}{2}$ hr period of observation, and no indication was found of a temperature gradient along the loops.

Similar measurements were made by Kundu, Schmahl & Gerassimenko (1980) for McMath region 12379, although no actual values of T_e and N_e for individual loops are given in their paper. Comparison of their Figs. 4 and 6 reveals a close similarity between the loop structure visible on a soft X-ray photograph and that seen on photographs taken in the three hot EUV lines Mg X $\lambda 62.5$, Si XIV $\lambda 41.7$ and Fe XVI $\lambda 33.5$ some two hours later. The EUV loops were found to have an electron density of about 2×10^{10} cm^{-3} (Section 3.6.3).

More recently, Webb *et al.* (1987) have determined T_e and N_e for six loops observed from rockets. The results (see their Table 2) are mostly concordant with the figures quoted above, although a few of their N_e values are somewhat higher.

Soft X-ray observations of other active regions have been analysed by Pallavicini *et al.* (1981) and Teske & Mayfield (1981), but no direct values of T_e or N_e for individual loops are given in these papers.

There is clearly a need for more precise and detailed information on the physical conditions in X-ray loops and on their spatial and temporal variations. But this will have to await not only the advent of better quality observations but also the development of more refined diagnostic methods.

3.6.5 *Analysis of observations at radio wavelengths*

In radioastronomy it is usual to express the emission from a radiating plasma in terms not of the emergent intensity but of the radio brightness temperature T_b. At a given point on the surface of the plasma and at a wave frequency ν, T_b is defined as the temperature of a black body having the same brightness at this frequency as the point in question. In the case of thermal emission the source function within the plasma in

Planckian and at radio frequencies ($hv < kT_e$) is given to sufficient accuracy by the Rayleigh–Jeans formula

$$B_v(T_e) \simeq \frac{2v^2 k T_e}{c^2}, \tag{3.11}$$

where T_e is the local electron temperature. Substituting from Eqn. (3.11) in the standard formula for the emergent intensity and assuming the refractive index of the plasma to be approximately unity, we find that the radio brightness temperature at the point of observation is

$$T_b(v) = \int_0^{\tau_v} T_e(t)\, e^{-t}\, dt, \tag{3.12}$$

where

$$\tau_v = \int_{s_0}^{\infty} \kappa_v\, ds. \tag{3.13}$$

Here κ_v is the linear absorption coefficient at frequency v and ds an element of length along the path of the ray to the observer. It should be noted that since the refractive index of the plasma varies with the electron density N_e (and with the magnetic field B), the ray will in general follow a curved trajectory. Consequently, the integration in Eqn. (3.13) must be performed along the actual trajectory, from the point of origin of the wave s_0 to infinity.

In the particular case of an *isothermal* plasma of optical thickness τ_v along the line of sight, Eqn. (3.12) reduces to

$$T_b = (1 - e^{-\tau_v})\, T_e. \tag{3.14}$$

The measured value of T_b is thus a *lower* limit to T_e: T_b is always smaller than T_e but approaches it when the plasma is optically thick ($\tau_v \gg 1$). Conversely, if the plasma is optically thin ($\tau_v \ll 1$), Eqn. (3.14) becomes

$$T_b = \tau_v\, T_e. \tag{3.15}$$

Furthermore, if L is the geometrical path length through the (isothermal) plasma, then by Eqn. (3.13) $\tau_v = \kappa_v L$ and we have

$$T_b = \kappa_v L T_e. \tag{3.16}$$

It is clear, therefore, that to infer the physical conditions within the plasma one needs to identify the emission and absorption mechanisms involved and to evaluate the relevant absorption coefficients for both the ordinary and extraordinary components of the radio emission.

As we have already remarked in Section 3.5.1, two processes contribute to the thermal emission at radio wavelengths of (non-flare) loops, namely bremsstrahlung and gyroresonance radiation. Their relative contributions depend in a complicated way on the physical conditions prevailing in the plasma.

(1) *Bremsstrahlung*

Emission by this mechanism is due to the acceleration of electrons (free–free transitions) in the electric field of positive ions, mainly protons. In the absence of a magnetic field in the plasma the dependence of the absorption coefficient κ_v on wave frequency v and on T_e and N_e is given by the formula

$$\kappa_v = \frac{0.023}{v^2} N_e^2 T_e^{-\frac{3}{2}}[3.23 + \log{(T_e N_e^{-\frac{1}{3}})}] \tag{3.17}$$

(see Bray & Loughhead, 1974, pp. 136–7, and references cited therein). The slowly varying logarithmic term can be evaluated by putting $T_e = 3 \times 10^6$ K, $N_e = 10^{9.5}$ cm^{-3} – figures typical of those found for radio loops (see below) – yielding

$$\kappa_v = \frac{0.15}{v^2} N_e^2 T_e^{-\frac{3}{2}}. \tag{3.18}$$

For the optically thin case, combining Eqns. (3.16) and (3.18) we obtain

$$T_b = \frac{0.15}{v^2} \cdot T_e^{-\frac{1}{2}} \cdot N_e^2 L. \tag{3.19}$$

This formula allows one to estimate T_e in an optically thin loop from the observed T_b when its emission measure $N_e^2 L$ is known from other data, e.g. hot EUV or X-ray observations. Formulae similar to Eqns. (3.18) and (3.19), differing only in the numerical coefficient, have been given by numerous authors.

In the presence of a magnetic field the absorption coefficients of the extraordinary and ordinary modes differ from one another and from the field-free value. Felli, Lang & Willson (1981) and Marsh & Hurford (1982) give the following expressions:

$$(\kappa_v)_e = \frac{(\kappa_v)_{B=0}}{[1 - (v_B \cos{\theta}/v)]^2}, \quad (\kappa_v)_o = \frac{(\kappa_v)_{B=0}}{[1 + (v_B \cos{\theta}/v)]^2}, \tag{3.20}$$

where θ is the angle between **B** and the direction of the ray and $v_B = eB/2\pi m_e c = 2.8 \times 10^6 B$ Hz is the gyrofrequency. The effect of the

Table 3.6. *Notional values of B (Gauss) satisfying resonance condition*

Wavelength (cm)	$s = 2$	$s = 3$	$s = 4$
6	892	594	446
20	268	178	134

field is thus to increase the opacity of the *e*-mode with respect to the *o*-mode, producing a circular polarization in the sense of the *e*-mode (cf. Dulk, 1985). Formulae for the degree of polarization have been given by Felli *et al.* (1981) and Dulk (1985).

(2)   *Gyroresonance radiation*

Gyroresonance (or cyclotron) emission and absorption, due to electrons spiralling around magnetic field lines, occurs at frequencies which depend on the field strength *B* and a harmonic number *s* according to the formula

$$v_s = \frac{e}{2\pi m_e c} \cdot s \cdot B = 2.8 \times 10^6 s B \text{ Hz}, \tag{3.21}$$

m_e being the electron mass.

Table 3.6 gives (notional) values for the field strength *B* which satisfy the resonance condition for various values of *s* at two wavelengths commonly used in observing radio loops and active regions, namely $\lambda = 6$ cm and $\lambda = 20$ cm.

When interpreting any given set of observations, it is not possible to determine in advance which particular harmonics are dominant. Accordingly it is necessary to use the general formulae for the ordinary and extraordinary absorption coefficients, treating the harmonic number *s* as yet one more (free) parameter. The formulae for $(\kappa_v)_o$ and $(\kappa_v)_e$ are extremely complicated, involving not only *s* but also N_e, T_e, θ, *B* and the scale length of variations in *B*. In their general form they have been quoted by a number of authors, including Alissandrakis, Kundu & Lantos (1980), Kundu, Schmahl & Gerassimenko (1980), McConnell & Kundu (1983), Dulk (1985), and Luo *et al.* (1985). Simplified formulae corresponding to particular values of *s* have been given by Felli, Lang & Willson (1981) and Pallavicini, Sakurai & Vaiana (1981).

When high values of the circular polarization are observed (up to 90% on occasion) it is usually taken as evidence that gyroresonance emission predominates over bremsstrahlung. However, low polarization values are consistent with the presence of either mechanism, or both, and many authors explore the influence of both mechanisms in their analyses of the data.

The analysis of radio observations of loops in order to obtain information about their physical conditions requires a complex modelling procedure. The steps involved include (a) establishing the relative contributions of bremsstrahlung and gyroresonance radiation to the observed emission; (b) applying the formulae for the absorption coefficients for the relevant processes; (c) setting up an appropriate *atmospheric* model. This may be derived from theory and may incorporate data (e.g., the emission measure) from other spectral regions; (d) setting up a *magnetic* model, i.e. a three-dimensional model of the field distribution; (e) integrating through the atmosphere to obtain the calculated radio emission and polarization; (f) comparing the results of (e) with the observations; and (g) varying the parameters until satisfactory agreement is obtained between observation and theory.

This procedure is evidently not capable of giving unique, independent results from radio data alone. On the other hand, the radio data can serve as a useful check on results obtained in other regions of the spectrum, or on the plausibility of proposed theoretical loop models. In one respect, moreover, the radio data is of great importance: it is virtually the only data now available that – in principle – is capable of yielding information about the three-dimensional structure of the magnetic field at various levels of the low corona.

Using essentially the procedure outlined above, Lang, Willson & Rayrole (1982) and McConnell & Kundu (1983) have obtained values for T_e, N_e, p and B consistent with radio observations of individual loops. Both sets of workers used the theoretical loop model of Rosner, Tucker & Vaiana (1978) (see Chapter 5). McConnell & Kundu found that the observed emission at 20 cm could be explained by assuming thermal gyroresonance radiation at the loop top with $T_e = 1.7 \times 10^6$ K, $N_e = 5 \times 10^8$ cm^{-3}, and $B = 130$–170 G and thermal bremsstrahlung at the loop feet. Lang *et al.* found that their observations could be wholly interpreted in terms of bremsstrahlung and obtained $T_e = 3 \times 10^6$ K (maximum), $N_e = 2.5 \times 10^9$ cm^{-3}, and $p = 2.0$ dyne cm^{-2}. Reviewing the physical conditions in loops, Kundu & Lang (1985) give the following figures as typical: $T_e = 2 \times 10^6$–4×10^6 K, $N_e = 10^9$–10^{10} cm^{-3} and $B = 200$ G.

A number of authors have applied similar procedures to radio observations, not of loops *per se*, but of active regions (perhaps including unresolved loops). These include Alissandrakis, Kundu & Lantos (1980), Kundu, Schmahl & Gerassimenko (1980), Felli, Lang & Willson (1981), Pallavicini, Sakurai & Vaiana (1981), Dulk & Gary (1983), Webb *et al.* (1983), Staude *et al.* (1984), Luo *et al.* (1985), Gary & Hurford (1987) and Webb *et al.* (1987).

Most of these investigations incorporate both bremsstrahlung and gyroresonance emission in their analyses. They have yielded values of B ranging from 20–70 G (Dulk & Gary) to 600 G (Felli *et al.*). The paper of Luo *et al.* is of special interest since the observations were carried out simultaneously at four wavelengths, thus imposing on the analysis a healthier number of constraints than is usually the case.

3.7 Relationship between loops observed in different spectral regions

3.7.1 *Introduction*

Little information is available about the detailed relationships between hot loops observed in different regions of the spectrum. As we have remarked in Section 3.4.1, the X-ray loops associated with an active region are similar in general appearance to those seen in hot EUV lines, but systematic comparisons are lacking. More attention has been paid to the complex relationship between radio and X-ray loops, summarized below.

The important question of the relationship between hot and *cool* loops observed in the EUV is discussed in Section 3.7.3. Do they coincide in position or do they exist as quite separate entities? Detailed studies have been made of this question and it has been firmly established that, while both hot and cool EUV loops may occur close to one another, they are *not* cospatial.

3.7.2 *Relationship between X-ray and radio loops*

The relationship between radio and X-ray loops is complex. They are frequently *assumed* to be the same phenomenon (e.g. Lang, Willson & Rayrole, 1982; Kundu & Lang, 1985), while Dulk & Gary (1983) have remarked that 20 cm radio maps of active regions on the disk taken with a spatial resolution of 40″ of arc are similar in appearance to pictures of comparable quality taken in soft X-rays. High-resolution observations (3″ × 6″ of arc) at 6 cm by Webb *et al.* (1983) have improved our understanding. These authors found that, in general, the areas of microwave emission on their maps were *not* associated with X-ray emission, although several of them were found to coincide with the apparent bases of short coronal X-ray loops and four of them with the tops of X-ray loops. Figures 2 and 3 of their paper show co-aligned maps of X-ray and 6 cm radio emission and clearly demonstrate the absence of a close correlation.

Webb *et al.* explain the apparent lack of association as a consequence of the strong dependence of the gyroresonance opacity on variations in the strength and direction of the magnetic field pervading the active region. In effect, a radio loop and an X-ray loop could in fact coincide in space but look very different on maps depicting the emission which escapes. This explains the observation of Lang & Willson (1982), also at 6 cm, of horseshoe-shaped regions of very high circular polarization (up to 95%) surrounding sunspots (see their Figs. 1B and 3B). This observation demonstrates, once again, the important role played by the strength and configuration of the magnetic field along the path to the observer. In a later paper, Webb *et al.* (1987) have again emphasized the absence of a detailed correspondence between X-ray and microwave images. No cases were found of cospatial X-ray and microwave emission outlining entire loops.

3.7.3   *Spatial relationship between hot and cool EUV loops*

The spatial relationship between hot and cool EUV loops was first investigated by Foukal (1975) using Skylab observations with a claimed spatial resolution of 5″ of arc. By comparing photometric tracings made across 14 individual loops he found that loops seen in the hot Mg X $\lambda 62.5$ and Si XII $\lambda 52.1$ lines always occur close to but significantly displaced from nearby cool EUV loops. Only in one case did the peak of the Mg X emission coincide with that of the emission in cool EUV lines to within the limit of resolution. In general, however, Foukal inferred from the tracings (see his Fig. 5) that the peak emission from a cool loop coincides with a *drop* in the level of hot EUV emission, which then rises to a low peak on either side. The lateral displacement of the Si XII peaks from the central axis of the cool loop is greater than that of the corresponding Mg X peaks. Bearing in mind that the formation temperature of Mg X is less than that of Si XII (1.0×10^6 K as opposed to 2.0×10^6 K) Foukal interpreted the observations as implying the existence of concentric sheaths of increasingly hotter material around a cool core, so that what is seen as a hot EUV loop is really a sheath of hot gas surrounding a cool EUV loop. Foukal's idea was subsequently elaborated by Levine & Withbroe (1977), who attempted a quantitative study of the variation in the physical conditions in a loop as a function of the distance from its axis.

However, this picture of the spatial relationship between hot and cool EUV loops has not been confirmed by subsequent workers. In particular,

Dere (1982) finds that, while cool Ne VII loops are located in the general vicinity of hot Fe XV loops, they are not coaxial. He concludes that the cool and hot loops are formed in quite separate locations and remarks that this result was confirmed by visual examination of active region structures on several hundred plates. A similar conclusion emerges from an examination of the Ne VII and Fe XVI maps of an active region published by Cheng *et al.* (1980, Fig. 2(*a*)) and from a study made by Cheng (1980) of another active region on the disk. Yet another study confirming the lack of cospatiality between hot and cool EUV loops has subsequently been reported by Habbal, Ronan & Withbroe (1985).

Hot and cool EUV loops, although obviously related, must therefore be regarded as separate physical structures. On the other hand, as Cheng (1980) has remarked, the existence of hot and cool loops side by side in active regions without any obvious direct spatial relationship to one another raises intriguing questions about the mechanism of formation and dynamical stability of coronal loops, problems discussed in Chapter 5.

In conclusion, it is of interest to note that hot and cool *flare* loops observed in Fe XIV λ530.3 and Hα respectively also show a lack of detailed correspondence (cf. Section 4.3.3). On the other hand, a systematic study of this question for the case of non-flare loops observed in the visible region of the spectrum does not appear to have been carried out.

3.8   Summary of data: hot loops

Tables 3.7 and 3.8 summarize the numerical data detailed in earlier sections of this chapter concerning the morphological properties of, and physical conditions within, hot loops. By contrast with cool loops, where the data comes almost exclusively from Hα and various EUV lines, here we are able to draw upon data from no less than four regions of the spectrum, namely the visible (largely λ530.3 of Fe XIV), the EUV, the (soft) X-ray, and the microwave. When we examine the tabulated figures for the various morphological and physical quantities, we see that – despite the rather large range of values in many cases – the picture is a consistent one. In fact, all hot loops appear to be basically similar in their properties regardless of the wavelength region in which they are observed.

One inconsistency in Table 3.7 may be noted: the figures given for the *lengths* of class II X-ray loops are not compatible with the figures quoted for their *heights*. Occasional apparent inconsistencies of this kind reflect the inhomogeneity of the original observational material. Some of the values in Tables 3.7 and 3.8 are in fact based on rather meagre material and, in order to find out the provenance of any particular figure, the reader should

Table 3.7. *Morphological properties of hot loops: summary of data*

Quantity	Value	Wavelength	Disk (D) or Limb (L)	Section
Height, true (km)	110000–130000	EUV	D[d]	3.3.3
	~45000	21 cm	D	3.5.2
Height, projected (km)	50000–100000	530.3 nm	L	3.2.2
	200000–250000	530.3 nm	L[d]	3.2.2
	5000–70000 (?)	EUV	D	3.3.3
	~100000–200000	X-ray	L[b]	3.4.3
	50000–100000	X-ray	D[c], L[c]	3.4.3
	25000–110000	X-ray	D[d], L[d]	3.4.4
	200000–500000	X-ray	L[e]	3.4.5
Length (km)	18000–29000	EUV	D	3.6.3
	7000–44000	X-ray	D[a]	3.4.3
	10000–100000	X-ray	D[b]	3.4.3
	50000–500000	X-ray	D[c]	3.4.3
	70000–100000	6, 20 cm	D	3.5.2
Separation of footpoints (km)	250000–500000	X-ray	L[e]	3.4.5
Diameter (km)	3000–22000	530.3 nm	L	3.2.2
	3000–18000	EUV	D, L	3.3.3, 3.6.3
	5000–11000	X-ray	D[a]	3.4.3
	5000–20000	X-ray	D[b]	3.4.3
	10000–30000	X-ray	D[c]	3.4.3
	15000	3.7–20 cm	D	3.5.2
Aspect ratio	0.03–0.18	EUV	D	3.3.3
	0.1	X-ray	D[a, b]	3.4.3
	0.04	X-ray	D[c]	3.4.3
	0.075	microwave	D	3.5.2
Inclination of loop plane to vertical β (deg)	7–25	EUV	D	3.3.3
Lifetime	hours (small loops)	530.3 nm	L	3.2.2
	days (large loops)	530.3 nm	L	3.2.2
	$\gtrsim 6$ hr	EUV	D	3.3.3
	>31 hr	EUV	D[d]	3.3.3
	>10 hr	X-ray	D[a]	3.4.3
	hours–days	X-ray	D[d]	3.4.2, 3.4.4
	hours–day	microwave	D	3.5.2

[a,b,c] Class I, II and III X-ray loops respectively; [d] interconnecting loop; [e] quiet region loop.

Table 3.8. *Physical conditions in hot loops: summary of data*

Quantity	Value	Wavelength	Disk (D) or Limb (L)	Section
Temperature T_e (K)	1.0×10^6–2.6×10^6	Various lines in visible region	L	3.6.2
	2.0×10^6–2.2×10^6	EUV	D	3.6.3
	2.0×10^6–3.2×10^6	X-ray	D	3.6.4
	1.7×10^6–3.0×10^6	20 cm	D	3.6.5
Electron density N_e (cm^{-3})	1.5×10^8–2.0×10^9	530.3, 637.4 nm	L	3.6.2
	1.8×10^9–2.0×10^{10}	EUV	D	3.6.3
	8.0×10^8–6.0×10^9	X-ray	D	3.6.4
	5.0×10^8–2.5×10^9	20 cm	D	3.6.5
Gas pressure p (dyne cm^{-2})	0.5	Visible	L[a]	3.6.2
	0.7–16.6	EUV	D	3.6.3
	2.4	X-ray	D[b]	3.6.4
	2.0	20 cm	D	3.6.5
Nonthermal velocity ξ (km s^{-1})	6–16	Various lines in visible region	L	3.6.2
Total magnetic field B (G)	130–200	20 cm	D	3.6.5

[a] Taking $p = 2kN_e T_e$, $T_e = 1.8 \times 10^6$ K, $N_e = 1.0 \times 10^9$ cm^{-3}.
[b] Taking $p = 2kN_e T_e$, $T_e = 2.6 \times 10^6$ K, $N_e = 3.4 \times 10^9$ cm^{-3}.

consult the relevant section of the chapter; for convenience, section numbers are given in the fifth column of each table.

By contrast with cool loops (Table 2.8), dynamical information is not yet available for hot loops. Furthermore, values for the various (true) geometrical parameters are sparse, since to date little work has been done on the geometrical reconstruction of hot loops.

The inferred temperatures and electron densities (Table 3.8) cover the ranges 1.0×10^6–3.0×10^6 K and 1.5×10^8–2.0×10^{10} cm^{-3} respectively, whereas the gas pressures cover a range of, at most, a single order of magnitude. This parallels a similar situation for cool loops (cf. Section 2.6) and, in fact, the actual values for p appear to be much the same for cool and hot loops. Once again, this suggests that, regardless of its temperature, a loop is in approximate pressure equilibrium with its surroundings (cf. Chapter 5).

In this book we have found it convenient and useful to divide (non-flare) loops into two classes, cool (Chapter 2) and hot (this chapter), depending on whether their temperature is less than or greater than about one million degrees. To what extent does this classification reflect a basic alteration in the physical processes occurring within cool and hot loops? One relevant and important fact has already been discussed (Section 3.7.3): in the case of EUV loops it has been conclusively established that, while both types may occur in the same vicinity, they are not cospatial. But we can now go further and ask whether this difference is accompanied by systematic differences in the morphological properties of, and physical conditions in, cool and hot loops. To answer this question we need to compare the numerical data for hot loops presented in Tables 3.7 and 3.8 with the corresponding data for cool loops given in Tables 2.8 and 2.9.

Hot loops, especially some of those observed in X-rays, can attain much greater heights than cool ones. Both hot and cool loops exhibit a wide range of (inferred) lengths, but certain classes of X-ray loops (interconnecting loops) are much longer than cool loops observed in $H\alpha$ or the cool EUV lines. Hot and cool EUV loops have much the same diameter (thickness), but cool $H\alpha$ and $L\alpha$ (Sections 2.3.3 and 2.4.3) loops appear to be much thinner than any other class. Measurement of the inclinations of loop planes to the vertical are sparse, but both hot and cool loops may be inclined at significant angles. In general, hot loops last longer than cool ones, and there appear to be no hot counterparts to the short-lived $H\alpha$ loops, which last for only a few minutes.

It is difficult to compare the electron densities in hot and cool loops, since both cover a large range, with considerable overlap. The gas pressures show a much smaller range, as pointed out above, and the values for hot and cool loops are much the same. This also applies to the non-thermal velocity ξ, but there are few measurements. Similarly, there are too few measurements of the total magnetic field B to permit meaningful comment.

To summarize, hot loops tend to be thicker, longer, higher, and longer-lived than cool loops, with which they are not cospatial. However, our present knowledge appears to indicate that their other morphological and physical properties (except temperature) are similar.

References

Alissandrakis, C.E., Kundu, M.R. & Lantos, P. (1980). A model for sunspot associated emission at 6 cm wavelength. *Astronomy and Astrophysics*, **82**, 30–40.

Aschenbach, B. (1985). X-ray telescopes. *Reports on Progress in Physics*, **48**, 579–629.

Athay, R.G. (1976). *The Solar Chromosphere and Corona: Quiet Sun*. Dordrecht: Reidel.

Berton, R. & Sakurai, T. (1985). Stereoscopic determination of the three-dimensional geometry of coronal magnetic loops. *Solar Physics*, **96**, 93–111.

Billings, D.E. (1966). *A Guide to the Solar Corona*. New York: Academic Press.

Bohlin, J.D. & Sheeley, N.R. (1978). Extreme ultraviolet observations of coronal holes. II. Association of holes with solar magnetic fields and a model for their formation during the solar cycle. *Solar Physics*, **56**, 125–151.

Bray, R.J. & Loughhead, R.E. (1974). *The Solar Chromosphere*. London: Chapman and Hall.

Bumba, V. & Howard, R. (1965). Large-scale distribution of solar magnetic fields. *Astrophysical Journal*, **141**, 1502–12.

Chase, R.C., Krieger, A.S., Svestka, Z. & Vaiana, G.S. (1976). Skylab observations of X-ray loops connecting separate active regions. *Space Research*, **XVI**, 917–22.

Cheng, C.-C. (1980). Physical properties of individual coronal loops in a solar active region observed in the XUV. *Astrophysical Journal*, **238**, 743–52.

Cheng, C.-C., Smith, J.B. & Tandberg-Hanssen, E. (1980). Morphology and spatial distribution of XUV and X-ray emissions in an active region observed from Skylab. *Solar Physics*, **67**, 259–65.

Culhane, J.L. & Acton, L.W. (1974). The solar X-ray spectrum. *Annual Review of Astronomy and Astrophysics*, **12**, 359–81.

Davis, J.M., Gerassimenko, M., Krieger, A.S. & Vaiana, G.S. (1975). Interpretation of simultaneous soft X-ray spectroscopic and imaging observations of an active region. *Solar Physics*, **45**, 393–410.

Davis, J.M. & Krieger, A.S. (1982). Properties of coronal arches. *Solar Physics*, **80**, 295–307.

Davis, J.M., Krieger, A.S., Silk, J.K. & Chase, R.C. (1979). Quest for ultrahigh resolution in X-ray optics. *Society of Photo-Optical Instrumentation Engineers*, **184**, 96–108.

Dere, K.P. (1982). The XUV structure of solar active regions. *Solar Physics*, **75**, 189–203.

Dere, K.P. & Mason, H.E. (1981). Spectroscopic diagnostics of the active region: transition zone and corona. In *Solar Active Regions*, ed. F.Q. Orral, pp. 129–64. Boulder: Colorado Associated University Press.

Dollfus, A. (1971). Investigations on coronal monochromatic emissions in the optical range. In *Physics of the Solar Corona*, ed. C.J. Macris, pp. 97–113. Dordrecht: Reidel.

Donnelly, R. (1977). Extreme ultraviolet and X-rays (> 10 Å). In *The Solar Output and its Variation*, ed. O.R. White, pp. 33–4. Boulder: Colorado Associated University Press.

Dulk, G.A. (1985). Radio emission from the Sun and stars. *Annual Review of Astronomy and Astrophysics*, **23**, 169–224.

Dulk, G.A. & Gary, D.E. (1983). The Sun at 1.4 GHz: intensity and polarization. *Astronomy and Astrophysics*, **124**, 103–7.

Dunn, R.B. (1971). Coronal events observed in 5303 A. In *Physics of the Solar Corona*, ed. C.J. Macris, pp. 114–29. Dordrecht: Reidel.

Felli, M., Lang, K.R. & Willson, R.F. (1981). VLA observations of solar active regions. I. The slowly varying component. *Astrophysical Journal*, **247**, 325–37.

Fort, B. & Martres, M.J. (1974). On the structure of the quiet corona near quiescent prominences. *Astronomy and Astrophysics*, **33**, 249–55.

Fort, B., Picat, J.P., Dantel, M. & Leroy, J.L. (1973). Coronal densities and temperatures derived from monochromatic images in the red and green lines. *Astronomy and Astrophysics*, **24**, 267–73.

Foukal, P. (1975). The temperature structure and pressure balance of magnetic loops in active regions. *Solar Physics*, **43**, 327–36.

Fuhr, J.R., Martin, G.A., Wiese, W.L. & Younger, S.M. (1981). Atomic transition probabilities for iron, cobalt, and nickel (a critical data compilation of allowed lines). *Journal of Physical and Chemical Reference Data*, **10**, 305–565.

Gaetz, T.J. & Salpeter, E.E. (1983). Line radiation from a hot, optically thin plasma: collision strengths and emissivities. *Astrophysical Journal Supplement Series*, **52**, 155–68.

Gary, D.E. & Hurford, G.J. (1987). Multifrequency observations of a solar active region during a partial eclipse. *Astrophysical Journal*, **317**, 522–33.

Gerassimenko, M. & Nolte, J.T. (1978). The quantitative interpretation of solar X-ray images. *Solar Physics*, **60**, 299–310.

Gerassimenko, M., Solodyna, C.V. & Nolte, J.T. (1978). Observational evidence of continual heating in X-ray emitting coronal loops. *Solar Physics*, **57**, 103–10.

Giacconi, R., Reidy, W.P., Vaiana, G.S., van Speyboeck, L.P. & Zehnpfennig, T.F. (1969). Grazing-incidence telescopes for X-ray astronomy. *Space Science Reviews*, **9**, 3–57.

Golub, L., Noci, G., Poletto, G. & Vaiana, G.S. (1982). Active region coronal evolution. *Astrophysical Journal*, **259**, 359–65.

Habbal, S.R., Ronan, R. & Withbroe, G.L. (1985). Spatial and temporal variations of solar coronal loops. *Solar Physics*, **98**, 323–40.

Haisch, B.M. & Claflin, E.S. (1985). X-ray resonance scattering in a spherically symmetric coronal model. *Solar Physics*, **99**, 101–9.

Harrison, R.A. (1987). Solar soft X-ray pulsations. *Astronomy and Astrophysics*, **182**, 337–47.

Hirayama, T. (1985). Modern observations of solar prominences. *Solar Physics*, **100**, 415–434.

Howard, R. & Svestka, Z. (1977). Development of a complex of activity in the solar corona. *Solar Physics*, **54**, 65–105.

Jacobs, V.L., Davis, J., Kepple, P.C. & Blaha, M. (1977a). The influence of autoionization accompanied by excitation on dielectronic recombination and ionization equilibrium. *Astrophysical Journal*, **211**, 605–16.

Jacobs, V.L., Davis, J., Kepple, P.C. & Blaha, M. (1977b). The influence of autoionization accompanied by excitation on the dielectronic recombination and the ionization equilibrium of silicon ions. *Astrophysical Journal*, **215**, 690–9.

Jacobs, V.L., Davis, J., Rogerson, J.E. & Blaha, M. (1979). Dielectronic recombination rates, ionization equilibrium, and radiative energy-loss rates for neon, magnesium, and sulfur ions in low-density plasmas. *Astrophysical Journal*, **230**, 627–38.

Jager, C. de (1959). Structure and dynamics of the solar atmosphere. In *Handbuch der Physik*, ed. S. Flügge, vol. LII, pp. 80–362. Berlin: Springer.

Kahler, S. (1976). Determination of the energy or pressure of a solar X-ray structure using X-ray filtergrams from a single filter. *Solar Physics*, **48**, 255–9.

Kato, T. (1976). Radiation from a hot, thin plasma from 1 to 250 Å. *Astrophysical Journal Supplement Series*, **30**, 397–449.

Kleczek, J. (1963). Regular structures in the green solar corona. *Publications of the Astronomical Society of the Pacific*, **75**, 9–14.

Koutchmy, S. (1977). Solar corona. In *Illustrated Glossary for Solar and Solar-Terrestrial Physics*, ed. A. Bruzek and C.J. Durrant, pp. 39–52. Dordrecht: Reidel.

Kramer, G. & 5 co-authors (1978). Analysis and interpretation of soft X-ray photographs of coronal active regions taken with Fresnel zone plates. I: Image Analysis. *Solar Physics*, **57**, 345–67.

Kreplin, R.W. (1961). Solar X-rays. *Annales de Géophysique*, **17**, 151–61.

Krieger, A.S., de Feiter, L.D. & Vaiana, G.S. (1976). Evidence for magnetic energy storage in coronal active regions. *Solar Physics*, **47**, 117–26.

Krieger, A.S., Vaiana, G.S. & van Speybroeck, L.P. (1971). The X-ray corona and the photospheric magnetic field. In *Solar Magnetic Fields*, IAU Symposium 43, ed. R. Howard, pp. 397–412. Dordrecht: Reidel.

Kuang, Y.-G. (1982). Optical design of a grazing-incidence X-ray imaging telescope. *Chinese Physics*, **2**, 1041–9.

Kundu, M.R. & Lang, K.R. (1985). The Sun and nearby stars: microwave observations at high resolution. *Science*, **228**, 9–15.

Kundu, M.R., Schmahl, E.J. & Gerassimenko, M. (1980). Microwave, EUV, and X-ray observations of active region loops: evidence for gyroresonance absorption in the corona. *Astronomy and Astrophysics*, **82**, 265–71.

Kundu, M.R. & Velusamy, T. (1980). Observation with the VLA of a stationary loop structure on the Sun at 6 centimeter wavelength. *Astrophysical Journal*, **240**, L63–7.

Landini, M. & Fossi, B.C. (1970). Solar radiation from 1 to 100 Å. *Astronomy and Astrophysics*, **6**, 468–75.

Landini, M., Fossi, B.C., Krieger, A. & Vaiana, G.S. (1975). The coronal structure of active regions. *Solar Physics*, **44**, 69–82.

Lang, K.R. & Willson, R.F. (1982). Polarized horseshoes around sunspots at 6 centimeter wavelength. *Astrophysical Journal*, **255**, L111–7.

Lang, K.R., Willson, R.F. & Rayrole, J. (1982). Very Large Array observations of coronal loops at 20 centimeter wavelength. *Astrophysical Journal*, **258**, 384–7.

Levine, R.H. (1976). Evidence for opposed currents in active region loops. *Solar Physics*, **46**, 159–70.

Levine, R.H. & Withbroe, G.L. (1977). Physics of an active region loop system. *Solar Physics*, **51**, 83–101.

Luo, X.-H., Yao, D.-Y., Ji, S.-C. & Zhao, R.-Y. (1985). A magnetic loop model for the slowly varying radio source observed during the solar eclipse of 1980 February 16. *Chinese Astronomy and Astrophysics*, **9**, 232–40.

Marsh, K.A. & Hurford, G.J. (1982). High spatial resolution solar microwave observations. *Annual Review of Astronomy and Astrophysics*, **20**, 497–516.

Martens, P.C.H., van den Oord, G.H.J. & Hoyng, P. (1985). Observations of steady anomalous magnetic heating in thin current sheets. *Solar Physics*, **96**, 253–75.

Maute, K. & Elwert, G. (1981). Quantitative analysis of Skylab X-ray pictures of the Sun by means of iterative deconvolution. *Solar Physics*, **70**, 273–91.

McConnell, D. & Kundu, M.R. (1983). VLA observations of a solar active region and coronal loops. *Astrophysical Journal*, **269**, 698–705.

McConnell, D. & Kundu, M.R. (1984). VLA observations of fine structures in a solar active region at 6 centimeter wavelength. *Astrophysical Journal*, **279**, 421–6.

McGuire, J.P. & 5 co-authors (1977). A long-lived coronal arch system observed in X-rays. *Solar Physics*, **52**, 91–100.

McIntosh, P.S., Krieger, A.S., Nolte, J.T. & Vaiana, G. (1976). Association of X-ray arches with chromospheric neutral lines. *Solar Physics*, **49**, 57–77.

Mewe, R. (1972). Calculated solar X-radiation from 1 to 60 Å. *Solar Physics*, **22**, 459–91.

Mewe, R. (1975). Calculated solar X-radiation. II: Spectrum between 61 and 220 Å. *Solar Physics*, **44**, 383–390.

Mewe, R. & Gronenschild, E.H.B.M. (1981). Calculated X-radiation from optically thin plasmas. IV. Atomic data and rate coefficients for spectra in the range 1–270 Å. *Astronomy and Astrophysics Supplement Series*, **45**, 11–52.

Noci, G. (1971). Atomic processes in the solar corona. In *Physics of the Solar Corona*, ed. C.J. Macris, pp. 13–28. Dordrecht: Reidel.

Nolte, J.T. & 5 co-authors (1977). Do changes in coronal emission structure imply magnetic reconnection? *Solar Physics*, **55**, 401–12.

Pallavicini, R., Peres, G., Serio, S., Vaiana, G.S., Golub, L. & Rosner, R. (1981). Closed coronal structures. III. Comparison of static models with X-ray, EUV, and radio observations. *Astrophysical Journal*, **247**, 692–706.

Pallavicini, R., Sakurai, T. & Vaiana, G.S. (1981). X-ray, EUV, and centimetric observations of solar active regions: an empirical model for bright radio sources. *Astronomy and Astrophysics*, **98**, 316–27.

Pallavicini, R., Vaiana, G.S., Tofani, G. & Felli, M. (1979). The coronal atmosphere above solar active regions: comparison of high spatial resolution soft X-ray and centimetric observations. *Astrophysical Journal*, **229**, 375–86.

Picat, J.P., Fort, B., Dantel, M. & Leroy, J.L. (1973). Photometric analysis of monochromatic photographs of the solar corona taken in the green line (5303 Å) and the red line (6374 Å). *Astronomy and Astrophysics*, **24**, 259–65.

Poletto, G., Vaiana, G.S., Zombeck, M.V., Krieger, A.S. & Timothy, A.F. (1975). A comparison of coronal X-ray structures of active regions with magnetic fields computed from photospheric observations. *Solar Physics*, **44**, 83–99.

Pye, J.P. & 6 co-authors (1978). The structure of the X-ray bright corona above active region McMath 12628 and derived implications for the description of equilibria in the solar atmosphere. *Astronomy and Astrophysics*, **65**, 123–38.

Ray, A. & van Hoven, G. (1982). Hydromagnetic stability of coronal arcade structures: the effects of photospheric line tying. *Solar Physics*, **79**, 353–64.

Raymond, J.C. & Foukal, P. (1982). The thermal structure of solar coronal loops and implications for physical models of coronae. *Astrophysical Journal*, **253**, 323–9.

Rosner, R., Tucker, W.H. & Vaiana, G.S. (1978). Dynamics of the quiescent solar corona. *Astrophysical Journal*, **220**, 643–65.

Rust, D.M. & Webb, D.F. (1977). Soft X-ray observations of large-scale coronal active region brightenings. *Solar Physics*, **54**, 403–17.

Saito, K. & Hyder, C.L. (1968). A concentric ellipse multiple-arch system in the solar corona. *Solar Physics*, **5**, 61–86.

Sakurai, T. & Uchida, Y. (1977). Magnetic field and current sheets in the corona above active regions. *Solar Physics*, **52**, 397–416.

Schmahl, E.J., Mouradian, Z., Martres, M.J. & Soru-Escaut, I. (1982). EUV arcades: signatures of filament instability. *Solar Physics*, **81**, 91–105.

Serio, S. & 5 co-authors (1978). Configuration and gradual dynamics of prominence-related X-ray coronal cavities. *Solar Physics*, **59**, 65–86.

Sheeley, N.R. (1980). Temporal variations of loop structures in the solar atmosphere. *Solar Physics*, **66**, 79–87.

Sheeley, N.R. (1981). The overall structure and evolution of active regions. In *Solar Active Regions*, Skylab Solar Workshop III, ed. F.Q. Orrall, pp. 17–42. Boulder: Colorado Associated University Press.

Sheeley, N.R. & 5 co-authors (1975a). XUV observations of coronal magnetic fields. *Solar Physics*, **40**, 103–21.

Sheeley, N.R. & 12 co-authors (1975b). Coronal changes associated with a disappearing filament. *Solar Physics*, **45**, 377–92.

Shull, J.M. (1981). The X-ray spectrum of a hot interstellar plasma. *Astrophysical Journal Supplement Series*, **46**, 27–40.

Staude, J. & 8 co-authors (1984). Observations in the X-ray, extreme ultraviolet, optical, and radio ranges. *Soviet Astronomy*, **28**, 557–63.

Stewart, R.T., Brueckner, G.E. & Dere, K.P. (1986). Culgoora radio and Skylab EUV observations of emerging magnetic flux in the lower corona. *Solar Physics*, **106**, 107–30.

Stewart, R.T. & Vorpahl, J. (1977). Radio and soft X-ray evidence for dense non-potential magnetic flux tubes in the solar corona. *Solar Physics*, **55**, 111–20.

Strong, K.T., Alissandrakis, C.E. & Kundu, M.R. (1984). Interpretation of microwave active region structures using SMM soft X-ray observations. *Astrophysical Journal*, **277**, 865–73.

Svestka, Z. & Howard, R. (1979). Transient brightenings of interconnecting loops. Morphology of the sudden brightenings. *Solar Physics*, **63**, 297–318.

Svestka, Z. & Howard, R. (1981). Transient brightenings of interconnecting loops. II. Dynamics of·the brightened loops. *Solar Physics*, **71**, 349–59.

Svestka, Z., Krieger, A.S., Chase, R.C. & Howard, R. (1977). Transequatorial loops interconnecting McMath regions 12472 and 12474. *Solar Physics*, **52**, 69–90.

Teske, R.G. & Mayfield, E.B. (1981). The distribution of maximum temperatures of coronal active region loops. *Astronomy and Astrophysics*, **93**, 228–34.

Tousey, R. & 8 co-authors (1973). A preliminary study of the extreme ultraviolet spectroheliograms from Skylab. *Solar Physics*, **33**, 265–80.

Tousey, R. & Koomen, M.J. (1971). The relation between the white light and XUV coronas on 7 March, 1970. *Solar Physics*, **21**, 401–7.

Tsubaki, T. (1975). Line profile analysis of a coronal formation observed near a quiescent prominence: intensities, temperatures and velocity fields. *Solar Physics*, **43**, 147–75.

Tucker, W.H. & Koren, M. (1971). Radiation from a high-temperature, low-density plasma: the X-ray spectrum of the solar corona. *Astrophysical Journal*, **168**, 283–311.

Underwood, J.H. & McKenzie, D.L. (1977). The analysis and interpretation of solar X-ray photographs. *Solar Physics*, **53**, 417–33.

Underwood, J.H. & McKenzie, D.L. (1978). Further remarks on the analysis and interpretation of solar X-ray photographs. *Solar Physics*, **60**, 311–14.

Underwood, J.H., Milligan, J.E., deLoach, A.C. & Hoover, R.B. (1977). S056 X-ray telescope experiment on the Skylab Apollo Telescope Mount. *Applied Optics*, **16**, 858–69.

Vaiana, G.S., Krieger, A.S. & Timothy, A.F. (1973). Identification and analysis of structures in the corona from X-ray photography. *Solar Physics*, **32**, 81–116.

Vaiana, G.S., Krieger, A.S., Timothy, A.F. & Zombeck, M. (1976). ATM observations, X-ray results. *Astrophysics and Space Science*, **39**, 75–101.

Vaiana, G.S. & 5 co-authors (1977). The S-054 X-ray telescope experiment on Skylab. *Space Science Instrumentation*, **3**, 19–76.

van Hoven, G. (1981). Simple-loop flares: magnetic instabilities. In *Solar Flare Magnetohydrodynamics*, ed. E.R. Priest, pp. 217–75. New York: Gordon & Breach.

Waldmeier, M. (1963). Slow variations of the solar corona. In *The Solar Corona*, IAU Symposium 16, ed. J.W. Evans, pp. 129–36. London: Academic Press.

Walker, A.B.C. (1972). The coronal X-spectrum: problems and prospects. *Space Science Reviews*, **13**, 672–730.

Walker, A.B.C. (1975). Interpretation of the X-ray spectra of solar active regions. In *Solar Gamma-, X-, and EUV Radiation*, IAU Symposium 68, ed. S.R. Kane, pp. 73–100. Dordrecht: Reidel.

Walker, A.B.C. (1977). Comment on the solar spectrum between 10 and 300 Å. In *The Solar Output and its Variation*, ed. O.R. White, pp. 279–86. Boulder: Colorado Associated University Press.

Webb, D.F. (1981). Active region structures in the transition region and corona. In *Solar Active Regions*, Skylab Solar Workshop III, ed. F.Q. Orrall, pp. 165–98. Boulder: Colorado Associated University Press.

Webb, D.F., Davis, J.M., Kundu, M.R. & Velusamy, T. (1983). X-ray and microwave observations of active regions. *Solar Physics*, **85**, 267–83.

Webb, D.F., Holman, G.D., Davis, J.M., Kundu, M.R. & Shevgaonkar, R.K. (1987). The plasma and magnetic field properties of coronal loops observed at high spatial resolution. *Astrophysical Journal*, **315**, 716–28.

Webb, D.F., Krieger, A.S. & Rust, D.M. (1976). Coronal X-ray enhancements associated with Hα filament disappearances. *Solar Physics*, **48**, 159–86.

Webb, D.F. & Zirin, H. (1981). Coronal loops and active region structure. *Solar Physics*, **69**, 99–118.

Wiese, W.L., Smith, M.W. & Miles, B.M. (1969). *Atomic Transition Probabilities*, vol. II. Washington: National Bureau of Standards.

Willson, R.F. (1985). VLA observations of solar active regions at closely spaced frequencies: evidence for thermal cyclotron line emission. *Astrophysical Journal*, **298**, 911–17.

Zirin, H. (1966). *The Solar Atmosphere*. Waltham: Blaisdell.

Zombeck, M.V. & 5 co-authors (1978). An atlas of soft X-ray images of the solar corona from Skylab. *Astrophysical Journal Supplement Series*, **38**, 69–85.

4

Flare loops: observed properties

4.1 Introduction

Solar flares are remarkably diverse and complicated phenomena involving the transient heating of localized regions of the corona and underlying chromosphere within an active region. The sudden release of energy is accompanied by the emission of electromagnetic radiation over a very wide span of the spectrum, ranging – in extreme cases – from γ-rays to kilometric radio waves. In almost all cases flares seen in the chromospheric Hα line also produce an increase in the flux of soft X-rays. Moreover, the variation in the soft X-ray flux with time roughly follows that of the Hα intensity at the brightest point of the flare (Svestka, 1981, p. 74). Both curves are characterized by a rapid rise to a comparatively short-lived maximum, followed by a much slower decay. The intensity of the soft X-ray emission generally increases with the optical importance of the flare, although individual flares may show marked deviations from this rule. For this reason it has become customary to assign both an optical (Hα) and an X-ray importance to each flare. Both the Hα and soft X-ray emission pertain to what is conventionally called the 'thermal' or 'quasi-thermal' component of the flare, i.e. they originate in plasmas where the distribution of electron velocities is believed to be Maxwellian.

Flares are classified optically according to their area at the time of maximum Hα brightness into the four categories listed in Table 4.1, adapted from Svestka (1981). Strictly speaking, a transient brightening in the chromosphere should be called a flare only if its area, corrected for foreshortening, exceeds 100 millionths of the visible hemisphere or 3×10^8 km^2; otherwise it is classified as a subflare and labelled with the symbol S. Account is taken of the brightness of the flare or subflare in Hα by adding the letter F = faint, N = normal or B = brilliant to the symbol denoting its area classification: thus the largest and brightest flares are classified as 4B and the faintest subflares as SF.

Table 4.1. *Optical classification of flares*

| Importance | Area at time of maximum Hα brightness[a] | |
	Millionths of hemisphere	Square degrees[b]
S	<100	<2.06
1	100–250	2.06–5.15
2	250–600	5.15–12.4
3	600–1200	12.4–24.7
4	>1200	>24.7

[a] Corrected for foreshortening; [b] 1 square degree = 1.476×10^8 km² of solar surface.

Table 4.2. *X-ray classification of flares*

| Class | Peak flux at Earth's distance | |
	erg cm⁻² s⁻¹	W m⁻²
C1–C9	$1–9 \times 10^{-3}$	$1–9 \times 10^{-6}$
M1–M9	$1–9 \times 10^{-2}$	$1–9 \times 10^{-5}$
X1–X9⁺	$1–>9 \times 10^{-1}$	$1–>9 \times 10^{-4}$

Under the scheme adopted by the United States National Oceanic and Atmospheric Administration (NOAA) the X-ray importance of a flare is assessed in terms of the peak flux of soft X-rays received at the distance of the Earth over the spectral band 0.1–0.8 nm, measured in watts per square metre. The letters C, M and X are used to represent powers of ten in the manner set out in Table 4.2 (Simon & McIntosh, 1972). The number following the letter acts as a multiplier: for example, M4 is the classification assigned to a flare giving rise to a peak soft X-ray flux at the Earth's orbit of 4×10^{-5} W m⁻².

Time variability is the essence of flare emission. All flares pass through at least three phases: rise, maximum and decay (Moore *et al.*, 1980, p. 341). These phases are easily recognizable in curves showing the variation of the Hα intensity and soft X-ray flux over the lifetime of a flare. The physical properties of a flare change markedly over its lifetime and, on occasion, we shall take advantage of the existence of the three phases common to all flares to divide our discussion of the available observational data into the four natural evolutionary stages: pre-flare, rise, maximum and decay (cf. Sections 4.4.2 and 4.5.2). Of course, this is possible only when sufficient data are available.

Some flares exhibit only a thermal component. Others, however, also display a highly energetic *impulsive* component superimposed on the rising phase of the thermal component. The impulsive phase is marked by the occurrence of short-lived 'bursts' of radiation, which are most prominent at hard X-ray, EUV, and some optical and radio (microwave) wavelengths (Kane, 1974; Svestka, 1976, pp. 302–3; Kane *et al.*, 1980). It takes place during what is sometimes called the 'flash' phase of an Hα flare and coincides with the period of the most energetic release of flare energy (Svestka, 1981, p. 55). Following Kane (1974), the evolution of impulsive flares is sometimes described on the basis of the following three phases: precursor, impulsive and gradual. Notwithstanding, we shall endeavour to keep the discussion of the material in this chapter within the framework of the four-stage sequence of evolutionary phases described above, which applies to all flares. Impulsive phenomena, when present, will be dealt with as part of the rising phase.

Calculation shows that explanation of impulsive hard X-ray and microwave bursts in terms of a purely thermal process implies plasma temperatures exceeding those characteristic of the hottest thermal flare regions ($\sim 1 \times 10^7$ K) by one or two orders of magnitude. For this reason such impulsive events are often referred to as the *non-thermal* component of the flare. However, some workers have challenged this view and the question of whether the impulsive component is thermal or non-thermal in origin has not been entirely closed.

Flare loops play an integral and decisive role in the origin and physics of flares and our aim in the present chapter is to give a comprehensive account of their properties. Our account is based on the best modern space and ground-based observations which have been obtained in the visible, EUV, soft and hard X-ray and microwave regions of the spectrum (Sections 4.3–4.6). The complex spatial and temporal relationships between flare loops observed in the various widely-separated regions of the spectrum are gathered together in Section 4.7; relationships between loops observed in Hα and λ530.3 and between hard and soft X-ray emission are described in Sections 4.3.3 and 4.5.4 respectively. Diagnostic techniques for determining physical conditions are only touched upon lightly, but the extensive numerical results obtained in the various spectral regions are outlined in Section 4.8. Finally, the numerical data pertaining to the morphological and dynamical properties of flare loops and the physical conditions within them are summarized in Section 4.9 (Tables 4.16 and 4.17). These results are compared with similar data for cool and hot non-flare loops tabulated in Sections 2.6 and 3.8.

4.2   The role of loops in flares: an overview

Before immersing outselves in a detailed description of the properties of flare loops observed in different spectral regions, it is worthwhile to present a simplified and abbreviated overview of the role of loops in flares.

Flares appear to form two basic types, *simple-loop* flares and *two-ribbon* flares (Priest, 1981, 1985; Svestka, 1981). Two-ribbon flares are perhaps the more interesting since (a) most large flares fall into this category so that – with a given spatial resolution – the structure is more easily observed; (b) they are highly dynamic events; (c) they contain numerous loops.

A simple-loop flare, also known as a compact flare or subflare, is a small flare which consists of a single loop or collection of loops which simply brightens and fades, without movement or change of shape (Priest, 1982). It is observed both in Hα and in soft X-rays. Compact flares comprise some of the best flare observations obtained from Skylab; their properties have been described in detail by Moore *et al.* (1980, pp. 386–99). The main difference between compact and large two-ribbon flares is in the relative absence of continued heating and of loop system growth of a compact flare during the decay phase.

Two-ribbon flares rank as the largest and most energetic of all solar flares; a number of authors have given clear descriptions of the phenomenon, including Dodson-Prince & Bruzek (1977), Priest (1981, 1982), Pneuman (1981), Svestka (1981) and Loughhead, Blows & Wang (1985). Photospheric magnetograms show that the two characteristic ribbons of bright Hα emission lie on either side of a line of zero longitudinal field strength ($B_\parallel = 0$) threading the active region. Prior to the onset of the flare, this inversion line is usually occupied by a dark filament. Sometimes the two bright ribbons are straight, parallel and very similar in appearance. A beautiful time sequence of high resolution photographs of such an event has been published by Martin (1979, Fig. 1). Often, however, the structure seen in the Hα line is much more irregular (see, for example, Dodson-Prince & Bruzek, 1977, Fig. 9.2; Dwivedi *et al.*, 1984, Fig. 3). In such cases, however, the true two-ribbon nature of the flare can be established with the aid of maps of the longitudinal (photospheric) field, where the neutral line will be observed to separate the ribbons of flare emission.

The process of the onset of a two-ribbon flare has been described by

Svestka (1981). The process starts, quite some time before the onset of the flare, with changes in the photospheric magnetic field. On the occasions when the $B_\parallel = 0$ line is marked by a filament, the rearrangement of the field is signalled by the activation and ultimately disruption of the filament. (The phenomenon of pre-flare filament activation has been described in detail by Svestka, 1976, pp. 216–20.) The disruption marks the appearance of newly-formed loops visible in Hα which form an *arcade* spanning the neutral line. The footpoints of the loops are located in the ribbons which constitute the Hα flare. As the flare proceeds, the two ribbons are often seen to move apart with a velocity of 2–10 km s^{-1} (Dodson-Prince & Bruzek, 1977). The number of Hα loops that are visible at any given time varies, but Martin (1979) was able to distinguish at least 16 loops or (unresolved) groups of loops, whose duration of visibility varied from 6 to 23 min.

Early in the course of a flare a soft X-ray loop system is also observed; this system can outlast the optical event by many hours (Moore *et al.*, 1980; Pneuman, 1981). As time proceeds new, progressively higher, Hα loops are formed, their footpoints remaining rooted to the ribbons (Svestka, 1981). This process can last for hours, which explains why the Hα loops are often called 'post-flare' loops. During the decay phase of the flare the X-ray loops are also formed at successively greater heights, greater than those of the Hα loops (Dodson-Prince & Bruzek, 1977); the height of X-ray loops can extend to beyond 100000 km (Nolte *et al.*, 1979). As has been stressed by Pneuman (1981), the Hα and X-ray loop systems do not appear to consist of single loops rising upwards but rather of 'newly formed or activated stationary loops appearing at successively higher levels'. At heights similar to those of Hα and X-ray loops, loops are also observed in the EUV lines (MacCombie & Rust, 1979; Pneuman, 1981; Table 4.16).

The physical connection between the footpoints of the Hα loops and the chromospheric ribbons has been known for a long time. More recent observations (see Pneuman, 1981) show that the Hα footpoints are located on the *insides* of the ribbons, while the X-ray footpoints are rooted in the middle and outer portions of the ribbons. As Pneuman points out, this is consistent with the concept of a system of cool loops nested within and lower than a system of hot loops, a picture also adopted by Moore *et al.* (1980, p. 365).

It is of considerable importance to our understanding of the role of loops in flares to establish the site of the primary energy release. MacCombie & Rust (1979) found that the tops ('apices') of soft X-ray loops were considerably hotter than the legs, and that in each case the temperature difference was maintained for at least 8 hr, indicating a continual heating. Nolte *et al.* (1979) also found that the brightest emission was near the top.

Fig. 4.1. Loop prominence system photographed simultaneously in Hα (upper) and λ530.3 of Fe XIV (lower) at the Mees Solar Observatory, Haleakala (McCabe, 1973). Despite an obvious similarity between the two photographs, the cold and hot loops are *not* co-spatial (see text). Note that loop detail appears sharper in Hα than in λ530.3.

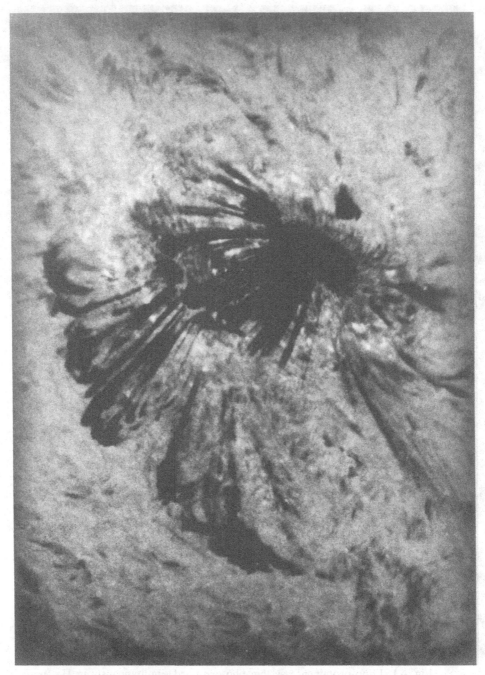

Fig. 4.2. Fine example of a cool post-flare loop system photographed during the late phase of the flare of 10 September 1974. This photograph was obtained at Sacramento Peak Observatory at Hα+0.08 nm (Svestka, 1976). Owing to the combined effects of geometry and the high axial velocity of the loop material, many of the numerous loops present appear incomplete.

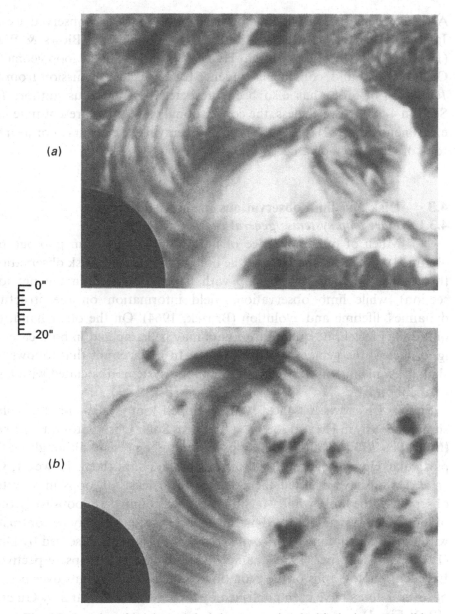

(a)

0"

20"

(b)

Fig. 4.3. Arcades of bright and dark loops associated with the large two-ribbon flare of 13 May 1981, photographed at Culgoora Solar Observatory (Loughhead, Wang & Blows, 1983). (a) Bright loops photographed at Hα line centre near their time of maximum development; their footpoints are lost in regions of intense flare emission. (b) Dark loops photographed at Hα+0.1 nm some 14 min later near their maximum development; note their termination in small bright patches ('kernels') marking the points of most intense emission.

A source height of 66000 km for hard X-ray emission observed by the Japanese Hinotori satellite was inferred by Loughhead, Blows & Wang (1985) from the location of bright Hα kernels and the likely loop geometry. On the other hand, convincing evidence for hard X-ray emission from the *footpoints* of loops has also been presented by numerous authors (see Section 4.5.3). Nevertheless, the conventional view at the present time (see, e.g., Pneuman, 1982) is that the *primary* energy release occurs at or near the tops of the flare loops, i.e. in the inner corona.

4.3 Disk and limb observations in the visible region
4.3.1 *Flare loop systems: general characteristics*

Much of our knowledge of flare loops comes from photographs obtained on the disk and beyond the limb in the Hα line. Disk observations throw light on their relationship with flares (discussed in the previous section), while limb observations yield information on the structure, dynamics, lifetime and evolution (Bruzek, 1964). On the other hand, the obtaining of accurate quantitative data may be hampered in both cases by ignorance of the geometry of the system. In the account that follows, we shall draw mainly on data pertaining to loop systems associated with large two-ribbon flares.

Figure 4.1 shows a typical well-developed loop system photographed simultaneously in Hα and λ530.3 at the Mees Solar Observatory, Haleakala (McCabe, 1973). A number of loops is seen to be present, although on this particular Hα photograph it is not possible to count them. However, Cui *et al.* (1985) have counted the presence of at least 10 loops in a system observed at the limb, while Martin (1979) has counted 16 loops (or groups of loops) in a system observed on the disk. These figures may be contrasted with the counts obtained for *compact* flare loop systems observed by Dizer (1978) and Kopriva (1976), namely 3 loops and 4–5 loops respectively. Excellent photographs of the evolution of such small systems over periods of several hours have been published by these two authors and by Gu *et al.* (1983, Fig. 1).

Good limb photographs such as Fig. 4.1 often reveal a marked blobby appearance ('knots') of the individual loops. This is well shown on limb photographs reproduced by Foukal (1978, Fig. 1), Albregtsen & Engvold (1979, Fig. 1), Chapman & Neupert (1979, Fig. 1), Ambastha & Bhatnagar (1985) and Cui *et al.* (1985, Plate 1), and on disk photographs of *bright* loops observed by Loughhead, Wang & Blows (1983) (see Fig. 4.3). The knots are very useful for measuring the component of the loop velocity in

Table 4.3. *Time of first appearance of a flare loop system*

Reference	t_1 (min)[a]	t_2 (min)[b]	Disk (D) or Limb (L)
Bruzek (1964)	~40[c]	—	L
	36–98	—	D
Roy (1972)	—	−13	L
Svestka *et al.* (1982)	31	14	D
Hiei, Okamoto & Tanaka (1983)	~30	—	L
Loughhead, Wang & Blows (1983)	51	4	D

[a] Measured from time of flare onset; [b] measured from time of flare maximum; [c] on occasion, small loops were observed as early as $t_1 = 8$–14 min.

the plane of the sky. They are less visible on disk photographs of *dark* loops.

A noticeable characteristic of many large loop systems is that the planes of the extreme members may be tilted at large angles to the solar vertical, e.g. up to 45° or so. This is well demonstrated by limb photographs published by a number of authors, including Jefferies & Orrall (1965, Fig. 1), Fisher (1971, Fig. 2), Roy (1972, Fig. 1g) and Svestka (1976, Fig. 9).

A remarkably fine off-band disk photograph (Hα+0.08 nm) of a large flare loop system is shown in Fig. 4.2, taken from Svestka (1976, Fig. 88). This photograph, obtained at Sacramento Peak Observatory, has also been reproduced by Pneuman (1981, Fig. 7.2) and Priest (1982, Fig. 10.18). At this wavelength the loops appear intensely dark against the chromospheric background. Most of them appear incomplete owing to the combined effects of their geometry and the high axial velocities present inside them (cf. Section 2.3.1). Once again it is hard to count the number of loops in the system; a rough estimate indicates at least 25. A photograph of the same event taken in the helium line D_3 is also reproduced by Svestka. It shows the bright flare ribbons to which a number of the dark loops are seen to be anchored.

Both on- and off-band (Hα+0.1 nm) photographs obtained at Culgoora of a loop system accompanying a large, two-ribbon flare are shown in Fig. 4.3 (Loughhead, Wang & Blows, 1983). The footpoints of the bright loops in Fig. 4.3(*a*) are lost in regions of intense flare emission, but the dark loops in Fig. 4.3(*b*) can be seen to terminate in small bright patches (kernels) marking the points of most intense flare emission.

Other photographs of (dark) flare loops – observed on the disk with varying degrees of spatial resolution – have been published by Bruzek

Table 4.4. *Time of maximum development of a flare loop system*

Reference	$t_1{}'$ (min)[a]	$t_2{}'$ (min)[b]	Disk (D) or Limb (L)
Bruzek (1964)	~120	—	L
Rust & Bar (1973)	282	—	D
Loughhead, Wang & Blows (1983)	74[c]	27[c]	D
	88[d]	41[d]	D

[a] Measured from time of flare onset; [b] measured from time of flare maximum; [c] bright loops.; [d] dark loops.

(1964, Fig. 3), Rust & Bar (1973, Fig. 3), Martin (1979, Fig. 1), Sturrock (1980, Frontispiece) and Kahler *et al.* (1984, Fig. 1). The high-resolution photograph reproduced by Sturrock, obtained at Hα line centre at Big Bear Solar Observatory, is interesting: it shows very thin low-contrast dark loops against the bright background of a large two-ribbon flare.

The time of the first appearance of a flare loop system and the time of its maximum development have been recorded by a number of observers using both disk and limb observations. These estimates are given in Tables 4.3 and 4.4 respectively. The figures are reasonably concordant, allowing us to reach the general conclusion that the first appearance occurs some 40 min after flare onset or, alternatively, within a few minutes before or after flare maximum. Maximum development of the loop system is achieved some 1 to 4 hr after flare onset, thus accounting for the (slightly misleading) terminology, 'post-flare' loop system.

The total lifetime of a large flare loop system is typically 12 hr, although considerably longer in some cases (Bruzek, 1964). Thus Cui *et al.* (1985) found that 14 hr elapsed between the eruption of a limb flare and the disappearance of the associated loops. Rust & Bar (1973) found that loops associated with a 2B flare observed at the limb persisted for 4 hr.

In a classic investigation, Bruzek (1964) studied 30 loop systems at the limb and 25 on the disk using Hα filtergrams from Sydney (CSIRO), Sacramento Peak Observatory and the Anacapri station of the Kiepenheuer Institute. An important characteristic property analysed with the help of the limb observations was the apparent rise of the system, which starts at the 'flash' phase of the associated flare. Bruzek emphasized that the loop system enlarges because successive loops appear at short intervals at ever-increasing heights, while the lower and older ones disappear: no individual loop is observed to expand. Figure 2 of Bruzek's paper gives height v. time curves for 6 systems. In all cases the initial (apparent) velocity is greater than at later times, Bruzek finding a velocity of >10 km s^{-1} in the first hour, then <5 km s^{-1}, and finally after several hours only 2 km s^{-1}. Similar height–time curves but displaying, in addition, the height variation of the lower envelope of the system have been published by Kureizumi *et al.* (1977, Fig. 3) and Martin (1979, Fig. 5).

Other workers have given figures for the apparent rise velocity which are broadly consistent with Bruzek's, although there is a considerable spread in individual values. For the initial velocity, figures have been quoted ranging from 3 km s^{-1} (Phillips & Zirker, 1977; Dizer, 1978) to 25 km s^{-1} (Kureizumi *et al.*, 1977). A maximum (not initial) velocity of ascent of 17.5 km s^{-1} has been measured by Gu *et al.* (1983). A system associated with a major white light flare observed by Deszo *et al.* (1980) also showed untypical behaviour, the velocity *increasing* from 5 to 68 km s^{-1} in the first 11 min. Most observers agree that after a few hours the rise velocity has reduced to 1–3 km s^{-1} (Bruzek, 1964; Roy, 1972; Phillips & Zirker, 1977). A good sequence illustrating the upward expansion of a typical flare loop system over a period of $\sim 2\frac{1}{2}$ hr has been published by Pneuman (1981, Fig. 7.3).

Closely related to the expansion of the loop system associated with a two-ribbon flare is the motion apart of the ribbons themselves (Roy, 1972; Svestka *et al.*, 1982). In one case observed by Svestka *et al.* the ribbons, after formation, moved apart with an initial velocity of 40 km s^{-1}. Thereafter the velocity declined to 10 km s^{-1} after 29 min and to 2 km s^{-1} after a further 2 hr or so. The separation of the ribbons is evidently consistent with the expansion of the associated loop system.

4.3.2 *Properties of individual loops*

The properties of individual loops in a flare loop system can be conveniently summarized under the following headings:

(1)   *Height, inclination of loop plane, tilt of axis of symmetry*

The height of loops observed at the limb in Hα has been measured by numerous authors. Although the exact geometry is usually unknown, the results have led to a consistent picture. Bruzek (1964) gives a range of values of 40000–150000 km, the most probable height being 60000–79000 km. Values agreeing with these figures have been quoted by numerous authors, including Fisher (1971) and Gu *et al.* (1983) (60000 km), Phillips & Zirker (1977) (40000 km), Engvold, Jensen & Andersen (1979) (60000–90000 km) and Ruzdjak (1981) (70000 km).

The occurrence of lower heights, on the other hand, is well attested in the literature, e.g. 34 000 km (Kureizumi *et al.*, 1977), 21 000 km (Dizer, 1978) and < 20 000 km (Martin, 1979). At the upper end of the range heights of 140 000–150 000 km have been measured in Hα by Cui *et al.* (1985) and in the λ530.3 line of Fe XIV by Roy (1972). These heights are characteristic of loops observed a number of hours after the onset of the associated flare.

It is of interest that in one loop system observed on the disk by Martin (1979) the loops developed through the same space as previously occupied by the associated *filament*, which ascended rapidly 5 min before the flare onset.

On the disk the height and other geometrical characteristics of three flare loops have been determined by Loughhead, Wang & Blows (1983) with the aid of a geometrical reconstruction technique. Besides heights in the range 45 000–60 000 km they found that the loop planes were inclined 1°–12° to the vertical. In a later paper, Loughhead, Chen & Wang (1984) took account of a possible tilt of the axis of symmetry to the vertical of one of their (symmetrical) loops. They found this to be 14°, requiring only small changes to the geometrical parameters deduced from their earlier reconstruction, which assumed no tilt. From limb observations in Hα (filtergrams) and in the H and K lines of Ca II (spectra), Engvold, Jensen & Andersen (1979) derived an inclination of the loop plane of 45° and a tilt of the axis of 5°–10°.

From disk observations Martin (1979) measured an azimuthal angle between the loops and the magnetic neutral line of 45°±8°, while Rust & Bar (1973) found that as the loops evolved they rotated progressively from a 20° angle to 90°–100°.

(2)   *Length, separation of footpoints, diameter*
Loops which reach a height characteristic of the upper end of the range can extend to a total length – measured along the central axis of the loop – of 290 000 km (Cui *et al.*, 1985). However, more typical lengths which have been quoted in the literature lie in the range 60 000–100 000 km (Phillips & Zirker, 1977; Gunkler *et al.*, 1984). Similarly, a typical footpoint separation lies in the range 35 000–55 000 km (Dizer, 1978; Loughhead, Wang & Blows, 1983) while an untypically large one might be as much as 180 000 km (Cui *et al.*, 1985).

The diameters (thickness) of loops have been estimated by a number of authors to be several thousand kilometres, but it is probable that these are overestimates resulting from inadequate spatial resolution and (unrecognized) merging of individual loops in the system. On high resolution Hα photographs (cf. Fig. 4.3) bright and dark flare loops are seen to have diameters in the range < 1500 km to ~2200 km (see also the Frontispiece to Sturrock, 1980). In one system observed by Cui *et al.* (1985), the average width was 3000 km.

(3)   *Lifetime*

On the disk Martin (1979) found that the average duration of individual loops (or unresolved groups of loops) observed in Hα ranged from 6 to 23 min, the average duration being 15 min. From his limb observations Bruzek (1964) found a lifetime for individual loops (and knots) of 33–83 min. The difference between the two sets of figures probably reflects the greater visibility of loops observed beyond the limb.

(4)   *Velocities*

It is characteristic of flare loops that the material motions along the central axis of the loop, i.e. the axial speed, are invariably observed to be directed downwards in both legs. Two methods are available in principle for determining its magnitude: (a) measurement of the motion of knots in either limb or disk loops, giving the component of the true velocity in the plane of the sky; and (b) measurement of Doppler shifts in appropriate lines of loops observed beyond the limb, giving the line-of-sight component of the axial speed.

Unfortunately, when the geometry of the loop is unknown (which is usually the case) it is not possible to derive an accurate value for the axial speed from such measurements. However, a number of authors have been able to conclude that their Doppler measurements are consistent with free fall under gravity from the loop apex along the field line of an (assumed) magnetic dipole (Engvold, Jensen & Andersen, 1979; Ruzdjak, 1981; Cui *et al.*, 1985). Ruzdjak's measurements imply a downward velocity at the photospheric level of 110 km s^{-1}.

From knots in loops observed on the disk, Martin (1979) deduced a (maximum) velocity of 90 km s^{-1} in the plane of the sky, while from Hγ spectra of loops observed at the limb Foukal (1978) – assuming the geometry – inferred an axial speed of $\gtrsim 45$ km s^{-1}.

The above figures imply that in flare loops, as in cool active region loops (cf. Table 2.8), the material motions are supersonic.

(5)   *Loop intensities*

It was remarked by Bruzek (1964) that the brightness of (bright) loops observed on the disk – especially their tops – was approximately flare brightness in most cases. This has been confirmed in one case observed at the limb by Malville & Schindler (1981), who compared absolute measurements of Hα loop intensity with the intensity of two disk flares near the loop footpoints, both as functions of time (see their Fig. 2). The loop intensity recorded by these authors, while still rising, was $I = 4.9 \times 10^4$ erg cm^{-2} s^{-1} sr^{-1}; Kureizumi *et al.* (1977, Fig. 7) measured a maximum Hα intensity across a limb loop of 4.0×10^5 erg cm^{-2} s^{-1} sr^{-1}. The latter authors also give simultaneous measurements of other Balmer lines.

4.3.3   *Relationship between loops observed in Hα and λ530.3*

A question of considerable interest is the spatial relationship between cool flare loops observed in Hα and hot flare loops observed in λ530.3 of Fe XIV. Fortunately there are a few observations bearing on this question. Fisher (1974) observed a flare loop system with the 40-cm coronagraph and universal spectrograph at Sacramento Peak Observatory, obtaining both graded height spectra and filtergrams in Hα and λ530.3 and also in λ637.4 of Fe X. He found that there were noticeable differences between Hα and λ530.3: for example, there was one region containing a bright λ530.3 loop but *no* Hα emission.

A special study of the question was carried out by McCabe (1973), using filtergrams and spectrograms obtained during the late phase of a flare loop system observed at Haleakala – the system illustrated in Fig. 4.1. She concluded that the hot and cool loops are not co-spatial. After measuring the positions of individual identifying features on her filtergrams and spectrograms, she arrived at a picture of '...a system of cool loops within and underlying a corresponding hot loop system, but closely related to it insofar as the individual loops are not entirely separate...'. Figure 5 of McCabe's paper shows a drawing of the Hα loop system superimposed on a λ530.3 filtergram. From excellent simultaneous eclipse photographs in Hα and λ637.4, Hanaoka & Kurokawa (1986) found that the hot loops were broader in diameter than the Hα loops and that they were located close to them, but slightly higher.

Loop detail seen in Hα usually appears sharper than in λ530.3, where the loops appear diffuse and more uniform, thus showing less structure than the characteristic knots visible in Hα. This difference is evident in Fig. 4.1 and in photographs published by Rust & Bar (1973, Fig. 5) and Chapman & Neupert (1979, Fig. 1). On the other hand, the exceptionally good λ530.3

photograph published by the latter authors shows hot loops that are no wider than the Hα loops belonging to the same system.

Loops were photographed in the He I D_3 line as well as in Hα and λ530.3 by Rust & Bar (1973). One 'family' of loops in the system appeared predominantly in D_3, not at all in Hα, and only faintly in λ530.3. A second family was mainly a λ530.3 phenomenon.

4.4 Observations in EUV lines
4.4.1 *Introduction*

A characteristic feature of the rising phase of an impulsive flare is the sudden enhancement in the level of EUV radiation (both line and continuum) – sometimes described as an 'EUV burst'. The increase occurs simultaneously over the whole EUV spectrum (Svestka, 1976, p. 164) and may affect lines ranging from cold ones like the Lα line of H I to intensely hot ones like the λ19.2 and λ25.5 lines of Fe XXIV (Widing, 1975). Many of the lines which have been used to obtain the EUV observations described in this chapter are listed in Table 4.5, where the third and fourth columns give respectively the ionization potential of the ion and the formation temperature of the line computed on the basis of the ionization equilibrium calculations of Shull & van Steenberg (1982). (The significance of the quantity G_{max} (fifth column) is discussed in Section 2.5.4.)

Three factors combine to make EUV observations of great importance to the study of the loop structure of flares. First, as shown in Table 4.5, EUV flare lines encompass almost the full range of temperatures encountered in flares, from as low as 20000 K to as high as 20000000 K. Second, the spatial resolution of a few arc seconds achieved by the best EUV spectroheliograms (Section 2.4.1) is much higher than that attained by any X-ray observations available at the present time (Section 3.4.1); Lα and UV continuum photographs taken from rockets have achieved even better resolution. Finally, by combining images taken simultaneously in lines spanning a wide temperature range one can sometimes obtain a more complete picture of the structure of a flare than that given by any one of the component images. This is important when studying flare loops because, as we shall see, the plasma near the top of a loop may be much hotter than that at the footpoints, particularly near the time of flare maximum and the early decay stage.

The Naval Research Laboratory EUV spectroheliograph on Skylab was designed with these considerations in mind (cf. Widing & Cheng, 1974) and yielded valuable information. For a variety of reasons, however, the flare observations from Skylab were mostly taken around or after the maximum

Table 4.5. *EUV lines used to obtain observations of flare loops*

Ion	Wavelength (nm)	Ionization potential (eV)	$\bar{T}_e$ (K)	G_{max}
H I	*121.57*	13.7	$\sim 2 \times 10^4$	
He II	25.63	54.4	$\sim 8 \times 10^4$	
	30.38			
C II	*133.45*	24.4	4.5×10^4	2.7×10^{-4}
	133.57			2.7×10^{-4}
C III	*97.70*	47.9	8.1×10^4	4.1×10^{-4}
O III	52.5	54.9	1.2×10^5	1.8×10^{-4}
O IV	*55.41*	77.4	1.9×10^5	3.7×10^{-4}
	140.12		1.8×10^5	8.7×10^{-4}
O V	137.13	113.9	2.5×10^5	5.7×10^{-4}
O VI	103.20	138.1	3.0×10^5	2.6×10^{-4}
Si IV	140.28	45.1	1.2×10^5	2.0×10^{-5}
Si XII	49.93	523.2	2.0×10^6	1.3×10^{-4}
	52.11			1.3×10^{-4}
Ne VI	43.32	157.9	4.3×10^5	3.5×10^{-4}
Ne VII	*46.52*	207.2	5.1×10^5	2.2×10^{-4}
Mg VII	*43.49*	224.9	6.5×10^5	3.7×10^{-4}
Mg VIII	*43.67*	266.0	8.1×10^5	3.1×10^{-4}
Mg IX	*36.81*	327.9	9.8×10^5	1.5×10^{-4}
Mg X	*62.53*	367.4	1.1×10^6	1.8×10^{-4}
Ca XVII	19.29	1087	4.9×10^6	6.5×10^{-5}
Fe XIV	*26.48*	392.0	2.7×10^6	2.1×10^{-4}
Fe XV	*28.42*	456.0	3.4×10^6	1.8×10^{-4}
Fe XVI	*33.54*	489.5	4.3×10^6	8.8×10^{-5}
Fe XXI	135.41	1685	9.8×10^6	7.7×10^{-5}
Fe XXIII	26.38	1950	1.3×10^7	7.4×10^{-5}
Fe XXIV	*19.20*	2045	1.6×10^7	9.3×10^{-5}
	25.51			9.4×10^{-5}

[a] Wavelengths in the second column referring to resonance lines (Section 2.5.4) are given in italic print; [b] For each line $\bar{T}_e$ is the temperature at which the function $G(T_e)$ defined by Eqn. (2.15) attains its maximum value, G_{max} (Section 2.5.4); [c] For more complete atomic data see Wiese, Smith & Glennon (1966), Wiese, Smith & Miles (1969) and Fuhr *et al.* (1981).

of the thermal phase (see, for example Cheng *et al.*, 1982; Poland *et al.*, 1982). More extensive information on the impulsive and pre-flare phases has come subsequently from the ultraviolet spectrometer and polarimeter (UVSP) on the Solar Maximum Mission satellite (SMM). This instrument was designed to operate over the wavelength range 115–360 nm and to have a spectral resolution of 0.002 nm in the second order of the grating (Woodgate *et al.*, 1980). Using pixel sizes ranging from 3″ × 3″ to 10″ × 10″, the UVSP obtained raster pictures of flares over areas on the Sun extending up to 256″ × 256″.

The UVSP was a modification of a unit built originally as a backup for the University of Colorado ultraviolet spectrometer on OSO 8, which also carried an ultraviolet and visible polychromator constructed by the French Centre National de la Recherche Scientifique (CNRS). However, observations with these instruments did not contribute significantly to our knowledge of the loop structure of EUV flares, partly because OSO 8 was flown during a period of low solar activity (June 1975 to October 1978) and partly because spatial resolution was sacrificed to compensate for a loss of instrumental sensitivity. On the other hand, OSO 8 added to our knowledge of the time variation of the EUV radiation from flares and its relationship to the corresponding variations of the emission in other spectral regions (Bonnet, 1981).

We shall now describe the properties of the EUV loops observed in flares and – to the extent that existing observations permit – explore their role during each of the four evolutionary stages of a flare: pre-flare, rise, maximum and decay.

4.4.2   *Loop structure of EUV flares*
(1)   *Pre-flare phase*

A question of considerable theoretical interest is the percentage of EUV flares which occur in a loop or loops visible before the event. The available data are far too meagre to allow a quantitative answer to this question, but present observations show that both cases occur.

An example of a low-lying loop becoming the site of a flare is shown in Fig. 4.4, due to Cheng *et al.* (1982). The figure shows raster pictures of the loop (footpoints '1' and '2') during the 20-min period just prior to the onset of the flare, taken in the cool EUV lines Si IV $\lambda 140.3$ and O IV $\lambda 140.1$ by the UVSP on SMM. The spatial resolution was 4″ of arc. At 02.47 UT the brightness of footpoint '2' doubled, whereas footpoint '1' was little affected. Five minutes later the whole loop began to brighten and by 02.59 UT was conspicuous in both the Si IV and O VI lines. The sudden onset of the importance 1B flare was marked by an eightfold increase in the brightness of footpoint '1' but there was little effect on '2' or on the two bright features labelled '3' and '4' in Fig. 4.4, which were visible throughout the period of observation. The pre-flare loop itself had an (apparent) width of 4″ and a length of 28″.

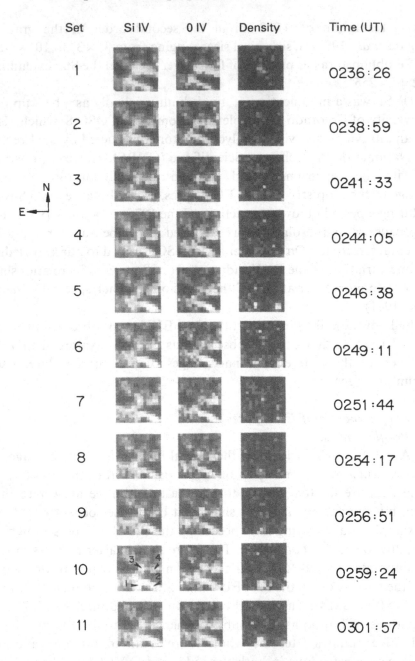

8 April, 1980 flare

Fig. 4.4. Low-lying loop observed in the cool EUV lines Si IV λ140.3 and O IV λ140.1 which became the site of a flare (Cheng *et al.*, 1982). The sequence of raster pictures illustrates the evolution of the loop over the 20-min period immediately prior to the onset of the flare at 03.02 UT. The numbers '1' and '2' identify the footpoints of the loop, while '3' and '4' label two other bright features visible throughout the period.

Using NRL spectroheliograms from Skylab Brueckner, Patterson & Scherrer (1976) have also studied pre-flare conditions in active regions, concentrating their attention on new emerging regions at the stage when they were producing many flares, albeit mostly small ones. They concluded that such highly flare productive regions are characterized by the presence of numerous, small, low-lying loops arranged in a chaotic pattern whose overlapping, however, makes it difficult to identify individual loops that may later become the site of a flare.

Turning to an active region with a lower rate of flare production, the same authors (Brueckner *et al.*, *loc. cit.*, Fig. 5) identified a small loop which became the site of a small flare of X-ray importance C2 and subsequently disappeared. There was little to distinguish this particular loop from the non-flaring loops in the same active region. Finally, mention may be made of a study by Schmahl *et al.* (1978) of the conditions preceding the onset of a SN,M1 flare which was intensively observed with the EUV and X-ray instruments on Skylab on 5 September 1973. They found that the flare appeared to have occurred in a pre-existing loop that was particularly prominent in the cool λ55.4 line of O IV.

In other cases the observations show no trace of a flare loop prior to the onset of the flare itself. For example, Cheng *et al.* (1981, Fig. 4) have illustrated the evolution of a small SN,C8 flare observed with the UVSP from the pre-flare to the post-maximum stages. Even a mere 18 s before flare onset there was no trace of impending activity at the location.

(2)   *Rise (impulsive) phase*

Most of our knowledge of the early development of EUV flares and, in particular, of the highly energetic impulsive phase has come from observations made with the UVSP instrument on SMM. The results are difficult to summarize because, in the words of Tandberg-Hanssen, Reichmann & Woodgate (1983), '...flares exhibit a nearly bewildering variety of impulsive-phase behaviour'. Moreover, because of differences in the light curves, a flare observed in a hot EUV line such as Fe XXI λ135.4 ($\bar{T}_{e} = 1 \times 10^{7}$ K) may look different when seen in a cool line such as O V λ137.1 ($\bar{T}_{e} = 2.5 \times 10^{5}$ K) (see Tandberg-Hanssen *et al.*, *loc. cit.*, Fig. 7).

The SMM observations show that, during the rise phase, the EUV emission from a flare originates mainly in small, localized bright areas called *kernels*. In some flares only a few kernels are seen, whereas in others many such bright points are visible. It is not uncommon for a flare kernel to appear in just one pixel of a raster picture and thus be only $\sim 3''$ or less in diameter (see, for example, Cheng, Tandberg-Hanssen & Orwig, 1984, Fig. 4).

Flare kernels observed in cool EUV lines are believed to mark the footpoints of loops or systems of loops joining areas of opposite magnetic polarity. This is supported by a comparison of EUV and Hα photographs of flares. One finds that the flare kernels observed in cool EUV lines are situated in areas of enhanced Hα emission lying on opposite sides of the neutral line in the underlying longitudinal magnetic field (see Tandberg-Hanssen *et al.*, 1984, Figs. 2(*a*) and 2(*b*); MacNeice *et al.*, 1985, Fig. 1). In most cases the footpoints are the only parts of the loops actually seen and the likely geometry of the loops has to be inferred from indirect evidence. Sometimes, however, the whole flare loop is visible, as in Fig. 4.4, a loop which spanned the neutral line in the photospheric field.

A second case of almost complete visibility has been described by Cheng (1977a), who analysed Skylab observations of a rather complicated SN flare near the west limb taken on 21 January 1974. The loop was conspicuous in the $\lambda 25.6$ line of He II and was also visible in other cool EUV lines such as Ne VI and Mg VII–VIII and, albeit only faintly, in the hotter Fe XIV and XVI lines. Its ends were anchored in bright He II flare kernels lying on either side of the neutral line separating two sunspots of opposite polarities. Over a period of about 4 min the loop underwent a 'wriggling' motion.

According to Levine (1978) the occurrence of localized bright patches of concentrated EUV flare emission is not restricted to the footpoints of loops. In the case of a small flare observed from Skylab on 28 November 1973 in Mg X $\lambda 62.5$ and other lines he found brightenings not only at the footpoints but also at various locations along the lengths of the loops involved (see, for example, Levine's Fig. 4(*b*)).

Woodgate *et al.* (1983) have combined observations of four small flares taken in the cool EUV line O V $\lambda 137.1$ and in Hα to estimate the lengths of the individual loops believed to have been involved. The inferred lengths range from 23 000 to 27 000 km, implying heights of between 7300 and 8600 km.

During the impulsive phase the brightness of individual kernels increases dramatically. In some cases all the kernels present reach peak intensity at the same time (cf. Cheng *et al.*, 1981), whereas in others the various kernels behave differently (Cheng *et al.*, 1985, Fig. 5). Moreover, the degree of enhancement may vary considerably from line to line. The light curves of a flare kernel recorded in cool EUV lines generally display a complicated and decidedly 'spiky' appearance which is marked by the presence of a

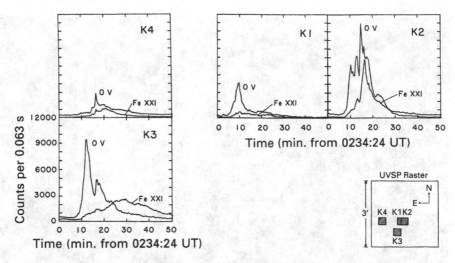

Fig. 4.5. Light curves of four bright kernels (labelled K1, K2, K3 and K4) observed in the EUV lines O V λ137.1 (cool) and Fe XXI λ135.4 (hot) during the impulsive phase of a 1B,M2 flare on 12 November 1980 (Cheng *et al.*, 1985). Note that the O V light curves display a decidedly 'spiky' appearance, whereas the Fe XXI curves show a more gradual rise and decay and reach their maxima later.

number of separate peaks. This is illustrated in Fig. 4.5, which shows the light curves obtained in the cool O V λ137.1 line by Cheng *et al.* (1985) for four bright kernels (labelled K1, K2, K3 and K4) in a 1B,M2 flare observed with the UVSP on 12 November 1980. As we shall see in Section 4.7.4, there is a close temporal correlation between the individual peaks in the EUV light curves of individual kernels and peaks in the curve of hard X-ray emission from the flare.

The situation changes when we turn from cool to hot EUV lines, such as the commonly used λ135.4 line of Fe XXI ($\bar{T}_e = 1 \times 10^7$ K). While the light curves in cool EUV lines tend to mimic the curve of hard X-ray emission, that recorded in Fe XXI tends to follow the rise and fall of the soft X-ray flux (Section 4.5.2). Compared to those obtained in cool lines, the Fe XXI light curves of individual kernels show a more gradual rise and decay and reach their maxima later. From the onset of a flare until at least midway into the impulsive phase the kernels observed in Fe XXI are found to be cospatial with those observed in cool EUV lines to within the accuracy of the available measurements. Later, however, the Fe XXI emission may come predominantly from the top of an (inferred) loop (Tandberg-Hanssen *et al.*, 1984).

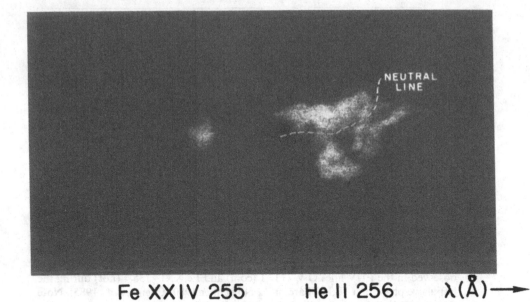

Fig. 4.6. Location of the very hot Fe XXIV $\lambda 25.5$ emission at the maximum phase of the large 1B,M3 flare of 15 June 1973 with respect to the ribbonlike pattern of cool He II $\lambda 25.6$ emission (Widing & Cheng, 1974). It can be seen that the Fe XXIV emission lies over the neutral line dividing the two He II flare ribbons, consistent with the view that the primary release of flare energy occurred at the top of a loop joining regions of opposite magnetic polarity.

(3)   *Maximum phase*

The NRL spectroheliograph on Skylab could obtain observations in resonance lines of iron at ionization stages ranging from Fe XVIII to Fe XXIV (Widing & Cheng, 1974). Of particular utility were the lines $\lambda 26.4$ of Fe XXIII and $\lambda\lambda 19.2$ and 25.5 of Fe XXIV, which have formation temperatures of 1.3×10^7 K (Fe XXIII) and 1.6×10^7 K (Fe XXIV) – temperatures characteristic of the very hot parts of flares. In fact, observations in Fe XXIV lines have provided useful information on the three-dimensional structure of EUV flares during the maximum phase.

The Skylab observations show that hot Fe XXIV emission is nearly always present at flare maximum and during the early part of the post-maximum (decay) phase. Figure 4.6 shows the large two-ribbon 1B,M3 flare of 15 June 1973, the ribbons being clearly delineated by emission in the cool $\lambda 25.6$ line of He II (Widing & Cheng, 1974; see also Cheng 1977b and Widing & Dere, 1977). Lying on the magnetic neutral line dividing the ribbons there is a small blob of Fe XXIV emission, consistent with the common view that the primary release of flare energy occurs at the top of a loop joining regions of opposite polarity. In this particular case the

Fe XXIV feature remained visible for at least 8 or 9 min (cf. Cheng, 1977b, Fig. 7). Its disappearance marked the time when the temperature of the hottest part of the flare dropped below 1×10^7 K.

Cheng & Widing (1975) reached similar conclusions in respect of a SN,M1 flare recorded on 5 September 1973. Once again the hot Fe XXIV plasma was situated over the neutral line while the He II emission was brightest where the longitudinal photospheric field was strongest. In this case the Fe XXIV (and Fe XXIII) emission vanished after 6.5 min. The space between the ribbons then became filled with radiation from cooler lines of He II and Fe XIV–XVI. The loop-like structure then became clearly apparent (see also Dere, Horan & Kreplin, 1977, Fig. 2).

Limb observations can provide further evidence of the loop structure of EUV flares; a good example is given in Fig. 4.7, which shows a small SN,C6 flare photographed at the west limb on 15 January 1973 (Cheng & Widing, 1975). The high temperature Fe XXIV plasma emission is concentrated in the small feature indicated by the arrow; it lies well below the tall active region loops visible in the (cooler) Fe XIV and Fe XVI lines. The Fe XXIV feature is some $3''$ above the image in the somewhat cooler line $\lambda 19.3$ of Ca XVII ($\bar{T}_e = 5 \times 10^6$ K). Cheng & Widing concluded that only the top of a low-lying flare loop is hot enough to radiate in Fe XXIV.

In all, these authors analysed Skylab observations of five flares and, in each case, concluded that the flare occurred in a low-lying loop joining regions of opposite magnetic polarity. The widths of the loops ranged from 4500 to 14000 km and their inferred heights from 4200 to 13000 km. Dere & Cook (1979) give figures of 2500 km and 10000 km for the widths and lengths respectively of loops observed in a number of EUV lines in a compact SN,M2 flare.

It should be pointed out that the flares studied by Cheng & Widing (1975) were all of (optical) importance 1B or less, so the results of their studies may not be applicable to larger flares. On the other hand, for the cases studied we can conclude that the energy release took place at or near the top of a low-lying loop and was marked by emission from the very hot Fe XXIII–XXIV lines.

(4)   *Decay phase: post-flare loops*

The fading of the very hot Fe XXIII and Fe XXIV emission from a flare is sometimes marked by the appearance of an apparently new system of loops, often referred to as 'post-flare' loops. However, as we have seen in Section 4.3, the term is a misnomer as the flare itself is well in progress when they start to appear. Moreover, as Brueckner (1976) has emphasized,

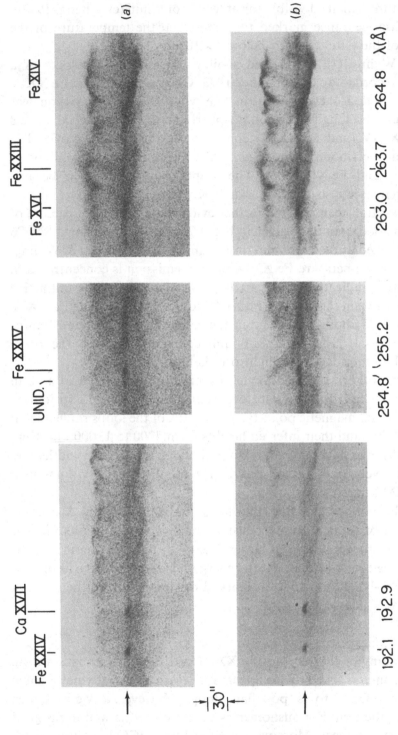

Fig. 4.7. Small SN,C6 flare observed at the west limb on 15 January 1973 in the hot EUV lines Fe XIV λ26.5, Fe XVI λ26.3, Fe XXIII λ26.4 and Fe XXIV λ19.2 (Cheng & Widing, 1975). The very hot Fe XXIV emission is concentrated in the small feature indicated by the arrow and lies some distance below the tall active region loops seen in the (cooler) Fe XIV and Fe XVI lines. Composite images obtained by mapping the Fe XXIII images in exposures (a) and (b) onto the adjacent Fe XIV and Fe XVI images reveal the presence of a low-lying flare loop, only the top of which is hot enough to radiate in Fe XXIV ($\bar{T}_e = 1.6 \times 10^7$ K).

1250 UT

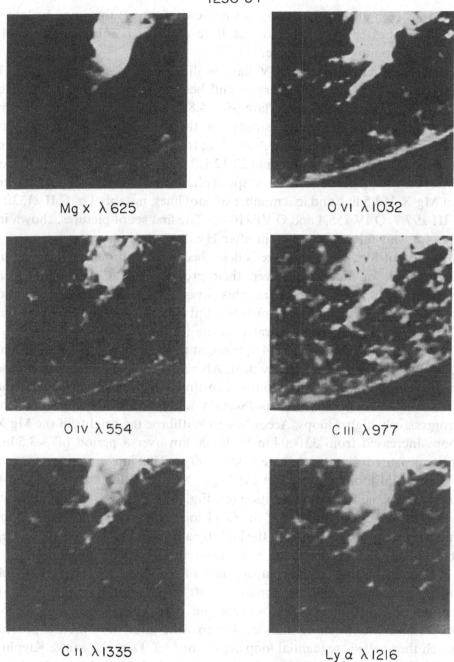

Fig. 4.8. Large 2B flare photographed from Skylab on 7 September 1973 (Withbroe, 1978). The pictures show the appearance of the flare during its decay phase in a number of cool EUV lines and in the hot λ62.5 line of Mg X. While some trace of the presence of flare loops can be seen in the O IV and O VI images, the emission in the cooler Lα, C II and C III lines is concentrated at the feet of the hot loops.

the loops seen at this phase may not be newly formed: they may be loops which are heated as a result of the flare and which subsequently cool, thereby becoming newly visible.

The decay phase of an EUV flare is illustrated by the 2B flare of 7 September 1973, one of the largest and best-observed during the Skylab period (Withbroe, 1978). This flare (Fig. 4.8) was of the classic two-ribbon type (Section 4.2) and emitted strongly in the radio, visible, EUV and soft X-ray regions of the spectrum. According to the Hα record the flare began at 11.41 UT, reached maximum at 12.12 UT and ended at 13.42 UT. It was observed with the Harvard EUV spectroheliometer (Section 2.4.1) in the hot Mg X λ62.5 line and in a number of cool lines, notably Lα, C II λ133.5, C III λ97.7, O IV λ55.4 and O VI λ103.2. The first set of pictures, shown in Fig. 4.8, was taken some 38 min after Hα flare maximum.

In the hot Mg X line the flare is described by Withbroe as consisting of bright loops or arches; however, their presence is not particularly well brought out in Fig. 4.8, taken from his paper. A trace of the loops can also be seen in the O IV and O VI pictures, but the emission in the cooler Lα, C II and C III lines is concentrated at the feet of the hot loops.

In the Mg X pictures of Fig. 4.9, taken at the later times of 14.11 UT and 15.45 UT, the loops are more evident. Also noticeable is a marked increase in the separation of the footpoints with time which, by analogy with the behaviour of Hα post-flare loops (Section 4.3.1), implies the appearance of progressively higher loops. According to Withbroe the heights of the Mg X loops increased from 33000 km to 46000 km over a period of ~ 3.5 hr; their measured diameters range from 3700 to 7500 km.

Individual loops observed in the Mg X and O VI lines appear to have nearly the same shape and position (cf. Fig. 4.9). By careful measurement, however, Withbroe found that the O VI loops lay within the Mg X loops and were displaced away from the limb by a few seconds of arc, suggesting that the Mg X loops extended a few thousand kilometres higher.

From their study of the compact flare of 9 August 1973 Dere & Cook (1979) concluded that the apparent shift in position of the region of maximum emission observed during the decay phase was due to the progressive activation of individual loops one after the other, a process which they called 'sequential loop activation' (cf. Dere, Horan & Kreplin, 1977; Widing & Dere, 1977). On this interpretation, well into the decay stage of the flare, new loops are being heated to temperatures high enough to radiate in Fe XXIII lines ($\bar{T}_e = 1.3 \times 10^7$ K).

A nice example of a long-lived, post-flare loop system has been described by Cheng (1980) and is illustrated in Fig. 4.10. The system was conspicuous

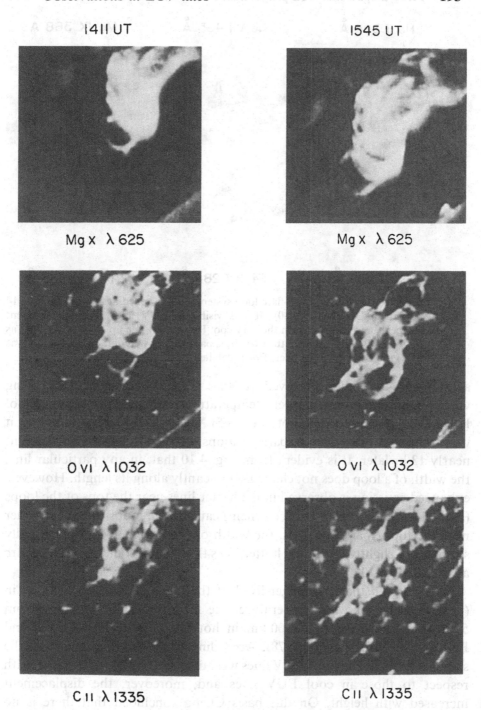

1411 UT

1545 UT

Mg x λ 625

Mg x λ 625

O vi λ 1032

O vi λ 1032

C ii λ 1335

C ii λ 1335

Fig. 4.9. Same flare as in Fig. 4.8, photographed at later times (Withbroe, 1978).
The presence of hot (post-) flare loops is now more apparent in the Mg X pictures.
Also evident is the marked increase in the separation of the footpoints with time,
implying the appearance of progressively higher loops (see text).

He II 304 Å   Ne VII 465 Å   Mg IX 368 Å

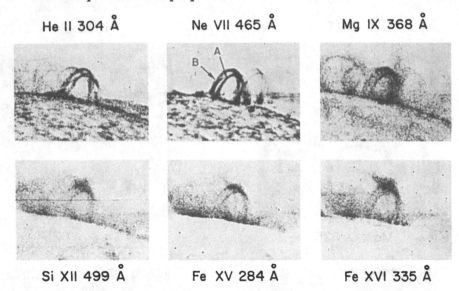

Si XII 499 Å   Fe XV 284 Å   Fe XVI 335 Å

Fig. 4.10. Long-lived, post-flare loop system photographed at the west limb on 14 August 1973 (Cheng, 1980). It was visible in EUV lines of highly different temperatures, ranging from the very cool He II λ30.4 line up to such hot lines as Si XII $\lambda\lambda$49.9, 52.1. One of the two large, parallel loops present on this picture was still visible nearly 13 hr later. For a detailed description see text.

at the west limb when observed at 00.51 UT on 14 August 1973, being visible in lines of highly different temperatures ranging from the very cool He II λ30.4 line up to such hot lines as Si XII $\lambda\lambda$49.9, 52.1. At this stage it was composed of two large, parallel loops, one of which was still present nearly 13 hr later. It is evident from Fig. 4.10 that, in any particular line, the width of a loop does not change significantly along its length. However, enhanced emission is observed in the hotter lines near the tops of the loops (as well as in the region between them) causing the loops to appear wider near the top. Apart from this, the width of the longer-lived loop actually seems to be slightly less in the hotter lines than in the cooler ones, 9″ of arc as opposed to 10″.

Cheng found that the longer-lived of the loops showed an increase in (projected) height with temperature, the measured values ranging from 57 000 km in Ne VII to 66 700 km in hot lines like Si XII, Fe XV and Fe XVI (see also Foukal, 1978). According to Cheng, the images of the same loop obtained in hot EUV lines were displaced by about 3″ to 10″ with respect to those in cool EUV lines and, moreover, the displacement increased with height. On this basis Cheng concluded that there is no evidence for the view advanced by Foukal (1975, 1978) that what is seen as a hot EUV loop is really a sheath of hot gas surrounding a cool EUV loop (cf. Section 3.7.3).

However, evidence contrary to Cheng's conclusion had earlier been published by Chapman & Neupert (1979) from an analysis of OSO 7 observations of a post-flare loop system. These showed clear loops in the cool $\lambda 36.8$ line of Mg IX but a diffuse blob seemingly filling the entire envelope of the loop system in the hot $\lambda 33.5$ line of Fe XVI. The same effect has already been noted (Section 3.3.2) in the case of non-flare active region loop systems.

Taken together, the EUV observations described above have added much to our knowledge of the properties of flare loops derived from data in Hα and other visible lines described in Section 4.3. Whereas visual observations refer primarily to the 'post-flare' loops characteristic of the maximum and decay phases, EUV observations are capable of yielding information about all four stages of a flare: pre-flare, rise, maximum and decay.

Active regions at the stage of producing many flares are characterized by the presence of numerous, small, low-lying loops arranged in a seemingly chaotic pattern. In rare instances it has been possible – after the event – to identify from the observations the individual loop or loops that became the site of a flare. In other cases no trace of the loop can be found prior to the flare.

Cool EUV loops about to become the site of a flare may show quite large brightness fluctuations on a time scale of a few minutes during the period immediately preceding the flare. The onset itself is marked by a sudden conversion (thermalization) of energy at one or more of the footpoints of the particular loops involved, giving rise to the bright flare kernels characteristic of the rising phase of a flare. Individual kernels may have diameters as small as 3″ of arc and the number present may vary markedly from flare to flare. Usually the kernels are the only parts of the loop actually seen at this initial stage, but occasionally virtually the whole of the flare loop can be discerned. In the latter case the loop is invariably found to link areas of opposite magnetic polarity within the active region.

As the flare approaches its maximum the brightness of individual kernels increases dramatically. At the same time extended areas of less intense emission appear on either side of the magnetic neutral line in EUV lines ranging from cool ones such as He II and Ne VII to hot ones such as Fe XV and Fe XVI. Sometimes all the kernels present attain peak intensity simultaneously, at other times they do not. Meanwhile emission in the hot EUV lines has also been increasing, albeit more gradually than in that of the cool ones. At the same time the source of the hot emission moves away from the footpoints up towards the top of the loop. The maximum and early post-maximum (decay) stages are almost always marked by the presence of emission in very hot lines of the ions Fe XXIII and Fe XXIV, implying temperatures of nearly 2×10^7 K. The site of the Fe XXIII and Fe XXIV radiation is found to lie over the neutral line of the active region, consistent with the view that the very hot emission arises near the top of the loop. By combining observations taken in a range of cool and hot EUV lines it is sometimes possible to construct composite pictures of the loop as a whole.

The fading of the very hot Fe XXIII and Fe XXIV emission from a flare is sometimes marked by the appearance of an apparently new system of loops, closely analogous to the Hα post-flare loops described in Sections 4.3.1 and 4.3.2. These loops are often best seen in moderately hot lines like Mg X λ62.5 and appear to lie above the loops visible in cool EUV lines. On the other hand, there is evidence of post-flare loops being heated to temperatures high enough to radiate in Fe XXIII well into the decay stage of the flare. As in the case of Hα post-flare loops, there is a marked increase in the separation of the footpoints of the EUV loops with time, implying the appearance of progressively higher loops. They attain heights of tens of thousands of kilometres and may persist long after the disappearance of the Hα flare.

The picture provided by the existing EUV observations may seem reasonably complete but is in fact deficient in at least two respects. First, the description is based on studies of a relatively small sample of flares, many of only minor importance (i.e. optical classification 1 or less). Their behaviour may by no means be typical of *all* flares. Second, EUV observations provide no information on regions of the flare hotter than 2×10^7 K, where the most energetic processes may be expected to occur. To probe these ultra-hot regions we now turn to observations made in the X-ray portion of the spectrum.

4.5 Observations in X-rays
4.5.1 *Introduction*
The first soft X-ray photographs of a flare of (relatively) high resolution were obtained by Vaiana and his colleagues at AS&E on 8 June 1968 (see Section 1.5). They used a grazing-incidence telescope carried aloft by a rocket launched a few minutes after the start of a two-ribbon Hα flare of importance 1N. In X-rays the flaring region was more than an order of magnitude brighter than any other active region present and, according to Vaiana & Giaconni (1969), showed detail on the original negative as small

as 2″ of arc. Their photographs indicated a close resemblance between the X-ray and Hα flare images; diffuse X-ray emission overlay the two Hα flare ribbons and tended to be brighter at places where the Hα intensity was greatest. But there was also an important difference: the presence in X-rays of a bright central 'core' linking the two ribbons across the magnetic neutral line (Vaiana & Giaconni, *loc. cit.*, Fig. 14).

Consistent with these results, Vorpahl *et al.* (1975) concluded that in almost all soft X-ray flares the bulk of the radiation comes from a compact core surrounded by a larger and less well-defined area of fainter emission from the associated active region. In many cases the core was visible on longer-exposure photographs taken before the start of the flare. It was found that generally the core does not change its shape significantly during the lifetime of the flare; rather, it evolves by first brightening and then fading and becoming more diffuse.

The question of whether or not the core of an X-ray flare consists entirely of arcades of loops or loop-like structures is subject to dispute in the literature. Basing their results on an examination of 132 soft X-ray flares observed with the Aerospace Corporation/Marshall Space Flight Center (AC/MSFC) telescope on Skylab, Vorpahl *et al.* (1975) concluded that '. . . Practically all core features with one dimension greater than 10″ had a loop-like or filamentary appearance, and many of those smaller than 10″ could clearly be resolved as tiny loops'. It was further claimed that some of the apparently 'linear' structures observed within flare cores, referred to as 'loops', turned out on closer examination to be '. . . an entire arcade of closely spaced loops or the superposition of several overlapping, partially aligned smaller loops'.

The claim that many flare loops are actually composed of arcades was disputed by Kahler (1978). After an examination of a sample of 45 flares observed during the rise phase with the AS&E X-ray telescope on Skylab, he reported that he had '. . . not found evidence that any of the events . . . are composed of the kinds of arcades shown in Figure 1a and 1b of Vorpahl *et al.*'. We also conclude that the photographs of Vorpahl *et al.*, as reproduced, do not adequately substantiate their claims. On the other hand, as Gibson (1977) has pointed out, there is no substitute for direct scrutiny of the first or second generation films which, because of their greater dynamic range, may well show structure that is lost in reproduction.

There is a large literature on X-ray flares. In the following section we shall proceed to outline the basic properties of flare loops observed in soft X-rays ($\lambda \simeq 0.1$–10 nm), basing our account mainly on those papers containing more or less clear evidence of the loop phenomenon. As in our

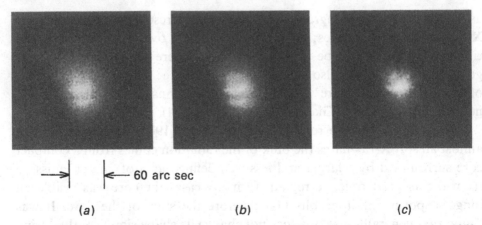

<- 60 arc sec

(a)   (b)   (c)

Fig. 4.11. Soft X-ray pre-flare loops observed from Skylab on 1 September 1973 (Petrasso *et al.*, 1975). (*a*) and (*b*) show the loops at two different times prior to the flare (22.47 UT and 23.01 UT respectively), while (*c*) (23.04 UT) – taken with a shorter exposure – brings up the core of the flare itself.

account of the loop structure of EUV flares (Section 4.4.2), we shall deal separately with the pre-flare, rise, maximum and decay phases.

4.5.2 Loop structure of soft X-ray flares

(1) *Pre-flare phase*

The conditions prevailing in active regions during the 20 min immediately preceding the onset of a flare were investigated by Kahler & Buratti (1976) and Kahler (1979) using observations obtained with the AS&E soft X-ray telescope on Skylab. At least one-half of the flares studied were preceded by transient brightenings, but only a small fraction of these were actually located at the site of the subsequent flare. These findings were confirmed and extended by Webb (1985), again using photographs obtained with the AS&E instrument. During the 30-min period prior to flare onset Webb observed 'definite' brightenings in 16 out of the 23 cases considered, but again only a small fraction were located at the flare site. In each of the papers referred to above the pre-flare brightenings are described as having mostly a loop-like or kernel-like appearance. However, it is sometimes hard to identify well-defined loops as such on the published photographs, perhaps owing to the low contrast of the loops and the small scale of the reproductions (cf. Kahler & Buratti, 1976, Figs. 2–5; Kahler, 1979, Figs. 1–3; Webb, 1985, Figs. 4, 5 and 7).

Figure 4.11 illustrates an interesting event described by Petrasso *et al.* (1975), in which three compact loops spanning the magnetic neutral line in a small bipolar active region became the site of a subflare. The three loops gradually increased in brightness over a period of approximately 10 min leading up to the start of the flare (23.04 UT, 1 September 1973). The left-hand and middle photographs show the loops at two different times prior to the flare, while the right-hand photograph, taken with a shorter exposure, brings up the core of the flare itself. Photometric measurements showed that the brightest point in the flare coincided with the brightest point in the pre-flare loop system to within the accuracy of registering the photographs, $\pm 3''$.

(2) *Rise phase*

An attempt to summarize the evolution of X-ray flares from the rise through to the decay phase was made by Kahler, Krieger & Vaiana (1975) on the basis of observations obtained with the AS&E soft X-ray telescope on Skylab. From an examination of a sample of 12 small flares ranging in X-ray importance from C1.2 to M1.3 (Table 4.1) they reached the conclusion that '... each of the 12 flares ... is morphologically very different from every other flare'! The truth of this remark is apparent when one scrutinizes Kahler *et al.*'s Table 1, which summarizes the evolutionary history of each event. As in the case of EUV flares (Section 4.4.2), the rise phase of soft X-ray flares is characterized by the appearance of a number of small, very bright kernels. At the same time one or more pre-existing loops may begin to brighten or, alternatively, one or more *new* loops may appear and increase in intensity. A kernel may either be an apparently isolated feature or may be seen as a footpoint of a flare loop. Photometric measurements performed on two of the flares indicated that the true width (FWHM) of a kernel could be $<6''$ of arc (4500 km).

Vorpahl (1976, Fig. 2) has published a sequence of photographs illustrating the evolution of a flare observed on 5 September 1973 in an active region which was a prolific producer of flares. During the rise phase the elongated core of the flare displayed non-uniformities in brightness and, in fact, appeared to consist of a set of crosswise striations running at right angles to the axis of the core. The striations were 4–5″ long and were interpreted as evidence for the presence of an arcade of loops. Individual striations (or loops) were observed to brighten sequentially, implying the movement of some form of excitation along the arcade at an apparent speed of between 180 and 280 km s^{-1}. Other evidence of sequential loop activation has been mentioned in our discussion of EUV flare loops (Section 4.4.2: decay phase).

Limb observations provide clear evidence of the loop structure of soft X-ray flares; descriptions of such observations, obtained with the AS&E and AC/MSFC telescopes on Skylab, have been given by Pallavicini, Serio &

Vaiana (1977) and Gibson (1977) respectively. In both studies the authors concluded that the majority of flares appear to consist essentially of a loop or system of loops, embedded in a background of faint diffuse emission, although their presence is often hard to detect on the published illustrations. However, Fig. 4.12 is an exception: here the photograph in the upper right-hand corner shows an exceptionally well-defined loop, photographed during the rising phase. Close examination of the fainter (northern) leg suggests that the loop may in fact consist of two closely spaced loops. According to Pallavicini *et al.* the loop (or loops) was $\sim 15\,000$ km high and its footpoints were separated by $\sim 30\,000$ km. A second illustration of a well-defined flare loop, this time located a short distance inside the limb, is to be found in the paper by Gibson (1977, Fig. 1).

Photographs obtained with the AS&E telescope on Skylab have been used by Kahler (1978) to measure the diameters and lengths (uncorrected for foreshortening) of 61 flare loops observed during the rise phase of 45 separate flares. Corrections were applied to the measured values to take account of the effect of the instrumental profile of the telescope. The results are summarized in Kahler's Fig. 2; they imply values of $\sim 4''$ (3000 km) and $\sim 20''$ (15\,000 km) for the typical diameter and length respectively of a flare loop during the rise phase.

(3)   *Maximum phase*

All authors agree that loops form the principal structure of soft X-ray flares observed at the maximum phase. For example, in describing his analysis of 26 limb flares Gibson (1977) remarks that '... One or more well-defined loops were the only structures of flare intensity observed during the rise phase and near flare maximum...'. The number of (resolved) loops varies. An M2 flare observed by Underwood *et al.* (1978) showed 2 to 3 loops; according to Vorpahl *et al.* (1975) the typical core consists of 2 resolved loops, with a single loop being the next most frequent case. Sub-flares also show a loop-like structure (Kahler *et al.*, 1975).

The maximum phase is characterized by an increase in brightness – to its maximum value – of the flare core and the loop structures comprising it. Vorpahl *et al.* (1975) found that normally an individual feature in the core varies only in *brightness* during the flare's evolution, maintaining its shape throughout the duration of the event. It follows that observations obtained during the period around maximum accurately describe the event. Gibson (1977) agreed that in many cases there is no major geometrical change in the loops during the evolution of the flare, while Pallavicini *et al.* (1975) suggested that the system of inner loops in the flare core seen at the

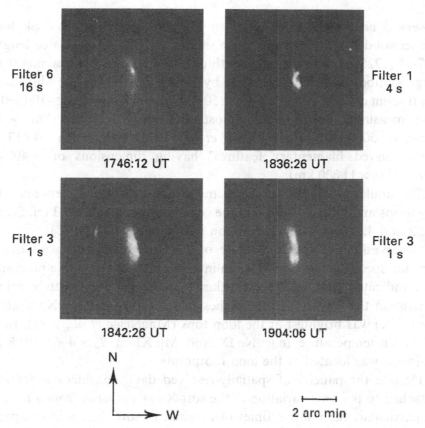

Fig. 4.12. Soft X-ray limb event photographed on 29 August 1973 (Pallavicini, Serio & Vaiana, 1977). The image at the top right shows an exceptionally well-defined loop, photographed during the rise phase. The image to its left shows the pre-flare phase of the event, while the lower right image shows the post-flare phase.

maximum phase is the same loop system as that present at the end of the impulsive phase.

In addition to loops, Gibson (1977) claimed to have seen thin vertical 'spikes' and triangular 'cusps'. However, Pallavicini, Serio & Vaiana (1977), also using limb observations, were not able to confirm these features. Photographs and contour maps showing with various degrees of clarity the loop structure of soft X-ray flares are to be found in the papers cited above and in the papers of Krall *et al.* (1978), Pallavicini & Vaiana (1980) and Cheng & Pallavicini (1984).

Several estimates are available for the lengths and widths of soft X-ray flare loops observed near the time of flare maximum. A single loop with footpoints anchored to the two brightest Hα regions present was observed from SMM by Cheng & Pallavicini (1984): this exceptionally large loop had a length of 100 000 km and a width of 20 000 km. The main loop observed near the maximum of an M2 flare observed from Skylab by Underwood *et al.* (1978) was much smaller, having an extimated length of 10″–15″ (7250–11 000 km) and width of only 1″–2″ (!). Measurements for a large number of flares were made by Vorpahl *et al.* (1975), who found a continuum of lengths; most were < 50 000 km but a few were > 100 000 km. The measured widths were almost exclusively < 15 000 km, with a mean of 5000 km. Finally, Krall *et al.* (1978) observed a 1B,M3 flare with 'curved filamentary features' having dimensions of ~ 40″ × 15″ (29 000 km × 11 000 km).

The similarity, both in geometry and in size distribution, between soft X-ray loops and loops observed in the coronal green line λ530.3 (cf. Sections 3.2.2 and 4.3.3) was remarked upon by Vorpahl *et al.* (1975).

Observations of the M1.4 flare of 12 November 1980 with the Flat Crystal Spectrometer on SMM, an imaging instrument with a pixel size of 15″, indicated that in the early flare phases and immediately prior to maximum the emission in the highest temperature line (Fe XXV at $\bar{T}_e \simeq 2 \times 10^7$ K) was brightest at the loop tops (MacNeice *et al.*, 1985). In lines of lower temperature (e.g. Ne IX and Mg XI at $\bar{T}_e \simeq 4$–6×10^6 K) the emission was located at the loop footpoints.

Despite the paucity of spatially-resolved data, considerable interest is attached to the time variation of the soft X-ray emission from a flare and, in particular, the relative times of maxima in soft X-rays in comparison with other spectral regions. Non-imaging flux data have been provided by the NRL satellite Solrad 9 over the bands 0.1–0.8 nm and 0.8–2.0 nm and the X-ray Event Analyser (X-REA) aboard Skylab (0.2–2.0 nm). The 'classical' curve shows a very steep rise to a relatively sharp maximum, followed by a smooth, slow decline (see, e.g., Pallavicini & Vaiana, 1980, Fig. 1; Vorpahl *et al.*, 1975, Fig. 4; Pallavicini *et al.*, 1975, Fig. 1). In other cases, due to different parts of the flare rising in brightness at different times or to the effect of other active events that may be present on the solar disk, the curves are irregular or show multiple peaks (e.g. Kahler *et al.*, 1975, Fig. 1; Vorpahl *et al.*, 1975, Fig. 7).

Figure 4.13 shows the light curves of four different 2″ × 2″ pixel elements of the 1B two-ribbon flare of 15 June 1973 observed from Skylab (Pallavicini *et al.*, 1975, Fig. 10). It will be noticed that there are differences in the times of maximum extending up to 10 min or so. Also shown is the total flux of the flaring region derived from the Skylab film; this curve is broadly similar to the simultaneous Solrad 9 flux curve (*loc. cit.*, Fig. 1).

Spatially-unresolved data obtained from Solrad 9, the X-REA aboard Skylab, and the Bent Crystal Spectrometer aboard SMM indicate that the

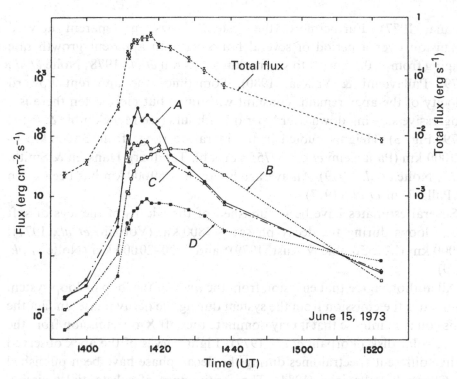

Fig. 4.13. Soft X-ray light curves corresponding to four different 2″ × 2″ elements of a 1B two-ribbon flare observed from Skylab (Pallavicini *et al.*, 1975). The curve at the top is the total flux integrated over the flaring region. Note the differences in the times of maximum for the various locations and the steepness of the rise to maximum.

soft X-ray maximum appears to lead the Hα maximum by periods ranging from zero to about 7 min (Pallavicini & Vaiana, 1980; Pallavicini *et al.*, 1975; Cheng & Pallavicini, 1984, Fig. 1; Krall *et al.*, 1978, Fig. 5). Observations from SMM obtained by MacNeice *et al.* (1985) and Cheng & Pallavicini (1984) indicate that the hard X-ray maximum precedes the soft X-ray maximum (lines of Ca XIX) by about 2–5 min. The maximum in the (hotter) lines of Fe XXV precedes that in Ca XIX by 1–2 min.

Light curves corresponding to individually-resolved points of soft X-ray flare loops are not available at the present time. It is likely that they would show some of the characteristics of the spatially-averaged curves described above.

(4)   *Decay phase: post-flare loops*

During the decay phase the loop system present at the rise and maximum phases retains its identity but becomes progressively fainter and more diffuse (Kahler *et al.*, 1975; Vorpahl *et al.*, 1975; Pallavicini, Serio &

Vaiana, 1977). Furthermore, the system shows an apparent upward expansion over a period of several hours with an apparent growth rate ranging from ~ 0.5 km s^{-1} to several km s^{-1} (Krall *et al.*, 1978; Nolte *et al.*, 1979; Pallavicini & Vaiana, 1980). Sometimes the apparent upward velocity of the apex remains constant with time, but more often there is a progressive slowing down over a period of hours (see MacCombie & Rust, 1979, Fig. 3). Heights quoted in the literature range from 35000 km to 180000 km (Pallavicini *et al.*, 1975; Vorpahl, Tandberg-Hanssen & Smith, 1977; Nolte *et al.*, 1979). An average figure of ~ 50000 km has been given by Pallavicini *et al.* (1977).

Several estimates have been published of the widths of the legs of soft X-ray loops during the decay phase: ~ 7800 km (Vorpahl *et al.*, 1977); 12000 km (MacCombie & Rust, 1979); and 7000–20000 km (Nolte *et al.*, 1979).

All authors agree that emission from the *apex* of the loop or loop system dominates the emission from the system during the decay phase. In fact the emission is so intense that it may dominate the soft X-ray emission from the entire solar disk (Vorpahl *et al.*, 1977)! Light curves of the apex observed in five different spectral lines during the decay phase have been published by Cheng & Pallavicini (1984). The continuance of substantial emission from the apex, in some cases for several hours following flare maximum, has led several authors to suggest that continuous heating of the apex occurs during the decay phase (e.g. MacCombie & Rust, 1979). Sometimes during this phase the apex material appears to spread out along the neutral line (Pallavicini *et al.*, 1975).

As in the earlier phases, the footpoints of the loops are anchored to the bright Hα ribbons of a two-ribbon flare or to points of opposite magnetic polarity on either side of the neutral line (Nolte *et al.*, 1979; Pallavicini & Vaiana, 1980). In the famous flare of 29 July 1973 – the best-observed of the Skylab flares – the footpoints of the X-ray loops were cospatial with the brightest patches of the Hα ribbons (Svestka *et al.*, 1982). Various authors have estimated the distance between the footpoints, but the results are very dependent on the (usually unknown) geometry and therefore not really worth quoting. Pallavicini *et al.* (1975) noticed a progressive increase in the footpoint separation following the maximum of the 1B two-ribbon flare of 15 June 1973; they attributed this to the appearance of successively higher loops. On the other hand, Vorpahl *et al.* (1977) observed a *decrease* in the separation of the legs (from 275000 km to 93000 km), which they explained as a probable change in perspective due to changing magnetic field geometry.

The decay phase of a flare loop system observed in soft X-rays can be very extended indeed. For example, the loops observed by Vorpahl *et al.* (1977) (flare of 13–14 August 1973) were visible for more than 50 hr following flare maximum, while Nolte *et al.* (1979) (flare of 29 July 1973) observed a decay phase lasting 15 hr. It has been suggested by Pallavicini *et al.* (1977) that systems that endure for more than 3 hr or so constitute a special class characterized by greater height as well as long lifetime. Four such events, with lifetimes up to 72 hr, have been studied by MacCombie & Rust (1979), who use the term 'long-decay enhancement' (LDE) in describing them.

Convincing disk and limb photographs of soft X-ray loops and loop systems obtained during the decay phase have been published by a number of authors, including Pallavicini & Vaiana (1976, Fig. 3), Vorpahl *et al.* (1977, Fig. 1), MacCombie & Rust (1979, Fig. 1), Nolte *et al.* (1979, Fig. 1) and Kahler *et al.* (1984, Fig. 1). The illustration of Kahler *et al.* is of particular interest because it shows, in addition to an X-ray loop, an Hα photograph of the flaring region, a map of the 6 cm microwave emission and a magnetogram. The Hα photograph shows a well-defined loop prominence system of the kind described in Section 4.3.1, which is broadly cospatial with the (broad) X-ray loop.

Figure 4.14 shows a reproduction of the limb photographs of MacCombie & Rust. The photograph on the right was taken with four times the exposure of that on the left and shows evidence of the presence of several loops, i.e. an arcade. Detailed contour maps of a very well-defined loop at the early decay phase of a 2N flare observed from SMM have been published by Cheng & Pallavicini (1984, Fig. 5).

All of the LDEs observed by MacCombie & Rust (1979) were associated with Hα loop prominence systems (Section 4.3.1). In reviewing the possible relationship between the two systems MacCombie & Rust suggest that an LDE is essentially the high-temperature/low-density X-ray counterpart of the corresponding system observed in Hα. A similar identification has been made by Pallavicini & Vaiana (1980).

4.5.3 *Properties of hard X-ray (HXR) bursts*

The observation of HXR bursts from the Sun has been dominated by two satellites, the U.S. Solar Maximum Mission (SMM) and the Japanese Hinotori satellite. SMM was launched on 14 February 1980 near the peak of cycle 21. Its instruments cover wavelengths ranging from γ-rays to the visible and it has been described as the most powerful package ever flown for studying solar flares (Pallavicini, 1984). After nine months of

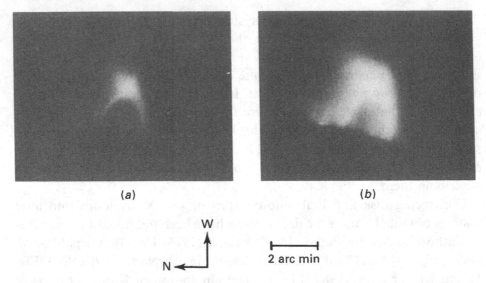

(a)   (b)

W

N

2 arc min

Fig. 4.14. Soft X-ray images of flare loops observed at the limb on 13 August 1973 during the decay phase (MacCombie & Rust, 1979). Photograph (*b*) was taken at the same time as (*a*), but with four times the exposure. (*a*) shows clearly only a single loop, whereas (*b*) reveals clear evidence of an arcade of loops. Note the brightness of the loop apex in (*a*) compared to the legs.

successful observations SMM's attitude control system failed, thus interrupting the operation of the fine-pointed instruments. However, in April 1984 a rescue mission carried out successful repairs and the restored spacecraft (SMM 2) again became operational. More than 300 flares were recorded during the Solar Maximum Year (1980) (Rust, Simnett & Smith, 1985).

There are seven experiments on SMM, three of which observe the full Sun: (a) Gamma-Ray Experiment (GRE); (b) Active Cavity Radiometer Irradiance Monitor (ACRIM); and (c) Hard X-ray Burst Spectrometer (HXRBS). The last of these covers the range 28–490 keV and is similar to instruments flown on OSO 5 and OSO 8 (cf. Sections 1.6 and 4.4.1).

The four imaging experiments are as follows: (a) Ultraviolet Spectrometer and Polarimeter (UVSP), which includes a polarimeter for measuring magnetic fields and is a modification of the unit on OSO 8; (b) Coronagraph/Polarimeter (C/P) for observations in the visible region (Hα, λ530.3 and 445–659 nm) with a resolution of 10″; (c) Soft X-ray Polychromator (XRP) consisting of two instruments – Bent Crystal Spectrometer (BCS) for the range 0.17–0.32 nm, and Flat Crystal Spectrometer (FCS) providing spectroheliograms in seven lines in the range 0.14–2.24 nm (resolution, 14″); and (d) Hard X-ray Imaging Spectrometer (HXIS), giving imaging in six bands in the range 3.5–30 keV with a spatial resolution of 8″ or 32″. HXIS was the first instrument in space to provide images in hard X-rays up to 30 keV.

The Japanese satellite Hinotori ('Firebird') was launched in February 1981 and is dedicated to X-ray observations of flares, no less than 675 of which were recorded during its first sixteen months of operation (Tanaka, 1983a). The Hard X-ray Imaging Telescope (SXT) on Hinotori operates in the range 17–40 keV; the effective resolution quoted by various workers who have analysed the observations ranges from 10″ to 15″. Other instruments on Hinotori include a Solar Gamma-ray Detector (SGR); two non-imaging Hard X-ray Monitors (HXM and FLM) covering the range 2–340 keV; and a Soft X-ray Crystal Spectrometer (SOX) covering the band 0.18–0.19 nm with high spectral resolution.

Virtually all of the observations of HXR bursts described in the present section were obtained with the HXIS and HXRBS instruments on SMM and the equivalent instruments on Hinotori. In view of the huge amount of material obtained from these satellites, it is not surprising that there is a large literature on the subject, including a number of reviews and workshop proceedings. Good reviews of the SMM results have been published by Simnett (1983) and Dennis (1985), while short reviews of the Hinotori results are to be found in Tanaka (1983b) and Takakura, Tanaka & Hiei (1984). A much lengthier survey by Tanaka (1987) covers both hard and soft X-ray observations. A more specialized review by Trottet (1986) deals with the relative timing of HXR bursts and microwave bursts and other radio phenomena observed at different phases of a flare. A short review by Hénoux (1984) discusses energy release in flares, with special reference to hard and soft X-rays, Hα profiles and microwave emission. An excellent review of pre-SMM and pre-Hinotori observations of HXR bursts and their interpretation was published some years ago by Brown (1975).

In the account that follows we start by systematically describing the basic properties of HXR bursts, dealing successively with location in (inferred) loop, height, size, time profiles, classification, relationship with magnetic field, and pre-cursor and post-maximum phenomena. We include a brief account of the physical processes involved. In the next section (Section 4.5.4) the spatial and temporal relationships between HXR bursts and soft X-ray emission are outlined. Relationships at other wavelengths – EUV, microwave and Hα – are to be found in Section 4.7. Finally, physical conditions derived from analyses of observations in hard and soft X-rays are discussed in Section 4.8.4.

(1)   *Location in loop; height, size*

The extensive body of observations of HXR flare bursts obtained from SMM and Hinotori have almost invariably been interpreted by writers on the subject in terms of a loop or a system of loops. On the other hand, actual images showing HXR emission from the *entire extent* of a loop are relatively rare. In some cases, of course, the associated loop is delineated by images at other wavelengths. Nevertheless, our use of the word 'loop' in this section may often imply no more than an *inferred* or *notional* loop.

The observations indicate that HXR bursts may occur at the footpoints of a loop, at or near its top, or at intermediate locations. Location at one or more footpoints – particularly during the impulsive phase – has been well established by a number of workers (Duijveman, Hoyng & Machado, 1982; Simnett & Strong, 1984; MacNeice *et al.*, 1985; Wang *et al.*, 1987) from both disk and limb observations for both large and small flares (Rust *et al.*, 1981). Typically, but not always, the HXR emission shifts to the top of the loop after the impulsive phase (Duijveman *et al.*, *loc. cit.*; Simnett & Strong, *loc. cit.*; Wang *et al.*, *loc. cit.*). On the other hand, reviewing Hinotori observations of many flares, Tanaka (1986) has remarked that a footpoint source is seen in only 10% of events. Similarly, Machado, Rovira & Sneibrun (1985) have emphasized that the great majority of SMM and Hinotori flares showed no HXR footpoints.

The Hinotori observations have led Japanese workers to conclude that, particularly in the case of very intense HXR bursts, the bursts are located along the ridge of a loop arcade (Takakura *et al.*, 1984a, b; Tsuneta *et al.*, 1984b). Tsuneta *et al.* (1983) observed a limb source which maintained a height of 50000 km for 17 min, while Hoyng *et al.* (1981a) observed HXR emission patches which changed position during the course of a flare. Pictures showing unmistakable loop-like structures in HXR have been published by Kosugi & Tsuneta (1983); one of these is reproduced in Fig. 4.15. Tanaka (1986) has in fact concluded that HXR emission over the whole of a flare loop is common in the Hinotori results.

A number of authors have measured or estimated the height of the HXR emission associated with flares, using both disk and limb observations. Excluding HXR footpoint emission, for which the height must evidently be <2000 km or so, the results show a wide variation, as the following figures demonstrate: 6000–10000 km (Haug, Elwert & Hoyng, 1984); 14000 km (Kosugi & Tsuneta, 1983); 14000 and 30000 km (Takakura *et al.*, 1983); 15000 and 30000 km (Sakurai, 1985); 20000 km (Takakura *et al.*, 1984b); 27000–45000 (Svestka & Poletto, 1985); 40000 km (Tsuneta *et al.*, 1984b);

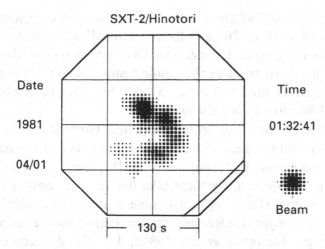

SXT-2/Hinotori

Date

1981

04/01

Time

01:32:41

Beam

─── 130 s ───

Fig. 4.15. Loop-like HXR burst observed from Hinotori in the energy range >17 keV (Kosugi & Tsuneta, 1983). This event is classified by the Japanese workers as an 'extended outburst'. Despite its appearance, the apparent single loop may be the top parts of several loops in an arcade.

and 50000 km (Tsuneta *et al.*, 1983). In compact flares the heights are of course smaller; figures of 3000 and 6000 km have been quoted by Tsuneta *et al.* (1984a).

The first measurements of the altitude structure of HXR sources at energies >100 keV were obtained by Kane *et al.* (1982), using a stereoscopic technique involving the International Sun Earth Explorer 3 (ISEE-3) and Pioneer Venus Orbiter (PVO). Comparison of light curves from instruments on these satellites showed that most of the high-energy emission came from heights of less than 2500 km.

There are relatively few estimates of the lengths of the loops associated with HXR flare emission. However, Rust, Simnett & Smith (1985, Table 1) give estimated lengths ranging from 45000 to 250000 km for eight flares whose X-ray classification ranged from C5 to M4 (their table also gives corresponding projected loop lengths). From limb observations Haug, Elwert & Hoyng (1984) found that following the last HXR peak, the *soft* X-ray emission came from a linear strip of length ∼35000 km.

The size of an HXR source is frequently comparable to that of a single pixel, so there are few reliable measurements for the smaller sources. An estimate of ∼6000 km for the thickness of a loop assumed to be associated with HXR bursts has been quoted by Duijveman, Hoyng & Machado (1982) – a figure comparable with the 8″ resolution of the HXIS instrument on SMM. From Hinotori observations of a 2B,X1.5 two-ribbon flare (claimed resolution, 15″), Tsuneta *et al.* (1984b) estimated an HXR source size of $\sim 70'' \times 30''$ ($\sim 51\,000 \times 22\,000$ km²). The source was coincident in size and position with a soft (5–10 keV) X-ray source. For the 2B,X1 flare of 21 May 1980 (described by de Jager & Svestka (1985) as one of 'the best observed and best analysed flares in the whole history of solar research') the footpoints or kernels observed at $\gtrsim 15$ keV – but also detectable at lower energy – had widths of 3000–10 000 km.

Disk and limb images obtained from SMM and Hinotori in the low HXR range (i.e. ~ 10–40 keV) are to be found in a number of the papers cited above and also in papers by Hoyng *et al.* (1981b), Kundu *et al.* (1984), and Takakura *et al.* (1986). Usually these take the form of contour maps of the HXR emission and show single or double sources in the form of a blob, which is often elongated. Occasionally the maps show a loop-like appearance (see, e.g., Takakura *et al.*, 1984b, Fig. 3). A particularly convincing example of this category – but, of course, not necessarily typical – is shown in Fig. 4.15.

The question of a single versus a double source in HXR flares is of considerable importance because of its bearing on identification of the physical processes responsible for the emission (see below). Discussions of this question with reference to earlier observations have been published by Takakura *et al.* (1984a) and Wang *et al.* (1987), who also describe their own observations of single sources. The weight of the evidence suggests that single or double sources can occur both at the footpoint or footpoints of a loop (and therefore at the chromospheric level) and at levels in a loop corresponding to the corona (see, e.g., Tsuneta *et al.*, 1983; Sakurai, 1985). A second fundamental question is whether flares occur in a single loop or whether several loops are involved. Existing observations are inadequate to provide an answer, but an interesting paper by Harrison & Simnett (1984) explores the theme that 'all flares occur within a hierarchy of magnetic loops'.

(2)   *Time profiles; classification*

The non-imaging Hard X-ray Burst Spectrometer (HXRBS) on SMM and the equivalent instrument on Hinotori (HXM) have provided a considerable amount of data on the variation with time of the hard X-ray emission from flares in various bands ranging up to several hundred keV. From the numerous time profiles published by various authors, the temporal characteristics of HXR burst are readily summarized.

All bursts show a steep rise to maximum, followed by a more gradual decay. A typical rise time is 3–4 min (Hoyng *et al.*, 1981a, Fig. 3; Hoyng *et al.*, 1981b, Fig. 3; Tsuneta *et al.*, 1983, Fig. 4; Tsuneta *et al.*, 1984b, Fig.

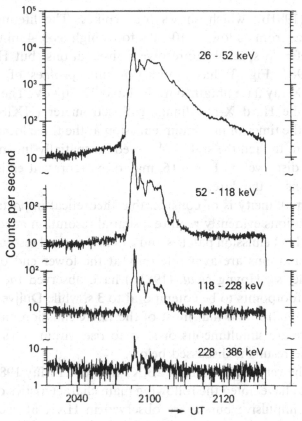

Fig. 4.16. Hard X-ray time profiles (non-imaging) for the flare of 21 May 1980, observed with the Hard X-ray Burst Spectrometer on SMM (Hoyng *et al.*, 1981b). Note the steep rise but slower decay and the presence of a number of distinct spikes.

1). However, values of $\lesssim 1$–2 min are also frequently observed (de Jager & Boelee, 1984, Fig. 2; Takakura *et al.*, 1984b, Fig. 1; Lin & Schwartz, 1987, Fig. 1 (balloon observations); Wang *et al.*, 1987, Fig. 2(*a*)). An exceptionally fast rise time of only ~ 40 s is exhibited by a curve covering the range 29–511 keV published by Simnett & Strong (1984, Fig. 1). The total duration of a burst or cluster of bursts, derived from the references cited above, varies from ~ 3 min to ~ 30 min. There is no marked correlation between the steepness of the rise and the total duration. The steepness of the decline from maximum appears to increase with energy in the HXR region of the spectrum.

Some time profiles show a relatively smooth rise, maximum and decay (cf. Hoyng *et al.*, 1981a; Tsuneta *et al.*, 1984b), but more often there are multiple peaks or spikes, which can occur both during near the maximum and during the decay phase. A good example is shown in Fig. 4.16, taken from Hoyng *et al.* (1981b), which shows four spikes. The lifetime of individual peaks varies from as low as 10–30 s to as high as 3–4 min.

The above data refer to spatially unresolved observations, but Haug, Elwert & Hoyng (1984, Fig. 3) have published time profiles of three separate points of an X-ray flare obtained in the band 22–30 keV. The data were obtained with the Hard X-ray Imaging Spectrometer (HXIS) on SMM and show that the times of maximum emission at the three locations differed by up to 1 min. In fact, the spikes observed on spatially-integrated curves, such as those displayed in Fig. 4.16, may often represent emission from different locations in the flare.

The question of simultaneity is of considerable theoretical importance. The necessary observations evidently require a spatial resolution adequate to resolve individual HXR emission patches and a high temporal resolution. At present such observations are available only at the lower end of the HXR range. Nevertheless, Hoyng *et al.* (1981b) have observed the time profiles of two HXR footpoints to be cotemporal to 3 s, while Duijveman, Hoyng & Machado (1982) found that most of the HXR emission patches associated with a flare are simultaneous or start to rise within ~ 5 s. The interpretation of these results is discussed below.

In a discussion of the very well-observed 2B,X1 flare of 21 May 1980, de Jager & Svestka (1985) have noted the following main characteristics of the impulsive phase: (a) impulsive bursts are observed in HXR at energies $\gtrsim 20$ keV and in microwaves at $\nu \gtrsim 10$ Ghz; (b) kernels or loop footpoints are present, best visible at energies $\gtrsim 15$ keV but also detectable at lower energies; (c) there are violent upward motions; (d) there is an increase in the thermal energy content of the plasma, a high-temperature component being present during the very first part of the impulsive phase.

Analysis of HXR images obtained with the imaging instrument on Hinotori (SXT) has led the Japanese workers to conclude that HXR flares can be divided into three categories (Sakurai, 1985). Type A are compact and show smoothly-varying HXR emission, attributed to a hot thermal plasma. Type B – impulsive flares – are in the majority and show several impulsive spikes in the time profile. Type C – gradual flares – show a less spiky time profile with a long duration; other names given to this category are 'gradual hard burst' and 'extended burst'. A detailed description of the three types has been published by Tanaka (1986), while characteristics of Types A and C have been listed by Tsuneta *et al.* (1984a) and Takakura *et al.* (1984b) respectively. A lengthy study of 10 gradual hard bursts has been published by Cliver *et al.* (1986). The Japanese classification has some interest in that it illustrates the diversity of the phenomena (and perhaps of their physical origin) associated with HXR flare emission. However, de Jager & Svestka (1985) – in describing a flare having both Type B and Type C characteristics – have questioned the usefulness of such a simplified classification scheme.

(3)   *Relationship with magnetic field*

Only meagre information is available about this potentially important topic. Hoyng *et al.* (1981a, Figs. 1 and 4) compared HXIS images of two flares with magnetograms from NASA's Marshall Space Flight Center, but made no definite conclusions. Also using observations from the MSFC vector magnetograph, Machado (1985) found that the site of the HXR emission in a highly flare-productive region observed from SMM showed a significant correlation with magnetic *shear*. He suggested that magnetic shear may be a necessary, but not always sufficient, condition for the high energy release which causes HXR bursts.

Finally, Sakurai (1985) computed three-dimensional configurations of the magnetic field and compared the resulting maps with HXR images of five flares. In one case – a 2B,X1.5 two-ribbon flare – he found that the field line calculations confirmed an earlier conclusion (based on Hα data) that the HXR source lay along the top of a magnetic arcade.

(4)   *Pre-cursor and post-maximum phenomena*

From a study of HXR images of flares obtained with the HXIS, Machado *et al.* (1984) found that in a majority of cases there was pre-HXR burst (excess) emission in the 11.5–30 keV band. They examined the period 5 min before the HXR burst – as defined by HXRBS time profiles – and found that the excess emission was highly localized ($\sim 8''$). Pre-cursor phenomena observed in soft X-ray are described in Section 4.5.2.

An excellent description of the post-maximum growth of an HXR loop system following the well-observed flare of 21 May 1980 has been given by Svestka & Poletto (1985). The growth was not a continuous process: on the contrary, the tops of the loops appeared to stay for a few minutes at a constant height, and then suddenly new loops appeared above them. The tops of the higher loops were first seen in the 22–30 keV channel, and only at lower energies minutes later. This suggested a new energy release in a relatively small volume high in the corona and late in the post-maximum phase, i.e. 16 min after HXR flare maximum. In a lengthy review of the same flare de Jager & Svestka (1985) draw the analogy with the discontinuous growth of post-flare loops observed in the Hα line (cf. Section 4.2).

(5)   *Electron beam precipitation and other physical processes*

Before proceeding to describe the observed relationships between HXR flare emission and emission in other spectral regions, it is useful to outline some of the physical processes which have been inferred from the HXR observations. Three possible sources for the HXR emission have been listed by Duijveman, Hoyng & Machado (1982) and Doyle *et al.* (1985): hot thermal plasma emission, thick target emission and thin target emission. Following a detailed discussion of the thick target mechanism, Duijveman *et al.* concluded that '...by far the most likely and simple explanation is that the observed impulsive hard X-ray emission from the footpoints is thick target bremsstrahlung from fast electrons precipitating downwards along the magnetic loops into the dense atmospheric layers'.

Evidence for energy deposition at the footpoints is provided by the observed temporal and spatial coincidence, or near-coincidence, of the HXR and Hα flare emission (Gunkler *et al.*, 1984; MacNeice *et al.*, 1985; Kurokawa *et al.*, 1986). This evidence is supported by observations in the D_3 line of He I and in certain EUV lines (Hoyng *et al.*, 1983; Tandberg-Hanssen *et al.*, 1984). Further support for the thick target model is provided by the stereoscopic observations of Kane *et al.* (1982), which located the HXR impulsive emission at heights $\lesssim 2500$ km. The generally accepted picture has been embodied in a useful schematic diagram by Tanaka (1986, Fig. 7) which shows a bi-directional electron beam resulting from energy release at the apex of a loop, together with the various emissions to be found at or near the footpoints.

Additional support for the thick target electron beam model has been provided by certain theoretical studies. For example, numerical simulations by Takakura (1986) of the effect of injecting non-thermal electrons with energy > 10 keV into a series of model loops were successful in reproducing some of the observed characteristics of HXR bursts. Moreover, McClymont & Canfield (1986) found that the model satisfactorily reproduced both the observed EUV/HXR flux ratio and the observed (small) time difference between the EUV and HXR peaks.

On the other hand, a detailed analysis of HXR footpoint spectra of three flares during the impulsive phase by MacKinnon, Brown & Hayward (1985) failed to support the conventional thick target beam interpretation: the footpoint photon and electron fluxes were too small and their spectra were too soft.

Using HXR imaging of three flares, Machado, Rovira & Sneibrun (1985) sought to discriminate between electron beam and thermal heating. In the case of one event they came to the firm conclusion that a large part of the emission was in fact due to thermal bremsstrahlung from a high temperature source or sources. They asserted that *all* flares should show a thermal component which, however, may be difficult to isolate from the mixture of processes occurring during the impulsive phase. The thermal component is supposed to originate in a region of magnetic reconnection.

During the decay phase soft X-ray observations sometimes clearly show the upward transfer of material into the top of a loop (Doyle *et al.*, 1985). The upward expansion of dense cool plasma is caused by energy deposition in the chromosphere. The process is usually given the name 'chromospheric evaporation' but the somewhat better word 'ablation' is used by Simnett & Strong (1984) and Karpen, Doschek & Seely (1986). In a flare studied by the latter authors electron beam-induced ablation was thought to play only a minor role. In agreement with this conclusion Gunkler *et al.* (1984) calculated that chromospheric ablation due to a non-thermal electron beam was 2–3 times less efficient than that caused by thermal *conduction*.

Rust (1984) has also concluded that most of the flare energy reaching chromospheric levels does so by thermal conduction rather than by energetic electron beams. Actual observations of fast-moving thermal conduction fronts on HXIS images of eight flares have been reported by Rust, Simnett & Smith (1985). They were generally found to occur during the rise phase and appeared to travel along (inferred) magnetic flux tubes with velocities ranging from 300 to 1700 km s ^{1}.

A well-balanced and critical review of the electron beam and other mechanisms for producing HXR emission has been published by Vilmer (1987).

4.5.4 Spatial and temporal relationships between hard and soft X-ray flare emission

Owing to limited spatial resolution and to difficulties in registering the images, no very clear picture emerges concerning the spatial relationship between hard and soft X-ray flare emission. Duijveman, Hoyng & Machado (1982) found that the soft and hard X-ray emission regions do *not* coincide, the former lying between patches of the latter which they identified with the footpoints of flare loops. Wang *et al.* (1987) also found that the core of the hard X-ray source was located at a footpoint, but only in the early phase. In the main and later phases it shifted to the top of the loop, where the soft X-ray emission was located. Studying a 2B,X1.5 flare, Tsuneta *et al.* (1984b) found that the soft and hard X-ray sources approximately coincided in position and size.

Simultaneous time profiles in hard and soft X-rays reproduced by a number of authors demonstrate that the peak of the former invariably leads that of the latter by a small but easily measurable amount. Measurement of the published curves indicates a lead varying from ~0.5 min to 3.5 min (e.g. Hoyng *et al.*, 1981a, Fig. 3(*a*); Duijveman *et al.*, *loc. cit.*, Figs. 1 and 3; Kundu *et al.*, 1984, Fig. 6; Tandberg-Hanssen *et al.*, 1984, Fig. 1(*a*); Doyle *et al.*, 1985, Fig. 1; Karpen, Doschek & Seely, 1986, Fig. 3).

The intensities of individual soft X-ray lines of Ca XIX, Ca XX and Fe XXV also show the characteristic lag behind the hard X-ray peak. However, it is significant that these lines also show a measurable blue shift which occurs well *before* the hard X-ray peak, indicating upward-moving material (Karpen *et al.*, 1986). Antonucci *et al.* (1982) had earlier observed large blue shifts in lines of Ca XIX and Fe XXV coincident in time with bursts observed over the range 52–386 keV.

Further relationships between flare and/or loop emission observed in different regions of the spectrum are to be found in Section 4.7.

4.6   Observations at radio wavelengths
4.6.1   *Introduction*
Observations in the microwave region of the radio spectrum enable us to approach closely, in space and in time, the primary flare event – the injection of energetic electrons into a magnetic loop at or near its apex. In fact, modern observations in the 2–20 cm region (15–1.5 GHz) have provided a considerable corpus of knowledge concerning the properties of individual impulsive bursts/loops (Section 4.6.2); the relationship between bursts and corresponding Hα and X-ray flare emission (Sections 4.7.3 and 4.7.5) and the physical conditions within the (radio) loop near the time of primary energy release (Section 4.8.5).

One- or two-dimensional observations of microwave bursts are listed in Tables 4.6 and 4.7. All of the two-dimensional observations were obtained with the Very Large Array at Socorro, New Mexico, a telescope which has also been successful in observing non-flare radio loops (cf. Section 3.5.1). The spatial resolution (beam shape) achieved, amounting to only a few sec of arc in most cases, is shown in the last column of Table 4.7. The one-dimensional observations listed in Table 4.6 have employed other radio interferometers in addition to the VLA and have achieved a time resolution as high as 0.1 s (Cornell *et al.*, 1984).

The quantities usually mapped in the case of two-dimensional observations are the total intensity $I = \frac{1}{2}(\text{LH} + \text{RH})$ and the 'polarization' $V = \frac{1}{2}(\text{LH} - \text{RH})$, where LH and RH are the left- and right-hand circularly

Table 4.6. *One-dimensional observations of microwave bursts*[e]

Reference	Telescope	Wavelength (cm)	Frequency (GHz)
Alissandrakis & Kundu (1978)	WSRT[a]	6	5
Marsh *et al.* (1979)	VLA[b, c]	6.1	4.9
Marsh *et al.* (1980)	VLA[b]	6.1	4.9
Willson (1983)	WSRT[a]	6	5
Cornell *et al.* (1984)	OVRO[d]	2.8	10.6

[a] Westerbork Synthesis Radio Telescope; [b] Very Large Array; [c] Partially completed; [d] Owens Valley Radio Observatory; [e] See also Kundu (1984).

Table 4.7. *Two-dimensional observations of microwave bursts*[a]

Reference	Wavelength (cm)	Frequency (GHz)	Resolution[b] (sec of arc)
Marsh & Hurford (1980)	1.3, 2	15.05, 22.5	0.75 × 1.0
Kundu *et al.* (1982)	6	5	—
Velusamy & Kundu (1982)	6	5	2 × 3
Hoyng *et al.* (1983)	2	15	0.6 × 0.6
Willson (1983)	2	15	2.2 × 3.1
	6	5	5.8 × 7.4
	6	5	4.2 × 5.5
	20	1.5	0.1–13.6
	20	1.5	0.1–6.4
Kahler *et al.* (1984)	6	5	15 × 19
Willson (1984)	17.6, 21.7	1.4, 1.7	3 × 4
Willson & Lang (1984)	17.6–21.7	1.4–1.7	3 × 4
Schmahl *et al.* (1985)	6	5	8 × 8
Shevgaonkar & Kundu (1985)	2	15	1.6 × 2
	6	5	3 × 5
Takakura *et al.* (1985)	6	5	6 × 7.5
Velusamy *et al.* (1987)	6	5	4 × 8
	20	1.5	12 × 18

[a] All observations obtained with the Very Large Array; [b] or synthesized beam shape.

polarized intensities respectively. The ratio $V/I = \rho_{CP}$ is the degree of circular polarization. 'Snapshot' maps of I and V with a time resolution of 10 s have been achieved by some of the observers listed in Table 4.7.

Three separate phases can be distinguished in impulsive microwave bursts (see, e.g., Kundu, 1983). The impulsive phase itself may contain several peaks, each of which may last for less than 1 min (see below). This phase is preceded by a slower rise in the emission intensity, called the 'pre-flare' or 'pre-burst' phase, and is followed by a decay called the 'post-flare', 'post-burst' or 'gradual' phase.

Two-dimensional maps of I and V obtained at 6 cm and/or 20 cm of both the pre-flare and impulsive phases have been published by a number of authors, including Kundu, Schmahl & Velusamy (1982), Velusamy & Kundu (1982), Willson (1983, 1984), Willson & Lang (1984) and Schmahl, Kundu & Dennis (1985). In some cases the I and V radio maps are accompanied by photospheric magnetograms and Hα photographs of the flaring region. Maps of the LH and RH circularly polarized intensity obtained at 2 cm during the impulsive phase have been compared with Hα photographs by Hoyng *et al.* (1983), while the post-flare phase has been mapped at 6 cm and 20 cm by Velusamy & Kundu (1981), Kahler *et al.* (1984) and Kundu (1986). A workshop summary and review of possible pre-flare effects observed at microwave and other wavelengths prior to the impulsive phase has been published by Van Hoven & Hurford (1984). In the account that follows, on the other hand, we shall concentrate mainly on the impulsive phase.

4.6.2 *Properties of individual bursts/loops*

The observations obtained with the VLA and other interferometers have given us a reasonably good picture of the properties of individual microwave bursts and their relationship to the loops in which they occur. These properties may be described under the following headings:

(1) *Location of burst in loop*

Observations at 6 and 20 cm show that bursts commonly occur near the top of a loop but may also occupy a significant portion of the legs (Kundu, Schmahl & Velusamy, 1982; Willson & Lang, 1984; Willson, 1984). A burst observed beyond the limb by Schmahl, Kundu & Dennis (1985) showed emission only along the legs; the absence of emission from the top was attributed to the circumstance that the orientation of the loop was such that at the apex the magnetic vector **B** was closer to the line of sight than in the legs.

Several authors have stressed that on theoretical grounds one may expect the appearance of a loop to depend on its orientation and, in particular, on the orientation of **B** with respect to the line of sight (Petrosian, 1982; Alissandrakis & Preka-Papadema, 1984). The latter authors have also

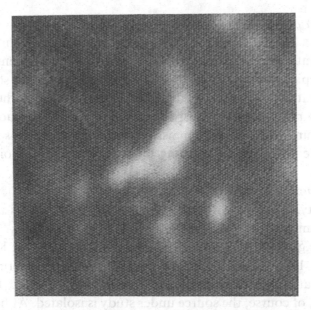

Fig. 4.17. Photographic representation of a 10 s 'snapshot' map of an impulsive burst observed at 6 cm (Velusamy & Kundu, 1982). The photograph depicts the total intensity I at a time close to the peak of the burst, which was associated with a 2B optical flare. Note that the apex of the loop-like structure is brighter than the legs.

explored the effect of wavelength, finding that at small λ emission can be expected only from the feet, at large λ from the entire loop. However, at the present time there are not enough multi-wavelength observations available to test this result.

An excellent example of a radio loop observed at the impulsive phase at 6 cm is shown in Fig. 4.17, taken from Velusamy & Kundu (1982, Fig. 4(g)). In this case the apex of the loop is noticeably brighter than the legs. The original paper reproduces a number of 10 s 'snapshots' of this event and includes maps of the circular polarization V as well as the intensity I, together with an Hα photograph of the same flare.

(2)   *Single or double sources*

According to Dulk & Dennis (1982), both single and double sources of emission are seen at 6 cm, but at 2 cm a double source has never been reported. At 20 cm Willson (1984) observed two oppositely polarized sources which he suggested were located near the footpoints of a loop activated by an adjacent, larger one. The limb loop of Schmahl et al. (1985), observed at 6 cm, showed two sources with the same polarization, interpreted as the legs of a single loop.

(3)   *Identification of the footpoints*
    In a complex system observed by Kundu, Schmahl & Velusamy (1982), the footpoints were identified as the emission components of opposite polarity depicted on the maps of V. On occasion the burst source itself can be located at the footpoints, rather than at the apex or on the legs (Kundu, 1983). Observing simultaneously at 6 cm and 2 cm, Shevgaonkar & Kundu (1985) found that the 2 cm emission could either be co-spatial with the 6 cm source at the top of a flaring loop or near the footpoints.

(4)   *Frequency maximum of flux density spectrum*
    This quantity, together with the slope of the spectrum, is a key observational parameter for comparison with the predictions of a theoretical model (e.g. Dulk & Dennis, 1982; Schmahl *et al.*, 1985; Klein, Trottet & Magun, 1986). The frequency spectrum can be determined without the high spatial resolution needed to obtain maps of I and V (cf. Table 4.7) provided, of course, the source under study is isolated. Although it depends, *inter alia*, on the viewing angle (Mok, 1983) the various determinations of the frequency maximum cover a relatively narrow range: ~ 5 GHz (Velusamy & Kundu, 1982); 5–8 GHz (4 sources) (Holman, Kundu & Dennis, 1984); 10 GHz (a limb flare) (Schmahl *et al.*, 1985); 9 GHz (Mackinnon *et al.*, 1986); and 5–9 GHz (Klein, Trottet & Magun, 1986, Fig. 4).

(5)   *Magnitude and sense of circular polarization*
    A high value of the degree of circular polarization $\rho_{\mathrm{CP}} = V/I$ is a characteristic feature of impulsive microwave bursts. For example, observing 8 bursts at wavelengths of 2, 6 and 20 cm, Willson (1983, Table II) measured maximum values of ρ_{CP} ranging from $< 10\%$ to $\sim 100\%$. Values at the lower, middle and upper parts of this range have been quoted by numerous authors, including Velusamy & Kundu (1982), Hoyng *et al.* (1983), Willson & Lang (1984), Willson (1984) and Schmahl *et al.* (1985). A common observation is two polarized sources of *opposite sign*, perhaps separated by an (inferred) neutral line in the photospheric magnetic field (e.g. Velusamy & Kundu, 1982; Hoyng *et al.*, 1983; Willson & Lang, 1984; Willson, 1984). The high value of ρ_{CP} is strong evidence that the microwave source, and hence the loop in which it is located, is pervaded by a significant magnetic field; furthermore, adequate modelling of real events might be expected to give us information about the magnitude and direction of the magnetic field vector **B** (see Section 4.8.5). Theoretical modelling shows that the sense of V depends strongly on the heliocentric

angle of the burst/loop and on the wavelength at which it is observed (Alissandrakis & Preka-Papadema, 1984) and on the properties of the surrounding medium (Schmahl *et al.*, 1985). Particularly for loops away from the disk centre, one must exercise caution in deducing the polarity of **B** from the sense of the observed polarization.

A change in the sign of ρ_{CP} a few minutes prior to the maximum phase has been noted by Van Hoven & Hurford (1984, Fig. 5), but these authors remark that such pre-flare effects during the pre-impulsive phase are relatively rare, occurring only in some 10% of cases. An increase in the magnitude of ρ_{CP} during the pre-maximum phase from $\leqslant 20\%$ to $\sim 60\%$ was recorded by Willson (1984).

(6)   *Length and height of radio loop*
        Since, at any one wavelength, a microwave burst source may occupy only a portion of the associated loop, it is not easy to determine the loop's total length. Estimates which are to be found in the literature include 30000 km (Willson & Lang, 1984); 65000 km (Willson, 1984); and 57000 km (Kahler *et al.*, 1984). It may be significant that all of these figures are smaller than those quoted for non-flare radio loops in Table 3.7, Chapter 3, for the same wavelengths.

The height is also hard to determine. Figures of 20000–30000 km and 15000 km have been given by Willson & Lang (1984) and Willson (1983) respectively on the basis of the apparent limbward displacement of the source with respect to associated Hα emission (kernels). Direct measurement of a source beyond the limb by Takakura *et al.* (1985) gave a height of 43000–50000 km. It should be noted that these figures only represent the height of the *apex* of the loop if the source itself occurred at the apex: otherwise these figures represent lower limits.

(7)   *Size of microwave source: diameter of radio loop*
        If we assume that a microwave burst occupies the entire width of its associated loop, then the smallest dimension of the source can reasonably be identified with the diameter of the loop. A number of measurements of this quantity are available, which are collected together in Table 4.8. Despite the relatively small number of values, these figures exhibit a definite trend for the diameter to increase with λ, the average diameters being 3″0 (2200 km) at 2 cm or less than 2 cm, 9″4 (6800 km) at 5 cm and 13″ (9400 km) at 20 cm. The same trend, i.e. for the source to appear smaller at higher frequencies, is predicted by the theoretical model of Petrosian (1982).

Table 4.8. *Measured sizes of microwave bursts: diameter of radio loop*

Reference	Wavelength (cm)	Frequency (GHz)	Size (sec of arc)
Marsh & Hurford (1980)	1.3, 2	23, 15	2
Dulk & Dennis (1982)	2	15	2–5
Hoyng *et al.* (1983)	2	15	1.3
Willson (1983)	2	15	5
Marsh *et al.* (1979)	6	5	2–4
Dulk & Dennis (1982)	6	5	10–15
Willson (1983)	6	5	5–15
Schmahl *et al.* (1985)	6	5	9–10
Takakura *et al.* (1985)	6	5	12
Willson (1983)	20	1.5	10–16

At all wavelengths the measured diameters appear to be significantly smaller than those of non-flare radio loops (Table 3.7).

As in the case of observations at other than radio wavelengths, the possibility of the existence of a 'micro'-structure cannot be ruled out. In fact, using the technique of VLBI (Very Large Baseline Interferometry) at 18 cm Tapping (1986) claims to have found strong evidence for multiple sources of very small size, viz. $\sim 0''07$ (50 km).

(8)   *Lifetime*

The lifetime of an individual impulsive burst peak can be remarkably short, of the order of one minute or even less. Flux v. time profiles obtained at 2 and 6 cm have been published by Shevgaonkar & Kundu (1985, Fig. 5). The total lifetime of the burst was about 10 min but it contained a number of peaks – six at 2 cm and five at 6 cm. The individual peaks lasted for 1 min or less. Similarly, a flux density curve published by Klein, Trottet & Magun (1986, Fig. 1) (3.6 cm) shows an irregular peak occupying some 1–2 min. Various portions of the curve, together with a simultaneous curve obtained in the hard X-ray region, were attributed by the authors to five successive injections into the loop of electrons. Other authors have given somewhat larger values for the lifetimes of bursts, e.g. a few minutes to greater than 30 minutes (Alissandrakis & Kundu, 1978) and 1–5 min (Lang & Willson, 1984).

4.7   Spatial and temporal relationships between flare emissions observed in different spectral regions

4.7.1   *Introduction*

Simultaneous observations of flares covering the full wavelength range, visible to hard X-ray, with adequate spatial resolution are at present relatively sparse. Nevertheless, there exists a considerable body of reliable information about significant aspects of the various relationships. The aim of the present section is to outline these relationships – spatial and temporal – which have been uncovered between Hα and hard X-ray flare emission (Section 4.7.2), Hα and microwave (Section 4.7.3), EUV and hard X-rays (Section 4.7.4), and hard X-rays and microwave (Section 4.7.5). Other relationships are described elsewhere in this chapter, viz. Hα and λ530.3 (Section 4.3.3), EUV and Hα (Section 4.4.2) and hard and soft X-rays (Section 4.5.4).

Some of the relationships described below are remarkably close. For example, the time profiles of impulsive microwave bursts from a flare and the corresponding hard X-ray emission may be very similar even on time scales of less than 1 s. Such relationships evidently have a profound bearing on our understanding of the physics of flare loops and on the role of loops in the flare event as a whole.

4.7.2   *Hα and hard X-rays*

Several of the authors who have compared simultaneous HXR images and on- or off-band Hα photographs have concluded that – particularly in the initial phase – the HXR emission coincides with the brightest Hα patches or kernels (Hoyng *et al.*, 1981a, b; Duijveman, Hoyng & Machado, 1982; Tsuneta *et al.*, 1983). Such coincidence has been cited as evidence for the view that during the impulsive phase there is significant energy deposition at the footpoints of the associated loop (MacNeice *et al.*, 1985). On the other hand, it must be admitted that the low spatial resolution of the HXR images, together with difficulties of registration, sometimes makes it difficult to draw reliable conclusions from such comparisons (see, for example, Kundu *et al.*, 1984, Fig. 5(*a*)). Furthermore, there is a fairly convincing case in which an HXR source appears to be located at the apex of an arcade of Hα post-flare loops seen in emission near the limb some 26 min later (Tsuneta *et al.*, 1984b).

The simultaneity or near-simultaneity between the peak of an HXR burst and the corresponding Hα intensity maximum is well established. For example, curves reproduced by Gunkler *et al.* (1984, Fig. 3) compare the HXR emission over the band 28–54 keV as a function of time with curves for the Hα *excess*, relative to pre-flare, of three separate strands of an Hα flare. Despite irregularities in the curves, the maxima are seen to be simultaneous to ~1 min. Kurokawa *et al.* (1986) made a special study of the question and found simultaneity to within the time resolution of the Hα observations used, 1–10 s (see their Table I). Furthermore, they found that individual impulsive spikes in the HXR (and microwave) time profiles were closely correlated with brightenings in the Hα kernels, in one case to < ±1 s. Not surprisingly, the sharp rise in HXR emission precedes the maximum in Hα *area* by up to several minutes (Wang *et al.*, 1987, Figs. 2 and 3).

Earlier references to the correspondence between HXR bursts and chromospheric flare emissions are to be found in Rust (1984).

4.7.3   *Hα and microwave*
From one-dimensional observations obtained at 6 cm with the Westerbork Radio Synthesis Telescope Alissandrakis & Kundu (1978, Figs. 2–3, 6–8) showed that the first peak of the microwave emission preceded the maximum of the associated Hα flare by intervals ranging from ~1 min to ~4 min. Their effective time resolution was about 30 s. In agreement with these figures, Kundu, Schmahl & Velusamy (1982) found that the peak of a 6 cm burst observed with the VLA preceded the maximum of the Hα flare by ~1 min.

Simultaneity to within ±10 s was observed by Marsh *et al.* (1980) between the brightening of Hα kernels and the impulsive increase in 6 cm flux; similar results were obtained at 1.8 cm by Kurokawa *et al.* (1986). Alissandrakis & Kundu (*loc. cit.*, Figs. 2–3, 7) found that the microwave peak followed or preceded the Hα flare *start* by intervals of the order of 1–2 min or less.

The spatial link between microwave bursts/loops and Hα flare emission is fairly close, the impulsive phase emission usually being located near to or over a magnetic neutral line (cf. Section 4.2) and between Hα flare kernels or ribbons (Dulk & Dennis, 1982; Lang & Willson, 1984; Shevgaonkar & Kundu, 1985). Impulsive bursts observed at various wavelengths and lying between or spanning Hα kernels have been illustrated by Hoyng *et al.* (1983, Fig. 2(*a*)) (2 cm), Marsh & Hurford (1980, Figs. 4 and 5) (2 cm and 1.3 cm) and Willson (1983, Fig. 1) (6 cm). Figure 5 of the last of these papers shows a 20 cm burst lying along the magnetic neutral line, while the corresponding relationship at 2 cm is well shown by Figs. 2(*a*) and 2(*b*) of Hoyng *et al.*

Two loops observed by Velusamy & Kundu (1982) at 6 cm (resolution,

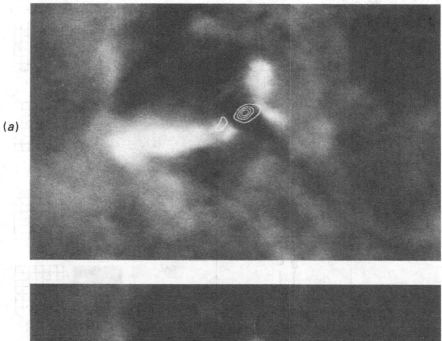

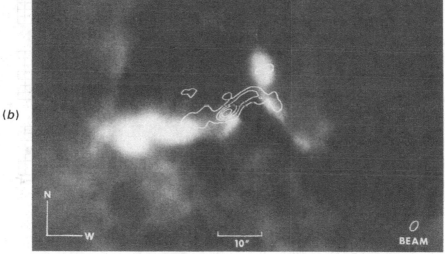

Fig. 4.18. (*a*) 2 cm intensity map of a compact, short-lived impulsive burst, superimposed on Hα photograph of associated optical flare (Marsh & Hurford, 1980). Note that the microwave emission is concentrated between two Hα kernels, strongly suggesting that the source is located at the top of a loop. (*b*) A more gradual burst, which followed the burst in (*a*) by a few minutes. In this case the burst shows a loop-like structure which extends between two of the Hα kernels. The peak brightness temperature of (*b*) was approximately twice that of (*a*).

2″ × 3″) were more clearly seen soon *after* the impulsive maximum and were observed to coincide roughly with faint Hα loops connecting the two flare ribbons.

Figure 4.18(*a*) shows a VLA map of a 2 cm burst at the peak of the

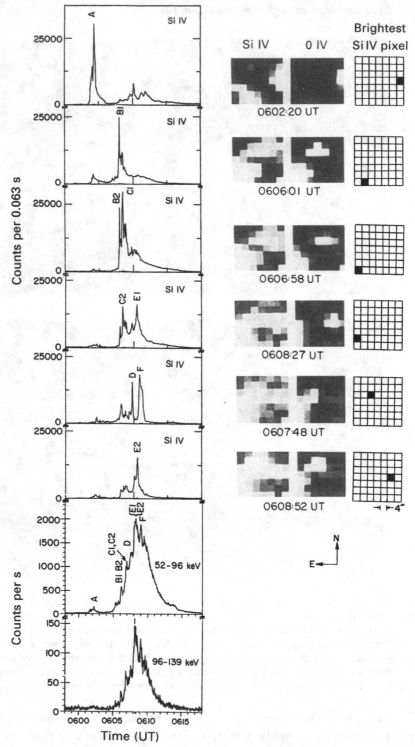

Fig. 4.19. Light curves in the cool EUV line Si IV λ140.3 of the brightest kernels in a 3B disk flare observed on 14 October 1980 (Cheng *et al.*, 1981). Shown for comparison is the time variation of the hard X-ray flux from the flare in the energy bands 52–96 keV and 96–139 keV. The peak Si IV emission at a particular time is labelled A, B, C, etc., while peaks associated with different kernels, but reaching maximum intensity at the same time, are distinguished by numerals – e.g. B1, B2, etc. The peaks in the hard X-ray flux are labelled with the same symbols as the Si IV peaks, with which they show a convincing temporal correlation.

impulsive phase, superimposed upon an Hα photograph of the associated optical flare (Marsh & Hurford, 1980, Fig. 4). This illustration strongly suggests that this short-lived compact burst was located at the top of a loop. It was followed a few minutes later by a more gradual (7 min) burst which took the form of a loop-like structure extending between two Hα kernels (Fig. 4.18(*b*)).

4.7.4 *EUV and hard X-rays*

In most cases the bursts of EUV and hard X-ray emission observed during the impulsive phase of a flare are closely correlated in time. As we have seen in Section 4.4.2, the impulsive EUV emission from a flare originates mainly in small, localized bright areas (kernels), believed to mark the footpoints of loops or systems of loops joining areas of opposite magnetic polarity. When light curves of kernels observed in cool EUV lines are compared with a plot of the hard X-ray emission from the flare, a close temporal correlation is found between individual peaks in the light curves of the individual kernels and peaks in the curve of the hard X-ray flux.

The correlation is well demonstrated by the results obtained by Cheng *et al.* (1981) from raster pictures of a large 3B disk flare observed on 14 October 1980 in the cool EUV lines Si IV λ140.2 and O IV λ140.1 with the ultraviolet spectrometer and polarimeter (UVSP) on SMM. Figure 4.19 compares the Si IV light curves of the brightest individual pixels (= kernels) with the time curves of the hard X-ray flux in the energy bands 52–96 keV and 96–139 keV. The peak Si IV emission at a particular time is labelled A, B, C, etc., while peaks associated with different pixels, but reaching maximum intensity at the same time, are distinguished by numerals – e.g. B1, B2, etc. The X-ray peaks are labelled with the same symbols as the Si IV peaks with which they are temporally correlated.

It is evident from Fig. 4.19 that the individual hard X-ray spikes are associated with different Si IV kernels; corresponding X-ray and Si IV peaks are coincident to within the time resolution of the UVSP observations, 4.6 s. Each hard X-ray spike is mainly associated with the brightest one or two of the EUV kernels, but others may brighten up to some degree at the same time. Similar results have been obtained by Poland *et al.* (1982) for the cool line λ137.1 of O V.

How closely does the time of a peak in a burst of EUV radiation from a flare coincide with the time of the corresponding peak in the flux of hard X-rays? Using UVSP observations of several impulsive flares made in the line λ137.1 of O V Woodgate *et al.* (1983) showed that any time difference between the O V and hard X-ray peaks *was less than 1 s*. Consistent with this Tandberg-Hanssen *et al.* (1984) found that the peaks of the O V and hard X-ray bursts from the 1B,M1.2 two-ribbon flare of 1 November 1980 were simultaneous to within ± 2 s.

More recently, Orwig & Woodgate (1986) have used UVSP observations of higher time resolution to study the correlation in the cases of two other flares, one of importance SB,M4.6 (20 May 1984) and the other of importance 1N,C8.8 (24 April 1985). In both cases high time resolution was achieved by restricting the field of view to a small area ($10'' \times 10''$ and $30'' \times 30''$ respectively) centred on the flare. The combined results showed that related peaks in the O V, continuum and hard X-ray emission were simultaneous to within 0.1 s.

The nearly simultaneous occurrence of peaks in the emission of hard X-rays from a flare with peaks in the light curves of individual EUV kernels supports the view that the X-rays actually originate in the kernels, i.e. at the footpoints of loops or systems of loops joining areas of opposite magnetic polarity. In a number of cases this has been confirmed by a comparison of EUV pictures of flares with spatially resolved hard X-ray observations of the same events (Duijveman, Hoyng & Machado, 1982; Poland *et al.*, 1982; MacNeice *et al.*, 1985). A theoretical model (thick-target) which satisfactorily predicts the observed smallness of the time difference between the HXR and EUV peaks – and also the observed EUV/HXR flux ratio – has been studied by McClymont & Canfield (1986).

4.7.5 *Hard X-rays and microwave*

There is generally a very close correlation between time variations in the flux from impulsive microwave bursts and the corresponding hard X-ray emission. The correlation is usually (but not always) so close that it is taken as evidence that both types of emission owe their origin to basically the same population of energetic electrons (Dulk & Dennis, 1982). Time profiles at various radio and X-ray wavelengths which illustrate this correlation have been published by a number of authors, including Klein, Trottet & Magun (1986, Fig. 1) (3.6 cm, 102–126 keV); Willson (1984, Fig. 1) (22 cm, 1.6–12.5 keV; 3.1–25 keV); and Cornell *et al.* (1984, Fig. 2) (2.8 cm, 32–55 keV). The results of the last of these are particularly interesting since these authors observed with a time resolution of only 0.1 s; studying five flares, they found that the average delay of the microwave peak behind the X-ray peak was only 0.2 s. Furthermore, the two profiles were closely similar even on sub-second time scales. Similarly, a limb event observed by Schmahl *et al.* (1985) was simultaneous with hard X-ray emission to within $\lesssim 1$ s.

Sometimes a correlation is absent: MacKinnon *et al.* (1986, Fig. 1) have published an interesting case of two bursts, one of which was well-correlated with 26–52 keV X-ray emission, the other hardly appearing on the X-ray record at all. Shevgaonkar & Kundu (1985, Fig. 9) have illustrated a case where the 2 cm emission peak is simultaneous with 30–60 keV X-rays, but the 6 cm peak occurs some 15 s later.

Despite the generally close correlation in time between microwave bursts and X-ray emission, the two are generally found to be spatially separated, there being a tendency for X-rays to come from the footpoints of a loop and microwaves from the top (Petrosian, 1982). Kundu (1984) has made a special study of the observations bearing on the question of the relative positions of microwave and hard X-ray burst sources. He finds that the common belief that the microwave and hard X-ray sources are located near the top and footpoints respectively is true in some but not all cases; more than one loop may be involved, thus complicating the picture.

A good example of non-coincidence between 6 cm and soft X-ray emission during the post-impulsive phase of a 1B,M3 flare has been published by Kahler *et al.* (1984, Fig. 1). The X-ray image showed a markedly loop-like appearance, not shown by the 6 cm emission. A limb flare observed at 6 cm and in hard X-rays discussed by Schmahl, Kundu & Dennis (1985) showed a single X-ray source, lower in the corona than the double 6 cm source; these observations were interpreted in terms of a single loop. Hoyng *et al.* (1983) observed a small microwave source during the impulsive phase which lay between locations showing excess HXR flux above 16 keV.

4.8   Physical conditions in flare loops
4.8.1   *Introduction*
The analysis of observations of flare loops for the purpose of inferring their physical parameters is described in the sections that follow. The diverse spectral regions are dealt with separately, starting with chromospheric and coronal line and continua in the visible region, continuing with the EUV and the soft and hard X-ray regions, and ending with microwave observations. Our main aim is to catalogue the quite extensive results which have been obtained in the different spectral regions, so that diagnostic methods are only lightly sketched. (In some cases the reader is referred back to discussions of diagnostics in earlier chapters.)

When one examines the individual figures for the electron density N_e and temperature T_e inferred from observations in widely separated spectral regions, one is struck by the consistency of the results. For example, various visual region data give values of N_e ranging from 1×10^{10} cm^{-3} to 7×10^{10} cm^{-3} (cf. Table 4.17), which is consistent with values derived from both soft and hard X-rays and microwave observations. The EUV data also give consistent values for N_e but stand out in giving a much larger apparent spread, 10^9–10^{12} cm^{-3}.

Similarly, all regions of the spectrum yield values of T_e in the relatively narrow range 5×10^6–7×10^7 K; this includes the figure for T_e derived for the so-called thermal component of hard X-ray and microwave bursts. Of course in the visible region Hα and other chromospheric lines, together with EUV lines of moderate ionization, give temperatures of only a few tens of thousand degrees. This plainly indicates that flare loops of widely differing temperature can co-exist in time while occupying different regions of space.

4.8.2   *Analysis of observations in the visible region*

The diagnostic methods available for determining the physical conditions in flare loops observed on the disk and beyond the limb in the visible and near-visible regions of the spectrum are the same as those used for cool and hot non-flare loops. Descriptions of these techniques are to be found in Sections 2.5.2, 2.5.3 and 3.6.2 and need not be repeated here. Table 4.9 lists values for the electron temperature T_e and non-thermal velocity ξ obtained by various authors, mostly by the line width method. Once again, we should point out that results obtained by this method must be viewed with caution: for example Kureizumi *et al.* (1977) found that the widths of the He I and He II lines that they measured were inconsistent with one another and with those of H I.

The temperature given by Hiei *et al.* (1983) for H I was derived from their measurement of the Balmer continuum, together with a value for N_e obtained from the widths of the higher Balmer lines (see below); their temperature for Fe XIV was assumed. The approximate value for Ca XV of Fisher (1974) was equated to the temperature at which the ionization curve for this ion reaches its peak. The figures in Table 4.9 show that, viewed in lines of H I, He I and Ca II, flare loops have temperatures and non-thermal velocities lying in the ranges 7600–21 000 K and 3–16 km s^{-1} respectively.

The principal methods which have been applied to the determination of the electron density N_e in flare loops observed beyond the limb are (a) measurement of the intensity of the electron scattering continuum (cf. Eqn. (3.2), Section 3.6.2), and (b) measurement of the widths of the higher Balmer lines (cf. Section 2.5.3). Moreover, Fisher (1974) has used the ratio

Table 4.9. *Temperature of flare loops: visible region*

Line	Ion	T_e (K)	ξ (km s^{-1})	Reference
Hα, H and K	H I, Ca II	12 500–21 000	5–7	Machado (1971)
Hγ, H and K	H I, Ca II	13 200 (top)	—	Machado *et al.* (1972)
		11 500 (legs)	7	Machado *et al.* (1972)
λ569.4, λ544.5	Ca XV	~5 × 10⁶	—	Fisher (1974)
H and K	Ca II	—	6–13	Kureizumi *et al.* (1977)
Var. metal lines	—	—	3–4	Kureizumi *et al.* (1977)
Hβ, D₃	H I, He I	15 400	16	Ruzdjak (1981)
Balmer lines + cont.	H I	7600	—	Hiei *et al.* (1983)
λ530.3	Fe XIV	2 × 10⁶ᵃ	30–40	Hiei *et al.* (1983)
Hα, K	H I, Ca II	12 000–13 000	5–15	Cui *et al.* (1985)
Hα	H I	21 400ᵇ	15	Yan *et al.* (1987)

[a] Assumed; [b] Excitation temperature.

Table 4.10. *Electron density in flare loops: visible region*

Line	Ion	N_e (cm^{-3})	Reference
Continuum	—	<1.9 × 10¹⁰–5.9 × 10¹⁰	Fisher (1971)
Continuum	—	<3.2 × 10¹⁰	Fisher (1974)
λ569.4/λ544.5	Ca XV	<1 × 10¹⁰	Fisher (1974)
H₁₀–H₁₅	H I	2 × 10¹²	Kureizumi *et al.* (1977)
Continuum	—	4 × 10¹⁰	Foukal (1978)
Continuum	—	3 × 10¹⁰–7 × 10¹⁰	Foukal *et al.* (1983)
H₂₂	H I	<7 × 10¹⁰ᵃ	Foukal *et al.* (1983)
Higher Balmer lines	H I	(2 × 10¹²)ᵇ	Foukal *et al.* (1983)
Higher Balmer lines	H I	1.6 × 10¹²–6.3 × 10¹²	Hiei *et al.* (1983)

[a] Assuming $T_e = 10\,000$ K; [b] regarded as 'unacceptable' by Foukal *et al.* (1983).

of the intensities of the lines $\lambda 569.4$ and $\lambda 544.5$ of Ca XV, while Foukal, Miller & Gilliam (1983) have derived N_e from an emission measure analysis of H_{22} in addition to applying methods (a) and (b) above. The results obtained by the various authors are listed in Table 4.10.

Excluding the values of N_e derived from measurements of the widths of the higher Balmer lines (Stark effect), we see from Table 4.10 that the figures all lie in the relatively narrow range $1 \times 10^{10} - 7 \times 10^{10}$ cm^{-3}. The line width measurements, on the other hand, yield values ~ 100 times larger. To explain this discrepancy in their own results, Foukal, Miller and Gilliam invoked an additional source of Stark effect broadening, viz. a hypothetical electric field E. The magnitude of the field required to obtain a fit with their own observations was $E \simeq 170$ V cm^{-1}.

The clear visibility in *emission* at Hα line centre of the tops of the flare loops shown in Fig. 4.3(*a*) has been used by Heinzel & Karlicky (1987) to estimate N_e. By non-LTE modelling they found that emission at Hα line centre required N_e to be of the order of $5 \times 10^{11} - 10^{12}$ cm^{-3}. Their models all gave zero contrast at H$\alpha + 0.1$ nm for the loop tops in agreement with the Culgoora observations (see Fig. 4.3(*b*)). A similar technique for estimating N_e, with similar results, has been applied by Svestka *et al.* (1987).

The total particle density of hydrogen N_H (i.e. neutral + ionized) can be derived by measuring the ratio of the central intensities of the K and H lines of Ca II in a loop observed beyond the limb, i.e. the quantity I_K/I_H. This method has been employed by Machado, Grossi Gallegos & Silva (1972), Ruzdjak (1981) and Cui *et al.* (1985). The ratio I_K/I_H is related to the optical depth at the centre of the K line by the formula

$$\tau_K = -2 \ln [(I_K/I_H) - 1]. \tag{4.1}$$

Having determined τ_K from Eqn. (4.1), one can then apply Eqns. (2.3) and (2.4) of Section 2.5.2 to determine the particle density $N_{Ca\,II}$, provided the geometrical path length through the loop along the line of sight can be estimated. It is now assumed that all the calcium is singly ionized: N_H can then be derived by making use of the known ratio of calcium to hydrogen atoms in the Sun's atmosphere. With this method the above-named authors obtained consistent values for N_H of $\sim 0.8 \times 10^{10}$, 1×10^{10} and $1.3 \times 10^{10} - 2.6 \times 10^{10}$ cm^{-3}. Finally, by applying the 'cloud' model (cf. Section 2.5.2) to Hα observations of a flare loop system observed on the disk, Yan, Ding & Xu (1987) derived the figure $N_H = 6.7 \times 10^{10}$ cm^{-3} as a lower limit to the hydrogen density at the bottom of a loop.

If hydrogen in flare loops can also be assumed to be almost completely ionized, then $N_H \approx N_e$ and these figures provide independent estimates of

N_e. Comparing them with the figures in Table 4.10 we see that they are indeed consistent with values derived from the electron scattering continuum. Actual values for the degree of ionization of hydrogen and helium x_H and x_{HeI} have been estimated by Kureizumi *et al.* (1977) to be 0.8–1.0 and 0.1–0.2 respectively.

An average longitudinal magnetic field in a flare loop of $B_\parallel = 76$ G has been measured by Malville & Schindler (1981), using the 512 channel magnetograph of the Kitt Peak National Observatory. By fitting hypothetical field lines to loops associated with a 2B flare, Rust & Bar (1973) have inferred total magnetic fields at the loop tops of $B = 40$–80 G.

4.8.3   *Analysis of observations in EUV lines*

It is generally assumed that EUV bursts are produced under *thermal* conditions even though the heating of the regions involved may itself be non-thermal in origin, e.g. the heating of the flare plasma as a result of the collisional thermalization of streams of accelerated electrons (Svestka, 1976, p. 167). This accounts for the term 'quasi-thermal' sometimes applied to the impulsive EUV emission from flares. Accordingly, the determination of the electron temperatures and densities in EUV flare loops hinges on the standard techniques discussed in earlier chapters:

(a) measurement of absolute line intensities to determine the emission measure $\int N_e^2 \cdot d\ell$ along the line of sight through the loop (Section 2.5.4);

(b) the use of observed values of suitably chosen temperature- or density-sensitive line intensity ratios to infer T_e or N_e; and

(c) measurement of the effective spectral line width from EUV slitless spectroheliograms, making the usual assumptions of optical thinness and Gaussian profile (Section 3.6.2).

(1)   *Electron temperatures and densities in loops and kernels*

The temperatures of EUV flare loops vary from as low as a few tens of thousands of degrees to as high as nearly twenty million degrees. Likewise, the published values of the electron density cover a wide range, from less than 10^9 cm^{-3} to greater than 10^{12} cm^{-3}. Moreover, the density may vary substantially over the lifetime of the flare (Withbroe, 1978) and there are marked differences in the physical conditions generally from flare to flare. A selection of modern determinations of T_e and N_e in flare loops and kernels is presented in Table 4.11. The seventh column indicates the phase of the flare pertaining to the observations in question, and the eighth the method of analysis employed.

Table 4.11. *Physical conditions in EUV flare loops and kernels*

Source	Spectral line	Temperature[a] (K)	Electron density (cm^{-3})	Gas pressure[b] (dyne cm^{-2})	Loop (L) or kernel (K)	Flare phase	Method/remarks
Dere, Horan & Kreplin (1977)	various	(2.2×10^6) (5.0×10^6)	3×10^{10} 2×10^{11}	— —	L	late decay	emission measure
Dere, Horan & Kreplin (1977)	various	$10^6 - 10^7$	$1 \times 10^{10} - 5 \times 10^{11}$	—	K	max + decay	intensity ratios
Foukal (1978)	various	—	—	0.02–1.06	L	decay	emission measure; for detailed results, see Table 4.12(a)
Withbroe (1978)	Mg X λ62.5	(1.1×10^6)	$2.4 \times 10^{10} - 5.3 \times 10^9$	7.0–1.5	L	decay	absolute intensities
Widing & Spicer (1980)	Fe XXIII λ26.4 Fe XXIV λ19.2, 25.5	1.4×10^7 1.5×10^7	— —	— —	K K	~max.	line widths
Widing & Spicer (1980)	various	$(6.9 \times 10^5 - 1.6 \times 10^7)$	$1 \times 10^{10} - 3.2 \times 10^{11}$	—	K	early decay	emission measure + intensity ratios
Cheng (1980)	various	$(0.5 \times 10^6 - 4.0 \times 10^6)$	$1.4 \times 10^9 - 1.0 \times 10^{10}$	0.2–4.0	L	decay	emission measure; for detailed results, see Table 4.13
Cheng et al. (1982)	Si IV λ140.2 O IV λ140.1	—	$2 \times 10^{11} - 3 \times 10^{12}$	—	L	pre-flare	intensity ratios

[a] Inferred loop and kernel temperatures are placed in parentheses.
[b] Calculated from the formula $p = 1.92 k N_e T_e$.

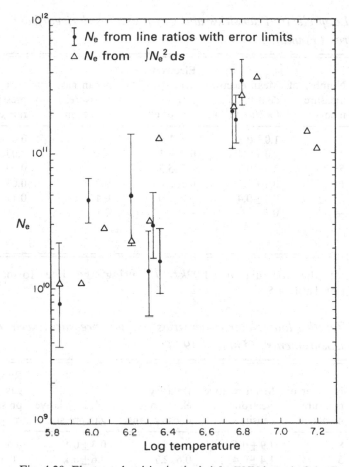

Fig. 4.20. Electron densities in the bright EUV kernel of the flare of 17 December 1973 during the early decay phase (Widing & Spicer, 1980). The electron density shows an increase of more than an order of magnitude over the temperature range 8×10^5 K–7×10^6 K. Note that with few exceptions the density derived by the emission measure technique agrees with that determined by the line ratio method to within the estimated error.

There is a clear tendency for the density of flare loops and kernels to increase with temperature. This is shown (for a kernel) in Fig. 4.20, the results referring to the early decay phase of a flare (Widing & Spicer, 1980). Over the temperature range 8×10^5–7×10^6 K, the electron density is seen to increase by more than an order of magnitude. The significance of the large pressure differences implied by these results cannot be understood without a detailed knowledge of the geometry of the flare region, but Widing and Spicer conclude that there is no strong tendency towards filamentation in the kernel. As Fig. 4.20 shows, with few exceptions the density derived from the emission measure agrees with that determined by the line intensity ratio method.

Table 4.12(a). *Loop electron densities and gas pressures derived from emission measures (Foukal, 1978)*

Spectral line	Number of measurements	Mean electron density, N_e ($\times 10^9$ cm^{-3})	Electron density range ($\times 10^9$ cm^{-3})	Mean gas pressure, p (dyne cm^{-2})[a]	Gas pressure range
O IV $\lambda 55.4$	5	1.0 ± 0.3	0.3–1.9	0.05	0.02–0.10
O VI $\lambda 103.2$	9	2.0 ± 0.4	0.4–4.1	0.16	0.03–0.33
Ne VII $\lambda 46.5$	6	2.3 ± 0.7	0.7–5.3	0.31	0.09–0.72
Mg X $\lambda 62.5$	11	0.8 ± 0.2	0.3–3.0	0.23	0.09–0.87
Si XII $\lambda 52.1$	5	0.8 ± 0.4	0.2–2.0	0.42	0.11–1.06
Background corona at 4×10^4 km	—	0.3	—	0.1	—

[a] Calculated from the formula $p = 1.92kN_e\bar{T}_e$, using the line formation temperatures ($\bar{T}_e$) of Table 4.5.

Table 4.12(b). *Relative loop electron densities and gas pressures derived from contrast measurements[b] (Foukal, 1978)*

Spectral line	Number of measurements	Mean relative electron density	Relative electron density range	Mean relative gas pressure	Relative gas pressure range
O VI $\lambda 103.2$	8	1.9 ± 0.6	0.4–5.6	0.3 ± 0.1	0.06–0.8
Ne VII $\lambda 46.5$	5	1.4 ± 0.4	0.6–2.7	0.6 ± 0.1	0.4–0.9
Mg X $\lambda 62.5$	9	8.1 ± 3	2.5–32	5.2 ± 2	1.5–19
Si XII $\lambda 52.1$	4	12 ± 3	6–18	11 ± 2	5–16

[b] Relative to the corresponding values in the surrounding corona.

The electron densities and gas pressures derived by Foukal (1978) are given in detail in Table 4.12; these values refer to high-density regions of a number of prominent post-flare loops observed from Skylab. For several of the lines Foukal used the method described in Section 2.5.5 to derive the mean electron density and gas pressure in the loops relative to the surrounding coronal background. These results are given in Table 4.12(b); note that the pressure in the loop ranges from substantially *less* to substantially *more* than that in the coronal surroundings.

The height variation of N_e in a long-lived post-flare loop is shown in Table 4.13, after Cheng (1980). It is evident that in any given line the density is almost constant with height and therefore constant along the

Table 4.13. *Electron densities in post-flare loop of 14 August 1973 (Cheng, 1980, loop 'A')*

		Electron density ($\times 10^9$ cm^{-3})			
Spectral line	Line formation temperature[a] ($\times 10^6$ K)	Height above limb along the dispersion			
		13".1	26".5	38".6	46".9
Ne VII λ46.5	0.51	1.5	1.4	1.3	1.3
Mg VIII λ43.7	0.81	2.3	1.6	1.9	—
Mg IX λ36.8	0.98	—	4.2	3.1	4.6
Fe XV λ28.4	3.4	—	—	3.5	3.6
Fe XVI λ33.5	4.3	—	7.0	7.0	—
Si XII λ49.9	2.0	—	10.0	10.0	10.0

[a] Taken from Table 4.5.

length of the loop, thus ruling out the presumption of hydrostatic equilibrium within the loop. Once again there is a marked increase of N_e with T_e. Cheng also found that, in any particular line, there was no observable variation in temperature along the loop. He calculated the gas pressure in the Ne VII loop to be 0.2 dyne cm^{-2} compared to a coronal background pressure of 0.6 dyne cm^{-2}, but 4.0 dyne cm^{-2} in the hot Fe XVI loop.

The electron temperatures and densities in the hot 'core' of a flare near the time of flare maximum do not appear to differ substantially from those of loops and kernels (cf. Widing & Dere, 1977; Cheng, 1977b). Non-thermal velocities in kernels and cores have been given values ranging from zero to 160 km s^{-1} (Cheng, 1977a; Widing & Spicer, 1980).

(2)   *Velocities*

Reliable data about mass flows throughout the bulk of an EUV flare loop is lacking at the present time, but there have been a few measurements of velocities at the lowest points of such loops, viz. flare kernels. Cheng & Tandberg-Hanssen (1986) measured the velocity at three kernels of a 2N,C9 flare in the cool Si IV λ140.3 line and found that the mass motions at all three points were directed downwards throughout the course of the flare. The largest downflow reached a magnitude of almost 30 km s^{-1}. Two other flares showed a similar behaviour.

A different picture emerges from the measurements of Mason *et al.* (1986), made in the hot line Fe XXI λ135.4 during the impulsive phase. They found evidence for *upflows*, the velocity magnitude sometimes exceeding

200 km s^{-1}. During the declining phase, on the other hand, profiles of the Fe XXI line gave no indication of motion.

4.8.4 *Analysis of observations in soft and hard X-rays*

(1) *Soft X-rays*

Our knowledge of the physical conditions in soft X-ray flare loops is based on extensive imaging and non-imaging observations obtained from various extra-terrestrial vehicles – rockets, Skylab, Solrad, SMM and Hinotori. The theoretical and practical aspects of the methods which have been used to derive T_e and N_e from soft X-ray data have been described at length in the previous chapter (Section 3.6.4) and need not be repeated here (short accounts specific to flares have been published by Kahler, Petrasso & Kane, 1976; Vorpahl, Tandberg-Hanssen & Smith, 1977; and Mac-Combie & Rust, 1979).

A selection of the results for the maximum and decay phases is presented in Table 4.14. For once, the numerical results are sufficiently abundant to make it meaningful to calculate average, or representative, values. For T_e at the maximum and decay phases respectively we thereby obtain 1.5×10^7 and 6.8×10^6 K and for N_e at the same phases, 4.7×10^{10} and 1.7×10^{10} cm^{-3}. In the case of T_e the spread of the values (for both phases) for the different authors cited is remarkably small. The spread is much larger in the case of N_e. Part of the spread is probably due to real differences in density between loops associated with different flares, but part is due to errors and uncertainties in the reduction procedures (cf. Section 3.6.4).

A comparison of the above values with figures for T_e and N_e derived for flare loops observed in other regions of the spectrum (visible, EUV, and microwave) is included in Section 4.8.1.

(2) *Hard X-rays*

As we have seen in Section 4.5.3(5), the thick target electron beam model for the production of hard X-rays from flares has found considerable support. If this model is accepted, it follows that the HXR spectrum observed, for example, by the Hard X-ray Burst Spectrometer on SMM can be used to determine the peak power and spectrum of the non-thermal electrons responsible (e.g. Gunkler *et al.*, 1984; Karpen, Doschek & Seely, 1986; Takakura, 1986). The observed HXR spectrum is customarily fitted to a single power law with a spectral index γ:

$$I(\varepsilon) = a\varepsilon^{-\gamma}; \tag{4.2}$$

in some cases a double power law has been found to provide a better fit with the observations (Lin & Schwartz, 1987; Smith & Orwig, 1988).

Table 4.14. *Physical conditions in soft X-ray flare loops*

Reference	T_e (K)	N_e (cm^{-3})	Flare phase	Remarks
Pallavicini	6.5×10^6	1.5×10^{10}	decay	—
et al. (1975)	1.1×10^7	5×10^{10}	~max.	—
Pallavicini &	8×10^6	2.7×10^{10}	decay	Single loop
Vaiana (1976)				
Vorpahl et al. (1977)	6.8×10^6	3.5×10^9	decay	Not a typical flare loop system
Krall et al. (1980)	4×10^6	$2 \times 10^9 - 4 \times 10^9$	decay	Long decay event (2 loops)
MacCombie & Rust (1979)	$4.6 \times 10^6 - 6.0 \times 10^6$	$5.0 \times 10^9 - 1.3 \times 10^{10}$	decay	Long decay events
Pallavicini &	8×10^6	$1.2 \times 10^{10} - 1.1 \times 10^{11}$	decay	—
Vaiana (1980)	1.4×10^7	—	max.	—
Svestka et al.	8×10^6	$9.3 \times 10^9 - 1.9 \times 10^{10}$	decay	See original paper
(1982)	1.1×10^7	—	~max.	(Figs. 5 and 6) for time variation
Kahler et al. (1984)	$6.5 \times 10^6 - 8.5 \times 10^6$	7×10^9	decay	—
Pallavicini	$6.5 \times 10^6 - 1.2 \times 10^7$	$7.1 \times 10^9 - 1.9 \times 10^{11}$	rise +	Range of values
et al. (1977)			decay	for 43 events
Krall et al. (1978)	1.4×10^7	3.5×10^{10}	max.	See original paper (Table II) for values at pre- and post-flare phases
Cheng & Pallavicini (1984)	$7 \times 10^6 - 8 \times 10^6$	2×10^{10}	max.	At apex
de Jager & Boelee (1984)	1.8×10^7	$7.9 \times 10^9 - 5.0 \times 10^{10}$	~max.	See original paper (Table I) for values at other times
Tsuneta et al. (1984b)	2×10^7	3×10^{10}	~max. (?)	—
Doyle et al. (1985)	2×10^7	—	max.	Compact flare
MacNeice et al. (1985)	1.6×10^7	1.2×10^{11}	max.	—
Kahler et al. (1976)	$6.0 \times 10^6 - 1.2 \times 10^7$	$\gtrsim 3 \times 10^{11}$	—	Kernels

The power law for the high energy non-thermal electrons responsible can be written in the form

$$n(E) \propto E^{-\delta}. \tag{4.3}$$

For the thick target mechanism, $\delta = \gamma + \frac{3}{2}$ (Hoyng et al., 1983; Takakura, 1986). It is important to realize that the electron spectrum deduced in this way is the spectrum of the high energy electrons within the X-ray emitting region (Brown, 1971) – e.g. the footpoints. Thus δ in Eqn. (4.3) may not be the same as δ in Eqn. (4.4) below (Section 4.8.5): the electrons responsible for the impulsive microwave bursts may be located higher up in the loop.

Numerous authors have inferred values of the spectral index γ from the observed HXR spectra. In fact, the published values are too numerous to quote individually. A typical average value would be $\gamma = 4.8$, but values as high as 7 and as low as 3 are common. It is of interest that Tsuneta *et al.* (1984b) observed a *hardening* of the spectrum with time, γ being ~ 4.5–5 at the start of the observation, ~ 3.5 at the peak and ~ 3 at the start of the decay phase. When observed with very high time resolution, γ may show marked variations in a period of only 3 min (Lin & Schwartz, 1987, Table I).

The fit of an observed HXR spectrum to a single power law is often extremely close. Excellent examples have been published by Tanaka (1986, Fig. 1), showing the fit from < 10 keV to 200 keV in the early phase of a flare. Nevertheless, there is evidence that at certain times during the development of a flare, the HXR spectrum may contain a strong thermal component. For example, Tanaka found that no less than 7 out of 13 large flares observed from Hinotori showed a thermal component hotter than 3×10^7 K, which became conspicuous during the middle phase. Takakura *et al.* (1986) analysed a thermal component ($T_e = 2.8 \times 10^7$ K) that occurred during the later part of the decay phase; they were able to fit the observed spectrum to the sum of the thermal and an impulsive component. Similar values of T_e were obtained for thermal components just following the main peak or during the decay phase by Hoyng *et al.* (1981a, b). On the other hand, much higher values were obtained at the maximum phase by Hoyng *et al.* (1983). These authors analysed two peaks and found values of γ on the basis of a power law fitting of 3.8 and 3.4; however, thermal bremsstrahlung also gave an acceptable fit, with $T_e = 2.9 \times 10^8$ K and 3.7×10^8 K respectively.

The analysis of HXR emission from flares does not, by itself, allow of a direct determination of the electron density N_e in a flaring loop. On the other hand, by a process of *modelling* it is possible to derive values which correctly reproduce the observed flux, wavelength variation, and perhaps height and time variation of the HXR emission. A key aspect of the physics involved is the ability of the high energy electrons accelerated by a postulated energy release mechanism at the top of a loop to penetrate down to the lower levels. This is qualitatively illustrated by a schematic diagram published by Tanaka (1986, Fig. 7) which shows 150 keV electrons penetrating to the photosphere ($N_e \simeq 10^{14}$ cm^{-3}), producing white-light and 150 keV emission; 50 keV electrons reaching the chromosphere ($N_e \simeq 10^{13}$ cm^{-3}), giving Hα, EUV and 50 keV X-ray emission; and 20 keV electrons being stopped while still in the corona ($N_e \simeq 10^{11}$–10^{12} cm^{-3}).

More exact modelling is needed to refine this picture. For example, Takakura *et al.* (1984b) calculated the number density of thermal electrons in a loop needed to stop the high energy electrons at the coronal level and thus produce a steady (observed) HXR source (13–29 keV) along the ridge of a loop arcade. The figure obtained was $N_e > 1.1 \times 10^{11}$ cm^{-3}. Takakura (1986) later performed detailed simulations of the effect of injecting electrons of energy > 10 keV into a model coronal loop. He carried out the calculations for a number of models whose physical parameters were allowed to vary: the density, for example, was varied from 1×10^{10} to 5×10^{10} cm^{-3}. The most important parameter influencing the spatial, temporal and spectral characteristics of the HXR emission was found to be the column density ('CD') of the plasma. At a critical CD of 10^{20} cm^{-2} the characteristics of the emission showed a marked change. With an assumed half-length of the loop of 3×10^9 cm, this value of CD yields a critical density of 3×10^{10} cm^{-3}.

The limiting density required to stop 16 keV electrons reaching the footpoints was calculated by Duijveman, Hoyng & Machado (1982) to be $\sim 10^{10}$ cm^{-3}. Using the same data but a more refined calculation, MacKinnon, Brown & Hayward (1985) found a limiting density of 4×10^{10} cm^{-3}.

A detailed simulation, together with comparison with three observed events, has also been carried out by Machado, Rovira & Sneibrun (1985). They used densities of 5×10^{10} and 10^{11} cm^{-3} for their analysis of two of the observed peaks – values similar to those used by Tanaka. They also estimated an upper limit to the density of the *beaming* electrons, assumed uniform over the footpoint area, during one of the peaks, $(N_e)_b < 2.5 \times 10^9$ cm^{-3} (average velocity of 10 keV electrons $= 6 \times 10^4$ km s^{-1}).

The peak power of the non-thermal electrons can be calculated from the flux and spectrum of (non-imaging) HXRBS data. Estimating the beam area from Hα images, Gunkler *et al.* (1984) calculated the flux of electrons with energy > 20 keV to be $F_{20} = 1.45 \times 10^{11}$ erg cm^{-2} s^{-1}. These authors also calculated the density of material evaporated at a footpoint due to (a) non-thermal electrons; and (b) thermal conduction. A calculation of beam power and flux has also been carried out by Duijveman *et al.* (*loc. cit.*); the results strongly depend on the observed value of γ.

4.8.5   *Analysis of observations at radio wavelengths*

A scenario for the production of microwave emission from flare loops, widely subscribed to at the present time, has been described and elaborated by a number of authors, including Velusamy & Kundu (1982), Kundu (1983), Mok (1983), Lang & Willson (1984), Schmahl, Kundu & Dennis (1985), Alissandrakis (1986) and Mackinnon *et al.* (1986). According to this picture, the microwave radiation from an impulsive burst is due to gyrosynchrotron emission from highly-energetic electrons trapped near the top of a closed magnetic structure – i.e., a loop. The hard X-rays from near the footpoints of the loop owe their origin to a beam of lower-energy electrons through the agency of thick-target bremsstrahlung (Section 4.5.3(5)); these also produce the Hα flare kernels and ribbons described in Section 4.3.1. The lower-energy electrons thus produce their effect as they precipitate onto the dense chromosphere, whereas the high-energy electrons emit in the diffuse (low) corona.

On this picture, the primary energy release takes place at the top of the loop. The microwave and X-ray emissions owe their origins to the same basic electron population, but the subsets involved differ in respect of (a) location (height); and (b) the portion of the electron energy distribution curve which is effective.

Observations of microwave bursts at 2, 6 and 20 cm show that a typical value for the brightness temperature T_B at the impulsive phase is $2 \times 10^7 - 2 \times 10^8$ K (Willson, 1983, 1984; Willson & Lang, 1984). On occasion, however, values as high as 10^9 K have been recorded (Velusamy & Kundu, 1982; Hoyng *et al.*, 1983). Brightness temperatures below about 10^7 K (e.g. Takakura *et al.*, 1985) are more characteristic of relatively small flares or of the pre- and post-impulsive phases of the larger flares.

A question of considerable importance is whether the energetic electrons involved are *thermal* (Maxwellian energy distribution) or *non-thermal* (e.g. power law energy distribution). If they are thermal, then the value of the electron temperature T_e must be very large indeed – at least as large as the measured values of T_B (cf. Section 3.6.5). Dulk (1985, p. 210) and Shevgaonkar & Kundu (1985) consider that this question is not yet fully resolved and, in fact, several authors have made use of purely thermal models in their analyses (e.g. Dulk & Dennis, 1982; Takakura *et al.*, 1985). Shevgaonkar & Kundu (1985) have modelled events observed simultaneously at 2 and 6 cm and concluded that the 6 cm and 2 cm emission could be accounted for by thermal and non-thermal electrons respectively.

Modelling of real events on the basis of an assumed non-thermal electron energy distribution has been carried out by a number of authors, including

Dulk & Dennis (1982), Hoyng *et al.* (1983), Holman, Kundu & Dennis (1984), Schmahl, Kundu & Dennis (1985), Shevgaonkar & Kundu (1985) and Klein, Trottet & Magun (1986). Among the observational parameters which such modelling seeks to reproduce are the size of a microwave source and the slope and peak frequency of its spectrum (Dulk & Dennis, 1982). In fact these authors concluded that more or less satisfactory models could be constructed either for a non-thermal electron distribution or for an inhomogeneous thermal distribution, at least for the optically thick case. Similarly, Hoyng *et al.* found that simultaneous X-ray and 2 cm bursts could be explained by means of a common non-thermal source of electrons. Holman *et al.* also concluded that their observed microwave and hard X-ray spectra of several sources were consistent with a non-thermal distribution, the microwave source being optically thick at 6 cm and longer wavelengths. In modelling their event, Klein *et al.* postulated a *multiple* injection process and sought to account for the temporal evolution of the observed flux at three microwave frequencies.

A purely theoretical calculation of the microwave signature of a single flaring magnetic loop was carried out by Mok (1983). He used a specific model for the non-thermal electrons, whose distribution needs to be known in 'configuration space' as well as 'velocity space'. Further theoretical modelling has been undertaken by Petrosian (1982), Alissandrakis & Preka-Papadema (1984) and Preka-Papadema & Alissandrakis (1988). The latter authors calculated the expected *appearance* of a loop observed at various wavelengths and orientations with respect to the observer and pointed out, *inter alia*, that the observed circular polarization could be greatly affected by propagation effects in the surrounding corona (see also Willson, 1983; Schmahl, Kundu & Dennis, 1985).

Finally, Shevgaonkar & Kundu (1985) sought to explain an observed time delay between maxima observed at 2 cm and 6 cm on the hypothesis that, while the 2 cm radiation was due to non-thermal electrons, the 6 cm emission was due to thermal gyrosynchrotron radiation from plasma at a temperature of $T_e = 3 \times 10^7 - 7 \times 10^7$ K. This conclusion was supported by a calculation of the time differences between the heating of the plasma and the acceleration of the particles in the two cases, thermal and non-thermal, which depends, *inter alia*, on the DC electric field in the plasma.

The energy band of the non-thermal electrons believed responsible for impulsive microwave bursts is $E > 100$ keV – perhaps extending as far as 500 keV or more (Petrosian, 1982; Kundu, 1983; Alissandrakis & Preka-Papadema, 1984; Lang & Willson, 1984; Dulk, 1985; and Schmahl, Kundu & Dennis, 1985). By contrast, the hard X-rays are due to (beamed) electrons of energy $E < 100$ keV, or perhaps < 50 keV (Petrosian, *loc. cit.*; Klein, Trottet & Magun, 1986; see also Section 4.8.4).

Model calculations of the radiation from a microwave burst assume that the energy of the non-thermal electrons is distributed according to a power law of the following form:

$$n(E) = KE^{-\delta}, \tag{4.4}$$

where
$$K = (\delta - 1) E_0^{\delta-1} N \tag{4.5}$$

and N is the number of electrons with $E > E_0$. For the purposes of normalization Dulk (1985) and Shevgaonkar & Kundu (1985) assume that $E_0 = 10$ keV, although electrons with energy E less than 50–100 keV contribute very little to the radiation. The value of δ can be determined from the observed spectrum and lies roughly in the range 3–4 (Schmahl *et al.*, 1985; Shevgaonkar & Kundu, *loc. cit.*).

As pointed out by Dulk (1985) the evidence is strong that the mechanism for microwave radiation from impulsive bursts is gyrosynchrotron emission (predominantly the extraordinary mode) from electrons in the energy range $E \simeq 100$–1000 keV, spiralling in a magnetic field of $B \simeq 100$–500 G, and involving harmonics $s \simeq 10$ to $s \simeq 50$ of the gyrofrequency ν_B (c.f. Eqn. (3.21)). In contrast to the (low-harmonic) gyroresonance mechanism, which accounts for part of the emission from non-flare hot loops (see Section 3.6.5), the gyrosynchrotron process is due to mildly relativistic particles, consistent with the higher effective temperatures of flare loops. In fact, the gyrosynchrotron mechanism is intermediate between the gyroresonance and synchrotron cases, the latter being characterized by highly relativistic electrons. Besides differing in respect of the range of harmonics s which are effective, the three processes differ markedly with regard to the directional properties of the emission (Dulk, *loc. cit.*, pp. 176–7).

The review cited above gives mathematical expressions for the absorption and emission coefficients κ_ν and η_ν, the effective temperature T_{eff}, the degree of circular polarization ρ_{CP} and peak emission frequency ν_{peak}, all for the case of power law electrons. These formulae exhibit the dependence of the various quantities on such physical parameters as B, θ (the viewing angle) and δ (the power law index) and are of course essential in modelling real events. Also given are expressions for κ_ν and ρ_{CP} for gyrosynchrotron emission from thermal electrons. Similar formulae for the various quantities, both for the thermal and non-thermal cases, have been given by a number of other authors, including Dulk & Dennis (1982), Petrosian (1982), Schmahl, Kundu & Dennis (1985) and Shevgaonkar & Kundu (1985).

Table 4.15. *Physical conditions in microwave bursts: inferred values*[a]

Reference	T_{eff} (K)	N_e (cm^{-3})	B (G)	Remarks
Velusamy & Kundu (1981)	—	—	120–176	Decay phase
Dulk & Dennis (1982)	1×10^9	2×10^8	220	Inhomogeneous thermal model
	—	3×10^8	500	Power law model (inhomogeneous B)
Velusamy & Kundu (1982)	—	—	500	—
Hoyng *et al.* (1983)	2.5×10^7–3×10^7	3.5×10^{10}	480–700	—
Kahler *et al.* (1984)	8×10^6	—	—	Decay phase
Willson (1984)	2.5×10^7	1×10^{10}	—	Pre-burst phase
Schmahl *et al.* (1985)	—	—	550	—
Shevgaonkar & Kundu (1985)	3×10^7–7×10^7	—	160–350	—
Klein *et al.* (1986)	5×10^7	7×10^{10}	175–240	Loop 'A'
	5×10^7	2×10^9	21–240	Loop 'B'

[a] Unless otherwise stated, the tabulated values refer to the impulsive phase; for details of the use made of any simultaneous X-ray data, the reader should consult the original papers.

Several authors have sought to infer the physical conditions in microwave bursts from model calculations. Their results are summarized in Table 4.15 and refer, unless otherwise noted, to the impulsive phase. Most of the analyses rely in part on simultaneous X-ray data; for details the reader should consult the original papers. The figures in Table 4.15 indicate a value for T_{eff} (see Dulk, 1985, pp. 171–2) of $\sim 5 \times 10^7$ K, some 20 times higher than the value of T_e for non-flare radio loops (Table 3.8). Furthermore, as in the case of the latter, the values of N_e appear to cover a large range. The inferred value of B for impulsive bursts appears to be systematically higher than that for non-flare loops, but the data is sparse.

4.9   Summary of data: flare loops

The morphological, dynamical and physical properties of flare loops observed in different regions of the spectrum are summarized in Tables 4.16 and 4.17. For many of the quantities there is a large range in values – due in part to real variations from flare to flare and from loop to loop. Nevertheless a satisfactory degree of consistency is apparent in most, but not all, cases.

Table 4.16. *Morphological and dynamical properties of flare loops: summary of data*

Quantity	Value	Wavelength	Flare phase[a]	Section
Height, true (km)	45000–60000	Hα	3	4.3.2
	4200–13000	EUV	1, 2	4.4.2
	<2000–50000	hard X-ray	~2	4.5.3
	15000–50000	microwave	~2	4.6.2
Height, projected	60000–79000	Hα	3	4.3.2
(km)	33000–67000	EUV	3	4.4.2
	15000	soft X-ray	1	4.5.2
	35000–180000	soft X-ray	3	4.5.2
Length, true (km)	60000–100000	Hα	3	4.3.2
	23000–27000	EUV	1	4.4.2
	30000–65000	microwave	~2	4.6.2
Length, projected	10000–20000	EUV	0, 2	4.4.2
or estimated (km)	15000	soft X-ray	1	4.5.2
	7250–>100000	soft X-ray	2	4.5.2
	45000–250000	hard X-ray	~2	4.5.3
Separation of	35000–55000	Hα	3	4.3.2
footpoints (km)	30000	soft X-ray	1	4.5.2
Diameter (km)	<1500–2200	Hα	3	4.3.2
	2500–14000	EUV	0, 2, 3	4.4.2
	3000	soft X-ray	1	4.5.2
	5000–20000	soft X-ray	2, 3	4.5.2
	6000	hard X-ray	~2	4.5.3
	2200–9400	microwave	~2	4.6.2
Diameter of	2200	EUV	1	4.4.2
kernel (km)	<4500	soft X-ray	1	4.5.2
	3000–10000	hard X-ray	~2	4.5.3
Inclination of loop	1–12	Hα	3	4.3.2
plane to vertical	45	Hα, H, K	3	4.3.2
β (deg)				
Tilt of axis of	14	Hα	3	4.3.2
symmetry σ (deg)	5–10	Hα, H, K	3	4.3.2
Lifetime of loop	12	Hα	3	4.3.1
system (hr)				
Lifetime of loop	0.25–1.5	Hα	3	4.3.2
(hr)	13	EUV	3	Fig. 4.10
	<72	soft X-ray	3	4.5.2
Lifetime of burst	3–30	hard X-ray	~2	4.5.3
or cluster of	1–30	microwave	~2	4.6.2
bursts (min)				
Axial flow speed	45–110	visible	3	4.3.2
V_0 (km s^{-1})	30 (downflow)	EUV	1–3	4.8.3
	200 (upflow)	EUV	~2	4.8.3
Velocity of bodily	2–10	Hα	3	4.3.1
motion (km s^{-1})	~0.5–'several'	soft X-ray	3	4.5.2

[a] 0 = preflare, 1 = rise, 2 = maximum, 3 = decay.

Table 4.17. *Physical conditions in flare loops: summary of data*

Quantity	Value	Wavelength	Flare phase[a]	Section
Temperature T_e (K)	7600–21 000	H I, He I, Ca II	—	4.8.2
	$\sim 5 \times 10^6$	Ca XV	—	4.8.2
	$2 \times 10^4 - 2 \times 10^7$ (range)	EUV	—	4.4.1
	$5 \times 10^5 - 5 \times 10^6$	EUV	3	4.8.3
	$6.9 \times 10^5 - 1.6 \times 10^7$ (kernel)	EUV	$\sim 2, 3$	4.8.3
	1.5×10^7	soft X-ray	2^b	4.8.4
	6.8×10^6	soft X-ray	3^b	4.8.4
	$\sim 3 \times 10^7$ (thermal comp.)	hard X-ray	$\sim 2-3$	4.8.4
	$3 \times 10^7 - 7 \times 10^7$ (thermal comp.)	microwave	2	4.8.5
Electron density N_e (cm^{-3})	$1 \times 10^{10} - 7 \times 10^{10}$	cont. + Ca XV	—	4.8.2
	$2 \times 10^{11} - 3 \times 10^{12}$	EUV	0	4.8.3
	$0.8 \times 10^9 - 2 \times 10^{11}$	EUV	3	4.8.3
	$1 \times 10^{10} - 5 \times 10^{11}$ (kernel)	EUV	2, 3	4.8.3
	4.7×10^{10}	soft X-ray	2^b	4.8.4
	1.7×10^{10}	soft X-ray	3^b	4.8.4
	$3 \times 10^{10} - > 1.1 \times 10^{11}$	hard X-ray	—	4.8.4
	1×10^{10}	microwave	0	4.8.5
	$2 \times 10^8 - 7 \times 10^{10}$	microwave	1–2	4.8.5
Gas pressure p ($dyne\ cm^{-2}$)	55	Ca XV	$-^c$	4.8.2
	0.02–7.0	EUV	3^d	4.8.3
	195	soft X-ray	2^e	4.8.4
	138	microwave	2^f	4.8.5
Non-thermal velocity ξ ($km\ s^{-1}$)	3–16	H I, He I, Ca II	—	4.8.2
Total magnetic field B (G)	40–80	—	—	4.8.2
	21–500	microwave	1–2	4.8.5
	120–176	microwave	3	4.8.5

[a] 0 = pre-flare, 1 = rise, 2 = maximum, 3 = decay.   [b] Average or representative values: for range of values see Table 4.14.   [c] Taking $p = 2kN_e T_e$, $T_e = 5 \times 10^6$ K, $N_e = 4 \times 10^{10}$ cm^{-3}.   [d] See Table 4.11; see also discussion in this section.   [e] Taking $p = 2kN_e T_e$, $T_e = 1.5 \times 10^7$ K, $N_e = 4.7 \times 10^{10}$ cm^{-3}.   [f] Taking $p = 2kN_e T_e$, $T_e = 5 \times 10^7$ K, $N_e = 1 \times 10^{10}$ cm^{-3}.

It is important to attempt a comparison between the figures in Tables 4.16 and 4.17 with the corresponding figures for cool and hot non-flare loops summarized in Sections 2.6 and 3.8 of earlier chapters. Let us consider, firstly, the morphological and dynamical properties. In the case of Hα loops, we find that the properties of flare and non-flare loops are the same or nearly the same. However, compared with non-flare loops, Hα flare loop systems appear to be slightly higher and to last longer, as do individual flare loops. Hα flare loops must therefore be distinguished from non-flare loops by their other properties, including their closer association with flares, brightness, direction of material flow, and number of loops in a system (cf. Section 4.3).

The morphological and dynamical properties of EUV flare loops are also similar to those of EUV non-flare loops. EUV flare loops appear to lie somewhat lower, but the ranges of values overlap. The microwave data are inadequate to make a comparison between flare and non-flare loops meaningful.

Finally, in soft X-rays both flare and non-flare loops reach very great heights, but the non-flare ones appear to reach greater heights. But again the ranges overlap. The footpoint separation for soft X-ray flare loops appears to be much smaller, but this particular quantity is very dependent on the geometry – usually unknown.

In general, therefore, we are led to the conclusion that the morphological properties of flare and non-flare loops are remarkably similar, regardless of the wavelength region in which they are observed.

The situation is very different when we compare physical conditions (Table 4.17). Comparing the results, firstly, with *hot* non-flare loops (Table 3.8) we find that independent of the wavelength region, flare loops are approximately an order of magnitude hotter than non-flare loops. Also, except in the EUV, the electron density is also an order of magnitude greater. Accordingly, we arrive at the important conclusion that the gas pressure in flare loops is approximately two orders of magnitude greater than in non-flare loops.

The values of p derived from EUV flare loop observations given in Table 4.17 appear to be anomalous. They refer to observations in the decay phase of flare loop systems (Table 4.11) and give figures similar to those quoted for non-flare EUV loops. On the other hand, if we take as typical or representative values of T_e and N_e the figures 5×10^6 K and 2×10^{11} cm^{-3} respectively (cf. Table 4.17) we obtain $p = 280$ dyne cm^{-2}, in rough agreement with the values given by the other spectral regions.

The value of the total magnetic field B given by the microwave observations appears somewhat bigger for flare than for non-flare loops, but in both cases the results are sparse.

Finally, unlike hot flare loops, cool flare loops observed in Hα and other visible region lines yield similar values of T_e (and also of non-thermal velocity ξ) to those of cool non-flare loops.

In summary, the extensive numerical data which has been derived for flare and non-flare loops leads to the conclusion that *hot* flare loops are distinguished from hot non-flare loops by their very different physical conditions rather than by morphological differences. By contrast, *cool* flare loops can be distinguished from their non-flare counterparts only by characteristics other than their morphological and physical properties.

The extensive data on the properties of flare loops summarized in Tables 4.16 and 4.17 and discussed in detail in earlier sections of the present chapter represent the product of an enormous observational and analytical effort over the past two decades or so. However, this must not blind us to the fact that much of the observational material is to some extent fragmentary. Further progress requires properly coordinated and simultaneous observations of the same events at visible, EUV, soft and hard X-ray, and microwave regions of the spectrum, together with improved imaging. Such a program will require the full panoply of ground-based, balloon, rocket and satellite instrumentation. At the time of writing (May 1989), an international program of this kind, scheduled to catch the peak of the next solar cycle (cycle 22), is at the discussion and planning stage. A new satellite, Solar-A, dedicated to improved soft and hard X-ray imaging, is due for launch in 1991 (Ogawara, 1987).

References

Albregtsen, F. & Engvold, O. (1979). High resolution Ca II K spectra of the September 8, 1977 loop system. In *Physics of Solar Prominences*, ed. E. Jensen, P. Maltby & F.Q. Orrall, pp. 246–9. Blindern-Oslo: Institute of Theoretical Astrophysics

Alissandrakis, C.E. (1986). Gyrosynchrotron emission of solar flares. *Solar Physics*, **104**, 207–21.

Alissandrakis, C.E. & Kundu, M.R. (1978). 6 centimeter observations of solar bursts with 6″ resolution. *Astrophysical Journal*, **222**, 342–56.

Alissandrakis, C.E. & Preka-Papadema, P. (1984). Microwave emission and polarization of a flaring loop. *Astronomy and Astrophysics*, **139**, 507–11.

Ambastha, A. & Bhatnagar, A. (1985). *Photographic Atlas of the Solar Chromosphere*. Udaipur: Udaipur Solar Observatory.

Antonucci, E. & 8 co-authors (1982). Impulsive phase of flares in soft X-ray emission. *Solar Physics*, **78**, 107–23.

Bonnet, R.M. (1981). The contribution of OSO-8 to our knowledge of the chromosphere and transition region. *Space Science Reviews*, **29**, 131–200.

Brown, J.C. (1971). The deduction of energy spectra of non-thermal electrons in flares from the observed dynamic spectra of hard X-ray bursts. *Solar Physics*, **18**, 489–502.

Brown, J.C. (1975). The interpretation of spectra, polarization, and directivity of solar hard X-rays. In *Solar Gamma-, X-, and EUV Radiation*, ed. S.R. Kane, pp. 245–82. Dordrecht: Reidel.

Brueckner, G.E. (1976). A.t.m. observations on the X.u.v. emission from solar flares. *Philosophical Transactions of the Royal Society of London*, Series A, **281**, 443–59.

Brueckner, G.E., Patterson, N.P. & Scherrer, V.E. (1976). Spectroscopic far ultraviolet observations of transition zone instabilities and their possible role in a pre-flare energy build-up. *Solar Physics*, **47**, 127–46.

Bruzek, A. (1964). On the association between loop prominences and flares. *Astrophysical Journal*, **140**, 746–59.

Chapman, R.D. & Neupert, W.M. (1979). The high-temperature corona associated with a loop prominence system. *Astrophysical Journal*, **229**, 799–811.

Cheng, C.-C. (1977a). Observation of a kink instability in a solar flare. *Astrophysical Journal*, **213**, 558–68.

Cheng, C.-C. (1977b). Evolution of the high-temperature plasma in the 15 June 1973 flare. *Solar Physics*, **55**, 413–29.

Cheng, C.-C. (1980). Spatial distribution of XUV emission and density in a loop prominence. *Solar Physics*, **65**, 347–56.

Cheng, C.-C. & Pallavicini, R. (1984). Analysis of the magnetic field configuration of a filament-associated flare from X-ray, UV, and optical observations. *Solar Physics*, **93**, 337–50.

Cheng, C.-C., Pallavicini, R., Acton, L.W. & Tandberg-Hanssen, E. (1985). Energy release topology in a multiple-loop solar flare. *Astrophysical Journal*, **298**, 887–97.

Cheng, C.-C. & Tandberg-Hanssen, E. (1986). Dynamic evolution of the transition zone plasma in solar flares and active region transients. *Astrophysical Journal*, **309**, 421–34.

Cheng, C.-C., Tandberg-Hanssen, E. & Orwig, L.E. (1984). Correlated observations of impulsive UV and hard X-ray bursts in solar flares from the Solar Maximum Mission. *Astrophysical Journal*, **278**, 853–62.

Cheng, C.-C. & Widing, K.G. (1975). Spatial distribution of XUV emission in solar flares. *Astrophysical Journal*, **201**, 735–9.

Cheng, C.-C. & 7 co-authors (1981). Spatial and temporal structures of impulsive bursts from solar flares observed in UV and hard X-rays. *Astrophysical Journal*, **248**, L39–43.

Cheng, C.-C. & 7 co-authors (1982). Observations of solar flare transition zone plasmas from the Solar Maximum Mission. *Astrophysical Journal*, **253**, 353–66.

Cliver, E.W. & 6 co-authors (1986). Solar gradual hard X-ray bursts and associated phenomena. *Astrophysical Journal*, **305**, 920–35.

Cornell, M.E., Hurford, G.J., Kiplinger, A.L. & Dennis, B.R. (1984). The relative timing of microwaves and hard X-rays in solar flares. *Astrophysical Journal*, **279**, 375–81.

Cui, L.-S. & 5 co-authors (1985). Simultaneous monochromatic and spectral observations of two large loop prominence groups. *Chinese Astronomy and Astrophysics*, **9**, 49–53.

Dennis, B.R. (1985). Solar hard X-ray bursts. *Solar Physics*, **100**, 465–90.

Dere, K.P. & Cook, J.W. (1979). The decay of the 1973 August 9 flare. *Astrophysical Journal*, **229**, 772–87.

Dere, K.P., Horan, D.M. & Kreplin, R.W. (1977). Structure and dynamics of a solar flare: X-ray and XUV observations. *Astrophysical Journal*, **217**, 976–87.

Dezso, L., Gesztelyi, L., Kondas, L., Kovacs, A. & Rostas, S. (1980). Motions in the solar atmosphere associated with the white light flare of 11 July 1978. *Solar Physics*, **67**, 317–38.

Dizer, M. (1978). A loop prominence system observed on May 24, 1972. *Solar Physics*, **59**, 357–60.

Dodson-Prince, H.W. & Bruzek, A. (1977). Flares and associated phenomena. In *Illustrated Glossary for Solar and Solar-Terrestrial Physics*, ed. A. Bruzek & C.J. Durrant, pp. 81–96. Dordrecht: Reidel.

Doyle, J.G. & 5 co-authors (1985). Energetics of a double flare on November 8, 1980. *Solar Physics*, **98**, 141–58.

Duijveman, A., Hoyng, P. & Machado, M.E. (1982). X-ray imaging of three flares during the impulsive phase. *Solar Physics*, **81**, 137–57.

Dulk, G.A. (1985). Radio emission from the Sun and stars. *Annual Review of Astronomy and Astrophysics*, **23**, 169–224.

Dulk, G.A. & Dennis, B.R. (1982). Microwaves and hard X-rays from solar flares: multithermal and nonthermal interpretations. *Astrophysical Journal*, **260**, 875–84.

Dwivedi, B.N., Hudson, H.S., Kane, S.R. & Svestka, Z. (1984). Hα and hard X-ray development in two-ribbon flares. *Solar Physics*, **90**, 331–41.

Engvold, O., Jensen, E. & Andersen, B.N. (1979). Kinematics of a loop prominence. *Solar Physics*, **62**, 331–41.

Fisher, R.R. (1971). On the distribution of material as a function of temperature in the post-flare loop system of 12 August 1970. *Solar Physics*, **19**, 440–50.

Fisher, R.R. (1974). The loop prominence of 11 August 1972: a coronal continuum event. *Solar Physics*, **35**, 401–8.

Foukal, P. (1975). The temperature structure and pressure balance of magnetic loops in active regions. *Solar Physics*, **43**, 327–36.

Foukal, P. (1978). Magnetic loops, downflows, and convection in the solar corona. *Astrophysical Journal*, **223**, 1046–57.

Foukal, P., Miller, P. & Gilliam, L. (1983). Electric fields in coronal magnetic loops. *Solar Physics*, **83**, 83–102.

Fuhr, J.R., Martin, G.A., Wiese, W.L. & Younger, S.M. (1981). Atomic transition probabilities for iron, cobalt, and nickel (a critical data compilation of allowed lines). *Journal of Physical and Chemical Reference Data*, **10**, 305–565.

Gibson, E.G. (1977). Observations of limb flares with a soft X-ray telescope. *Solar Physics*, **53**, 123–38.

Gu, X.M., Li, B.S., Ding, Y.J., Li, S.C. & Li, Z. (1983). The Hα-cyclonic spectra of a flare loop system on 1981 April 27. *Solar Physics*, **87**, 155–64.

Gunkler, T.A., Canfield, R.C., Acton, L.W. & Kiplinger, A.L. (1984). A consistent picture of coronal and chromospheric processes in a well-observed solar flare. *Astrophysical Journal*, **285**, 835–42.

Hanaoka, Y. & Kurokawa, H. (1986). The post flare loops observed at the total eclipse of February 16, 1980. *Solar Physics*, **105**, 133–48.

Harrison, R.A. & Simnet, G.M. (1984). Do all flares occur within a hierarchy of magnetic loops? In *Solar Maximum Analysis*, ed. P.A. Simon, pp.199–202. (*Advances in Space Research*, vol. 4, no. 7.) Oxford: Pergamon.

Haug, E., Elwert, G. & Hoyng, P. (1984). Analysis of the 1980 November 18 limb flare observed by the hard X-ray imaging spectrometer (HXIS). In *Solar Maximum Analysis*, ed. P.A. Simon, pp. 211–13. (*Advances in Space Research*, vol. 4, no. 7.) Oxford: Pergamon.

Heinzel, P. & Karlicky, M. (1987). Hα diagnostics of (post)-flare loops based on narrow-band filtergram observations. *Solar Physics*, **110**, 343–57.

Hénoux, J.-C. (1984). Study of energy release in flares. In *Solar Maximum Analysis*, ed. P.A. Simon, pp. 227–37. (*Advances in Space Research*, vol. 4, no. 7.) Oxford: Pergamon.

Hiei, E., Okamoto, T. & Tanaka, K. (1983). Observation of the flare of 12

June 1982 by Norikura coronagraph and Hinotori. *Solar Physics*, **86**, 185–91.

Holman, G.D., Kundu, M.R. & Dennis, B.R. (1984). A study of the evolution of energetic electrons in a solar flare. *Astrophysical Journal*, **276**, 761–5.

Hoyng, P., Marsh, K.A., Zirin, H. & Dennis, B.R. (1983). Microwave and hard X-ray imaging of a solar flare on 1980 November 5. *Astrophysical Journal*, **268**, 865–79.

Hoyng, P. & 23 co-authors (1981a). Hard X-ray imaging of two flares in active region 2372. *Astrophysical Journal*, **244**, L153–6.

Hoyng, P. & 11 co-authors (1981b). Origin and location of the hard X-ray emission in a two-ribbon flare. *Astrophysical Journal*, **246**, L155–9.

de Jager, C. & Boelee, A. (1984). Impulsive phase heating and a coronal explosion in a solar flare. *Solar Physics*, **92**, 227–43.

de Jager, C. & Svestka, Z. (1985). 21 May 1980 flare review. *Solar Physics*, **100**, 435–63.

Jefferies, J.T. & Orrall, F.Q. (1965). Loop prominences and coronal condensations. I. Non-thermal velocities within loop prominences. *Astrophysical Journal*, **141**, 505–25.

Kahler, S.W. (1978). The dependence of solar flare energetics on flare volumes. *Solar Physics*, **59**, 87–104.

Kahler, S.W. (1979). Preflare characteristics of active regions observed in soft X-rays. *Solar Physics*, **62**, 347–57.

Kahler, S.W. & Buratti, B.J. (1976). Preflare X-ray morphology of active regions observed with the AS&E telescope on Skylab. *Solar Physics*, **47**, 157–65.

Kahler, S.W., Krieger, A.S. & Vaiana, G.S. (1975). Morphological evolution of X-ray flare structures from the rise through the decay phase. *Astrophysical Journal*, **199**, L57–61.

Kahler, S.W., Petrasso, R.D. & Kane, S.R. (1976). The quantitative properties of three soft X-ray flare kernels observed with the AS&E X-ray telescope on Skylab. *Solar Physics*, **50**, 179–96.

Kahler, S.W., Webb, D.F., Davis, J.M. & Kundu, M.R. (1984). The spatial distribution of 6 centimeter gyroresonance emission from a flaring X-ray loop. *Solar Physics*, **92**, 271–81.

Kane, S.R. (1974). Impulsive (flash) phase of solar flares: hard X-ray, microwave, EUV and optical observations. In *Coronal Disturbances*, ed. G. Newkirk, pp. 105–41. Dordrecht: Reidel.

Kane, S.R., Fenimore, E.E., Klebesadel, R.W. & Laros, J.G. (1982). Spatial structure of ≥ 100 keV X-ray sources in solar flares. *Astrophysical Journal*, **254**, L53–7.

Kane, S.R. & 10 co-authors (1980). Impulsive phase of solar flares. In *Solar Flares*, ed. P.A. Sturrock, pp. 187–229. Boulder: Colorado Associated University Press.

Karpen, J.T., Doschek, G.A. & Seely, J.F. (1986). High-resolution X-ray spectra of solar flares. VIII. Mass upflow in the large flare of 1980 November 7. *Astrophysical Journal*, **306**, 327–39.

Klein, K.-L., Trottet, G. & Magin, A. (1986). Microwave diagnostics of energetic electrons in flares. *Solar Physics*, **104**, 243–52.

Kopriva, D. (1976). Frontispiece. *Solar Physics*, **50**, 2.

Kosugi, T. & Tsuneta, S. (1983). Time variations of hard X-ray bursts observed with the solar X-ray telescope aboard Hinotori (with a movie). *Solar Physics*, **86**, 333–8.

Krall, K.R., Reichmann, E.J., Wilson, R.M., Henze, W. & Smith, J.B. (1978). Analysis of X-ray observations of the 15 June 1973 flare in active region NOAA 131. *Solar Physics*, **56**, 383–404.

Krall, K.R., Smith, J.B. & McGuire, J.P. (1980). On the physics of a long decay X-ray event. *Solar Physics*, **66**, 371–91.

Kundu, M.R. (1983). Spatial characteristics of microwave bursts. *Solar Physics*, **86**, 205–17.

Kundu, M.R. (1984). Relative positions of microwave and hard X-ray burst sources. In *Solar Maximum Analysis*, ed. P.A. Simon, pp. 157–62. (*Advances in Space Research*, vol. 4, no. 7.) Oxford: Pergamon.

Kundu, M.R. (1986). Decimeter continuum radio emission from a post-flare loop. *Solar Physics*, **104**, 223–6.

Kundu, M.R., Machado, M.E., Erskine, F.T., Rovira, M.G. & Schmahl, E.J. (1984). Microwave, soft and hard X-ray imaging observations of two solar flares. *Astronomy and Astrophysics*, **132**, 241–52.

Kundu, M.R., Schmahl, E.J. & Velusamy, T. (1982). Magnetic structure of a flaring region producing impulsive microwave and hard X-ray bursts. *Astrophysical Journal*, **253**, 963–74.

Kureizumi, T. & 5 co-authors (1977). An interpretation of hydrogen and helium line spectra of the loop prominence observed on November 3, 1973. *Publications of the Astronomical Society of Japan*, **29**, 129–48.

Kurokawa, H., Kitahara, T., Nakai, Y., Funakoshi, Y. & Ichimoto, K. (1986). High resolution observations of Hα solar flares and temporal relations between Hα and X-ray, microwave emissions. *Astrophysics and Space Science*, **118**, 149–52.

Lang, K.R. & Willson, R.F. (1984). V.L.A. observations of flare build-up in coronal loops. In *Solar Maximum Analysis*, ed. P.A. Simon, pp. 105–10. (*Advances in Space Research*, vol. 4, no. 7.) Oxford: Pergamon.

Levine, R.H. (1978). EUV structure of a small flare. *Solar Physics*, **56**, 185–203.

Lin, R.P. & Schwartz, R.A. (1987). High spectral resolution measurements of a solar flare hard X-ray burst. *Astrophysical Journal*, **312**, 462–74.

Loughhead, R.E., Blows, G. & Wang, J.-L. (1985). High-resolution photography of the solar chromosphere. XXII. Relationship between the Hα and hard X-ray emission from the 3B/X1.5 flare of 13 May 1981. *Publications of the Astronomical Society of Japan*, **37**, 619–31.

Loughhead, R.E., Chen, C.-L. & Wang, J.L. (1984). High-resolution photography of the solar chromosphere. XVIII. Axial tilt of Hα loops observed on the disk. *Solar Physics*, **92**, 53–65.

Loughhead, R.E., Wang, J.-L. & Blows, G. (1983). High-resolution photography of the solar chromosphere. XVII. Geometry of Hα flare loops observed on the disk. *Astrophysical Journal*, **274**, 883–99.

McCabe, M.K. (1973). Spatial relationships between λ5303 and Hα components of a loop prominence system. *Solar Physics*, **30**, 439–48.

McClymont, A.N. & Canfield, R.C. (1986). The solar flare extreme ultraviolet to hard X-ray ratio. *Astrophysical Journal*, **305**, 936–46.

MacCombie, W.J. & Rust, D.M. (1979). Physical parameters in long-decay coronal enhancements. *Solar Physics*, **61**, 69–88.

Machado, M.E. (1971). Thermal, turbulent and macroscopic motions in the loop prominence of May 4, 1960. *Bulletin of the Astronomical Institutes of Czechoslovakia*, **22**, 117–21.

Machado, M.E. (1985). An evidence of flare energy buildup and release related to magnetic shear and reconnection. *Solar Physics*, **99**, 159–66.

Machado, M.E., Grossi Gallegos, H. & Silva, A.F. (1972). The loop prominence of May 13, 1971 and its associated effects. *Solar Physics*, **25**, 402–12.

Machado, M.E., Hernandez, A.M., Rovira, M.G. & Sneibrun, C.V. (1984). Study of combined soft and hard X-ray images of solar flares. In *Solar Maximum Analysis*, ed. P.A. Simon, pp. 91–4. (*Advances in Space Research*, vol. 4, no. 7.) Oxford: Pergamon.

Machado, M.E., Rovira, M.G. & Sneibrun, C.V. (1985). Hard X-ray imaging evidence of nonthermal and thermal burst components. *Solar Physics*, **99**, 189–217.

MacKinnon, A.L., Brown, J.C. & Hayward, J. (1985). Quantitative analysis of hard X-ray 'footprint' flares observed by the Solar Maximum Mission. *Solar Physics*, **99**, 231–62.

MacKinnon, A.L., Costa, J.E.R., Kaufmann, P. & Dennis, B.R. (1986). Interpretation of temporal features in an unusual X-ray and microwave burst. *Solar Physics*, **104**, 191–8.

MacNeice, P. & 8 co-authors (1985). Multiwavelength analysis of a well observed flare from SMM. *Solar Physics*, **99**, 167–88.

Malville, J.M. & Schindler, M. (1981). Oscillations of a loop prominence preceding a limb flare. *Solar Physics*, **70**, 115–28.

Marsh, K.A. & Hurford, G.J. (1980). Two-dimensional VLA maps of solar bursts at 15 and 23 GHz with arcsec resolution. *Astrophysical Journal*, **240**, L111–14.

Marsh, K.A., Hurford, G.J., Zirin, H. & Hjellming, R.M. (1980). VLA observations of impulsive solar flares at 4.9 GHz. *Astrophysical Journal*, **242**, 352–8.

Marsh, K.A., Zirin, H. & Hurford, G.J. (1979). VLA observations of solar flares, interpreted with optical, X-ray, and other microwave data. *Astrophysical Journal*, **228**, 610–15.

Martin, S.F. (1979). Study of the post-flare loops on 29 July 1973. III. Dynamics of the Hα loops. *Solar Physics*, **64**, 165–76.

Mason, H.E., Shine, R.A., Gurman, J.B. & Harrison, R.A. (1986). Spectral line profiles of Fe XXI 1345.1 Å from the Solar Maximum Mission. *Astrophysical Journal*, **309**, 435–48.

Mok, Y. (1983). Microwave signatures from a reconnecting plasma pinch, with application to loop flares. *Astrophysical Journal*, **275**, 901–15.

Moore, R. & 15 co-authors (1980). The thermal X-ray flare plasma. In *Solar Flares*, ed. P.A. Sturrock, pp. 341–409. Boulder: Colarado Associated University Press.

Nolte, J.T., Gerassimenko, M., Krieger, A.S., Petrasso, R.D. & Svestka, Z. (1979). Study of the post-flare loops on 29 July 1973. I. Dynamics of the X-ray loops. *Solar Physics*, **62**, 123–32.

Ogawara, Y. (1987). The Solar-A Mission. *Solar Physics*, **113**, 361–70.

Orwig, L.E. & Woodgate, B.E. (1986). Comparison of solar hard X-ray and UV line and continuum bursts with high time resolution. In *The Lower Atmosphere of Solar Flares*, ed. D.F. Neidig, pp. 306–17. Sunspot: National Solar Observatory/Sacramento Peak.

Pallavicini, R. (1984). The Solar Maximum Mission. *Memorie della Società Astronomica Italiana*, **55**, 633–52.

Pallavicini, R., Serio, S. & Vaiana, G.S. (1977). A survey of soft X-ray limb flare images: the relation between their structure in the corona and other physical parameters. *Astrophysical Journal*, **216**, 108–22.

Pallavicini, R. & Vaiana, G.S. (1976). The spatial structure of a solar flare in soft X-rays and centimetric wavelengths. *Solar Physics*, **49**, 297–313.

Pallavicini, R. & Vaiana, G.S. (1980). The flare of September 7, 1973: a typical example of a newly recognized class of solar transients. *Solar Physics*, **67**, 127–42.

Pallavicini, R., Vaiana, G.S., Kahler, S.W. & Krieger, A.S. (1975). Spatial structure and temporal development of a solar X-ray flare observed from Skylab on June 15, 1973. *Solar Physics*, **45**, 411–33.

Petrasso, R.D., Kahler, S.W., Krieger, A.S., Silk, J.K. & Vaiana, G.S. (1975). The location of the site of energy release in a solar X-ray subflare. *Astrophysical Journal*, **199**, L127–30.

Petrosian, V. (1982). Structure of the impulsive phase of solar flares from microwave observations. *Astrophysical Journal*, **255**, L85–9.

Phillips, K.J.H. & Zirker, J.B. (1977). A model for X-ray emission from loop prominences. *Solar Physics*, **53**, 41–58.

Pneuman, G.W. (1981). Two-ribbon flares: (post-)flare loops. In *Solar Flare Magnetohydrodynamics*, ed. E.R. Priest, pp. 379–428. New York: Gordon and Breach.

Pneuman, G.W. (1982). Energetics of two-ribbon solar flares. *Solar Physics*, **78**, 229–41.

Poland, A.I. & 10 co-authors (1982). The impulsive and gradual phases of a solar limb flare as observed from the Solar Maximum Mission satellite. *Solar Physics*, **78**, 201–13.

Preka-Papadema, P. & Alissandrakis, C.E. (1988). Spatial and spectral structure of a solar flaring loop at centimeter wavelengths. *Astronomy and Astrophysics*, **191**, 365–73.

Priest, E.R. (1981). Introduction. In *Solar Flare Magnetohydrodynamics*, ed. E.R. Priest, pp. 1–46. New York: Gordon & Breach.

Priest, E.R. (1982). *Solar Magnetohydrodynamics*. Dordrecht: Reidel.

Priest, E.R. (1985). Theory of flares. *Transactions of the International Astronomical Union*, XIX A, 90–6.

Roy, J.-R. (1972). The magnetic configuration of the November 18, 1968 loop prominence system. *Solar Physics*, **26**, 418–30.

Rust, D.M. (1984). Energy-transfer processes in flares. In *Solar Maximum*

Analysis, ed. P.A. Simon, pp. 191–8. (*Advances in Space Research*, vol. 4, no. 7.) Oxford: Pergamon.

Rust, D.M. & Bar, V. (1973). Magnetic fields, loop prominences and the great flares of August, 1972. *Solar Physics*, **33**, 445–59.

Rust, D.M., Simnett, G.M. & Smith, D.F. (1985). Observational evidence for thermal wave fronts in solar flares. *Astrophysical Journal*, **288**, 401–9.

Rust, D.M. & 5 co-authors (1981). Optical and radio observations of the 1980 March 29, April 30, and June 7 flares. *Astrophysical Journal*, **244**, L179–83.

Ruzdjak, V. (1981). The loop prominence of September 26th, 1963. *Bulletin of the Astronomical Institutes of Czechoslovakia*, **32**, 144–7.

Sakurai, T. (1985). Magnetic field structures of hard X-ray flares observed by Hinotori spacecraft. *Solar Physics*, **95**, 311–21.

Schmahl, E.J., Kundu, M.R. & Dennis, B.R. (1985). Microwave imaging of a solar limb flare: comparison of spectra and spatial geometry with hard X-rays. *Astrophysical Journal*, **299**, 1017–26.

Schmahl, E.J., Solodyna, C.V., Smith, J.B. & Cheng, C.-C. (1978). The pre-onset morphology of the 5 September 1973 flare. *Solar Physics*, **60**, 323–39.

Shevgaonkar, R.K. & Kundu, M.R. (1985). Dual frequency observations of solar microwave bursts using the VLA. *Astrophysical Journal*, **292**, 733–51.

Shull, J.M. & Van Steeberg, M. (1982). The ionization equilibrium of astrophysically abundant elements. *Astrophysical Journal Supplement Series*, **48**, 95–107.

Simnett, G.M. (1983). Flares on the Sun: selected results from SMM. In *Activity in Red-Dwarf Stars*, ed. P.B. Byrne & M. Rodono, pp. 289–305. Dordrecht: Reidel.

Simnett, G.M. & Strong, K.T. (1984). The impulsive phase of a solar limb flare. *Astrophysical Journal*, **284**, 839–47.

Simon, P. & McIntosh, P.S. (1972). Survey of current solar forecast centers. In *Solar Activity: Observations and Predictions*, ed. P.S. McIntosh & M. Dryer, pp. 343–57. Cambridge: MIT Press.

Smith, D.F. & Orwig, L.E. (1988). Evidence for two hard X-ray components in double power-law fits to the 1980 June 7 flare. *Astrophysical Journal*, **327**, 466–72.

Sturrock, P.A. (ed.) (1980). *Solar Flares*. Boulder: Colorado Associated University Press.

Svestka, Z. (1976). *Solar Flares*. Dordrecht: Reidel.

Svestka, Z. (1981). Flare observations. In *Solar Flare Magnetohydrodynamics*, ed. E.R. Priest, pp. 47–137. New York: Gordon & Breach.

Svestka, Z. & Poletto, G. (1985). Hard X-ray images of possible reconnection in the flare of 21 May, 1980. *Solar Physics*, **97**, 113–29.

Svestka, Z.F. & 5 co-authors (1987). Multi-thermal observations of newly formed loops in a dynamic flare. *Solar Physics*, **108**, 237–50.

Svestka, Z. & 6 co-authors (1982). Study of the post-flare loops on 29 July 1973. IV. Revision of T and n_e values and comparison with the flare of 21 May 1980. *Solar Physics*, **78**, 271–85.

Takakura, T. (1986). Dynamics of electron beams in a coronal loop and the hard X-ray burst. *Solar Physics*, **104**, 363–89.

Takakura, T., Kundu, M.R., McConnell, D. & Ohki, K. (1985). Simultaneous observations of hard X-ray and microwave burst sources in a limb flare. *Astrophysical Journal*, **298**, 431–6.

Takakura, T., Ohki, K., Nitta, N. & Wang, J.L. (1984a). Hard X-ray imaging of a solar two-ribbon flare on 1981 August 21. *Astrophysical Journal*, **281**, L51–3.

Takakura, T., Tanaka, K. & Hiei, E. (1984). High-temperature phenomena in flares. In *Solar Maximum Analysis*, ed. P.A. Simon, pp. 143–52. (*Advances in Space Research*, vol. 4, no. 7.) Oxford: Pergamon.

Takakura, T., Tanaka, K., Nitta, N., Kai, K. & Ohki, K. (1986). X-ray imaging of a solar limb flare on 1982 January 22. *Solar Physics*, **107**, 109–21.

Takakura, T. & 6 co-authors (1984b). Hard X-ray imaging of a solar gradual hard X-ray burst on April 1, 1981. *Solar Physics*, **94**, 359–67.

Takakura, T. & 8 co-authors (1983). Hard X-ray imaging of a solar limb flare with the X-ray telescope aboard the Hinotori satellite. *Astrophysical Journal*, **270**, L83–7.

Tanaka, K. (1983b). Flares on the Sun: selected results from Hinotori. In *Activity in Red-Dwarf Stars*, ed. P.B. Byrne & M. Rodono, pp. 307–20.

Tanaka, K. (1986). High-energy observations of solar flares. *Astrophysics and Space Science*, **118**, 101–13.

Tanaka, K. (1987). Impact of X-ray observations from the Hinotori satellite on solar flare research. *Publications of the Astronomical Society of Japan*, **39**, 1–45.

Tanaka, Y. (1983a). Introduction to Hinotori. *Solar Physics*, **86**, 3–6.

Tandberg-Hanssen, E., Reichmann, E. & Woodgate, B. (1983). Behaviour of transition-region lines during impulsive solar flares. *Solar Physics*, **86**, 159–71.

Tandberg-Hanssen, E. & 6 co-authors (1984). Observation of the impulsive phase of a simple flare. *Solar Physics*, **90**, 41–62.

Tapping, K.F. (1986). Solar VLBI of compact transient sources. *Solar Physics*, **104**, 199–206.

Trottet, G. (1986). Relative timing of hard X-rays and radio emissions during the different phases of solar flares: consequences for the electron acceleration. *Solar Physics*, **104**, 145–63.

Tsuneta, S. & 7 co-authors (1983). Vertical structure of hard X-ray flare. *Solar Physics*, **86**, 313–21.

Tsuneta, S. & 8 co-authors (1984a). Hard X-ray imaging observations of solar hot thermal flares with the Hinotori spacecraft. *Astrophysical Journal*, **284**, 827–32.

Tsuneta, S. & 8 co-authors (1984b). Hard X-ray imaging of the solar flare on 1981 May 13 with the Hinotori spacecraft. *Astrophysical Journal*, **280**, 887–91.

Underwood, J.H., Antiochos, S.K., Feldman, U. & Dere, K.P. (1978). Evolution of the coronal and transition-zone plasma in a compact flare: the event of 1973 August 9. *Astrophysical Journal*, **224**, 1017–27.

Vaiana, G.S. & Giaconni, R. (1969). Observation of an X-ray flare: spatial distribution and physical parameters. In *Plasma Instabilities in Astrophysics*, ed. D.G. Wentzel & D.A. Tidman, pp. 91–118. New York: Gordon & Breach.

Van Hoven, G. & Hurford, G.J. (1984). Flare precursors and onset. In *Solar Maximum Analysis*, ed. P.A. Simon, pp. 95–103. (*Advances in Space Research*, vol. 4, no. 7.) Oxford: Pergamon.

Velusamy, T. & Kundu, M.R. (1981). VLA observations of postflare loops at 20 centimeter wavelength. *Astrophysical Journal*, **243**, L103–7.

Velusamy, T. & Kundu, M.R. (1982). VLA observations of the evolution of a solar burst source structure at 6 centimeter wavelength. *Astrophysical Journal*, **258**, 388–92.

Velusamy, T., Kundu, M.R., Schmahl, E.J. & McCabe, M. (1987). Simultaneous VLA observations of a flare at 6 and 20 centimeter wavelengths. *Astrophysical Journal*, **319**, 984–92.

Vilmer, N. (1987). Hard X-ray emission processes in solar flares. *Solar Physics*, **111**, 207–23.

Vorpahl, J.A. (1976). The triggering and subsequent development of a solar flare. *Astrophysical Journal*, **205**, 868–73.

Vorpahl, J.A., Gibson, E.G., Landecker, P.B., McKenzie, D.L. & Underwood, J.H. (1975). Observations of the structure and evolution of solar flares with a soft X-ray telescope. *Solar Physics*, **45**, 199–216.

Vorpahl, J.A., Tandberg-Hanssen, E. & Smith, J.B. (1977). Coronal plasma parameters in a long-duration X-ray event observed by Skylab. *Astrophysical Journal*, **212**, 550–60.

Wang, J.L. & 5 co-authors (1987). An observational study of the 2B/X2.8 flare of 30 March, 1982 in optical, radio, and X-ray ranges. *Solar Physics*, **108**, 169–81.

Webb, D.F. (1985). Coronal X-ray activity preceding solar flares. *Solar Physics*, **97**, 321–44.

Widing, K.G. (1975). Fe XXIV emission in solar flares observed with the NRL/ATM XUV slitless spectrograph. In *Solar Gamma-, X-, and EUV Radiation*, ed. S.R. Kane, pp. 153–63. Dordrecht: Reidel.

Widing, K.G. & Cheng, C.-C. (1974). On the Fe XXIV emission in the solar flare of 1973 June 15. *Astrophysical Journal*, **194**, L111–13.

Widing, K.G. & Dere, K.P. (1977). Multiple loop activations and continuous energy release in the solar flare of June 15, 1973. *Solar Physics*, **55**, 431–53.

Widing, K.G. & Spicer, D.S. (1980). XUV observations of a dense, compact flare. *Astrophysical Journal*, **242**, 1243–56.

Wiese, W.L., Smith, M.W. & Glennon, B.M. (1966). *Atomic Transition Probabilities. Volume I. Hydrogen Through Neon.* NSRDS-NBS 4. Washington: National Bureau of Standards.

Wiese, W.L., Smith, M.W. & Miles, B.M. (1969). *Atomic Transition Probabilities. Volume II. Sodium Through Calcium.* NSRDS-NBS 22. Washington: National Bureau of Standards.

Willson, R.F. (1983). High-resolution observations of solar radio bursts at 2, 6, and 20 cm wavelength. *Solar Physics*, **83**, 285–303.

Willson, R.F. (1984). Observations of preburst heating and magnetic field changes in a coronal loop at 20 cm wavelength. *Solar Physics*, **92**, 189–98.

Willson, R.F. & Lang, K.R. (1984). Very Large Array observations of solar active regions. IV. Structure and evolution of radio bursts from 20 centimeter loops. *Astrophysical Journal*, **279**, 427–37.

Withbroe, G.L. (1978). The thermal phase of a large solar flare. *Astrophysical Journal*, **225**, 641–9.

Woodgate, B.E., Shine, R.A., Poland, A.I. & Orwig, L.E. (1983). Simultaneous ultraviolet line and hard X-ray bursts in the impulsive phase of solar flares. *Astrophysical Journal*, **265**, 530–4.

Woodgate, B.E. & 13 co-authors (1980). The ultraviolet spectrometer and polarimeter on the Solar Maximum Mission. *Solar Physics*, **65**, 73–90.

Yan, G.-Y., Ding, Y.-J. & Xu, A.-A. (1987). Computation and analysis of the physical fields of a disk flare loop. *Chinese Astronomy and Astrophysics*, **11**, 57–63.

5

Structure, dynamics and heating of loops

5.1   Introduction

In the preceding chapters we have summarized the observed properties of coronal loop structures; we now turn to the interpretation of these properties and provide an account of the physics of coronal loops. In this chapter we shall describe the models of individual loops and in the following chapter we shall consider systems of loops in the context of global models of the solar corona and of stellar coronae in general.

As a preliminary, though, we need to define more precisely than before what is meant by a coronal *loop*. So far we have employed the intuitive definition of a *plasma loop* as a continuous structure traceable from a point near the photospheric surface along an arc in the corona to another point at which it returns to the surface. Generally the limited spatial resolution offered by the observations precludes us from saying whether a loop is simple, with a more or less uniform curvature, or whether it possesses a more complicated topology with knots or braids. We have also adopted the view that the visible structures of the corona – the plasma loops – act as tracers of an underlying magnetic field structure, although we have seen that there is little direct evidence of the connection. Thus the first question which we must address in this chapter is the relation between the observed loops and magnetic loops. At this point, the meaning of the term *loop* becomes a matter of some subtlety.

Since the work of Faraday and Maxwell, physicists have found it convenient to depict a magnetic field in terms of the mathematical construction of a line of force, also known as a field line; through each point of space a line may be traced having the direction of the magnetic induction or magnetic flux density $\mathbf{B}$ at each successive step. The number of lines drawn crossing a unit area normal to $\mathbf{B}$ can be used to represent the magnitude of $\mathbf{B}$. The topological properties of field lines stem pri-

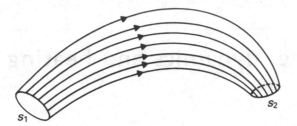

Fig. 5.1. A section of a magnetic flux loop. The side walls are delineated by the magnetic field lines passing through the perimeters of the end faces S_1 and S_2. No field line may cross from the interior to the exterior of the loop.

marily from the first of the Maxwell equations governing the electromagnetic field:

$$\nabla \cdot \mathbf{B} = 0. \tag{5.1}$$

The zero on the right-hand side of this equation expresses the fact that there are no static sources of magnetic field in the absence of magnetic monopoles. As a consequence, lines of force must be continuous and cannot end at any point of space. In the simplest geometries this means that all field lines are closed loops which may be termed *magnetic field loops*.

The coronal magnetic field is generally believed to be generated in the interior of the Sun from whence segments rise to the surface in bundles of field lines (cf. Section 6.2.1). The protruding bundles fill the coronal volume, where they compose the coronal field. The portions of the bundles remaining below the surface anchor the coronal field to the Sun. The topology of a bundle of field lines is characterized by the fact that the field lines passing through a compact area about any particular point retain a compact cross section throughout the coronal region. Figure 5.1 illustrates how the superficial field lines of the bundle define a volume which is called a *magnetic flux loop*. The end faces S_1 and S_2 are the photospheric cross-sections – the footpoints – where the field lines return to the interior. All the field lines passing through the end face S_1 emerge through the other end face S_2. Most, if not all, models of the solar magnetic field assume that the coronal field largely preserves this character at all times subsequent to emergence.

However, as Parker (1979) stresses, other topologies are possible. After emergence, magnetic flux loops are forced to adjust continually both to subsurface motions which change the locations of their footpoints, and to the emergence of new flux loops. Field lines belonging to adjacent flux loops may then link, destroying the topological isolation of each loop. Linkage is achieved by the process of reconnection in which contiguous field lines are severed and rejoined each to the other. Ultimately, lines of force may be made to wander endlessly through space. In this situation, a field line may return an indefinite number of times to thread any small area in the corona. Moreover, the field lines threading this area may follow quite independent paths as they move away from it, so that they form ever thinner and more tangled filaments. Individual magnetic field loops may then be traced, but no coherent magnetic flux loops emerge from the chaotic mass of field lines.

The validity of the flux loop picture of coronal magnetic fields has been hotly debated in recent years. Some authors stress the difficulty of finding general three-dimensional static configurations. Parker (1979), for example, suggested that most structural changes forced on the coronal field by subsurface evolutionary processes would lead to *non-equilibrium*, a state which could be resolved only by continual reconnection and change of topology. Since energy dissipation is inevitably associated with re-connection (cf. Section 5.7.3), this has come to be known as *topological dissipation*. Others claim that possible equilibrium states are more universal than Parker's analysis would suggest. According to van Ballegooijan (1985), a flux loop topology can change under only three circumstances: if the coronal field has a pre-existing singular point, i.e. a so-called neutral point where the field vanishes, or if the ends of the tube are discontinuous bundles of flux or are subjected to a discontinuous distortion. Whilst it is generally believed that the photospheric magnetic field is concentrated into small-scale elements (cf. Section 6.2.2), van Ballegooijan discounts the possibility that discontinuities and singular points are of general enough occurrence for topological dissipation to be significant. This controversy is discussed more fully in Section 6.3.1, but theory at present does not allow us to decide whether flux loops are quasi-permanent features of the coronal field structure or whether they are merely transient simplifications of a much more complicated magnetic configuration.

In Section 5.2 we look at the empirical evidence for the existence of magnetic flux loops. We describe how magnetic field structures may be calculated from observations of the magnetic field at the footpoints in the photosphere. The techniques assume implicitly that the field is not chaotic. In Section 5.2.3 we demonstrate that a reasonable correspondence exists between the computed magnetic structure and the observed plasma structure. The flux-loop picture is therefore consistent, but neither the theoretical nor empirical arguments are conclusive. In the remainder of the chapter we shall follow the vast majority of authors and adopt the premise that plasma loops are equivalent to magnetic flux loops as a working hypothesis.

We then outline, in Section 5.3.1, how we may describe the behaviour of a plasma under conditions pertaining to the solar corona. These conditions are characterized by the dominance of the magnetic pressure over the gas pressure, and the equations of fluid motion in the presence of a strong magnetic field are developed in Section 5.3.2. These are the *magneto-hydrodynamic* or MHD equations. We also introduce the *ideal* MHD approximation in which all sources and sinks of energy and momentum are neglected. This is an important simplification which can be applied to dynamic phenomena in the corona. To be valid, the time scales must be much shorter than those of the neglected terms, which are of the order of minutes. The use of the ideal equations avoids a major difficulty associated with the full equations, namely, the fact that we cannot specify the source of thermal energy. The identification of the principal mechanism responsible for heating the corona is an outstanding problem (cf. Section 5.7). In the case of slower dynamical processes, the ideal MHD approximation is not appropriate and we shall be forced to use arbitrary parametrized forms for the heating source in the energy equation in much of the discussion.

Having established the physical and mathematical bases, we then consider in detail coronal loop models, beginning in Section 5.4 with one-dimensional models. By taking only the component of the momentum equation in the direction of the field at each point along a loop, explicit reference to the magnetic field can be eliminated. Of course, it is the field which governs the geometry of the loop, so that the shape of the loop must be assumed in these models. In Section 5.4.2, we examine static loop models and show that their gross properties can be described very well by relations known as *scaling laws*. A simple analytical model reveals that the thermal structure of hot loops results from a balance between heating and radiation losses, with thermal conduction serving only to transport the heat from where it is released in the coronal part of the loop to where it is radiated away in the transition region at the base. In cool loops there is a local balance between heat input and radiation losses at each point. The balance between input and output allows a quite detailed discussion of the thermal structure of loops without a knowledge of the heating process because the processes of heat conduction and of radiative losses are well known. We here compare model calculations of emission measure with observations. In Section 5.4.3 we consider the effects of asymmetries of loop geometry, heating and footpoint behaviour. These can all cause steady flows to develop in coronal loops and we examine their influence on the thermal structure.

In Section 5.5 we relax the assumption of a steady state in the models and look at time-dependent behaviour. We first investigate the stability of the equilibrium models. In Section 5.5.1 we describe the local linear analysis of the one-dimensional models. This reveals two modes of response to a perturbation, an almost stationary thermal condensation mode and travelling acoustic modes. These modes appear to be unstable in the corona. However, the local analysis does not take account of the contrasting responses of the hot coronal part of a loop and the cooler basal parts. If these are both correctly incorporated in a global analysis (Section 5.5.2), the thermal stability of coronal loops seems to be assured. We look more closely at the processes by which equilibrium is established in Section 5.5.3 and discuss numerical simulations of time-dependent flows. Special attention is given to the flow characteristics of the cooler plasma, which may be compared with observations of emission measure and Doppler shift of EUV lines formed at about 10^5 K. The simulations and observations prove hard to reconcile, leaving the suspicion that the physics of the transition region are not correctly modelled at present.

One-dimensional models, which ignore magnetic effects, establish only necessary tests for stability. A general disturbance of a coronal loop would perturb also the magnetic field so that a proper discussion of loop stability requires the full three-dimensional structure to be included. In Section 5.6.1, we introduce some multi-dimensional equilibrium models of the magnetic and thermodynamic structures. These are all simplified models in which the curvature of the longitudinal axis of the loop is ignored. We shall refer to a loop that is treated as a straight cylindrically symmetric object embedded in a magnetic medium as a *magnetic flux tube*. In Section 5.6.2 we describe the various techniques employed to analyse the MHD stability of these tubes. The results depend critically on the boundary conditions imposed at the ends of the tube. In reality, the subsurface field is 'trapped' between convection cells and cannot move freely (cf. Section 6.2.1). This immobilization of the boundary field is known as *line-tying* and most authors agree that it can stabilize global MHD motions if the flux tube is not twisted too much. However, local instabilities might still develop along special magnetic surfaces within the loop. The role such instabilities might play in breaking down the thermal and dynamical isolation of coronal loops is discussed briefly in this section, but their role in heating the plasma is deferred to Section 5.7.

We move next to the response of stable structures to perturbations. High-frequency waves can be analysed in the ideal MHD limit. In Section 5.6.3 the discrete spectrum of Alfvén and magnetoacoustic waves is derived for uniform media and for structures consisting of uniform segments joined discontinuously together. The spectrum consists of its normal modes. Each mode can exist independently of the others and, as a consequence of the ideal approximation, each propagates energy at a constant rate. When there is a continuous variation of properties between structures, a disturbance excites not a set of normal modes but a large-scale collective response which is coupled energetically to a set of waves in a continuous frequency band, localized to the interface region between the structures. The process by which energy is transferred from the collective disturbance to the interface waves is known as *resonant absorption*. If it is argued that the flow of energy into the waves must be balanced in the steady state by the dissipation of those waves, the absorption rate can be interpreted as a dissipation rate, even though the dissipation itself is not described by the ideal MHD equations. The dissipation processes that result when there are departures from ideal MHD behaviour and when the MHD approximation itself breaks down are the subject of Section 5.6.4.

Resonant absorption of mechanical disturbances is undoubtedly one mechanism that must contribute to the heating of the corona, and has been studied in great detail in recent years. However, there are other sources of energy and other dissipation mechanisms that can also contribute to the heating. In Section 5.7.1 we shall see that the *ultimate* source of the energy is thought to reside in the subsurface layers where the gas-dynamical forces are strong enough to transmit stresses to the magnetic field. We may picture the energy as being transported to the corona by mechanical disturbances propagating along the field or by currents flowing along the field. In either case, the process is electrodynamic in character, at least in the corona where the magnetic field is everywhere dominant. The various heating mechanisms can be grouped according to the time scale of the electrodynamic disturbances. Those of short time scale have rapidly fluctuating currents that flow predominantly perpendicular to the field. They are known as *AC mechanisms* (Section 5.7.2). The wave heating theories fall into this class. Those of long time scale are associated with almost steady electric currents and are therefore known as *DC mechanisms* (Section 5.7.3). Heating by filamentary currents flowing along coronal loops belongs to this class. We are unable to assess the relative importance of these mechanisms at present. Indeed, it is likely that different mechanisms are important in different regions and at different times.

However, this gap in our knowledge of the basic physics of coronal loops is not as serious as might at first appear. In one of the most significant developments of recent years, Ionson (1983) showed that one could devise a *global* description of loop electrodynamics in terms of an electric circuit analogy. The energy fed into this circuit by the subphotospheric disturbances can be calculated. In a steady state, the thermal losses are fixed simply by the need to balance this input, the mechanism being irrelevant. However, this approach requires the coronal portion of the loop to be properly coupled to its photospheric and subsurface continuation. The treatment of loops as part of a global system is the subject of Section 6.3.2 of the following chapter.

This chapter concentrates on the commonest varieties of coronal loop, the hot and cool quiescent loops. We expect these loops to display the physics of the solar corona in its simplest manifestation, and a thorough understanding of these phenomena is a necessary prerequisite to any discussion of more exotic varieties of loops. In the present state of knowledge, a detailed treatment of flare loops is not warranted. Flares in loops are essentially dynamical phenomena with time scales ranging from a few seconds to hours, resulting from the rapid injection of large amounts of energy. The major source of flare energy, like the source of quiescent coronal heating, has not yet been identified, but many of the ideas concerning the mechanisms by which the energy is released in flare loops are similar to those forming the basis of theories of normal, quiescent coronal loops. Furthermore, the thermal and dynamical evolution of a flaring loop – at least in a compact flare – is similar, although on a grander scale, to the response of normal loops to smaller changes in energy input. We shall therefore refer to flare loops (in Section 5.8) only insofar as to note how their models relate to those of quiescent loops described in the preceding pages. For comprehensive reviews of flare theory *per se*, the interested reader is referred to Priest (1980), Sturrock (1980) and Spicer & Brown (1981).

We close this chapter in Section 5.9 with a brief summary of the current state of coronal loop modelling.

5.2 Coronal magnetic field models
5.2.1 *Introduction*

We have little empirical knowledge of the strength of the coronal field and almost none of its topology, so that a discussion of the relationship between the plasma loops described in the preceding chapters and the coronal magnetic field must rest heavily on models based on theory. In this section we shall outline the theory governing the structure of magnetic fields under the conditions obtaining in the solar corona, and describe the models that may be inferred from measurements of the magnetic field made in the solar photosphere, the only region of the Sun where such measurements are at all reliable. The aim of this procedure is to provide a basis for comparing the structure of the observed plasma morphology with that of the inferred fields.

Despite our inability to measure the field in the corona with any precision, the significance of its presence as a factor in the physics of the solar corona may be demonstrated by a simple calculation. The spatially averaged magnetic flux density in the solar photosphere ranges from about 0.4 mT in the quietest regions to over 30 mT in active regions. The *minimum* mean-square field strength is obtained by assuming that the field is uniform; hence the magnetic pressure $B^2/2\mu_0$ ranges upwards from about 0.1 Pa in quiet regions to well over 400 Pa in active regions. The estimates will not be significantly different in the low corona at heights of 2000 km or so, since this distance is small compared to the radius of the Sun ($1R_\odot = 7 \times 10^5$ km). On the other hand, empirical estimates of the gas pressure in coronal active regions lie in the range of 0.1–1 Pa. The magnetic pressure therefore greatly exceeds the gas pressure.

The ratio of the gas pressure to the magnetic pressure is denoted conventionally by β. The solar corona is thus an example of a low-β gas. The subsurface regions, on the other hand, comprise high-β gas. In the latter the gas dynamics largely control the field (cf. Section 6.2), but the field controls the gas in the former. In low-β systems, the field either can simply expand in response to unbalanced magnetic pressures or can adopt a static configuration in which magnetic stresses balance one another, a situation in which the field is said to be 'force-free'. However, as a consequence of the virial theorem (see, for example, Priest, 1982), no field – except the trivial case of a vanishing field – can be force-free everywhere. A field which is force-free throughout a given volume must experience stresses on some bounding surface in order to maintain it. In the case of the solar corona, the field cannot simply expand away because it is anchored by the gas-dynamical stresses exerted on the subsurface portion of the field. These stresses are, of course, continually varying and produce in the corona a state of constant evolution. However, the *observed* changes to the overall structure of coronal loop systems are generally slow, which suggests that we may, as a first step, ignore the variations and assume a static magnetic structure. This assumption will be made throughout this chapter. The evolution of coronal loop systems is discussed in Chapter 6.

5.2.2   *Force-free fields*

In order to focus on the geometric properties of the magnetic field, we shall adopt a model in which the field is static and is determined solely by the distribution of its own stresses, divorced from any consideration of the gas that must be present. This reduces the problem to that of finding solutions of the Maxwell equations for which the Lorentz force vanishes everywhere within the coronal volume, i.e.

$$\mathbf{j} \times \mathbf{B} = 0, \tag{5.2}$$

subject to appropriate conditions at the lower bounding surface that reflect the determining influence of the subsurface field and current distributions.

Looking first at the condition expressed by Eqn. (5.2), we see that it requires the current density $\mathbf{j}$ either to be parallel to the magnetic induction $\mathbf{B}$ or to vanish. Thus we may write

$$\mu_0 \mathbf{j} = \alpha(\mathbf{x}, t)\, \mathbf{B}, \tag{5.3}$$

where μ_0 is the permeability of free space. The current-free case corresponds to $\alpha = 0$. If we can assume that the field has achieved a static configuration, the parameter α becomes a function of position alone. Then using the Maxwell equation for a stationary system,

$$\nabla \times \mathbf{B} = \mu_0 \mathbf{j}, \tag{5.4}$$

we obtain
$$\nabla \times \mathbf{B} = \alpha(\mathbf{x})\, \mathbf{B}. \tag{5.5}$$

The physical significance of α is revealed by the integral form of this equation,

$$\int_S (\nabla \times \mathbf{B}) \cdot d\mathbf{S} = \int_S \alpha \mathbf{B} \cdot d\mathbf{S},$$

where S is any surface within the volume. This can be transformed by means of Stokes' theorem to give

$$\oint_C \mathbf{B} \cdot d\mathbf{s} = \int_S \alpha \mathbf{B} \cdot d\mathbf{S}, \tag{5.6}$$

where C is the curve bounding the surface S. If we take S to be a flat disk and C to be its perimeter, the left-hand side of this equation involves the component of field around the circumference and the right-hand side the component of field normal to the disk. The quantity α thus determines the relative sizes of these components, i.e. it measures the degree of twist of the field.

When $\alpha = 0$ there is no current and no twist. Such a field configuration is known as a potential field and will be discussed in detail in Section 5.2.3.

By taking the divergence of Eqn. (5.4), we find in the steady state that

$$\nabla \cdot \mathbf{j} = 0. \tag{5.7}$$

Thus a steady current, like the magnetic field, cannot end in space..

Furthermore, taking the divergence of Eqn. (5.3) or Eqn. (5.5) produces the relation

$$(\mathbf{B} \cdot \nabla) \alpha + \alpha \nabla \cdot \mathbf{B} = (\mathbf{B} \cdot \nabla) \alpha = 0, \tag{5.8}$$

which implies that the value of α does not change in the direction of the field and so remains constant along a field line. If a field line crosses the boundary into a force-free region, it maintains the value of α set by the boundary conditions throughout the volume.

This seemingly innocuous result has profound implications when we turn to the question of the boundary conditions necessary for the construction of solutions to Eqns. (5.1) and (5.5). Their general form has been studied by Grad & Rubin (1958). Since the equations are neither hyperbolic nor elliptic, the problem is neither completely of initial-value nor of boundary-value type; one scalar quantity may be specified at each boundary point – this might be the normal component of the magnetic induction B_n – together with one scalar quantity at just *one* boundary point of each field line – α, for example. If the vector field $\mathbf{B}$ is known everywhere on the boundary, the value of α there may be obtained from the normal component of Eqn. (5.5). However, this overprescribes the problem and a solution will exist only if the vector field on the boundary satisfies a consistency condition.

Moreover, the problem of finding solutions to

$$(\nabla \times \mathbf{B}) \times \mathbf{B} = 0, \tag{5.9}$$

is non-linear. Such problems are not guaranteed to have solutions and, if they exist, they are not necessarily unique. The construction of general force-free models for the coronal magnetic field is a mathematically intractable problem. Aly (1989) summarizes the situation, provides constraints on the boundary vector field $\mathbf{B}$ that must be satisfied in order that a force-free extrapolation is possible and discusses critically two techniques for constructing solutions. Neither technique addresses the questions of existence and uniqueness. Under these circumstances, reliable information can be gleaned only from simpler models. We shall first describe these models and then return to the question of their validity.

Both of the difficulties associated with general models are circumvented if α is deemed to be constant within the volume under consideration; then the invariance of α between footpoints is automatically ensured. The solutions of Eqn. (5.5) when α is a constant are the eigenfunctions of the curl operator. Instead of solving this equation directly, Chandrasekhar & Kendall (1957) took the curl of both sides to produce the Helmholtz equation for each Cartesian component of $\mathbf{B}$,

$$(\nabla^2 + \alpha^2) \mathbf{B} = 0. \tag{5.10}$$

This equation is linear in **B** for given α so that the solutions are said to describe *linear* force-free fields. It should be noted that all solutions of Eqn. (5.5) with constant α satisfy Eqn. (5.10), but the converse is not true.

The properties of the Helmholtz equation are well known (cf. Koshlyakov, Smirnov & Gliner, 1964). If the problem is posed in a volume totally enclosed within boundaries upon which B_n, for example, is specified, any solution is unique but its existence is not guaranteed. If the problem refers to the volume exterior to some surface on which B_n is specified, a solution always exists but is not unique unless some further restriction – such as the 'radiation' condition – is imposed. The form of the restriction that is most suitable in the case of coronal modelling has received little attention (Durrant, 1989); that employed by Levine & Altschuler (1974) to generate models of the global coronal field has no physical justification. The problem may equally well be posed with alternative boundary conditions on the lower surface – the cases in which the tangential field component or an oblique field component is specified are discussed by Hannakam, Gary & Teubner (1984) and Semel (1988).

Most authors (e.g. Seehafer, 1978; Alissandrakis, 1981; Chen & Chen, 1989) have discussed the application of linear force-free fields to the modelling of regions of finite horizontal extent, such as an active region. However, the solution is still not unique if B_n is specified only on the lower boundary (Chiu & Hilton, 1977). Since most observations furnish nothing but the normal component of the field at the photospheric surface, the models contain a degree of arbitrariness which makes it difficult to draw any firm conclusions from the many published comparisons of their structure with observations of the morpology of active regions. Such comparisons are, in any case, based upon models in which the value of α is adjusted to provide the best agreement between the observed morphology and the field, i.e. it is assumed that the morphology traces the magnetic structure and the latter is tailored accordingly. Observations alone give no indication whether this procedure is justified and so cannot reveal whether a constant-α model is a valid model of the coronal field.

We can, instead, ask whether theory provides us with a reason to expect an approximation to a constant-α force-free field in the solar corona, but the answer is again not conclusive. Heyvaerts & Priest (1984) provide the following argument.

We have already assumed that the coronal magnetic field evolves through quasi-static equilibrium structures as the field at the photospheric boundary changes slowly. During the time interval over which equilibrium is achieved, the changes in the boundary conditions may be neglected and we may suppose that the field structure changes only within the volume where the field must be force-free. Under these circumstances, Woltjer (1958) proved that in the limit of infinite electrical conductivity (ideal MHD), the quantity known as the *magnetic helicity* is conserved for each field line as the field evolves within the volume V. The helicity is defined as $\mathbf{A} \cdot \mathbf{B}$, where $\mathbf{A}$ is the vector magnetic potential ($\mathbf{B} = \nabla \times \mathbf{A}$). The helicity is, like α, a measure of the twist of the field. Unlike α, though, the helicity of a potential field does not necessarily vanish.

It might be noted that Eqn. (5.1) does not define $\mathbf{A}$ uniquely; there are a class of transformations, the so-called gauge transformations, which change the magnetic potential but leave the magnetic and electric fields unaltered. The helicity as defined here is *not* invariant under such transformations. To overcome this drawback, Berger (1985) suggests working with the relative helicity, the difference between the total helicity of the force-free field and the total helicity of the potential field having the same normal field components on the boundary. This is both gauge-invariant and a true measure of the twist of the field. However, Browning & Priest (1986a) demonstrate that the dependence of helicity on the gauge is of no consequence so long as one works consistently within one gauge.

Returning now to the evolution of the force-free system, we may note that, if helicity is conserved for each field line, the *total* helicity in the volume,

$$K = \int_V \mathbf{A} \cdot \mathbf{B}\, dV, \qquad (5.11)$$

will be constant throughout. The total magnetic energy

$$E_M = \int_V \frac{B^2}{2\mu_0}\, dV, \qquad (5.12)$$

will, however, change. The lowest possible value occurs when the field adopts precisely that constant-α force-free configuration (if it exists) having the prescribed normal component B_n on the boundary of V (Sakurai, 1979).

Although a minimum energy state exists, the system has no means of evolving to this state when there is no dissipation, i.e. when the electrical conductivity is infinite. If dissipation is introduced by allowing the conductivity to be finite, helicity will no longer be conserved on each field line. Field changes within the volume will often involve *magnetic reconnection*, the process whereby neighbouring field lines are severed at some point and rejoined each to the other. By this means, new connections are established between the (invariant) distribution of B_n over the bounding surface. Heyvaerts & Priest now conjecture that the total helicity will be almost unchanged as the process of reconnection proceeds, although the helicity will be redistributed within the volume V. This conjecture finds support in laboratory experiments, in which context it is known as Taylor's hypothesis. If it also applies in the solar corona, the field should evolve through almost minimal energy states – almost constant-α configurations – provided the boundary changes are slow compared to the rate of coronal reconnection. These arguments have renewed interest in linear force-free fields in the solar corona, an interest which has been intensified by the improvement in the techniques for measuring the full vector magnetic field in the photosphere. An extensive review of the properties of linear force-free fields has been published recently by Gary (1989).

Finite conductivity has additional consequences; it allows the dissipation of the coronal currents by ohmic losses and the diffusion of the magnetic field through the body of the gas to proceed on similar time scales. These time scales are generally much longer than the time scale for local reconnection, but if the boundary changes are sufficiently slow any currents *confined within the corona* may be eliminated from it altogether. However, we expect most currents to be generated in the high-β subsurface regions by dynamical interaction of the gas and the field; these currents will flow in a circuit only part of which extends into the coronal volume. The time constant for changes to these currents is therefore not the coronal diffusion time scale but $\tau \sim L/R$, where L is the total inductance of the circuit and R is its total resistance (cf. Section 5.7.3). These time scales are much longer than the time scales typical of boundary field changes. Therefore the magnitude of the currents passing through the surface is unlikely to change. Only if the currents are everywhere very small or are highly localized, will the magnetic field configuration in the corona adopt, on average, the *absolute* minimum energy state consistent with the prescribed normal component B_n on the photospheric boundary. This state is the current-free case in which $\alpha = 0$ (Sakurai, 1979), i.e.

$$\nabla \times \mathbf{B} = 0. \tag{5.13}$$

If this is the case, the magnetic induction $\mathbf{B}$ is derivable from a scalar potential ϕ:

$$\mathbf{B} = -\nabla\phi, \tag{5.14}$$

and the fields, as noted above, are known as potential fields.

The theory of potential fields is far more straightforward than that of general force-free fields. As a consequence, potential fields have been assumed in the majority of studies of the coronal magnetic configuration. The validity of the assumption that the corona can approximate a current-free state, though, requires critical examination.

5.2.3 Potential fields

Observations of the photospheric magnetic field were first used as a boundary condition for calculating potential field models of the coronal magnetic structure by Schmidt (1964), who considered the semi-infinite volume above a finite plane boundary region. The Schmidt procedure has an inherent difficulty posed by the unknown influence of the photospheric field outside this region, as discussed by Seehafer (1982). This difficulty is avoided in the alternative procedure due to Schatten, Wilcox & Ness (1969) and Altschuler & Newkirk (1969), who solved for the potential field exterior to the whole spherical photospheric surface. Since the solar wind is known to produce currents above $2R_\odot$, these authors introduced an outer boundary to the potential field region at a radial distance of $r = R_s \sim 2.5R_\odot$. The field was forced to be radial on the outer boundary – the so-called source surface – by making it an equipotential surface. As we shall see, this approach leads to a generally satisfactory correspondence between the shapes of the magnetic field lines and the observed structures of the outer corona.

The potential field model is based on the Laplace equation, obtained by combining Eqns. (5.1) and (5.14),

$$\nabla^2 \phi = 0, \quad R_\odot < r < R_s, \tag{5.15}$$

which has a unique solution if we impose the boundary conditions

$$\frac{\partial \phi}{\partial r} = B_r \quad \text{on} \quad r = R_\odot$$
$$\tag{5.16}$$
$$\phi = 0 \quad \text{on} \quad r = R_s.$$

The general solution in spherical coordinates (r, θ, φ) may be written in terms of the spherical harmonic functions $Y_{lm}(\theta, \varphi)$ as

$$\phi = \sum_{l=0}^{\infty} \sum_{m=-l}^{l} \left(A_{lm} r^l + \frac{B_{lm}}{r^{l+1}} \right) Y_{lm}(\theta, \varphi), \tag{5.17}$$

and the coefficients A_{lm} and B_{lm} found by the means described by Altschuler & Newkirk (1969) and Altschuler, Levine, Stix & Harvey (1977). Alternatively, the equation may be solved by a finite-difference numerical scheme (Adams & Pneuman, 1976).

The problem is also well-posed if either the radial component in cylindrical coordinates is specified on the lower boundary (relevant to line-of-sight measurements at central meridian passage) or the line-of-sight component in a fixed direction (Aly, 1987). In the latter case, Bogdan (1986) gives a method of constructing the potential field which is appropriate when there is no source surface.

Schulz, Frazier & Boucher (1978) have suggested improvements to the representation of the source surface, but the manner in which the solar wind effects are incorporated must remain *ad hoc* unless the full solution of the MHD equations is undertaken. Yeh & Pneuman (1977) provide an exact, although greatly idealized, MHD model from which it may be seen that the wind has no gross effect on the magnetic structure for $r \leqslant 2R_\odot$. Potential field models should therefore be reliable in this region.

Newkirk & Altschuler (1970) compared the potential model deduced from magnetograph observations in the photosphere with contemporaneous structures seen in the visible corona and concluded that 'the current-free model for the magnetic field below $\sim 2.5R_\odot$ gives a generally satisfactory correspondence between the shapes of the magnetic field lines and the shapes of such correspondingly fine scale density structures in the corona as arches, rays, and plumes'. An example of a potential field model employing magnetic observations with better spatial resolution (allowing the use of spherical harmonics up to $l = 90$), taken from Altschuler *et al.* (1977), is shown in Fig. 5.2. The right-hand panels show the field emanating from the 400 photospheric grid points where the field was strongest when averaged over 3.7×10^8 km² (upper, high-resolution panel) and over 2.3×10^9 km² (lower, medium-resolution panel). The strong active region fields seen with high resolution form generally low compact flux loops, whilst the less strong fields towards the edges of the active regions contribute larger-scale interconnecting loops linking the active regions. One of the most striking features is the behaviour of the 'open' field lines – those traced back from points on the source surface – depicted in the lower left panel. These lead back to a few spatially restricted areas of the photosphere outside the polar regions, but mainly to the polar regions themselves. The open field lines occupy the total coronal volume above $r \sim 2R_\odot$. However, the inner corona is almost wholly occupied by loops.

Regions composed of flux loops are known as *closed field* regions or *magnetically closed* regions. Active regions are the prime examples of closed regions, but it should be noted that active regions are not exclusively closed. Indeed, careful inspection of Fig. 5.2 reveals that when active regions form part of a complex, the field emerging from a single polarity patch in one region can separate into two domains, one arching towards the opposite polarity area of the same active region and the other diverting to a similar area in another active region. The two domains are separated by a narrow wedge of open field cutting across the active region, as noted by Svestka, Solodyna, Howard & Levine (1977). The topology of active region complexes can be complicated. Areas of a single magnetic polarity in the photosphere are separated from one another by so-called *neutral*

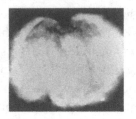

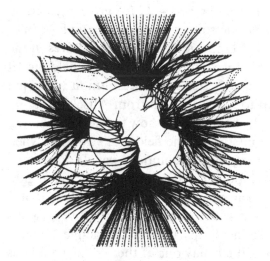

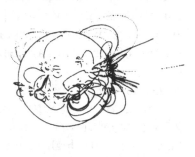

Fig. 5.2. An American Science and Engineering photograph of the X-ray emitting regions in the corona, compared with potential field calculations of the coronal magnetic field structure. The top-right panel shows the field lines emanating from the strong-field regions of the photosphere, mainly within active regions. The bottom-right panel shows the field lines emanating from weaker field regions, mainly around active regions. The bottom-left panel shows the open field lines traced back from the source surface placed at the top of the coronal field model. (From Altschuler *et al.*, 1977.)

lines, the locus of points where the field is horizontal. Yeh (1987) has used these loci to decompose coronal magnetic structures into hierarchies of nested magnetic cells.

The general correspondence between flux loop regions and X-ray enhancements on the one hand, and between open field regions and coronal holes on the other, has been noted by many authors (e.g. Altschuler *et al.*, 1977). Around active regions, though, a more critical comparison may be made. On the basis of potential models (calculated with the Schmidt procedure) and X-ray photographs, Poletto *et al.* (1975) could establish an overall correspondence between X-ray features and the calculated field lines both with regard to their topology and the number density of lines and the brightness of the X-ray emission. Specifically, they could identify the bright cores in X-ray images with field lines connecting adjacent areas of opposite polarity, high X-ray loops within active regions with interconnections between widely separated areas, and weak X-ray interconnecting loops with magnetic interconnections between adjacent active regions. Examples of each type may be found in Fig. 5.2.

The potential field models provide not only the direction of the field at each point but also its strength. On the basis of a series of models, Poletto *et al.* (1975) quote typical values for the magnitude of the magnetic induction B in loops of each type. In the cores of active regions they find that the flux loops can rise to maximum heights of 7500 km with B in the range 2–20 mT at the tops. Surrounding active regions, the flux loops rise to between 7500 and 75000 km with B in the range 0.2–2 mT. The extended (interconnecting) flux loops extend from 75000 up to 180000 km in height, where B falls to between 0.02 and 0.6 mT.

The decrease of field strength with height and scale is a consequence of the dominance of the B_{lm} terms in the expression for φ (Eqn. (5.17)). These are generated by the current sources within the Sun. The A_{lm} terms are small perturbations introduced to account for the solar wind current sources in the outer corona. Eqn. (5.17) shows that the small-scale components (with high values of l) of the dominant field decay rapidly with height, so that the outer coronal structure reflects the large-scale organization of the photospheric field.

The general correspondences established by these comparisons give some credence to the two assumptions underlying the comparison, namely, that the observed features trace coronal field lines and that the coronal field lines can be represented qualitatively by potential models. However, Levine & Altschuler (1974) point out that significant electric currents may be added to the potential models without changing the topology noticeably. Hence the presence of currents throughout the corona is not precluded by the results of the comparisons. Furthermore, a detailed comparison of the morphology of individual active regions with the geometry of the predicted potential field reveals significant discrepancies (e.g. Krieger, de Feiter & Vaiana, 1976; Levine, 1976), attributable to the presence of strong currents. However, plasma loops are found in these regions as in the surrounding corona.

These qualifications of the picture of coronal magnetic fields as potential field configurations are not serious as regards our present purpose. We are seeking merely to establish that plasma loops trace flux loops as a basis for further discussion. On the evidence available, this is a reasonable conclusion and we shall identify plasma loops with magnetic flux loops from now on.

5.3 Model equations for the coronal plasma
5.3.1 The MHD approximation

At the high temperatures and low densities characteristic of the corona, the atoms of the coronal gas are almost all ionized. The long-range electrostatic forces between the charges is then the most important factor governing the *small-scale* motion of the particles. Any medium (not necessarily a fluid) in which this is the case is said to be a *plasma*. It is a basic property of plasmas that the strength of the electrostatic interaction precludes any permanent large-scale separation of opposite charges. The average charge density is effectively zero everywhere, so that the *large-scale* dynamics of a plasma is controlled by the magnetic field.

In the presence of a uniform magnetic field, a single charged particle executes a helical motion along a field line, the radius of the circular projection of the path, the *Larmor radius*, being

$$r_{\mathrm{L}} = \frac{vm}{ZeB},$$ (5.18)

and the frequency of revolution, the *Larmor frequency* or *gyrofrequency*, being

$$\Omega = \frac{ZeB}{m},$$ (5.19)

where v is the speed of the particle in the plane of the circle and m and Ze are its mass and charge respectively. The quantity e is the charge on an electron. The motion in the direction of the field, i.e. along the field line, is uninfluenced by the field; the particle drifts freely. If other particles are present, the drift continues only until a 'collision' – a near encounter with another particle – scatters the particle from its helical path. These collisions occur in a fully ionized plasma at a frequency v_{c} given roughly by

$$v_{\mathrm{c}} \simeq 1.3 \times 10^{-20} \frac{n}{m^{\frac{1}{2}} T^{\frac{3}{2}}},$$ (5.20)

where n is the particle number density (the number of particles per unit volume) and T is the temperature of the system. The distance drifted between collisions is the *mean-free-path* λ_{c} and is given by $\lambda_{\mathrm{c}} = v_{\mathrm{th}}/v_{\mathrm{c}}$. Here $v_{\mathrm{th}} = \sqrt{(kT/m)}$ is the typical thermal speed of the particles, k being the Boltzmann constant.

If we adopt the canonical hot coronal loop values from Table 5.1, *viz.* a temperature of 2×10^6 K, an electron density of 10^{15} m^{-3} and a magnetic field of strength 1 mT, we may estimate $r_\perp$ to be ~ 1 m for ions and ~ 20 mm for electrons. The corresponding values for the gyrofrequencies are 10^5 rad s^{-1} and 10^7 rad s^{-1}, respectively. The collision frequency is of the order of 5 s^{-1} for electrons and 0.1 s^{-1} for ions, giving in each case a mean-free-path of the order of 1000 km. In cool loops, the collision frequency is much higher and the mean-free-path much smaller, as a consequence of the lower temperature and higher density.

These length scales are smaller than those so far resolved in plasma loop structures (Sections 2.4.3, 3.3.3, 3.4), and the time scales are shorter than those of all observed coronal phenomena except some transient flare stages (Section 4.5.3). Under these circumstances, the plasma may be treated as a fluid and the *magnetohydrodynamic* (or *MHD*) description is appropriate. In a fluid, the so-called *transport processes* of diffusion, viscosity, heat conduction and electrical resistance can all be modelled in terms of the local thermal and dynamical properties of the gas – the temperature T, pressure p and bulk velocity $\mathbf{v}$ – together with the macroscopic magnetic induction **B**. Since there is a large difference between the typical distances that particles can carry information in the direction of the field (measured by λ_c) and across the field (measured by $r_\perp$), these transport processes are generally anisotropic, the cross-field transport being smaller than the parallel transport by a factor of at least Ω/ν. This product is greater than 5×10^4 for electrons and greater than 10^3 for positive ions, under coronal conditions.

The fact that energy and momentum are transported preferentially along the field, i.e. along a magnetic flux loop rather than across it, allows great simplifications to be made in the modelling of coronal plasma loops, as we shall see shortly. However, a note of warning should be sounded at this stage. The particles are here assumed to be scattered by direct interactions with the random fields of the other charged particles. This assumption yields *classical* values for the transport processes, and is valid when the state of the plasma is close to thermal equilibrium. Given the low collision frequencies of the corona, thermalization can be a lengthy process. In a non-thermal plasma, groups of particles can organize their motion so as to produce a coherent collective electric field, rather than a random field. The collective field then sustains the group motion. Other charged particles are scattered much more effectively by the collective field than the random fields, resulting in a much higher effective collision frequency and correspondingly enhanced *anomalous* transport of matter, energy and

Table 5.1. *The magnitudes of various physical quantities for the canonical hot loop model*

The subscript TR refers to transition region conditions at $T = 10^5$ K.

Quantity	Symbol	Value
Length	$2L$	5×10^7 m
Maximum temperature	T_m	2×10^6 K
Electron (number) density	n_e	10^{15} m^{-3}
Hydrogen (number) density	n_H	10^{15} m^{-3}
(Mass) density	ρ	2×10^{-12} kg m^{-3}
Pressure	p	6×10^{-2} Pa
Magnetic flux density	B	10^{-3} T
Sound speed	v_a	2×10^5 m s^{-1}
Alfvén speed	v_A	6×10^5 m s^{-1}
Radiative loss rate at T_m	$\mathscr{L}_m$	2×10^{-5} W m^{-3}
Radiative loss rate at T_{TR}	$\mathscr{L}_{TR}$	2×10^{-2} W m^{-3}
Conductive flux divergence	$\kappa T/L^2$	2×10^{-4} W m^{-3}
Acoustic time scale	L/v_a	10^2 s
Radiative time scale at T_m	$p/\mathscr{L}_m$	3×10^3 s
Radiative time scale at T_{RT}	$p/\mathscr{L}_{TR}$	3 s
Conductive time scale	$pL^2/(\kappa T)$	2×10^2 s

momentum. Anomalous conditions are an essential ingredient of some coronal heating theories (cf. Section 5.7.3) but they are thought to obtain only in highly localized regions of the corona. In the following section we therefore develop the equations of motion of a plasma in the presence of a strong magnetic field in the case of semi-classical transport only.

5.3.2 The MHD equations

McWhirter, Thonemann & Wilson (1975) calculate that some 87% of the positively charged particles in the hot corona are protons (ionized hydrogen) and that these contribute 77% of the electrons. Similar figures appear also to apply to cool loops (cf. Table 2.9). Thus an appropriate point of departure is the two-component plasma model. Expressions for the transport processes in a plasma close to thermal equilibrium may be found in texts on plasma physics (e.g. Clemmow & Dougherty, 1969). They give rise to an electric current density

$$\mathbf{j} = \sigma \cdot \mathbf{E}'' + \boldsymbol{\alpha} \cdot \nabla T, \tag{5.21}$$

a conductive heat flux,

$$\mathbf{q} = -\boldsymbol{\beta} \cdot \mathbf{E}'' - \boldsymbol{\kappa} \cdot \nabla T, \tag{5.22}$$

and a viscous force,

$$\mathbf{F}_v = \nabla \cdot \mathbf{P}. \tag{5.23}$$

Here $\mathbf{E}''$ is the electric field strength in the frame moving with the plasma, $\mathbf{E}'$,

$$\mathbf{E}' = \mathbf{E} + \mathbf{v} \times \mathbf{B}, \tag{5.24}$$

with the addition of a term depending on the gradient of the electron pressure p_e:

$$\mathbf{E}'' = \mathbf{E}' + \frac{kT}{ep_e} \nabla p_e. \tag{5.25}$$

The transport coefficients σ, α, β and κ are second-order tensors (σ_{ij}, etc.). The dyadic notation is employed here in order to give the equations greater clarity; the expression $\sigma \cdot \mathbf{E}''$ should be interpreted as the vector $\sigma_{ij} E_j''$, etc. In the expression for the viscous force, $\mathbf{P}$ is the traceless part of the stress tensor and is related to the traceless part of the rate-of-strain tensor

$$S_{ij} = \tfrac{1}{2}\left(\frac{\partial v_i}{\partial x_j} + \frac{\partial v_j}{\partial x_i}\right) - \tfrac{1}{3}\delta_{ij} \nabla \cdot \mathbf{v}, \tag{5.26}$$

by

$$P_{ij} = 2\mu_{ijkl} S_{kl}, \tag{5.27}$$

where μ_{ijkl} is the tensor coefficient of viscosity.

To be consistent with the fluid description of the plasma, we must exclude any phenomena in which the displacement current in the Maxwell equations is important. This requires that Eqn. (5.4),

$$\mu_0 \mathbf{j} = \nabla \times \mathbf{B},$$

be satisfied at all times. Eliminating $\mathbf{j}$ between this equation and Eqn. (5.21), we find the electric field strength to be given by

$$\mathbf{E} = -\mathbf{v} \times \mathbf{B} - \frac{1}{n_e e} \nabla p_e + \sigma^{-1} \cdot \left(\frac{1}{\mu_0} \nabla \times \mathbf{B}\right) - (\sigma^{-1} \cdot \alpha) \cdot \nabla T. \tag{5.28}$$

If we now substitute this expression for $\mathbf{E}$ in the Maxwell equation,

$$\frac{\partial \mathbf{B}}{\partial t} = -\nabla \times \mathbf{E}, \tag{5.29}$$

we obtain the *induction equation*

$$\frac{\partial \mathbf{B}}{\partial t} = \nabla \times (\mathbf{v} \times \mathbf{B}) - \nabla \times [\eta \cdot (\nabla \times \mathbf{B})] + \nabla \times \left[(\sigma^{-1} \cdot \alpha) \cdot \nabla T + \frac{kT}{ep_e} \nabla p_e\right]. \tag{5.30}$$

The first term on the right-hand side describes advection of the magnetic flux by the flow as if the field were 'frozen' into the plasma (see below). The second is a diffusion term with $\eta = \sigma^{-1}/\mu_0$ being the magnetic diffusivity, which governs the slippage of the field through the plasma. The final term is a battery term by which thermal and pressure gradients can generate magnetic flux.

Before we can write down, with the aid of these results, the equations governing the motion of the plasma, we need to define the plasma density. The mass density ρ of a fully ionized gas is

$$\rho = m_e n_e + \overline{m_i} n_i, \tag{5.31}$$

where n_e and n_i are the electron and ion number densities, and m_e and $\overline{m_i}$ are the electron and mean ion masses. Under coronal conditions, $n_i \simeq n_e$ very closely and the mean relative molecular mass is 1.3, i.e. $\overline{m_i} = 1.3 m_H$ where m_H is the mass of a hydrogen atom, so that

$$\rho \simeq 1.3 m_H n_e = 2.2 \times 10^{-27} n_e.$$

The equations of motion are then:

continuity

$$\frac{D\rho}{Dt} + \rho \nabla \cdot \mathbf{v} = 0, \tag{5.32}$$

momentum

$$\rho \frac{D\mathbf{v}}{Dt} = -\nabla p + \rho \nabla \Phi + \mathbf{j} \times \mathbf{B} - \nabla \cdot \mathbf{P}, \tag{5.33}$$

energy

$$\frac{\rho^\gamma}{\gamma - 1} \frac{D}{Dt} (p\rho^{-\gamma})$$

$$= -\nabla \cdot \mathbf{q} + \mathbf{j} \cdot \mathbf{E}' - \mathbf{P} : \nabla \mathbf{v} - \nabla \cdot \mathbf{q}_r$$

$$= \nabla \cdot (\kappa^{eff} \cdot \nabla T) - \mathbf{j} \cdot \left[\left(\frac{\beta_\parallel}{\sigma_\parallel T} \right) \nabla T + (\sigma^{-1} \cdot \alpha) \cdot \nabla T + \frac{1}{n_e e} \nabla p_e \right]$$

$$+ \frac{j^2}{\sigma_\parallel} - \mathbf{P} : \nabla \mathbf{v} - \nabla \cdot \mathbf{q}_r. \tag{5.34}$$

In the energy equation, the quantity $\mathbf{P} : \nabla \mathbf{v}$ is to be interpreted as the scalar $P_{ij}(\partial v_i / \partial x_j)$ and the subscript $\parallel$ denotes the component in the direction of the magnetic field. The operator D/Dt is the time derivative following the motion,

$$\frac{D}{Dt} \equiv \frac{\partial}{\partial t} + \mathbf{v} \cdot \nabla. \tag{5.35}$$

The second term of this operator describes the changes at each point due to the advection of each quantity – $\rho, \mathbf{v}$ or $p\rho^{-\gamma}$ – by the flow.

In the momentum equation, the advection term $\rho(\mathbf{v} \cdot \nabla)\mathbf{v}$ is also called the *inertial force*. The terms on the right-hand side represent the pressure gradient force, the gravitational force (derived from a gravitational potential Φ), the Lorentz force and the viscous force.

The energy equation can be written in various forms; Priest (1982) gives several alternatives, some of which are given below. Equation (5.34) involves the entropy per unit mass. This is defined as $c_V \log(p\rho^{-\gamma})$, where c_V is the specific heat at constant volume and is equal to $k/(\gamma-1)\bar{m}$, $\bar{m}$ being the mean mass of *all* the particles, including both ions and electrons. In the corona, $\bar{m} = \rho/(n_i + n_e) \simeq \rho/2n_e \simeq 1.0 \times 10^{-27}$ kg. The value of the ratio of specific heats γ is almost exactly 5/3 in the corona, because the hydrogen is very nearly completely ionized.

The terms on the right-hand side of the energy equation represent conductive heat transport with an effective tensor coefficient of conduction

$$\kappa^{eff} = \kappa - \frac{\beta_\parallel}{\sigma_\parallel}\alpha, \tag{5.36}$$

a current associated energy transport (noted by Hinata, 1981), Joule heating, viscous heating and the rate of energy exchange with the radiation field. In the case of optically thin plasmas in which the absorption of radiation is negligible, the latter is an energy loss term and is given by

$$\nabla \cdot \mathbf{q}_r = 4\pi j - \frac{1}{c}\frac{\partial J}{\partial t}, \tag{5.37}$$

where j is the volume emission coefficient of the plasma integrated over wavelength, J is the mean radiative intensity integrated over wavelength and c is the speed of light (cf. Bray, Loughhead & Durrant, 1984). If the coronal time scales are longer than those required to establish ionization equilibrium (a matter of minutes), the second term on the right-hand side can be neglected and the emission coefficient evaluated at the local temperature and pressure. The radiative losses are then given simply by

$$\nabla \cdot \mathbf{q}_r = 4\pi j(T, \rho). \tag{5.38}$$

The quantity on the right-hand side has been calculated as a function of temperature and density for a gas of solar composition by several authors. It can be written as

$$\nabla \cdot \mathbf{q}_r = \mathscr{L} = n_e n_H Q(T), \tag{5.39}$$

where n_H is the number density of hydrogen (in the form of either neutral hydrogen or protons) and $Q(T)$ is a function that must be calculated numerically (cf. Raymond & Smith, 1977). The function increases rapidly to a broad maximum between temperatures of 5×10^4 and 3×10^5 K and then decreases as the temperature increases up to 10^7 K approximately as

$$Q(T) = 10^{-31.66} T^{-\frac{1}{2}} \quad \text{Wm}^3. \tag{5.40}$$

The function $Q(T)/T^2$ is sketched in Fig. 5.5. It must be realized, though, that this is a very simplistic representation of the effects of radiative energy transport. In dynamical models, the local evaluation of optically thin losses may become invalid (cf. Section 5.4.3). Moreover, the basal loop segments in the chromosphere are not optically thin and a proper treatment is exceedingly difficult (cf. Section 5.4.2).

The other transport processes, however, assume a much simpler form when the magnetic field is strong. If we take local axes with the z axis in the direction of the magnetic field, the viscous stress tensor can be approximated by

$$P_{xx} = P_{yy} = -\mu_\|(s_{xx} + s_{yy}), \quad P_{zz} = -2\mu_\| s_{zz}, \tag{5.41}$$

with all the other components vanishing. Kinetic theory gives the form of μ as

$$\mu_\| \simeq 0.5 \frac{n_e kT}{v_i}$$

$$\simeq 2.2 \times 10^{-17} T^{\frac{5}{2}} \text{ kg m}^{-1} \text{ s}^{-1}, \tag{5.42}$$

where v_i is the ion collision frequency. This component of the viscosity is equal to the value of the scalar coefficient of viscosity in the absence of a magnetic field. The property of being unchanged in value in the direction of the field is shared by all the transport coefficients. The form of the stress tensor under conditions applying in the solar corona is discussed by Hollweg (1986).

The energy equation can also be simplified to read

$$\frac{\rho^\gamma}{\gamma - 1} \frac{D}{Dt} (p\rho^{-\gamma})$$

$$= \nabla \cdot (\kappa_\|^{eff} \nabla_\| T) - 3.7 \times 10^{-4} \mathbf{j} \cdot \nabla T - 6.2 \times 10^{-5} j_\| \nabla_\| T$$

$$- 8.6 \times 10^{-5} \frac{T}{n_e} \mathbf{j} \cdot \nabla n_e + \frac{j^2}{\sigma_\|} - \mathbf{P} : \nabla \mathbf{v} - \nabla \cdot \mathbf{q}_r, \tag{5.43}$$

where

$$\kappa_\|^{eff} \simeq 2 \frac{n_e k^2 T}{m_e v_e}$$

$$\simeq 1.8 \times 10^{-11} T^{\frac{5}{2}} \quad \text{W m}^{-1} \text{ K}^{-1}$$

$$\simeq \kappa_0 T^{\frac{5}{2}}, \tag{5.44}$$

$$\sigma_\| \simeq \frac{n_e e}{m_e v_e}$$

$$\simeq 1.5 \times 10^{-3} T^{\frac{3}{2}} \quad \text{S m}^{-1}, \tag{5.45}$$

the quantity v_e being the electron collision frequency.

Supplementing the equations of motion is the constitutive relation between pressure, density and temperature. This takes the form of the perfect gas equation throughout the Sun,

$$p = (n_e + n_i)kT = \frac{\rho}{\bar{m}}kT = R_* \rho T. \qquad (5.46)$$

These equations, together with the induction equation (5.30) – which yields **B** – the Maxwell equation (5.4) – which yields **j** – and the relation (5.28) – which yields **E** – form a complete closed set.

The system of equations is still closed if Eqn. (5.28) is omitted. Because of the vanishing net electric charge, the electric field **E** plays no role in the large-scale dynamics of the plasma. It is simply evaluated *a posteriori*, but is needed in order to check the consistency of the MHD assumption.

If we examine the orders of magnitude of some of the remaining terms in the equations of motion, characterizing the typical length scale by L and the typical velocity by v, we discover that the ratio of inertial force to viscous force – a ratio known as the Reynolds number $Re = Lv/(\mu/\rho)$ – is reasonably large (of the order of 100) and the typical ratio of the advection term to the diffusion term in the induction equation governing the evolution of the magnetic field – the magnetic Reynolds number $Rm = Lv/\eta$ – is extremely large (of the order of 10^{13}) in the solar atmosphere. This suggests a much more drastic simplification of the equations that is frequently adopted when studying short time scale effects or large spatial scale structures (cf. Section 5.6). This is the *ideal* MHD approximation in which all the effects of viscosity, heat conduction and electrical resistance are ignored. For future reference, the governing equations are then

$$\frac{D\rho}{Dt} + \rho \nabla \cdot \mathbf{v} = 0, \qquad (5.47)$$

$$\rho \frac{D\mathbf{v}}{Dt} = -\nabla p + \rho \nabla \Phi + \frac{1}{\mu_0}(\nabla \times \mathbf{B}) \times \mathbf{B}, \qquad (5.48)$$

$$\frac{D}{Dt}(p\rho^{-\gamma}) = 0, \qquad (5.49)$$

$$\frac{\partial \mathbf{B}}{\partial t} = \nabla \times (\mathbf{v} \times \mathbf{B}). \qquad (5.50)$$

The ideal induction equation (5.50) implies that the magnetic flux N threading a surface S which moves with the fluid does not change, i.e.

$$\frac{dN}{dt} = \frac{d}{dt}\int_{S(t)} \mathbf{B} \cdot d\mathbf{S} = 0. \qquad (5.51)$$

This property is referred to as 'flux-freezing'. As a consequence, the structure of the magnetic field may be deformed as the fluid moves around but field loops may be neither created nor destroyed.

It may also be noted that whilst currents are allowed to flow in an ideal fluid ($j \neq 0$), there is no ohmic dissipation of these currents. Furthermore, it can be seen from Eqn. (5.28) that in the limit as $\sigma \to \infty$ the fluid motion generates an electric field which is perpendicular to the magnetic field. This electric field drives a cross-field drift of electrons until the motion is resisted by the build-up in the electron pressure gradient.

The other ideal equation which calls for some comment is the energy equation. When transport effects and radiation are ignored, the energy equation reduces to the adiabatic condition; the entropy of each volume element of plasma is preserved throughout the dynamical evolution. The equations do not define the *initial* distribution of entropy, which is needed to construct the temperature structure of the corona. The ideal equations can be used only to follow the dynamical evolution of a structure generated by other means.

However, even the full energy equation is not adequate for this purpose. If we examine, as before, the orders of magnitude of the various terms on the right-hand side of Eqn. (5.43), we find that the only significant terms are the conductive heat flux and the radiative loss terms. There appears to be no source of thermal energy to produce the high temperatures of the corona. We are thus forced to the immediate conclusion that our assessment of the coronal scales is not correct. The corona must contain unresolved, smaller-scale regions with higher dissipation rates. Whether the unresolved motions pervade the whole coronal volume and can be described consistently within the MHD approximation or whether they are localized to such an extent that fluid behaviour must break down, is an open question at present. The alternative scenarios are discussed more fully in Section 5.7. In order to make progress at this stage, the effects of the unresolved processes must be incorporated into the MHD equations by introducing *ad hoc* momentum and energy source terms to replace the viscous and volume current terms derived above, *viz.*

$$\rho \frac{D\mathbf{v}}{Dt} = -\nabla p + \rho \nabla \Phi + \mathbf{j} \times \mathbf{B} + \mathbf{K}, \tag{5.52}$$

$$\frac{\rho^\gamma}{\gamma - 1} \frac{D}{Dt}(p\rho^{-\gamma}) = \nabla \cdot (\kappa_\parallel^{eff} \nabla_\parallel T) - \mathscr{L} + \mathscr{H}, \tag{5.53}$$

where $\mathbf{K}$ and $\mathscr{H}$ represent the net rates of transfer of momentum and energy per unit volume from the unresolved regions (or unresolved scales) and $\mathscr{L} = \nabla \cdot \mathbf{q}_r$ is the rate of radiative energy loss per unit volume.

An alternative form of the energy equation is often useful. It is obtained by expanding the derivatives on the left-hand side and using the continuity equation (5.32) to eliminate the derivatives of ρ. This produces

$$\frac{1}{\gamma-1}\frac{Dp}{Dt}+\frac{\gamma p}{\gamma-1}\nabla\cdot\mathbf{v} = \nabla\cdot(\kappa_\parallel^{eff}\nabla_\parallel T)-\mathscr{L}+\mathscr{H}, \tag{5.54}$$

Still other commonly used forms are obtained by introducing the temperature via the perfect gas law (5.46),

$$\frac{\gamma}{\gamma-1}\frac{p}{T}\frac{DT}{Dt}-\frac{Dp}{Dt} = \nabla\cdot(\kappa_\parallel^{eff}\nabla_\parallel T)-\mathscr{L}+\mathscr{H}, \tag{5.55}$$

and the specific heat at constant pressure $c_p = c_V + R_*$,

$$c_p\rho\frac{DT}{Dt}-\frac{Dp}{Dt} = \nabla\cdot(\kappa_\parallel^{eff}\nabla_\parallel T)-\mathscr{L}+\mathscr{H}. \tag{5.56}$$

This last form is particularly relevant because we shall see that there is little pressure variation in coronal loops and that many processes are essentially isobaric. The second term on the left-hand side can then be dropped and the right-hand side balanced against the rate of change of heat energy acquired when the temperature and volume change at constant pressure. This quantity is known as the *enthalpy*, and comprises the heat energy acquired at constant volume plus the work done against the pressure force when the volume changes.

It might be noted that these equations of motion are referred to an inertial frame of reference. Of more obvious value would have been the equations referred to axes co-rotating with the solar plasma. The significance of rotational effects is then measured by the Rossby number *Ro*, the ratio of the inertial term to the Coriolis term in the momentum equation (see Priest, 1982). The lower the Rossby number, the more important it is to include rotational effects. For the largest-scale loops with dimensions of the order of $R_\odot$, this number is typically 50 for flows at the sound speed. For smaller-scale loops, the Rossby number is much greater. Rotation will have no significant effect on their structure and it will be ignored in the models of individual loops to which we now turn.

5.4   Steady-state structure of loops
5.4.1   *Introduction*
Having established the appropriate formalism for describing the physics of plasma loops, we can now assess which factors are most important in influencing their behaviour. In order to simplify the initial discussion, we may note that the observed lifetimes of coronal loops summarized in Tables 2.8 and 3.7 suggest that they exist in a quasi-steady state. We shall justify this claim in more detail when we look at time-dependent phenomena in Section 5.5. In this section, we shall concentrate on the steady-state structure of the coronal loop models.

Let us first focus attention on the momentum equation (Eqn. (5.33) or (5.52)). When there is a steady state ($\partial/\partial t \equiv 0$), the remaining terms represent the inertial force, the pressure force, the gravitational force, the Lorentz force and the viscous force. The Lorentz force may be written with the help of Eqn. (5.4) in the form

$$\mathbf{j} \times \mathbf{B} = \frac{1}{\mu_0}(\nabla \times \mathbf{B}) \times \mathbf{B} = \nabla\left(\frac{B^2}{2\mu_0}\right) + \frac{1}{\mu_0}(\mathbf{B} \cdot \nabla)\mathbf{B}, \qquad (5.57)$$

which demonstrates that it has two contributions, one from the magnetic pressure gradient and the other a curvature (or magnetic tension) force. We can now see that the Lorentz term will dominate the gas pressure term when the coronal conditions described in Section 5.2.1 apply.

The only potential competitor with the magnetic force is the inertial term, $\rho(\mathbf{v} \cdot \nabla)\mathbf{v}$. This is equivalent dimensionally to the gradient of a dynamic pressure, $\frac{1}{2}\rho v^2$. The various pressures are in the ratio

$$\rho v^2/2 : p : B^2/2\mu_0 = v^2 : v_a{}^2 : v_A{}^2$$

where v_a and v_A are the speeds of the acoustic (sound) wave and Alfvén wave respectively. For our canonical coronal conditions (cf. Table 5.1), we can see that $v_A > v_a$ and what observational evidence there is suggests that $v \sim v_a$ in loops. However, in open-field regions such as coronal holes and high above active regions, v_A may drop sufficiently for the flow speed to exceed the Alfvén speed. The inertial term then becomes dominant and we enter the domain of the solar wind. We exclude such regions here.

Even in closed-field regions there is an important exception to the rule that the Lorentz term dominates the force balance. In the direction of the (local) magnetic field, the component of the Lorentz force vanishes identically since $\mathbf{B} \cdot (\mathbf{j} \times \mathbf{B}) = 0$. When the flow speeds are sufficiently subsonic for the inertial term to be neglected, the force balance equation becomes

$$\mathbf{B} \cdot (-\nabla p + \rho \nabla \Phi) = 0. \qquad (5.58)$$

In other words, *hydrostatic* equilibrium applies along the field lines. The material at any point on a field line is in hydrostatic balance with its footpoint. If we choose a particular field line and label it α, the solution of Eqn. (5.58) is

$$p(\alpha, z) = p(\alpha, 0) \exp\left\{ -\int_0^z \frac{1}{\Lambda(\alpha, \xi)} d\xi \right\}, \tag{5.59}$$

where $\Lambda(\alpha, z)$ is the pressure scale height at a height z along that field line and is given by

$$\Lambda(\alpha, z) = \frac{p}{\rho |\nabla \Phi|} = \frac{kT(\alpha, z)}{\overline{m}|d\Phi/dz|}. \tag{5.60}$$

Since the mass of the corona is negligible in comparison to the mass of the body of the Sun, the gravitational potential is to a very good approximation,

$$\Phi = \frac{GM_\odot}{r}, \tag{5.61}$$

and the acceleration due to gravity acting vertically downwards is

$$g = -\frac{d\Phi}{dz} = \frac{GM_\odot}{(R_\odot + z)^2}. \tag{5.62}$$

For $0 < z < R_\odot$ this quantity varies between 274 m s^{-2} and 69 m s^{-2}. The scale height Λ ranges correspondingly from 50000 km to 200000 km (5×10^7–2×10^8 m) for the canonical coronal temperature 2×10^6 K. The scale height is thus comparable to the greatest height to which plasma loops are seen to rise, and there is consequently little gravitational stratification in the pressure distribution of hot loops. The density along the isothermal portion of a flux loop will show equally little variation.

The most natural starting point for a theoretical discussion of plasma loops is to build on the previous force-free model of the field alone by including the coronal plasma in a magnetostatic configuration. This requires us to solve

$$-\nabla p + \rho \nabla \Phi + \mathbf{j} \times \mathbf{B} = 0, \tag{5.63}$$

allowing a small Lorentz force to be balanced by equally small pressure and gravitational forces. We then require also Eqns. (5.1) and (5.4) and an energy equation. This set of equations, however, defeats mathematical analysis. This remains the case even when the need for an energy equation is obviated by invoking an idealized – and physically unrealistic – model in which the density is constant. Equation (5.29) then takes the simpler form

$$\nabla(-p + \rho \Phi) + \mathbf{j} \times \mathbf{B} = 0. \tag{5.64}$$

Nevertheless, the introduction of the single scalar function $p + \rho \Phi$ into the force balance equation renders the system intractable. Suitable boundary conditions have been defined by Grad & Rubin (1958), but the only known solutions possess a degree of symmetry unlikely to be found in Nature. When the density is treated as an independent scalar field, it is possible to find some fully three-dimensional solutions that have the geometry of a system of loops (Low, 1982) or of large-scale coronal structures (Bogdan & Low, 1986). However, the temperature can only be obtained *a posteriori* and there is no guarantee that it will satisfy a realistic energy equation. Indeed, Priest (1978) has surmized that the separate requirements of force and energy balance are probably incompatible, and that a system of loops cannot be in magnetostatic equilibrium. This remains another open question.

On the other hand, magnetostatic solutions can be found for the simplified – but not fully self-consistent – problem in which a thin force-free loop is embedded in an ambient potential field. A thin loop is one in which the diameter is much less than both its length and its radius of curvature. The variation of field and plasma properties across the loop can then be either approximated with an axisymmetric model (Lakshmi & Gokhale, 1987) or neglected altogether. Browning & Priest (1986b) adopted the latter strategy, assuming an isothermal loop with uniform axial field and uniform internal pressure surrounded by a similarly isothermal medium pervaded by a two-dimensional potential field. The force balance along the loop is described by Eqn. (5.58). In the thin loop approximation, this is supplemented by a transverse force balance equation which balances the magnetic pressure force against the magnetic tension and the differential gravitational force between the loop and its surroundings (the buoyancy force). The joint solution of these two equations yields the direction of **B** at each point along the loop, thus enabling its shape to be generated.

The models have three parameters, the ratio of the pressure scale height to the ambient field scale height (which measures the strength of gravity), the density decrement in the loop (which measures the buoyancy of the loop) and the separation of the footpoints. The numerical solutions show that the shapes are close to those of the ambient field lines. When gravity dominates, the height of the loop increases as both the footpoint separation and the buoyancy of the loop are increased. However, beyond a certain separation the tension becomes insufficient to balance the buoyancy and no further equilibrium solutions are possible. When gravity is weak, the buoyancy force produced by the density decrement is smaller than the increase in magnetic tension resulting from the increase in the internal field strength. The loop is then pulled down until the external magnetic pressure gradient can balance the tension. For large density increases no equilibrium is possible. In the weak gravity case there is also a critical footpoint separation beyond which equilibrium cannot be found. The authors conclude that this separation is proportional to the larger of the gravitational and magnetic length scales.

Most authors, though, have elected to treat a flux loop not as an entity within a system of loops but as an isolated feature, which can – hopefully – be embedded in the surrounding structure at a later stage. That this is conceptually possible was pointed out by Durney & Pneuman (1975). According to the definition of the flux loop illustrated in Fig. 5.1, no field line crosses the walls. They then argue that, because of the suppression of the cross-field transport processes noted above, no momentum or energy can be transferred from one loop to the next across their surfaces. Each flux loop is therefore thermally and dynamically isolated from its neighbours, so that the calculation of the dynamical equilibrium can be split into two steps. In the first step, only the component of the force balance equation along the field lines, usually with unresolved viscous effects ignored,

$$\mathbf{B} \cdot (\mathbf{v} \cdot \nabla \mathbf{v} + \nabla p - \rho \nabla \Phi) = 0, \tag{5.65}$$

is employed to find the pressure, temperature and flow velocity *along* the various flux loops. In the second step, the loci of the flux loops are adjusted to achieve a lateral force balance between the neighbouring loops. On the evidence available, we should be cautious in assuming that this second step is always, if ever, possible. Nevertheless, we are forced to adopt this approach at this stage in order to make any progress at all. Throughout the following discussion of the structure and dynamics of individual loops, this caveat should be borne in mind.

In the following sections we describe one-dimensional models of coronal flux loops which account only for mass, momentum and energy balance along the field lines. The geometry of the field lines is assumed to be given, so that the dynamic and thermal properties of the loop can be analysed without further reference to the field. The question of the three-dimensional structure of coronal loops is taken up again in Section 5.6.

5.4.2 *Static models of individual loops*

Steady-state models fall into two classes, those which are truly static, i.e. without any gas flows, and those which allow a steady, time-independent flow along the loop. We shall look first at the structure of

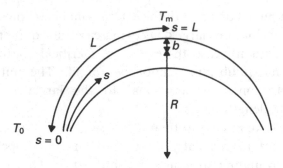

Fig. 5.3. The parameters describing a symmetric coronal loop. The length is $2L$, the diameter $2b$, and the radius of curvature R. Distance s is measured along the loop from the base, where the temperature is T_0, to the apex, where the temperature is T_m.

static loops and then turn to loops with steady flows in the following section.

Let us suppose that the geometry of the magnetic field is given, either empirically or by some theoretical approximation such as a potential field structure. This enables us to establish the coordinate system depicted in Fig. 5.3. The loop is taken to be symmetric about its apex, the half-length being L. The quantity s measures the distance along the arc of the loop, $A(s)$ measures the cross-sectional area of the loop at each point along its length and $\theta(s)$ measures the inclination of the axis of the loop to the vertical. A static loop is then described by Eqn. (5.65) without the inertial term,

$$-\frac{dp}{ds} - \rho g \cos \theta = 0, \tag{5.66}$$

where g is the local acceleration due to gravity (Eqn. (5.62)) and is sensibly constant for hot low-lying loops. The gas law Eqn. (5.46) can then be used to eliminate the density, producing

$$-\frac{dp}{ds} - \frac{\bar{m}g \cos \theta}{k} \frac{p}{T} = 0. \tag{5.67}$$

The only other equation is the *ad hoc* energy equation (5.53)

$$0 = \frac{1}{A}\frac{d}{ds}\left(A\kappa_{\parallel}^{eff}\frac{dT}{ds}\right) - \frac{p^2}{4k^2}\frac{Q(T)}{T^2} + \mathcal{H}. \tag{5.68}$$

Here the radiative loss term has been rewritten using the fact that the number densities of electrons and protons are roughly equal in the corona (McWhirter *et al.*, 1975), so that their partial pressures are each half the total pressure, $p_e \sim p_H \simeq p/2$.

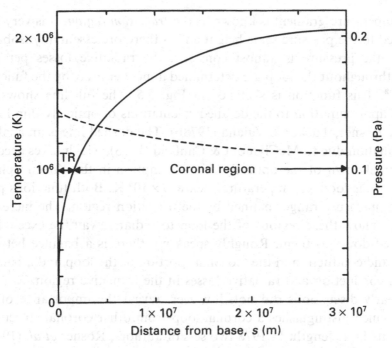

Fig. 5.4. The one-dimensional structure of a hot coronal loop of uniform cross section and uniform heating. The full line shows the temperature, the dashed line the pressure as a function of distance from the base. The basal temperature is 7000 K and the uniform heating factor $H_0 = 2 \times 10^{-4}$ W m^{-2}.

Solutions of these equations are known from earlier plane-parallel models of the corona, which take $\cos \theta = 1$ and $A(s) = 1$ throughout, but the equations were first interpreted in terms of active region loop models by Landini & Monsignori Fossi (1975) and Jordan (1976).

The main features of the loop solutions can be seen in Fig. 5.4. This shows the result of the numerical integration of Eqns. (5.67) and (5.68) for a symmetric semicircular loop of uniform cross-section and half-length $L = 3 \times 10^7$ m. The heating is assumed to be constant throughout and is set equal to $\mathcal{H} = 2 \times 10^{-4}$ W m^{-3}. As expected, the pressure, which was set at 0.14 Pa at the footpoint where the temperature is 7000 K, drops only 20% up to the apex. Moreover, the temperature shows only a similarly small variation over the upper 90% of the loop. At high temperatures the radiative loss rate ($\simeq 10^{-4}$ W m^{-3}) is lower than the heating rate, but the conductivity is so high that conduction can convey the excess heat input to the lower part of the loop by means of a very shallow temperature gradient. This is the 'coronal' part of the loop. However, when the temperature drops below 10^6 K, the conductivity drops sharply and the temperature gradient steepens in order to maintain the conductive flux. This region of steep temperature gradient is known as the *transition region*. It is very thin compared to the pressure scale height and is therefore essentially isobaric.

When the pressure is almost constant, the radiative losses per unit volume throughout the loop are determined almost entirely by the function $Q(T)/T^2$. This function is sketched in Fig. 5.5. The full line shows the analytic approximation to the detailed calculations of optically thin losses given by Rosner, Tucker & Vaiana (1978b). The dashed line is an analytic approximation due to McClymont & Canfield (1983), which takes account of the breakdown of the optically thin assumption in the chromospheric portion of the loop at temperatures below 3×10^4 K. Both functions peak in the temperature range spanned by the transition region. The increased loss rates allow these regions of the loop to radiate away the excess heat conducted down to them. Roughly speaking, there is a balance between heating and conduction in the 'coronal' portion of the loop and a balance between conduction and radiative losses in the transitive region.

The early discussions did not, however, reveal the importance of the feature which distinguishes a coronal loop from other coronal structures, namely, its finite length. In 1978 two sets of authors, Rosner *et al.* (1978b) and Craig, McClymont & Underwood (1978), demonstrated that a relationship must exist between the thermodynamic properties of a loop and its length. These relationships are known as *scaling laws*. The scaling laws allow the properties of loops to be systematized and described in very general terms. For instance, one scaling law given by Rosner *et al.* (1978b) involves only measurable quantities, relating the maximum temperature occurring at the apex of the loop, T_m, to the product of its pressure p and half-length L,

$$T_m = 1.4 \times 10^4 (pL)^{\frac{1}{3}}. \tag{5.69}$$

The value of this relation became apparent when it was shown by Vesecky, Antiochus & Underwood (1979) and Wragg & Priest (1981) to be very insensitive to changes in the assumed geometry of the field, i.e. to the basic uncertainty of the model.

The form of the scaling laws can be obtained from an analytic model of a constant pressure loop in which we suppose that the temperature is a monotonic function of position up to the apex of the loop. We may then parametrize the cross-section A and the heating function $\mathcal{H}$ in terms of T alone; let us take a general functional form (cf. Rosner *et al.*, 1978b; Levine & Pye, 1980),

$$A = A_0 T^\alpha, \quad \mathcal{H} = H_0 T^\beta, \tag{5.70}$$

and write

$$Q(T) = \chi T^\gamma. \tag{5.71}$$

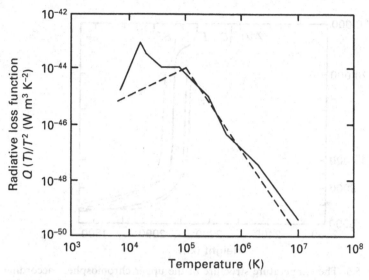

Fig. 5.5. The radiative loss function $Q(T)/T^2$. The full line is the analytic fit of Rosner *et al.* (1978b) to detailed computations of optically thin emission from a plasma of solar composition. The broken line is the analytic approximation of McClymont & Canfield (1983) to their detailed calculations incorporating optically thick transfer in the chromosphere.

Utilizing the classical form for the effective thermal conductivity (Eqn. (5.44)), the energy equation becomes

$$0 = T^{-\alpha}\frac{\mathrm{d}}{\mathrm{d}s}\left(\kappa_0\,T^{\alpha+\frac{5}{2}}\frac{\mathrm{d}T}{\mathrm{d}s}\right) - \frac{p^2\chi}{4k^2}\,T^{\gamma-2} + H_0\,T^\beta. \qquad (5.72)$$

For a constant pressure loop, the hydrostatic equilibrium equation is redundant. The thermal structure of the loop is determined entirely by the energy equation (5.72), together with appropriate boundary conditions.

The model of the coronal loop must match a model of the chromospheric portion of the magnetic flux loop. The lower boundary conditions are therefore set by the empirical models of the atmospheric structures in the chromosphere. Those of Basri *et al.* (1979) and of Vernazza, Avrett & Loeser (1981), shown in Fig. 5.6, indicate that the quiet Sun (the curves marked *C* and *AV*) possesses two temperature plateaux, one at a temperature of around 7000 K and the other at 20 000–25 000 K, where the temperature gradient, and hence also the conductive flux, becomes very small. The pressure at the upper plateau is about 0.015 Pa. In network elements (the curves marked *F* and *BP*), which are the footpoints of loops outside active regions, the higher temperature plateau shrinks in extent and moves deeper into the atmosphere where the pressure is somewhat higher, around 0.03 Pa. In active regions, the upper plateau appears to be absent

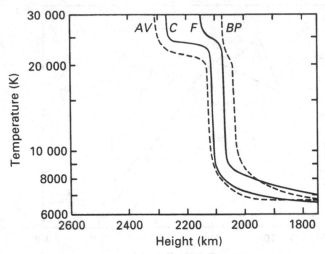

Fig. 5.6. The temperature structure of the upper chromosphere according to the empirical models of Basri *et al.* (1979) and Vernazza *et al.* (1981). Models of the average chromosphere are labelled *AV* and *C*, respectively. Models of bright network, thought to be the footpoints of coronal loops, are labelled *BP* and *F*. (After Vernazza *et al.*, 1981.)

and the conductive flux first becomes negligible at about 7000 K where the pressure is considerably higher, of the order of 0.6 Pa. This suggests that as a lower boundary condition we should set $s = 0$ and $dT/ds = 0$ when $T = T_0$, where $T_0 = 20\,000$ K in low pressure loops and $T_0 = 7000$ K in high pressure loops. The upper boundary condition is straightforward. If the loop is symmetrical about its apex, we must have $dT/ds = 0$ at $s = L$. By assumption, the maximum temperature T_m occurs at this point. These boundary conditions define what is known as a *thermally isolated loop* because no heat is conducted in or out at either end. The imposition of three boundary conditions on a second-order system means that a constraint is placed on the parameters describing the system. Since these are the length L, the heating factor H_0, and the pressure p, a scaling law of some form is inevitable.

The energy equation (5.72) may be integrated directly by rewriting it in the form

$$\frac{d}{ds}\left(\kappa_0\, T^{\alpha+\frac{5}{2}}\frac{dT}{ds}\right)^2 = 2\kappa_0\left(\frac{p^2\chi}{4k^2}\,T^{2\alpha+\gamma+\frac{1}{2}} - H_0\,T^{2\alpha+\beta+\frac{5}{2}}\right)\frac{dT}{ds}. \qquad (5.73)$$

Integrating this equation from $s = 0$ to $s = L$ and employing the boundary conditions $dT/ds = 0$ at each end as well as the assumption that $T_0 \ll T_m$, leads to the scaling law relating the heating rate to the maximum temperature of the loop,

$$H_0 \simeq \frac{p^2 \chi}{4k^2} \frac{2\alpha+\beta+\frac{7}{2}}{2\alpha+\gamma+\frac{3}{2}} T_m^{\gamma-\beta-2}. \tag{5.74}$$

This merely states that the overall heating must balance the overall radiation losses in a thermally isolated loop.

With this value for H_0 substituted, the indefinite integral of Eqn. (5.73) yields the temperature gradient at each point. For a loop whose temperature increases with height, the temperature gradient is given by

$$\frac{dT}{ds} = \left[\frac{p^2\chi}{2k^2\kappa_0(2\alpha+\gamma+\frac{3}{2})}\right]^{\frac{1}{2}} T^{\gamma/2-\frac{7}{4}}[1-(T/T_m)^{\beta-\gamma+2}]^{\frac{1}{2}}. \tag{5.75}$$

This can now be integrated along the loop to produce a generalized version of the earlier scaling law (cf. Landini & Monsignori Fossi, 1981)

$$pL = \left[\frac{2k^2\kappa_0(2\alpha+\gamma+\frac{3}{2})}{\chi}\right]^{\frac{1}{2}} T_m^{\frac{11}{4}-\gamma/2} I_{\beta,\gamma} \tag{5.76}$$

where

$$I_{\beta,\gamma} = \int_0^1 x^{\frac{7}{4}-\gamma/2}(1-x^{\beta-\gamma+2})^{-\frac{1}{2}} dx$$

$$= \frac{\sqrt{\pi}}{2+\beta-\gamma} \frac{\Gamma\left(\frac{11-2\gamma}{8+4\beta-4\gamma}\right)}{\Gamma\left(\frac{15+2\beta-4\gamma}{8+4\beta-4\gamma}\right)}. \tag{5.77}$$

When the gamma functions Γ are evaluated, this integral is found to lie between 0.5 and 1.0 for all realistic values of γ and β.

For hot loops with temperatures in excess of 5×10^5 K for most of their length we may take $\gamma = -\frac{1}{2}$ (cf. Eqn. (5.40)) and then recover the scaling law of Rosner *et al.* (1978b),

$$pL \simeq 0.75 \left[\frac{2k^2\kappa_0(2\alpha+1)}{\chi}\right]^{\frac{1}{2}} T_m^3. \tag{5.78}$$

It may be seen from the more general expression (5.76) that the scaling law is relatively insensitive to the form of the loss function. The integrals are dominated by the extended length of hot loops where the temperature is in excess of 5×10^5 K, so that the results change only slightly when the exact form of $Q(T)$ is used (Levine & Pye, 1980; Chiuderi, Einaudi & Torricelli-Ciamponi, 1981). Moreover, the geometric factor describing the change of cross-section along the loop enters only weakly in the coefficient of proportionality.

It might seem surprising that the scaling law is almost totally independent of the form of the heating function. The reason lies in the efficiency with which conduction spreads out the excess heat in the corona – irrespective of the point of input – and transfers it down to the transition region. The derivation of the scaling laws ignores the processes in this region, simply assuming that the radiative losses there are sufficient to balance the conductive input. Of course, numerical solutions cannot be found when this is not possible but, under conditions typical of observed loops, solutions exist and follow the scaling laws very closely.

Some authors (e.g. Hood & Priest, 1979a; Roberts & Frankenthal, 1980) have criticized models of thermally isolated loops on the grounds that the energetics of the lower transition region and chromosphere, where the lower boundary condition is imposed, are not described correctly. They suggested avoiding this uncertainty by terminating the loop higher in the atmosphere at a temperature of 10^6 K, say. Then the downward conductive heat flux at the base cannot be ignored and must be treated as a further parameter in the model. However, Chiuderi *et al.* (1981) found empirically that basal fluxes less than 5×10^2 W m^{-2} have only a very small influence on the energy balance and the scaling laws. An upper limit on the actual flux at this point in the solar atmosphere is provided by estimates of the losses by Lα radiation from the 20000 K plateau. This figure is about 10^3 W m^{-2} (Chapman, 1981), so that the basal flux in coronal loops does not appear to be significant.

Furthermore, any oversimplification of the physical processes occurring in the lower segment of a thermally isolated loop model does not appear to be damaging. McClymont & Canfield (1983) have made a detailed study of the radiative transfer in the optically thick chromospheric regime and find that their results can be reproduced roughly by suppressing the contribution of the Lα line to the low temperature end of the optically thin loss function $Q(T)/T^2$. They claim that this function is given to a good approximation by

$$\frac{Q(T)}{T^2} = \begin{cases} 10^{-49}T & [\text{W m}^3] \quad T < 10^5 \text{ K} \\ 10^{-29}T^{-3} & [\text{W m}^3] \quad T > 10^5 \text{ K.} \end{cases} \tag{5.79}$$

It is shown in Fig. 5.5 in comparison to the loss function for optically thin transfer. The change has very little effect on the thermal structure of equilibrium models, but is highly important as far as their stability is concerned (Section 5.5.2).

As well as being optically thick, the upper chromosphere is observed to be highly dynamic on small scales. This is even more true of the lower temperature regime in the transition region. Wallenhorst (1982) has modelled the effects of unresolved motions by an enthalpy flux term in the energy equation. The form of the term was derived empirically by comparison with observed emission measures. The scaling law Equation (5.69) again survived, with only a 15% modification.

The scaling law is, indeed, remarkably robust. High loops in which the pressure stratification may not be neglected require the numerical solution of the coupled Eqns. (5.67) and (5.68). Landini & Monsignori Fossi (1981) discuss the general characteristics of the solutions and conclude that the trends displayed by the constant-pressure scaling laws are unchanged. This is confirmed by the more detailed numerical results of Serio *et al.* (1981) for loops in which the pressure distribution was allowed to possess an arbitrary scale height Λ. They found that the scaling law took the form

$$T^3 \propto p_0 L \exp(-L/8\Lambda), \tag{5.80}$$

where p_0 is the pressure at the base of the loop. When $L \sim 8\Lambda$, the temperature becomes almost independent of the length of the loop and varies only with basal pressure (cf. Wragg & Priest, 1981, 1982b). Since Λ is of the order of 10^8 m for hot loops, this effect should be barely discernable even in the longest interconnecting loops.

It is therefore not surprising that the lengths of 'hot' loops derived from Eqn. (5.69) using empirical estimates of T_m and p agree quite well with the observed lengths. Figure 5.7, taken from Pallavicini *et al.* (1981), demonstrates that the predicted and observed lengths of virtually all the examples agree with a factor of 2, whether the loops are active-region loops, interconnecting loops or even flare loops. Since the temperatures of active region loops are systematically higher than those of quiet region loops, the relation correctly predicts that the densities of active region loops are higher than those of quiet region loops of the same length. It also follows, from Eqn. (5.74), that the heating rates are higher in active region loops.

The physical interpretation of this trend was provided by Pye *et al.* (1978). They suggested that an increase of heat input to a hot loop would initially increase the temperature, because the radiative loss rate falls (Eqn. (5.40)) and cannot compensate the heating increment. The temperature gradient must therefore steepen and conduct heat to the denser basal regions which will heat and expand, thereby 'evaporating' material up into the coronal part of the loop. This process will continue until the increased density of the loop raises the radiative loss rate sufficiently to balance the new heat input. In other words, coronal loops possess a mechanism acting as a thermostat. This mechanism will be described in detail in Section 5.5.

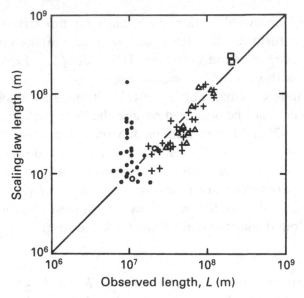

Fig. 5.7. Comparison of the observed lengths of coronal loops with lengths predicted on the basis of the Rosner *et al.* (1978b) scaling law from observed values of the maximum temperature and pressure. Different structures are denoted by different symbols: ○, coronal bright points; +, active region loops; □, interconnecting loops; △, prominence-related flare loops; ●, compact flare loops. (After Pallavicini *et al.*, 1981.)

The numerical calculations of Serio *et al.* (1981) and the earlier work of Hood & Priest (1979a) revealed a feature of the model that is not apparent in the simplified analytic treatment. As well as solutions with a monotonic increase of temperature up to the apex, there may exist other solutions in which the temperature first increases with height and then decreases to form a local minimum at the apex. These are the Class II solutions of Serio *et al.* and require a greater overall heating rate than the monotonic, or Class I, solutions because of the increased radiative losses from the cool apex. The drop in temperature at the top of the loop is generally accompanied by a rise in density. However, a density inversion is unstable unless the heavy material is supported by magnetic forces – by a sag in the field, for instance. Thus these 'high-energy' states describe the structure of a prominence rather than a coronal loop (see, for example, She, Malherbe & Raadu, 1986), and they will not be pursued further here.

On the other hand, there are solutions which are cool throughout. Their significance was first appreciated by Martens & Kuin (1982). They pointed out that 'cool' loops arise quite simply if the radiative losses balance the heating all along the loop, i.e., from (5.68),

$$\frac{p^2}{4k^2}\frac{Q(T)}{T^2} = \mathcal{H}. \tag{5.81}$$

Reference to Fig. 5.5 shows that Q/T^2 is greater at cool temperatures of around 10^4 K than at hot temperatures of 10^6 K. Thus cool loops must have lower pressures than hot loops with the same heat input. The energy balance enforces a relationship between the temperature and pressure at each point of the loop (instead of overall, as in hot loops); the temperature and pressure distributions in the loop are then determined solely by the hydrostatic equilibrium equation. With low temperatures, the pressure scale height will be small and the pressure at the top of a large-scale loop will be significantly lower than that at the base. The radiative loss rate will be maintained only if the $Q(T)/T^2$ factor can increase to compensate the decrease in p^2. Martens & Kuin argued, on the basis of the Rosner *et al.* (1978b) curve, that this will limit the temperatures of cool loops to about 20000 K. Above this value, local energy balance can no longer be maintained and the temperature must climb steeply to 'coronal' values, at which conduction can remove the excess heat; a hot loop results. Their trial models suggested that the maximum size that a cool loop can attain before becoming unstable is relatively insensitive to the form of the heating function and is of the order of the scale height corresponding to the maximum allowable temperature, i.e. about 10^6 m. The conclusion that cool loops are short and of low pressure is in accord with the general scaling law, Eqn. (5.69).

Confirmation is to be found in the work of Antiochus & Noci (1986), who investigated the limiting height of cool loops assuming various forms of heat input functions $\mathscr{H}$. Note that by employing the McClymont & Canfield (1983) loss function $Q(T)/T^2$, which has a positive gradient up to 10^5 K, it is possible to generate 'cool' loops of higher temperature than those found by Martens & Kuin. The properties of cool *versus* hot loops can be found from the appropriate scaling laws (5.76). For loops of the same length, the ratio of the basal pressures p_c and p_h of cool and hot loops with apex temperatures T_c and T_h respectively is

$$\frac{p_c}{p_h} \simeq 10^{-\frac{9}{2}} \left(\frac{T_c}{10^4}\right)^{\frac{5}{4}} \left(\frac{10^6}{T_h}\right)^{\frac{13}{4}}. \tag{5.82}$$

The cool loop pressure (and density) is orders of magnitude lower than a hot loop of the same length. Static cool and hot loops of equal length are unlikely to coexist. On the other hand, the scaling laws provide exactly the same relation between the half-lengths of cool and hot loops L_c and L_h having the same basal pressure,

$$\frac{L_c}{L_h} \simeq 10^{-\frac{9}{2}} \left(\frac{T_c}{10^4}\right)^{\frac{5}{4}} \left(\frac{10^6}{T_h}\right)^{\frac{13}{4}}. \tag{5.83}$$

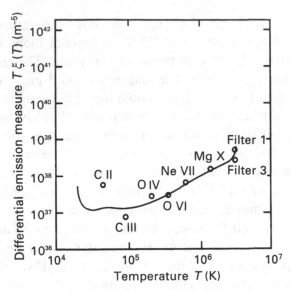

Fig. 5.8. Comparison of the product of temperature and differential emission measure $T\xi(T)$ observed from a 'composite' loop (see text) with that of a static model with $L = 9 \times 10^7$ m, $p_0 = 0.15$ Pa, $\alpha = 1$, $\beta = 0$. The values measured for each ion (or filter in the X-ray band) are plotted at the temperature of maximum ionization abundance. (After Pallavicini *et al.*, 1981.)

Thus very short cool loops, rising to correspondingly small heights, may exist together with larger scale hot loops.

An illuminating way of systematically mapping the cool and hot solutions corresponding to different choices of boundary conditions was suggested by Hood & Anzer (1988). They plot the conductive flux against the temperature in a 'phase-plane' diagram. Thermally isolated models start from the flux axis and return to it. Other models can be read off from arbitrary starting points.

Of course, the very insensitivity of the scaling laws to the forms of the unknown functions endows them with little power of discrimination. Of more diagnostic value is the distribution of emission along the loop, as characterized by the differential emission measure

$$\xi(T) = A(s)\frac{p^2}{4k^2T^2}\left(\frac{\mathrm{d}T}{\mathrm{d}s}\right)^{-1}. \tag{5.84}$$

In the simple model introduced above, this takes the form (cf. Rosner *et al.*, 1987b)

$$\xi(T) = A_0\left[\frac{\kappa_0(2\alpha+\gamma+\frac{3}{2})}{8\chi k^2}\right]^{\frac{1}{2}}pT^{\alpha-\gamma/2-\frac{1}{4}}[1-(T/T_{\mathrm{m}})^{\beta-\gamma+2}]^{-\frac{1}{2}}. \tag{5.85}$$

For hot loops ($\gamma \simeq -\frac{1}{2}$) of constant cross-section and uniform heating, this function decreases initially as temperature increases, then displays a broad and flat minimum for $10^5 < T < 10^6$ K followed by a sharp rise as the temperature approaches its maximum value T_m. As can be seen from the formula, the shape of the minimum depends strongly on the way that the cross section of the loop changes with height, which is parametrized by α. The importance of this factor was noted by Vesecky *et al.* (1979) and Levine & Pye (1980).

The predictions may be compared with empirical determinations of emission measure. These were first obtained from spatially unresolved observations of solar features and they revealed the remarkable fact that the differential emission measure in transition region material with $T < 10^5$ K had the same form

$$\xi(T) \simeq T^{-\frac{9}{2}} \qquad (5.86)$$

irrespective of the feature, whether quiet Sun, active Sun, flare or coronal hole (Jordan, 1980). Antiochus & Noci (1986) remark that a similar behaviour is found in other late-type stars. They represent the general form by

$$\xi(T) \begin{cases} \propto T^{-4} & 2 \times 10^4 \leqslant T \leqslant 10^5 \\ \simeq constant & 10^5 \leqslant T \leqslant T_m. \end{cases} \qquad (5.87)$$

The initial drop and the broad flat minimum of the observations thus broadly resemble the model. However, a more detailed fit between the two is more difficult to achieve.

Pallavicini *et al.* (1981) report observations of the emission measure in *individual* active region loops but do not publish any comparison with a model. Instead, they construct a *composite* loop by averaging the brightest points in an active region in each ion or filter wavelength band. This empirical composite loop is compared with a loop model with constant heating but steadily increasing cross section in Fig. 5.8. The rather flat curve between temperatures of 10^5 and 3×10^6 K agrees quite well with the 'observed' points. Note that Pallavicini *et al.* plot the quantity $T\xi(T)$ against T for the sake of clarity. (It should be noted that some authors, e.g. Antiochus & Noci (1986), define this quantity to be the differential emission measure.) Pallavicini *et al.* found that static, thermally isolated loop models could also match both a composite 'core' loop, derived from the bright centre of the active region, and large-scale interconnecting loops around the active region. However, the representation is not unique. Torricelli-Ciamponi, Einaudi & Chiuderi (1982) achieved an equally good fit between the composite active region loop and a quite different loop model, one which was not thermally isolated and which had a heating function which decreased rapidly with temperature ($\beta \sim -2$).

To obtain a fit beyond this temperature range is more difficult. Unresolved observations of an active region by Levine & Pye (1980) show that the emission measure drops steeply at temperatures above 3×10^6 K. These authors were able to reproduce this behaviour only by assuming an ensemble of loops at different maximum temperatures. However, they stress that the problem of deriving the characteristics of the loop population from spatially unresolved observations is badly conditioned and does not produce unique results.

Antiochus & Noci (1986) argue that the presence of an ensemble of loops is masked at intermediate temperatures by the fact that the emission throughout that range increases with increasing T_m, as may be seen from Eqn. (5.85). Thus composite observations will be dominated by the hottest loops present and a reasonable fit to a hot loop model is not surprising. However, at the low temperature end, the hot loop models provide far too little transition region emission ($T < 10^5$ K) to match the observations. They therefore appeal to the presence of cool coronal loops to provide the extra component. For such loops $\gamma = 3$, so that $\xi(T) \propto T^{-\frac{7}{4}}$ when we take $T \ll T_m$ and $\alpha \simeq 0$. The coolest loops thus dominate the low temperature emission, and a suitable ensemble could produce the T^{-4} dependence of unresolved observations of emission measure. It seems unlikely, though, that any properties of the loop ensemble can be safely derived from such observations.

We may therefore conclude that observations of emission measure from loops and ensembles of loops are *consistent* with simple loop models. Whilst the comparison provides very little constraint on the free parameters of the model, and almost none on the population characteristics of the ensemble in active regions, there is little doubt that the highly simplified models described in this section capture the essential processes governing the behaviour of coronal loops and allow insight to be gained into more complicated structures.

5.4.3 *Steady-flow models of individual loops*

An obvious limitation of the models of coronal loops presented in the previous section is the assumption that they are static. This conflicts with the (admittedly limited) observational material available. However, the apparent lifetimes of loops suggest that they are maintained for times at least comparable to the time for a sound wave or Alfvén wave to propagate along its length (cf. Section 5.5.1). To a first approximation, then, the flows within coronal loops may be considered to be steady. We shall return to the question of unsteady flows below; here we examine the modifications that are necessary to the picture drawn above when time-independent motions along the field lines are allowed.

Two types of steady flow have been considered. In the first, there is an upflow in one leg of the loop and downflow in the other. In a strictly isolated loop, mass conservation requires the same flux of material to enter at one footpoint as leaves at the other. In a loop which can exchange material with its surroundings this need not be the case. In the second type of flow, there is a downflow in each leg. This type clearly requires there to be a source of material distributed along the loop. The third possibility, an upflow in both legs, could be produced if there were a sink of material along the loop, but observation provides little motivation for pursuing this variant.

Let us look first at the case of the strictly isolated loop which possesses a steady mass-conserving flow from one end to the other. This model was first studied by Meyer & Schmidt (1968) in the context of the Evershed flow in sunspots. It was adopted as a coronal loop model by Glencross (1980) and Cargill & Priest (1980), and followed by Noci (1981) and Landini & Monsignori Fossi (1981). These authors differ only in the treatment of the energy equation. However, this makes little difference to the results because the most significant features of flow models are produced by the momentum equation.

The momentum equation for a steady flow is

$$\rho v \frac{dv}{ds} = -\frac{dp}{ds} - \rho g \cos \theta. \tag{5.88}$$

This must now be supplemented by the continuity equation, which reads

$$\frac{d}{ds}(\rho v A) = 0, \tag{5.89}$$

where $A(s)$ is the cross-sectional area of the loop. It simply expresses the fact that the mass flux $q = \rho v A$ is conserved along the loop.

Cargill & Priest (1980) adopted the adiabatic condition as the energy equation (5.53), namely,

$$\frac{d}{ds}(p\rho^{-\gamma}) = 0, \tag{5.90}$$

with γ being the ratio of specific heats. Noci (1981) assumed the same form for the energy equation but interpreted γ as a polytropic index, whilst Landini & Monsignori Fossi (1981) used the full equation (5.53). For a loop with almost constant pressure, the isobaric form of Eqn. (5.56) can be used,

$$c_p \rho v \frac{dT}{ds} = \frac{1}{A} \frac{d}{ds} \left(A \kappa^{eff} \frac{dT}{ds} \right) - \mathcal{L} + \mathcal{H}, \qquad (5.91)$$

which may be rewritten using the constancy of the mass flux as

$$\frac{1}{A} \frac{d}{ds} (c_p q T) = \frac{1}{A} \frac{d}{ds} \left(A \kappa^{eff} \frac{dT}{ds} \right) - \mathcal{L} + \mathcal{H}. \qquad (5.92)$$

This expression (cf. Craig & McClymont, 1986) balances the change in enthalpy flux against the change in conductive flux and the *net* heating rate, $\mathcal{H} - \mathcal{L}$.

In the adiabatic or polytropic case, expansion of the derivative leads to the relation

$$\frac{dp}{ds} = \frac{\gamma p}{\rho} \frac{d\rho}{ds} = v_a^2 \frac{d\rho}{ds}, \qquad (5.93)$$

where $v_a = \sqrt{(\gamma p / \rho)}$ is the local sound speed.

Now ρ and p can be eliminated from the momentum equation using Eqns. (5.89) and (5.93), to produce an equation for the flow speed v,

$$\left(\frac{v^2 - v_a^2}{v} \right) \frac{dv}{ds} = \frac{v_a^2}{A} \frac{dA}{ds} - g \cos \theta. \qquad (5.94)$$

From this equation it can be seen that, as the flow speed increases, the factor multiplying the derivative on the left-hand side increases, but remains negative whilst the flow speed remains less than the speed of sound. The derivative dv/ds is thus well-defined if the speed is everywhere subsonic. Models of coronal loops with subsonic flows have a structure very similar to those of static models and will be considered in more detail shortly.

When the flow exceeds the sound speed, the situation is very different. From Eqn. (5.94) it is clear that the only point at which the flow speed can change from being subsonic to supersonic, i.e. $v = v_a$, is the point at which the right-hand side happens to vanish. This critical point is determined mainly by the geometry of the loop. If we start with a subsonic flow everywhere, dv/ds must vanish at the critical point, so that the flow speed must reach its maximum value here. If we gradually increase the speed of the flow (by whatever means) a stage will be reached when the speed at this critical point will equal the sound speed, whereupon it becomes possible to satisfy Eqn. (5.94) without the flow speed having a turning point, i.e. the speed may continue to increase above the sound speed. The flow upstream of the critical point is subsonic and the flow downstream is supersonic.

However, a model with a continuous variation of velocity and thermo-dynamic state variables is then not generally possible. The structure of the flow upstream of the critical point cannot change; it is fixed by the need to maintain the flow speed matching the sound speed at that point. The flow downstream can still change but only by means of a discontinuous jump from the fixed solution through the critical point to whatever solution is demanded by the conditions enforced at the other footpoint. In other words, further solutions must contain a shock front.

Meyer & Schmidt (1968) suggested that the mechanism responsible for Evershed flows is a siphon maintained by a pressure difference between the footpoints. This model was developed for coronal loops by Glencross (1980), Cargill & Priest (1980) and Noci (1981) and their conclusions may be summarized as follows.

When the loop is perfectly symmetric the equations are symmetric in ρ, p, T and v^2. This has the rather surprising consequence that steady subsonic flows antisymmetric about the apex of the loop are possible without any pressure difference between the footpoints. However, Cargill & Priest show that any slight asymmetry of the loop requires a finite pressure difference to initiate the flow. Increasing the pressure difference causes the speed at the critical point to increase. Eventually it reaches the sound speed. As the pressure difference is increased further, a shock develops just downstream of the critical point. The structure of the flow upsteam then does not change and the critical point remains stationary. Further increases of the pressure difference between the footpoints causes the shock to move downstream and to increase in strength. Eventually the shock migrates to the other footpoint and no further change can be accommodated.

This general behaviour is found in both symmetric flows, in which case the critical point occurs at the apex, and asymmetric flows. The latter are the more physically plausible. A difference in gas pressure between the footpoints would require to be compensated by a complementary difference in magnetic pressure in order to maintain pressure balance with the surrounding material. The high pressure end would thus possess a lower magnetic pressure, i.e. a smaller magnetic induction, than the low pressure end. Since magnetic flux must be conserved along the loop, this can be achieved only by decreasing the cross-sectional area from the high pressure footpoint to the low pressure footpoint. The pressure-driven flow would therefore follow a converging duct.

This geometric asymmetry produces higher speeds at the low pressure end as a direct result of the continuity equation, and the critical point is moved into the leg of the loop with downflow. The greater the degree of convergence, the more difficult it is to achieve supersonic flow. The speed of the flow depends also on the length of the loop. The longer the loop, the greater the pressure difference that must be imposed to produce the same flow speed.

Whilst loops with flows are certainly observed, the concept of a steady siphon flow driven by a pressure difference between the footpoints has been criticized by Mariska & Boris (1983). They modelled the one-dimensional motion along a hot loop of relatively short length, $L = 1.2 \times 10^7$ m, with footpoints embedded in the chromosphere. This required the numerical solution of the time-dependent equations derived from Eqn. (5.32),

$$\frac{\partial p}{\partial t} + \frac{1}{A}\frac{\partial}{\partial s}(A\rho v) = 0, \tag{5.95}$$

from Eqn. (5.52)

$$\rho\frac{\partial v}{\partial t} + \rho v\frac{\partial v}{\partial s} = -\frac{\partial p}{\partial s} - \rho g\cos\theta, \tag{5.96}$$

and from Eqn. (5.54)

$$\frac{1}{\gamma-1}\left(\frac{\partial p}{\partial t} + v\frac{\partial p}{\partial s}\right) + \frac{\gamma p}{\gamma-1}\frac{1}{A}\frac{\partial}{\partial s}(Av) = \frac{1}{A}\frac{\partial}{\partial s}\left(A\kappa^{eff}\frac{\partial T}{\partial s}\right) - \mathscr{L} + \mathscr{H}. \tag{5.97}$$

One simulation followed the motion that ensued when a pressure difference was enforced between the footpoints. They found that a *static* balance was re-established after pressure waves had travelled back and forth along the tube several times. During this time, a flow occurred which transferred mass from the high pressure end to the low pressure end, expanding the chromosphere on the high-pressure side and compressing that on the low-pressure side. A uniform pressure was thereby created throughout the corona and transition region, and the motions died out.

However, these calculations prevented flows through the footpoints, thus allowing the material transferred from one side to the other to accumulate where it could eventually choke the flow. The picture of a siphon flow is rather different. The flow is envisaged as continuing through the chromosphere and photosphere and into the subsurface regions, where it is maintained. It is an open question whether such boundary conditions would lead to a steady coronal flow.

The question is in any case academic. The asymmetric geometries considered by Cargill & Priest (1980) immediately suggest that the heating function should also be asymmetric. Mariska & Boris (1983) simulated a loop with an asymmetric perturbation of the heating rate, taken originally to be constant per unit volume. In this case, they found that static equilibrium could not be established because an enthalpy flux is required to balance the energy budget in the transition region and chromosphere in the presence of different conductive fluxes in the two legs of the loop. The flow is upward on the side with excess heating, where the conductive heat flux down through the transition zone is greatest. However, the steady-state flows speeds were low and had little effect on the temperature and pressure distributions in the corona. The corona remained in almost hydrostatic balance. Craig & McClymont (1986) reached similar conclusions using a model in which the basic model had a constant heating rate per unit mass.

Little work has been done on flows in loops which are not isolated. If plasma injection or extraction is allowed to take place along the loop, a great number of further assumptions must be made in order to construct a model. So far there has been no systematic analysis of these. Noci & Zuccarello (1983) have explored a restricted class of models, in which only the continuity equation is deemed to be altered by the injection of plasma to the loop. This involves replacing the cross-sectional area A by a quantity σ, which describes both the changing area and the changing particle flux. There is then no necessity for σ to be positive. The solutions may be classified exactly as before, though they may be interpreted in terms of a *condensing* loop – one having a steady downflow in each leg – or an *evaporating* loop – one having a steady upflow in each leg.

In order to relate the theoretical dynamics to observations, the thermodynamic structure which governs the emission must be analysed using the full energy equation. Landini & Monsignori Fossi (1981) and Craig & McClymont (1986) discuss the thermodynamic effects of the flows in terms of modified scaling laws.

If we follow Craig & McClymont in approximating dT/ds by T_m/L, the scaling law Eqn. (5.74) becomes

$$H_0 = (1-\varepsilon)\frac{p^2\chi}{4k^2}\left(\frac{2\alpha+\beta+\frac{7}{2}}{2\alpha+\gamma+\frac{3}{2}}\right)T_m^{\gamma-\beta-2},\tag{5.98}$$

where ε measures the ratio of the enthalpy flux at the apex, $c_p q T_m$, to the rate of heating over the coronal part of the loop:

$$\varepsilon = \left(\frac{2\alpha+\gamma+\frac{3}{2}}{\alpha+\frac{7}{2}}\right)\frac{c_p q T_m}{(p^2\chi T_m^{\gamma-2}/4k^2)(A_0 T_m^\alpha L)}.\tag{5.99}$$

The other scaling law, Eqn. (5.76), becomes

$$pL = \left[\frac{2k^2\kappa_0(2\alpha+\gamma+\frac{3}{2})}{\chi}\right]^{\frac{1}{2}}T_m^{\frac{11}{4}-\gamma/2}I_{\alpha,\beta,\gamma}(\varepsilon),\tag{5.100}$$

where

$$I_{\alpha,\beta,\gamma}(\varepsilon) = \int_0^1 x^{\frac{7}{4}-\gamma/2}[1-x^{\beta-\gamma+2}+\varepsilon x^{2-\alpha-\gamma}(1-x^{\alpha+\beta})]^{-\frac{1}{2}}dx.\tag{5.101}$$

In this approximation, a leg of a loop with an upward flow ($\varepsilon > 0$) requires an increased heating rate to achieve the same maximum temperature as a static loop of the same length and pressure; since the enthalpy flux $c_p qT$ varies only with temperature along the loop, the outflow at the apex dominates and creates a net energy loss. By the same token, a leg with a downflow requires a lower heating rate.

The effect on the thermodynamic structure is more subtle. The new scaling law, Eqn. (5.100), allows three cases to be distinguished.

1. If $\alpha + \beta > 0$, i.e. if the loop expands faster than the heating rate drops off with temperature (or height), a leg of a loop with an upflow will have a lower pressure and density than a static leg of the same length and maximum temperature. Or a loop leg with an upflow will have a higher maximum temperature than a static leg of the same length and pressure. A downflow leads to the opposite result.
2. If $\alpha + \beta = 0$, i.e. the loop expands at the same rate as the heating drops off, the flow has no effect on the global thermodynamic structure.
3. If $\alpha + \beta < 0$, i.e. the loop expands more slowly than the heating drops off, an upflow will occur in loop legs which have higher pressure and density than static loops of the same length and maximum temperature, and so on.

In particular, a loop leg of constant cross section ($\alpha = 0$) and a constant heating rate per unit volume ($\beta = 0$) has the same scaling law as that of a static loop, whatever the flow. A loop of constant cross section but constant heating rate per unit mass ($\alpha = 0$ and $\beta = -1$) will have higher densities associated with upflows and lower densities with downflows. For loops with upflow in one leg and downflow in the other, the scaling laws are harder to apply because the maximum temperature does not occur at the geometrical centre and L is no longer the half-length of the loop. According to Landini & Monsignori Fossi, a steady flow around a loop reduces the temperature and density at the apex. For subsonic flows, the reduction is at most a factor of two. For supersonic flows, the drop in temperature and density can become much greater just upstream of the shockfront. Downstream of the shock, the temperature and density regain almost their undisturbed values.

Noci (1981), Borrini & Noci (1982) and Craig & McClymont (1986) have all calculated the effects of these structural changes on the appearance of loops in hot EUV lines. In the earliest paper, Noci found that the drop in temperature and density in supersonic flows led to much lower emission in such lines, the loop becoming almost invisible just upstream of a strong shock. However, in the second paper, account was taken of the departures from equilibrium in the ionization balance of the high-speed flows. The typical ionization and recombination rates are $(10^{-17}–10^{-16})\,n_e$ per second. For densities of 10^{15} m^{-3} and flow speeds of 10^5 m s^{-1}, an ion would move a distance of $10^7–10^8$ m before ionization equilibrium could be attained. This distance is comparable to the length of coronal loops, so that the ionization is frozen into the plasma to a large degree and the ionization temperature before the shock does not fall to anything like the same degree as the electron temperature. The asymmetry in the emission is thereby greatly reduced.

Since most velocity measurements are obtained in the EUV lines of relatively cool plasma, the structure of the coronal sections of hot loops is of less diagnostic value than that of the transition zone at the base of hot loops or throughout cool loops. Here a detailed numerical model employing a realistic energy equation is required. Most authors have had difficulty reconciling the models with observations.

Kopp, Poletto, Noci & Bruner (1985) concluded that neither a siphon flow model nor a steady downflow in each leg could reproduce their observations of C IV lines in a system of loops near the limb. Nor could a time-dependent variant in which a peculiar steady downflow was adopted in each leg, chosen so that the density everywhere had the same exponential decay.

Mariska & Boris (1983) and Craig & McClymont (1986) had no more success in reproducing the high downward velocities generally associated with plasma at around 10^5 K. The average Doppler shift of the C IV lines over the disk is about 7 km s^{-1} downwards. As noted above, in hot loop models the steady flow speeds are generally low. The maximum occurs around the apex of the loop, where the temperature is too high for emission in transition region lines. Thus attention must be concentrated on the legs of the loop.

Craig & McClymont injected a point source of excess heating into one leg of a loop. In this case, the temperature maximum is moved towards the point of heat injection, i.e. into the evaporating leg. Consequently, the evaporating leg has a larger temperature gradient than the condensing leg and therefore a lower emission measure (cf. Eqn. (5.84)). But the *excess* emission arising from downward moving plasma is smaller than observed, and the velocities are no higher than about 1 km s^{-1} in the transition region, too low to explain the observed Doppler shifts. Higher velocities, up to 10 km s^{-1}, were-found to occur whilst the loop adjusted to the change of heating, but the transients were too short-lived to contribute significantly to the overall emission.

However, the problem appears to be resolved in cool loops. McClymont & Craig (1987a) point out that conservation of mass and momentum require the flow speed v at temperature T to be related to the maximum possible flow speed (the isothermal sound speed) at the maximum temperature in the loop by

$$v \simeq 6.4 \left(\frac{10^6}{T_\mathrm{m}}\right)^{\frac{1}{2}} \left(\frac{T}{10^5}\right) \text{ [km s}^{-1}\text{]}. \tag{5.102}$$

Thus the speed in plasma at $T \simeq 10^5$ K will increase as the maximum temperature of the loop is decreased. Large asymmetries in the heating rate of a cool loop can produce steady flow speeds in the range of 5–10 km s^{-1}. When the excess heating is concentrated at a point in one leg, the excess emission is concentrated in the other, condensing leg. Klimchuk & Mariska (1988) confirm this behaviour. These authors were able to generate flow speeds up to 20 km s^{-1}. They also demonstrated that the results were sensitive to the distribution of excess heat. Heat conduction is much less effective at smoothing out temperature fluctuations at cool loop temperatures than hot. In particular, an excess heat source which decreases monotonically from one footpoint to the other produces a temperature maximum in the *condensing leg*, the side on which the heat input increases with height. In such cases, the emission measure is greater in the evaporating leg and the plasma would show a net blue shift, albeit a small one. This is contrary to observation, suggesting that, at least in cool loops, heat is injected in localized regions.

In summary, it would seem that when the conditions of perfect symmetry in the geometry of the magnetic field are violated, steady flows from one footpoint of a loop to the other are inevitable. As far as hot loops are concerned, the flows are relatively slow and cause little change to the overall structure. But as the maximum temperature of the loop decreases, the flow speeds tend to increase throughout the loop and the asymmetry between the two legs becomes much more pronounced. In cool loops, the emission characteristics are more in line with observations.

5.5   Thermal stability and dynamics
5.5.1   *Introduction*

In the previous section we looked at the steady-state structure of coronal loops. In hot loops, heat is added throughout the extended coronal section and is removed by radiation in a thin basal region, the heat being transferred from the point of deposition to the point of loss by means of thermal conduction. Conduction maintains conditions in the coronal portion almost uniform and the details of the physical processes operating in the transition region are immaterial so long as the required rate of radiative loss is achieved there. The observed thermodynamic properties of hot loops therefore provide little significant information about the physics of the corona.

Nevertheless, there is one observed feature of coronal loops which is much more revealing, namely, their lifetimes. It was claimed in Section 5.4 that the length of time that loops survive without significant change implies that the loops are stable. We shall now quantify that statement.

A loop is dynamically stable if it returns to its original state when the hydrostatic balance is disturbed. It is well known that hydrostatic equilibrium is stable if the density does not ever increase with height. A pressure perturbation is then nullified on a time scale governed by the time taken by a pressure (acoustic) wave to cross the system. This is of the order of L/v_a and, as may be seen from Table 5.1, is much shorter than observed loop lifetimes. Hence we may conclude that coronal loops are dynamically stable. Valid coronal models must therefore also be stable. This is true of all models with temperature increasing monotonically upwards. However, the Class II models of Serio *et al.* (1981) can have a density inversion and develop a Rayleigh–Taylor instability, unless supported by some other agency such as the magnetic field. As noted before, these solutions are more relevant as prominence models than as coronal loop models and do not concern us here.

Thermal stability requires energy balance to be restored following a perturbation. There are two means by which this can be achieved, by radiation losses or thermal conduction. The time scales may be estimated from Eqn. (5.55) as $\tau_r = p/\mathscr{L}$ and $\tau_c = pL^2/(\kappa T)$, respectively. From Table 5.1 we see that their typical coronal values are 3000 s for radiation and 300 s for conduction (cf. Kuin & Martens, 1982). Again, these are much shorter than observed loop lifetimes, so that coronal loops appear to be thermally stable. However, models of loops are not so unequivocably stable to thermal perturbations, and have been the subject of much discussion.

The simplest possible treatment reveals the problem. The coronal portion of a hot loop occupies almost all the length of the loop and dominates the global structure. In this part of the loop the conditions vary little, and we may take the plasma to be uniform and of infinite extent, as a first approximation. Since the thermal time scales are much longer than the dynamical scales, we may also assume that disturbances are isobaric. The energy equation then becomes Eqn. (5.56)

$$c_p \rho \frac{\partial T}{\partial t} = \kappa \frac{\partial^2 T}{\partial s^2} - \mathcal{L} + \mathcal{H}. \tag{5.103}$$

By assumption, the initial state has uniform temperature $T = T_0$. We now follow Parker (1953) and perform a local linear stability analysis by allowing a small temperature perturbation $T'(s, t)$ of the form

$$T' \propto \exp{(nt)} \exp{(iks)}, \tag{5.104}$$

where n is the growth rate and k the wavenumber. We assume that the perturbation is small enough for quadratic and higher-order terms in T' to be negligible in the equations of motion. The linearized energy equation is then

$$c_p \rho \frac{\partial T'}{\partial t} = \kappa \frac{\partial^2 T'}{\partial s^2} - \left(\frac{\partial \mathcal{L}}{\partial T}\right)_p T' + \left(\frac{\partial \mathcal{H}}{\partial T}\right)_p T', \tag{5.105}$$

the partial derivatives of $\mathcal{L}$ and $\mathcal{H}$ being calculated with the initial equilibrium values. The *linearized* equation describes only the initial development of a perturbation if it grows in amplitude.

Substituting the form of Eqn. (5.104) shows that

$$n = -\frac{1}{c_p \rho}\left[\kappa k^2 + \left(\frac{\partial \mathcal{L}}{\partial T}\right)_p - \left(\frac{\partial \mathcal{H}}{\partial T}\right)_p\right]. \tag{5.106}$$

From this relation we can see that the perturbation will decay away $(n < 0)$ if

$$\left(\frac{\partial \mathcal{H}}{\partial T}\right)_p < \left(\frac{\partial \mathcal{L}}{\partial T}\right)_p + \kappa k^2. \tag{5.107}$$

This will always be assured for large enough k, i.e. for perturbations that are of small enough wavelength for conduction to be able to erase them effectively, as discussed by Priest (1978). However, a disturbance in which the temperature perturbation is of the same sign throughout, cannot be stabilized by conduction from hotter to cooler regions. This is a characteristic of the *fundamental* mode for which $k \sim 0$; its stability is governed solely by the relative sizes of the temperature dependence of the heating and radiative loss rates. Under canonical coronal conditions, $\mathcal{L} \propto p^2 T^{-\frac{5}{2}}$ so that $(\partial \mathcal{L}/\partial T)_p$ is negative. We would thus conclude that a loop heated at a constant rate, i.e. with $(\partial \mathcal{H}/\partial T)_p = 0$, is *unstable*. The thermal instability arises simply from the fact that the radiative loss function $Q(T)/T^2$ in the high temperature range decreases with temperature. Adding heat produces an increment in temperature which makes the plasma *less* capable of radiating away that heat. This is clearly an unstable situation.

Allowing for plasma flow and compression does not change the conclusion. If we ignore gravity and start from a uniform static state (cf. Oran, Mariska & Boris, 1982), the linearized equations become

$$\frac{\partial \rho'}{\partial t} + \rho_0 \frac{\partial v'}{\partial s} = 0 \qquad (5.108)$$

$$\rho_0 \frac{\partial v'}{\partial t} = -\frac{\partial p'}{\partial s} \qquad (5.109)$$

$$c_p \rho_0 \frac{\partial T'}{\partial t} - \frac{\partial p'}{\partial t} = \kappa \frac{\partial^2 T'}{\partial s^2} + \left(\frac{\partial \mathscr{E}}{\partial T}\right)_p T' + \left(\frac{\partial \mathscr{E}}{\partial p}\right)_T p', \qquad (5.110)$$

where $\mathscr{E} = \mathscr{H} - \mathscr{L}$ is the *net* heating rate. We can again look for solutions of the form (5.104). Substituting for each perturbed variable and eliminating with the help of the perfect gas equation, yields a cubic equation for the growth rate n:

$$\frac{n^3}{\gamma - 1} + \left[\left(\kappa k^2 - \frac{\partial \mathscr{E}}{\partial T}\right) \frac{T_0}{p_0} - \frac{\partial \mathscr{E}}{\partial p}\right] n^2 + c_p T_0 k^2 n + \left(\kappa k^2 - \frac{\partial \mathscr{E}}{\partial T}\right) \frac{T_0}{p_0} k^2 = 0. \quad (5.111)$$

A first estimate of the roots n_0 can be found by neglecting conduction and heating fluctuations. Then we find

$$n_0 = 0 \quad \text{and} \quad n_0^2 = -(\gamma - 1) c_p T_0 k^2 = -\frac{\gamma p_0}{\rho_0} k^2. \qquad (5.112)$$

The first root corresponds to a neutrally stable mode. The perturbed system does not move back to the original state, but finds a new equilibrium in the perturbed state. The other two roots correspond to oscillatory modes having a frequency given by

$$\omega^2 = -n_0^2 = \frac{\gamma p_0}{\rho_0} k^2. \qquad (5.113)$$

These are acoustic waves travelling at the sound speed $v_a = \sqrt{(\gamma p_0/\rho_0)}$, and are responsible for transmitting the pressure perturbations. Waves will be treated in detail in Section 5.6.3.

If we now reintroduce conduction and heating rate fluctuations as *small* terms, we can calculate the corrections to the growth rates by writing $n = n_0 + n_1$, where n_1 is small. The growth rate of the first mode then becomes

$$n = n_0 + n_1 = -\frac{1}{c_p \rho_0} \left(\kappa k^2 - \frac{\partial \mathscr{E}}{\partial T}\right). \qquad (5.114)$$

This is the thermal mode discussed previously. It is known as a *condensation mode*. The growth rates of the other modes, the *acoustic* modes, are

$$n = n_1 \pm i\omega = -\frac{1}{2c_p \rho_0}\left[(\gamma-1)\left(\kappa k^2 - \frac{\partial \mathcal{E}}{\partial T}\right) - \frac{\gamma p_0}{T_0}\frac{\partial \mathcal{E}}{\partial p}\right] \pm i\sqrt{\left(\frac{\gamma p_0}{\rho_0}\right)}k. \quad (5.115)$$

The real part of this expression gives rise to an exponential decay (when negative) or growth (when positive) of the amplitude of the wave. Conduction damps the acoustic wave, as it does the condensation mode, whilst positive dependences of the net heating rate on temperature and pressure cause the wave to grow. In the latter case, we obtain an *overstable* acoustic wave. It should be noted that an unstable condensation mode is not necessarily accompanied by overstable (i.e. unstable) acoustic modes if $\partial \mathcal{E}/\partial p < 0$. However, the criterion for instability of the condensation mode is *unaffected* by the behaviour of the other modes. This analysis still suggests that the corona is unstable to thermal perturbations.

This conclusion is nevertheless incorrect. It ignores the fact that energy deposition occurs mainly in the coronal portion of the loop and energy loss in the transition region, where the physical conditions are quite different. A local analysis assuming uniform coronal conditions is quite inappropriate. In the following section we shall look at global stability, taking account of the loop as a whole.

5.5.2 Global thermal stability

The stability of a complete coronal loop can be analysed by subjecting the equilibrium model to a one-dimensional perturbation in which the temporal and spatial dependences are separated, e.g.

$$T' = T_1(s)\exp(nt), \quad v' = v_1(s)\exp(nt), \text{ etc.,} \quad (5.116)$$

but the assumption of uniformity along the loop is dropped. The linearized perturbation equations for a loop initially in static equilibrium are then

$$n\rho_1 + \frac{d}{ds}(\rho_0 v_1) = 0, \quad (5.117)$$

$$n\rho_0 v_1 = -\frac{dp_1}{ds} - g\rho_1, \quad (5.118)$$

$$nc_V \rho_0 T_1 - (\gamma-1)c_V \rho_0 \frac{dv_1}{ds} =$$

$$\frac{d}{ds}\left[\kappa_0 \frac{dT_1}{ds} + \left(\frac{d\kappa}{dT}\right)_0 \frac{dT_0}{ds}T_1\right] + \left(\frac{\partial \mathcal{E}}{\partial T}\right)_p T_1 + \left(\frac{\partial \mathcal{E}}{\partial p}\right)_T p_1. \quad (5.119)$$

These equations assume that the loop has uniform cross section. A varying cross section will affect the divergence terms, such as the conductive heat flux. Dowdy, Moore & Wu (1985) have investigated the inhibition of conduction at the constriction experienced at the boundary between the transition region and the chromosphere, but the effects have yet to be incorporated into dynamical models of loops.

The linearized equations form a set of ordinary differential equations for the spatial parts of the perturbations. With suitable boundary conditions, to be specified below, the growth rate n is an eigenvalue of the system. The eigenvalues may be real, as for the condensation mode discussed in the previous section, or complex, as for the acoustic modes. (Purely imaginary eigenvalues, corresponding to undamped waves, are found when the system is adiabatic, but is not the case here.) The eigensolutions corresponding to each eigenvalue are the *normal modes* of the system.

Stability requires that *all* perturbations of the equilibrium system die away with time. This in turn means that all the eigenvalues must be either real and negative or complex with negative real parts. If any eigenvalue is zero, that mode consists of displacements that yield further equilibrium states and so is neutrally stable. If any eigenvalue (or any real part of a complex eigenvalue) is positive, that mode will lead the system further and further away from its equilibrium state, and the system will be unstable.

In the following we shall concentrate on the condensation mode. This is almost static because the modulus of n is small. The other modes have a large imaginary part to n and are basically oscillatory. These are discussed under the heading of wave motion in Section 5.6.3.

However, before we can analyse the equations, the specification of the model must be completed. The radiative loss function $\mathscr{L}$ is given by Eqn. (5.39), but the heating function $\mathscr{H}$ is, as yet, unknown. Little attention has been given to self-consistent models in which a heating *mechanism* is proposed and the appropriate form of $\mathscr{H}$ is evaluated (an exception being the work of Zweibel (1980) discussed in Section 5.7.2). As Hollweg (1981b) emphasizes, some heating functions are difficult to prescribe. In the case of wave heating, for example, the energy dissipation at any particular point depends on the whole wave propagation path and not just on local conditions. In the following it must be remembered that only simple heating functions, parametrized in terms of the local values of p and T, are considered and the results may be of limited value.

The final requirement for a global stability analysis is a set of boundary conditions. These require careful consideration, because the results of the analysis of coronal loop models are critically dependent on the choice of lower boundary condition.

The first global analysis, due to Antiochus (1979), was based on the standard static equilibrium model described in Section 5.4.2. Antiochus adopted a rigid lower boundary ($v_1 = 0$) at the base of the transition region, where the conductive heat flux was taken always to vanish, i.e. $dT_0/ds = 0$ and $dT_1/ds = 0$. Like all subsequent authors, Antiochus assumed dynamical stability and imposed the condition of constant pressure at all times. This procedure removes the pressure force from the momentum equation and suppresses the short-period acoustic waves. Their effect is not neglected, though; it is implicit in the assumption of constant p_1. Antiochus then made the further assumption that $p_1 = 0$. This had the advantage of reducing the system of equations to one of Sturm–Liouville type, so that the stability can be ascertained by looking at the fundamental spatial modes alone. On the other hand, this choice has the unwelcome property of decoupling the coronal loop both thermally and dynamically from the chromosphere below.

In general, a symmetric equilibrium loop possesses two sets of modes of disturbance. Symmetric modes have mirror symmetry about the apex. Antisymmetric modes have perturbations which change sign at the apex, i.e. the perturbations must vanish there. In this case, the constancy of p_1 requires the pressure perturbation to vanish everywhere, as assumed by Antiochus. This is not true of the symmetric modes. Moreover, the rigid boundary condition is defective in regard to antisymmetric modes because it excludes the fundamental (least stable) mode, which has an inflow at one end and an outflow at the other. This was not critical in the analysis of Antiochus because he found that all his models, using a variety of parametrized loss and heating functions (Eqns. (5.71) and (5.70)), were in any case unstable. The growth time was found to be that of the radiative loss time. Hood & Priest (1980) performed a similar analysis, but including the fundamental antisymmetric mode, and confirmed the results of Antiochus.

The lower boundary conditions are clearly responsible for this instability. By placing the boundary at the immediate base of the transition region, as in the equilibrium models, there is not sufficient cool material in the system to radiate away the increased heat flux produced when the coronal temperature rises. Instability results from a breakdown in the radiation budget. If the lower boundary is placed *within* the transition region and heat is allowed to be conducted away through the base, the loops may be stabilized (Hood & Priest, 1979a; Chiuderi *et al.*, 1981). However, this gambit does not resolve the problem of how the heat flux is ultimately dissipated.

In fact, the difficulty was substantially resolved by Habbal & Rosner

(1979), who set the lower boundary at a fixed temperature (rather than at a fixed point of space) and chose p_1 to be a nonzero constant. These authors calculated the pressure fluctuation by (arbitrarily) forcing the perturbed loop to obey the scaling law (Eqn. (5.74)). Although *ad hoc*, this approach had the essential feature of allowing the loop and the underlying chromosphere to interact, with material being transferred into a loop. In other words, the transition region at the base of the loop was allowed to migrate down into the chromosphere. The chromosphere provided the extra material that is required to radiate away the heat input when the coronal temperature rises. Not surprisingly, the model of Habbal & Rosner indicated that loops are thermally stable.

A full analysis of this model was given by McClymont & Craig (1985a, b). A fixed lower boundary was employed in this analysis, but was placed deep enough in the chromosphere to allow for heat to be conducted out of the base of the transition region, whilst remaining within the system. With this placement, it was found to be immaterial whether the temperature fluctuation or the flux fluctuation was assumed to vanish at the boundary. By the same token, a rigid lower boundary condition, $v_1 = 0$, could also be employed.

For antisymmetric modes, the remaining boundary condition was $p_1 = 0$. For symmetric modes, the constant value of p_1 was determined by matching the pressure at the base of the loop to that of the chromosphere, the point of match being fixed by the requirement that the mass of the loop *plus* chromosphere remain constant. The relative change in pressure in the loop is then related to the *change in length* of the loop δL by

$$\frac{p_1}{p_0} \simeq \frac{\delta L}{\varLambda}, \tag{5.120}$$

where $\varLambda$ is the scale height in the chromosphere. Craig & McClymont (1987) stress the essential role played by gravity in producing a pressure stratification *in the chromosphere*. If gravity is neglected here, i.e. we set $\varLambda \rightarrow \infty$, the pressure change resulting from material transfer into the loop is suppressed. On the other hand, gravitational effects may be safely neglected in the coronal portion of the loop.

Of course, moving the boundary into the chromosphere presents a new problem, that of calculating the radiative losses in a medium which is not optically thin. Chromospheric losses have been studied in detail by McClymont & Canfield (1983). They concluded that their results were mimicked very well by simply omitting the optically thick Lα radiation from the optically thin loss function $Q(T)$. The loss function has a roughly triangular shape, increasing with temperature up to 10^5 K and then decreasing with temperature (cf. Fig. 5.5).

The results may be summarized as follows. When gravity is neglected in the loop – allowing the assumption of constant pressure – the fundamental antisymmetric mode is neutrally stable whenever the heating function depends only on the local temperature and pressure. Translating the plasma around the loop does not change it in any way. If, however, the heating function depends on position in space, the loop is not invariant to translations. The fundamental antisymmetric mode is then unstable if the heating function decreases with height, and is stable if it increases with height.

The stability of the symmetric modes is governed mostly by the pressure dependence of $\mathcal{H}$, when the heating function does not decrease with height. The authors conclude from a variety of models that loops are stable if $\mathcal{H} \propto p^\delta$ where $\delta < 1\frac{1}{2}$–2. Little constraint is placed on the temperature dependence of $\mathcal{H}$. In particular, the loop with uniform heating per unit volume and the loop with uniform heating per unit mass are both stable. Antiochus, Shoub, An & Emslie (1985) contest this conclusion, but Craig & McClymont (1987) claim that their results are invalidated by the neglect of gravity in the chromosphere (see above).

Gravity in the coronal portion of the loop is a stabilizing factor, as argued by Wragg & Priest (1982a), but McClymont & Craig (1985b) found that its inclusion had little effect on their conclusions.

These authors explain their results by noting that if heat is supplied mainly to the coronal portion of the loop, any increase will raise the temperatures there and the reduced radiative losses will be compensated by an increased conductive heat flux down through the transition region and *into the chromosphere*. The chromosphere rises in temperature but here the temperature increase results in greater radiative losses, which eventually balance the excess heat input. The temperature of some material increases beyond 10^5 K, at which point it ceases to be able to radiate effectively and the temperature climbs steeply to coronal values. The transition zone thus migrates downwards and the loop lengthens. If, on the other hand, the lower parts of the loop are subject to the greatest increase of heat input the temperature will rise to a maximum there and the conductive coupling between the corona and the chromosphere will be cut. This destabilizes the loop.

This interpretation of the global stability analyses is strongly supported by the numerical simulations of loop dynamics, which are described in the next section.

5.5.3 Loop dynamics

The assessment of stability may be approached not only by global analysis but also by numerical simulation. Here the aim is to solve the full set of (non-linear) equations numerically. An equilibrium loop is first constructed and is then disturbed by a transient perturbation to the heating function. The subsequent development of the system is followed by stepping forward in time. This procedure allows the response of the system to large-amplitude (non-linear) perturbations to be examined, and thus complements the analytical treatment of small-amplitude (linear) perturbations described in the previous section. The stability criteria are, of course, the same. If the initial loop represents a stable state, the system will return to that state, otherwise it will evolve to a new static or dynamic state. For example, Oran *et al.* (1982) follow the evolution of a linearly unstable constant-pressure coronal loop into the non-linear regime by numerical solution of the full equations. Cool condensations formed and migrated along the loop until a stable configuration was achieved with a cool 'chromosphere' and 'transition region' at each end. However, most authors are concerned with the dynamics of loops which are stable to small amplitude perturbations, in accord with the observational evidence.

Although simple in principle to apply to any given loop model, the approach faces major difficulties, associated again with the basal regions. The very fine spatial grid required to resolve the rapidly varying structure of the transition regions demands an unacceptably small time step – i.e. prohibitive computing times – unless a systematic implicit numerical scheme is developed. Inadequate resolution leads to decoupling of the coronal regions from the chromosphere (Craig & McClymont, 1981), and the artificial truncation of the model above the steep transition region forces the use of very *ad hoc* boundary conditions (Krall & Antiochus, 1980).

The first satisfactory simulation was presented by Craig, Robb & Rollo (1982). The model has the same features as that subjected to global analysis by McClymont & Craig (1985a). A rigid lower boundary ($v = 0$) was placed in the chromosphere at $T = 30\,000$ K. Again, it was found to make little difference whether the temperature was held constant or whether the temperature gradient was made to vanish at this point. The authors investigated a model of a symmetric loop in which the heating function per unit mass was constant. Similar results were obtained by Peres, Rosner, Serio & Vaiana (1982) for a simulation in which a uniform heating function per unit volume was adopted. According to the linear analysis of McClymont & Craig (1985b), both models should be stable. Pakkert,

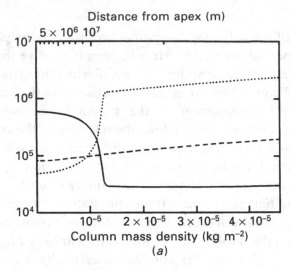

(a)

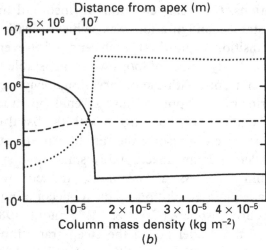

(b)

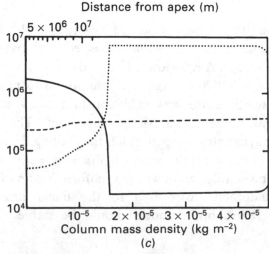

(c)

Fig. 5.9 (a)–(c). For legend see facing page.

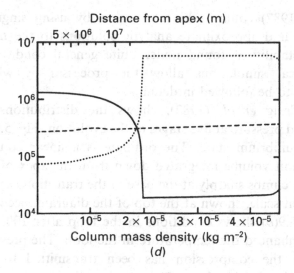

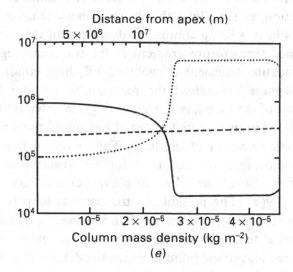

Fig. 5.9. The dynamical evolution of a model hot coronal loop subjected to a transient heating increment. In each panel are shown the temperature (in units of K) as a solid line, pressure (in units of 10^{-7} Pa) as a dashed line and density (in units of 10^{10} m^{-3}) as a dotted line. All are plotted as functions of the column mass density measured down from the apex (lower scale) and distance from the apex (top scale). (*a*) The initial static structure. (*b*) After 100 s, whilst the transient is still heating the coronal portion of the loop. (*c*) After 150 s, when the transient has just ceased. (*d*) After 250 s. (*e*) After 1000 s. (After Craig *et al.*, 1982.)

Martens & Verhulst (1987) confirm this conclusion by using singular perturbation theory to find approximate analytic solutions to the *time-dependent* equations. Stability is found under quite general conditions. However, the numerical simulations allow the processes by which equilibrium is regained to be followed in detail.

Figure 5.9, from Craig *et al.* (1982), shows the distributions of temperature, density and pressure at five stages of the evolution. Fig. 5.9(*a*) presents the initial, equilibrium state. The ordinate is a linear column density, the mass per unit volume integrated down from the apex of the loop. Where the density climbs sharply at the base of the transition region, the corresponding height scale, shown at the top of the diagram, becomes vastly expanded. Fig. 5.9(*b*) shows conditions in the loop after 100 s. A heating transient has enhanced the temperature in the loop. The pressure has followed suit and the compression has been transmitted to the chromosphere, raising the density there. The pressure rise enhances the chromospheric loss function, which allows the chromospheric temperature to *fall* somewhat. Excess heat is being conducted down out of the corona by means of the steepened temperature gradient in the transition region. After 150 s, when the heating transient is switched off, heat conducted down into the chromosphere has raised the material to coronal temperatures, so that the base of the loop has started to migrate down into the chromosphere, Fig. 5.9(*c*). The evaporated material has raised the density and pressure in the coronal portion of the loop. The increased pressure transmitted to the chromosphere has led to a further lowering of the temperature there. After 250 s, the chromospheric evaporation has proceeded further, Fig. 5.9(*d*). The pressure in the coronal loop is now sufficient for the radiative loss function to more than balance the quiescent heat input and the coronal temperature starts to sink. The mass of the chromosphere has been reduced to the point at which the density is starting to drop and the temperature has to rise again to maintain pressure balance. After 1000 s, evaporation ceases, Fig. 5.9(*e*). The migration of the base of the loop into the chromosphere is at its maximum at this time. The loop continues to cool and the chromosphere to heat up. The conductive heat flux from the corona drops and the gradient in the transition region flattens out accordingly. The material cooling at the base of the loop condenses back into the chromosphere. As the loop drains, the base of the loop migrates back upwards until the original state is regained. This relaxation phase is very slow, lasting of the order of 10^4 s.

These simulations display precisely the thermostat mechanism proposed by Pye *et al.* (1978) to account for the evolution of loops in accordance with the scaling laws.

There remains, however, the question as to how far this model reproduces processes operating in the Sun. A comparison of the model temperature structure in Fig. 5.9 with that of a semi-empirical model (Fig. 5.6) reveals that the former fails to show the high chromospheric temperature plateau. McClymont & Canfield (1983) analysed a theoretical loop model which had been fitted smoothly on to the model *F* shown in Fig. 5.6 by trial-and-error adjustment of the height dependence of the heating function. The empirical model then demanded a heating function which had a peak in the transition region and the increase in heating with height was sufficient to make the system *unstable* (cf. Section 5.5.2).

An, Canfield, Fisher & McClymont (1983) followed the development of this instability in a numerical simulation, whilst maintaining the ambient (destabilizing) heating rate. The instability predicted by the global linearized analysis led to a rapid growth of the perturbation in the transition region, but this soon saturated and the loop subsequently evolved by chromospheric evaporation and condensation on a time scale typical of coronal conduction. The speeds of the associated mass flows peaked at values of some 3 km s^{-1} in the upper transition region, and decreased rapidly in lower temperature material. The simulation showed no indication that the loop was relaxing to a new stable equilibrium, suggesting that a *dynamic* state would be maintained. Whilst this is more in accord with observations of the state of the transition region material, the model showed significant discrepancies in time scales and the authors suspected that they had not captured the essential physics.

The same criticism has been levelled at the simplified treatment of the chromosphere–corona coupling in the model of Kuin & Martens (1982). This model was unstable and in the non-linear regime showed a limit-cycle behaviour in which fast heating, slow evaporation, catastrophic cooling and quick draining occurred in succession. Craig & Schulkes (1985) later analysed the coupling more rigorously. They confirmed that limit-cycle behaviour is possible, but only below a critical value of the coupling constant. For larger values, stability is recovered. Since stability according to the linear analysis requires the coupling constant to greatly exceed the critical value, Craig & Schulkes conclude that coronal loops cannot go through limit cycles in response to large-amplitude disturbances and, at the same time, be stable to small-amplitude disturbances. The previous analyses then suggest that coronal loops are both linearly and non-linearly stable.

Recent work on loop dynamics has been concerned with the transient flows that develop in response to fluctuations in the coronal heating rate. As discussed in Section 5.4.3, the motivation is provided by observations of large downward velocities in transition region plasma. Hot loop models are not very promising. Very large changes in $\mathcal{H}$ are required to generate significant flow speeds. Mariska (1987) simulated the effect of reducing $\mathcal{H}$ to 10% and 1% of its original value over time scales of 100–2000 s. The velocity evolution was found to be similar in all cases. A condensation flow develops which takes over from conduction as the primary energy transport mechanism, but the flow speeds are still generally low. With a 90% reduction in heating, a maximum speed of 8 km s^{-1} can be generated in the transition region, and this figure is increased to only 15 km s^{-1} when the heating is reduced by 99%. In either case, the amount of high-speed material emitting at 10^5 K is too small to account for the observations.

Mariska also considered increases in the heating function. Adding a large excess heat input to rarified material produces a flare-like response (cf. Section 5.8). The temperature rises sharply where the heat is deposited, increasing the temperature gradient at the edge of the input volume. Heat is conducted out of the volume and rapidly raises the temperature of the surrounding material. The conduction front thereby migrates down the loop until it hits the chromosphere. There, chromospheric material is instantly evaporated and expands up into the loop at speeds reaching 100 km s^{-1}. This material meets material rising up the other leg at the apex and weak shocks develop that pass on down both legs. The passage of the shock decelerates the still rising material to about 20 km s^{-1}. However, the speeds in the transition region material remain low, less than 6 km s^{-1}, all the time. Mariska therefore concluded that 'a loop which is cycling between cooling and heating has difficulty in reproducing the mass motions observed in the quiet solar transition region'.

As suggested earlier in Section 5.4, this difficulty may be resolved in cool loops. There we saw that both hot conduction-dominated and cool radiation-dominated solutions are possible for short, low-lying loops. Klimchuk, Antiochus & Mariska (1987) studied their stability by solving numerically the non-linear equations when the equilibrium is perturbed. They included gravity but represented the chromosphere only crudely by means of a slab at 2×10^4 K. They found that loops with apex heights greater than 5×10^3 km only exist in the hot state, and that these are stable. Loops with apex heights less than 10^3 km are unstable in the hot state, and collapse into cool loops with essentially chromospheric temperatures, i.e. much less than 10^5 K. The cool states are stable. However, both cool and hot loops are stable if the apex heights are in the range 10^3–5×10^3 km. It is not possible to produce a 'hot' cool loop with a temperature around 10^5 K as a result of a thermal instability in a hot loop. The mechanisms that produce the loops seen in C III and C IV, for example, are not clear. However, they are seen to exist (cf. Fig. 2.13), and Antiochus & Noci (1986) suggest that they make an essential contribution to the cool emission of active regions.

In a later paper, Klimchuk & Mariska (1988) investigated whether time-dependent, transient flows in cool loops could explain the high velocities seen in cool emission lines (cf. McClymont & Craig, 1987a and Section 5.4.3). They simulated a *very short* cool loop of half-length 2400 km and maximum temperature 7.4×10^4 K. The short length is of course necessary to achieve stability and a realistic basal pressure of 0.02 Pa. They also assumed a uniform initial heating rate. This loop was then subjected to an excess heating rate in one leg. A flow was established from the hot side to the cool side within 200 s, and became steady after about 1000 s. The largest flow speeds occurred about midway along the loop and depended little on the location and extent of the excess heat input. Flow speeds were typically 5 km s^{-1}, but could be increased to 20 km s^{-1} with very large heating differentials. However, the speeds of the upflows and downflows were very similar and no clear excess emission arose from the condensing leg *throughout the evolution*. This model produced *no large transient velocities*; the speeds simply built up monotonically to their steady values.

When, on the other hand, the heat input was dropped symmetrically along the loop, the loop again adjusted on the time scale of radiative cooling, about 200 s. The material in the loop initially flowed down as hydrostatic support was lost. But the drop in coronal pressure was communicated to the chromospheric boundary, allowing chromospheric material to expand up into the loop. The upflow then dominated, with the highest speeds of about 4 km s^{-1} occurring near the chromospheric boundary (at $T \sim 2 \times 10^4$ K). A new equilibrium was achieved after some 2000 s. Faster flows could be produced by making the loop taller. This was achieved by adopting a background heating function that decreased with height. Draining flows could then accelerate up to 17 km s^{-1}, the highest speeds occurring at temperatures greater than the chromospheric value. However, they were transient and the material had already cooled beyond the 10^5 K required by observation.

Clearly, the physical processes in cool coronal loops have not yet been fully established. Such loops can certainly provide quasi-steady downflows in substantial amounts of material with temperatures below 10^5 K, which hot loops cannot. However, the prevalence of the observed downflows would suggest that cool loops are essentially dynamical systems, rather than static systems. Moreover, the magnitude of the observed flows is uncomfortably close to the upper limit found in the models. It would be premature to conclude that cool loop models now allow a satisfactory picture of the solar transition region to be drawn.

As regards hot loops, firmer conclusions are possible. One-dimensional loop models appear to be stable not only to infinitesimal perturbations but also to very large finite-amplitude disturbances. However, a one-dimensional analysis allows only necessary conditions for stability to be established. Sufficient conditions can be found only by examining *all possible* perturbations, including those that produce transverse disturbances of the loop. For this purpose we require a full multi-dimensional model of the loop structure and this introduces the magnetic field into the dynamical system. We look at magnetohydrodynamic models of coronal loops in the following section.

5.6   MHD stability and dynamics
5.6.1   *Introduction*

In the previous sections we have reviewed the extensive literature concerning one-dimensional models of coronal loops. In these models the thermodynamic structure and plasma motion in the direction of the field lines may be analysed without further regard to the magnetic field. However, loops have transverse structure as well as longitudinal. The properties of loops vary over their cross-section and vary one from another. The discussion of these features requires consideration of the structure of the magnetic field and of its effect on the plasma confined within it. The paucity of our empirical knowledge of the magnetic structure and an almost total lack of self-consistent theoretical models force us to proceed on the basis of *ad hoc* models chosen mainly for mathematical convenience. This approach has the merit, though, of clearly exposing the physical mechanisms which must be considered when seeking to understand the more complicated situation met with in solar active regions.

The basis for nearly all these models is provided by idealization of a long, thin loop as a *linear* structure with circular symmetry about its axis. Some justification is provided by the work of Xue & Chen (1983). These authors took into account longitudinal curvature in a simple equilibrium toroidal magnetic structure embedded in a field-free plasma, but found that it had little effect in *thin* loops. When the curvature is neglected, the models are better described as tubes rather than loops. The tube is taken to have a finite length $2L$ and is generally assumed to be uniform along that length. When the surroundings are taken into consideration, they are supposed (less legitimately) to possess the same symmetry and uniformity. If we adopt a cylindrical coordinate scheme with the z-axis along the loop axis, r as the radial coordinate and θ the azimuth, the magnetic field may then be written as

$$\mathbf{B} = (0, B_\theta(r), B_z(r)). \tag{5.121}$$

We have already seen that the terms due to the gas pressure and gravity in the momentum equation are very small in and around active region loops. The natural first approximation is therefore to assume that the magnetic field is force-free, i.e. one in which $\mathbf{j} \times \mathbf{B}$ vanishes. This leads to a unique solution for $\mathbf{B}$ in two cases.

The first is the potential field case for which

$$\mathbf{j} = \frac{1}{\mu_0} \nabla \times \mathbf{B} = 0.$$

Then we have

$$B_z = constant, \quad B_\theta = \frac{constant}{r}. \tag{5.122}$$

If we require the potential field to apply within the tube, we must take the second constant to be zero in order to avoid a singularity on the axis. An internal potential field is therefore purely longitudinal.

The second case is that of the linear force-free field (cf. Section 5.2.2) for which

$$\nabla \times \mathbf{B} = \alpha \mathbf{B},$$

where α is constant. This yields the Lundquist solution

$$B_\theta = B_0 J_1(\alpha r), \quad B_z = B_0 J_0(\alpha r), \tag{5.123}$$

where J_0 and J_1 are the Bessel functions of zero and first order respectively. It describes tubes with an azimuthal component of field, i.e. twisted tubes. The total angle through which a field line winds from one end of the tube to the other is known as the *twist*. It is given by the expression

$$\Phi = \frac{2LB_\theta(r)}{rB_z(r)}, \tag{5.124}$$

and varies in general with the radial position r.

If we admit non-linear force-free models, we require only that $\mathbf{j} \times \mathbf{B} = 0$. In our radially symmetric approximation this becomes

$$\frac{d}{dr}\left(\frac{B_\theta^2 + B_z^2}{\mu_0}\right) + \frac{B_\theta^2}{\mu_0 r} = 0. \tag{5.125}$$

Several simple *ad hoc* analytic solutions are known which satisfy this condition, including the case in which Φ is independent of r. In this uniform twist model we have

$$B_\theta = \frac{B_0(r/b)}{1+r^2/b^2}, \quad B_z = \frac{B_0}{1+r^2/b^2}, \tag{5.126}$$

where the parameter b is a measure of the transverse scale (or 'radius') of the loop and the twist is the constant $\Phi = 2L/b$. Another simple analytic solution is obtained when there is a constant axial current. Then

$$B_\theta = \frac{B_0 r}{b}, \quad B_z = B_0\left(1-\frac{2r^2}{b^2}\right)^{\frac{1}{2}}. \tag{5.127}$$

Zweibel & Boozer (1985) devised an elegant method for generating more general force-free field configurations in a cylinder as the ends are twisted or compressed, whilst Craig & Sneyd (1986) have shown how even more general loop configurations may be constructed numerically.

If we include the gas pressure as the next approximation, the *static* force balance equation becomes

$$\frac{dp}{dr}+\frac{d}{dr}\left(\frac{B_\theta^2+B_z^2}{\mu_0}\right)+\frac{B_\theta^2}{\mu_0 r}=0. \tag{5.128}$$

This has an extra degree of freedom in the unprescribed $p(r)$ and there are correspondingly more *ad hoc* solutions (Chiuderi, Giachetti & van Hoven, 1977; Hood & Priest (1979a, b); Chiuderi & Einaudi, 1981; Krishan, 1985).

The rationale for employing models of twisted tubes has been provided by Parker (1977) in a series of papers devoted to twisted magnetic field structures embedded in a medium maintaining a constant external pressure. He showed that if the tube radius is expanded with the twist maintained, the ratio of the azimuthal field B_θ to the axial field B_z *increases*. This effect is apparent when a tube expands along its length due to a decrease in the confining pressure along the tube. The relative strength of the azimuthal field is greater in the expanded portion. In the case of coronal magnetic fields, we must expect twists to be delivered to the photospheric footpoints of the coronal magnetic loops wherever there is locally a vertical component of vorticity. Since the viscosity of the plasma is so low, vertical vorticity will be present throughout regions where convection generates horizontal vorticity (cf. Bray *et al.*, 1984). The magnitude of the azimuthal field is probably small deep in the atmosphere, but Parker's results suggest that the expansion of the flux tube through the photosphere and chromosphere, where the confining gas pressure drops (cf. Section 6.2.3), will lead to the azimuthal field in the corona becoming much more significant. Most authors therefore try to account for it in multi-dimensional models of coronal loops.

Several authors also try to include the thermodynamic state of the plasma in the model by adopting various forms for $B_\theta(r)$, $B_z(r)$ and $p(r)$ consistent with Eqn. (5.128), and then using the one-dimensional model of Section 5.3 to derive the temperature structure along each constant-pressure field line. With suitable choices, any thermodynamic structure is possible. Hood & Priest (1979a) investigated fields with constant twist and with uniform axial strength and a pressure field that was a maximum on the axis. These authors then adopted a uniform heating function and generated a grid of models. However, if H, p and L are prescribed, the scaling law will not in general be satisfied and the loop cannot be thermally isolated. Einaudi, Torricelli-Ciamponi & Chiuderi (1984), on the other hand, assumed that the field lines were thermally isolated and that the pressure was uniform across the tube. The scaling laws then require the outer field lines, which twist and follow a longer path from footpoint to apex than the central field lines, to have correspondingly higher apex temperatures. A loop with a cool core would then result. However, it is clear from Eqn. (5.69) that a very large degree of twist would be necessary in order to produce the lengthening required to cause a substantial variation in temperature across the tube. Moreover, a variation in basal pressure across the tube, considered earlier by Priest (1978), would be equally effective.

Krishan (1983) adopted a different approach. In a *dynamic* system, the ideal MHD equations applied to an incompressible (uniform density) medium require the total energy, magnetic plus kinetic, to be conserved together with the magnetic helicity (Section 5.2.2) and a quantity known as the cross helicity

$$\int_V \mathbf{v} \cdot \mathbf{B} \, dV. \tag{5.129}$$

The state which minimizes the kinetic energy of the system, subject to these constraints, is a linear force-free field. The solutions for **B** also prescribe the flow field **v**. In a subsequent paper, Krishan (1985) modelled a tube by an arbitrary superposition of two force-free states, thereby producing a total field which was not force-free. The MHD equations were then used to find the thermodynamic structure consistent with the assumed magnetic and velocity fields. Much more work is required, however, to justify the assumption that the solar corona can achieve this statistical equilibrium state (cf. the discussion of the plasma-free case in Section 5.2.2).

Observations place little constraint on the possible static equilibrium structures. However, we have seen earlier, in Section 5.5.2, that the seeming stability of hot coronal loops provides the only real discriminant between one-dimensional models. The same is true of multi-dimensional models. We therefore devote the next section to a detailed discussion of stability problems, and we then look at the dynamical properties of stable loops in Section 5.6.3.

5.6.2 MHD stability

In Section 5.5 we discussed the stability of one-dimensional models of coronal loops. Because these models are governed by the hydrodynamic equations, the time scales of instabilities are those of hydrodynamic processes, as determined by the propagation speed of sound waves and the rate of energy loss by radiation. When the transverse structure of the loop is taken into account, the magnetic field comes into play and a whole new array of processes on both the macroscopic (MHD) and microscopic (kinetic) levels become possible. These can drive an enormous variety of instabilities. We cannot aim to give an exhaustive catalogue here, so we shall base our selection on the review provided by Priest (1978).

Priest assessed the importance of several classes of MHD instability in the context of solar coronal loops. The geometry of the disturbances is illustrated in Fig. 5.10. The first (Fig. 5.10(a)) is the kink instability, in which a sinuous displacement grows along the tube. It arises because the azimuthal field is compressed on the inside of the buckle and expanded on the outside. The pressure gradient due to the *azimuthal* field is thus directed outwards and drives the disturbance further.

The second is the interchange instability, in which ripples develop either along the tube or around it. The former (Fig. 5.10(b)) may be caused by a sausage instability in the azimuthal field or by the hydromagnetic equivalents of the Rayleigh–Taylor instability – arising from an inverted density gradient – or the Kelvin–Helmholtz instability – arising from a strong velocity shear. According to Priest's order of magnitude calculations, these last two processes are unlikely to be important, but this conclusion may be invalid if the gradients are confined to small-scale regions (cf. Section 5.9). Ripples around the tube (Fig. 5.10(c)) can develop when plasma is confined by a magnetic field, i.e. when the internal tube gas pressure exceeds the external.

The kink instability is a large-scale disturbance which completely disrupts the loop and develops on the time scale for the propagation of magnetic disturbances. This is the Alfvén time scale and is much shorter

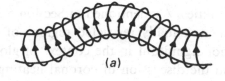

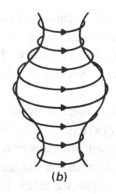

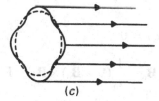

Fig. 5.10. Schematic diagrams of MHD instabilities. (*a*) The kink mode ($m = 1$, $k > 0$). (*b*) The sausage (interchange) mode ($m = 0$, $k > 0$). (*c*) The ripple (interchange) mode ($m > 1$, $k = 0$).

than the hydrodynamic time scales (cf. Table 5.1). For this reason, the kink instability has excited interest in the context of solar flares. By the same token, effects arising from the finite resistivity of the plasma and heat transfer are negligible, so that the ideal MHD equations with an adiabatic energy equation may be used in its study. Interchange instabilities tend to have much smaller scales. The ripples in the field at the edge of a loop carry material from inside the loop to outside and *vice versa*. Hence their global effect is to produce a cross-field transport of plasma, breaking down the isolation of the loop material. Heat conduction can be important on these scales. Its effect is to smooth out the temperature fluctuations. The limiting case of isothermal disturbances is simply modelled by the ideal MHD equations by setting the ratio of specific heats γ equal to unity. The time scales of these instabilities may be long enough to be influenced by resistive effects and other energy loss mechanisms such as radiation. On still smaller scales, the fluid description breaks down, and the kinetic equations are required to analyse the kinetic instabilities. As mentioned in Section 5.3.1, these might well be very important in spatially restricted regions of the coronal volume but probably are not significant in the context of global stability. We shall return to them in the discussion of coronal heating in Section 5.7.

In the following, we shall look in detail at the stability of coronal loops only in the MHD limit. We are, of course, interested in the stability of the loop as a feature embedded in some background structure. The stability analysis must therefore be global in scope, as in Section 5.5.2. It proceeds along the same lines.

The initial magnetostatic equilibrium is subjected to a small time-dependent displacement of the form $\delta \mathbf{x} = \xi(\mathbf{x}) \exp nt$. The equations of motion are the continuity equation, the momentum equation, the energy equation and the equation describing the changes in the magnetic field, the induction equation. In the ideal MHD limit, these are Eqns. (5.47), (5.48), (5.49) and (5.50). As before, we neglect the effects of gravity. The linearized versions of the equations are then

$$\rho_1 + \nabla \cdot (\rho_0 \xi) = 0 \tag{5.130}$$

$$\rho_0 n^2 \xi = -\nabla p_1 + \frac{1}{\mu_0}(\nabla \times \mathbf{B}_1) \times \mathbf{B}_0 + \frac{1}{\mu_0}(\nabla \times \mathbf{B}_0) \times \mathbf{B}_1 = \mathbf{F} \tag{5.131}$$

$$p_1 + \xi \cdot \nabla p_0 = \frac{\gamma p_0}{\rho_0}(\rho_1 + \xi \cdot \nabla \rho_0) \tag{5.132}$$

$$\mathbf{B}_1 = \nabla \times (\xi \times \mathbf{B}_0). \tag{5.133}$$

These can now be combined into a single equation governing the evolution of the displacement,

$$\rho_0 n^2 \xi = \nabla(\xi \cdot \nabla p_0 + \gamma p_0 \nabla \cdot \xi) + \frac{1}{\mu_0}\{\nabla \times [\nabla \times (\xi \times \mathbf{B}_0)]\} \times \mathbf{B}_0$$

$$+ \frac{1}{\mu_0}(\nabla \times \mathbf{B}_0) \times [\nabla \times (\xi \times \mathbf{B}_0)]. \tag{5.134}$$

Berstein, Frieman, Kruskal & Karlsrud (1958) have shown that this equation, together with appropriate boundary conditions on ξ, has only real eigenvalues n^2. If $n^2 > 0$, the perturbations $\delta \mathbf{x}$ have one mode growing in time and the other decaying in time. The system is then unstable. If $n^2 < 0$, the disturbance is an oscillation whose amplitude does not change with time. Stability is assured only if there are no eigenvalues for which $n^2 > 0$.

In order to proceed further, the form of the displacement ξ must be specified more exactly and this requires knowledge of the boundary conditions to be imposed at the outer surface of the loop and at the ends of the loop. The first is to a large extent arbitrary because there is little empirical evidence to suggest what constitutes the edge of a flux tube. Nearly all authors assume that there is a discontinuity in the current distribution, usually with the current system being confined to the tube. Any jump in the magnetic field at the edge can be accommodated with a suitable current sheet enveloping the tube. Some models also include a jump in the thermodynamic quantities (usually the density and pressure) at the boundary. Such discontinuities would be the natural consequence of different basal conditions or different heating rates in contiguous tubes. Neighbouring tubes cannot be properly treated in a cylindrically symmetric model, so most authors choose to embed the tubes in a simple force-free structure such as a linear force-free field or a potential field.

The other important boundary condition is that to be imposed at the ends of the tube at $z = \pm L$. Although the model is that of a straight tube, it is (somewhat inconsistently) assumed that both ends are rooted in the photosphere where the high-β external plasma holds the footpoints rigidly. This condition is known as *line-tying*. The physical processes involved in line-tying are not completely clear. Gibons & Spicer (1981) discuss the microscopic processes inherent in assuming that the field is frozen into a perfectly conducting photospheric plasma. However, it is not the increase in conductivity in the photosphere (in fact the conductivity has a minimum in the photosphere according to Kovitya & Cram, 1983) but the increase in mean kinetic energy density relative to the mean magnetic energy density which underlies the idea of line-tying. As far as the coronal loop is concerned, the effect of line-tying is to ensure that *all* coronal displacements with a component perpendicular to the axis must produce a curvature of the field, at least near the footpoints where the field cannot move. The magnetic tension associated with this curvature acts to restore the tube by straightening the field. It is thus a stabilizing factor, especially for the kink mode of instability. However, it should be noted that the curvature forces are forced on the perturbed field in the region where, in the Sun, the *equilibrium* field has strong curvature due to the funnelling of the large diameter coronal flux tube into a narrow photospheric flux tube (cf. Section 6.2.3). The curvature of the equilibrium structure has never been incorporated in the models, so that its effect on line-tying remains unknown.

Line-tying can be implemented in several ways. The weakest constraint, advocated by van Hoven, Ma & Einaudi (1981), is to demand that the displacement perpendicular to the field, $\xi_\perp$, vanish (I). This choice has been criticized by An (1984b), who argues that although the motion parallel to the field is not restricted by line-tying, it is governed by hydrodynamic processes which are slow compared to MHD processes. Since we are interested in the latter time scales, the plasma at the footpoint must be effectively static, i.e. the boundary condition should read $\xi = 0$ (II). This version is supported by the model analysis of Hood (1986a). A still stronger constraint, used by An (1982), is to demand that the fluctuation in the magnetic field $\mathbf{B_1}$ vanish (III). The latter two boundary conditions can be satisfied by writing

$$\xi = \xi(r) f(z) \exp i(m\theta + kz), \tag{5.135}$$

where

$$f(z) = 0 \quad \text{when} \quad z = \pm L \tag{5.136}$$

in the first case or

$$f(z) = 0 \quad \text{and} \quad \frac{df}{dz} = 0 \quad \text{when} \quad z = \pm L \tag{5.137}$$

in the second. When the gas pressure is neglected, we may take $f = \cos(\pi z / 2L)$ and satisfy the equations of motion and boundary condition (II). Such a simple function f does not seem to exist for the more general case. With $f(z)$ chosen, the $\xi_\theta(r)$ and $\xi_z(r)$ components may be eliminated from the equations, leaving a second-order differential equation governing $\xi_r(r)$ alone. This is to be solved subject to the boundary conditions imposed at $r = 0$ and at infinity.

Since we are interested mainly in the conditions under which stability turns into instability, only the marginally stable case, that for which $n^2 = 0$, need be considered. Setting $n = 0$, the required values of k are eigenvalues of the system for each choice of m. The determination is quite straightforward, thanks to the structure of the equations. For these equations it may be shown that, if the marginal stability equation is integrated outwards for any given k and a zero in ξ_r occurs before the outer boundary of the model is reached, then the corresponding value of n^2 is positive and the mode is unstable at that k value. Conversely, if no zero appears, then the corresponding n^2 is negative and the mode is stable.

A complication can arise when the differential equation has singular points. According to Giachetti, van Hoven & Chiuderi (1977), these occur when the perturbation propagates directly across the equilibrium magnetic field. Since the 'wavenumbers' in the expression (5.135) are k in the z-direction and m/r in the azimuthal direction, the singularity condition is

$$\mathbf{k} \cdot \mathbf{B_0} = kB_z + \frac{m}{r} B_\theta = 0. \tag{5.138}$$

In general there will be a set of radial surfaces at $r = r_s$ where this occurs; these are known as *mode rational surfaces*. Local solutions near these singularities must be obtained by asymptotic methods, and these solutions must then be tested for stability using a local (not global) analysis. The resulting test for stability in an incompressible medium is the Suydam criterion

$$\frac{\mathrm{d}p_0}{\mathrm{d}r} \geqslant -\frac{rB_z^2}{8\mu_0}\left[\frac{\mathrm{d}}{\mathrm{d}r}\left(\frac{rB_z}{B_\theta}\right)\right]^2, \qquad (5.139)$$

which involves the twist (5.124). When this condition is not met a local interchange instability, called a ballooning mode, develops which is driven by the pressure gradient. An (1984a) showed that compressibility adds another stabilizing term to this criterion. Hood (1986b) has further modified the Suydam criterion to take account of line-tying and this was used by Velli & Hood (1987) to obtain necessary (but not sufficient) conditions for the stability of various coronal loop models with cylindrically symmetric fields. They found that if $\mathrm{d}p_0/\mathrm{d}r < 0$ at any point, the model is unstable, the growth rate of the ballooning mode increasing with wavenumber k. However, the addition of viscosity (van der Linden, Goosens & Hood, 1988) may stabilize this mode completely. If the loop is locally stable, the global stability may be tested separately within each pair of singularities by examining the zeros in the solutions of the marginal stability equation as before.

The method of determining stability *via* the differential equations is formally identical to the approach based on the energy principle (Newcomb, 1960). If the momentum equation (5.131) is multiplied throughout by $\mathrm{d}\boldsymbol{\xi}/\mathrm{d}t$ and integrated over the volume of the system, the rate of change of kinetic energy is obtained,

$$\frac{\mathrm{d}}{\mathrm{d}t}\int_V \frac{1}{2}\rho_0\left(\frac{\mathrm{d}\boldsymbol{\xi}}{\mathrm{d}t}\right)^2 \mathrm{d}V = \int_V \mathbf{F}\cdot\frac{\mathrm{d}\boldsymbol{\xi}}{\mathrm{d}t}\mathrm{d}V. \qquad (5.140)$$

If the system is displaced from equilibrium over a small interval of time δt, the change in kinetic energy is

$$\delta W_{\mathrm{KE}} = \int_0^{\delta t}\left(\int_V \mathbf{F}\cdot\frac{\mathrm{d}\boldsymbol{\xi}}{\mathrm{d}t}\mathrm{d}V\right)\mathrm{d}t \simeq \frac{1}{2}\int_V \mathbf{F}\cdot\boldsymbol{\xi}(\delta t)\,\mathrm{d}V. \qquad (5.141)$$

Now the right-hand side is the work done by the force $\mathbf{F}$ throughout the displacement of the system to its final position $\boldsymbol{\xi}(\delta t)$ and this is, by definition, the negative of the change of potential energy, i.e.

$$\delta W_{\mathrm{PE}} = -\frac{1}{2}\int_V \boldsymbol{\xi}\cdot\mathbf{F}\,\mathrm{d}V. \qquad (5.142)$$

The equilibrium will be stable if the potential energy is *increased* by all possible displacements of the system. An increase in potential energy means that the system may return to equilibrium, gaining kinetic energy as it does so. If the potential energy decreases for any displacement, return to equilibrium is not possible and the system can gain kinetic energy only by moving away from equilibrium. Thus the condition for stability is that

$$\delta W_{\rm PE} > 0 \quad \text{for all possible displacements } \xi. \qquad (5.143)$$

This requirement can be restated in terms of the minimum value that the integral can take for arbitrary displacements. If the minimum value is positive, the equilibrium is stable.

In order to find the energy integral as a function of the displacement ξ, we use the linearized continuity, energy and induction equations, (5.130), (5.132) and (5.133), to substitute in **F** given by (5.131). This yields

$$\delta W_{\rm PE} = -\frac{1}{2}\int_V \xi \cdot \left(\gamma p_0 \nabla \cdot \xi + \xi \cdot \nabla p_0 + \frac{1}{\mu_0}\{\nabla \times [\nabla \times (\xi \times \mathbf{B}_0)]\} \times \mathbf{B}_0 \right.$$

$$\left. + \frac{1}{\mu_0}(\nabla \times \mathbf{B}_0 \times [\nabla \times (\xi \times \mathbf{B}_0)] \right) \mathrm{d}V. \quad (5.144)$$

This expression can be rearranged by use of the divergence theorem and integration by parts (assuming the fluctuations vanish at the boundary of the volume V) to give

$$\delta W_{\rm PE} = \frac{1}{2}\int_V \left(\gamma p_0 (\nabla \cdot \xi)^2 + (\xi \cdot \nabla p_0)(\nabla \cdot \xi) + \frac{1}{\mu_0}|\nabla \times (\xi \times \mathbf{B}_0)|^2 \right.$$

$$\left. - \frac{1}{\mu_0}(\nabla \times \mathbf{B}_0) \times \{[\nabla \times (\xi \times \mathbf{B}_0)] \times \xi\} \right) \mathrm{d}V. \quad (5.145)$$

We can now see that two of the terms are always positive, i.e. are always stabilizing. These terms describe the effect of compression of the plasma and the magnetic field. Such compressions drive stable oscillations, i.e. wave motions, in the absence of dissipation. A description of these modes is deferred to the following section. Spicer & Brown (1981) give another form, one in which each distinct mode of motion is isolated in the energy integral. Those terms which may be either positive (stabilizing) or negative (destabilizing) are identified there as the kink modes and the interchange modes. However, this anticipates the result of the analysis.

The analysis proceeds by minimizing the integral with respect to the vector displacement ξ. It is first minimized with respect to the components ξ_θ and ξ_z, so that the change in energy involves only ξ_r. According to the calculus of variations, the function ξ_r that minimizes the final integral must satisfy the Euler–Lagrange differential equation. This equation is identical to the marginal stability equation described above, thereby demonstrating the equivalence of the two approaches. Stability theory provides the proof that the system is stable (i.e. $\delta W_{\mathrm{PE}} > 0$) if the solution of this equation has no zeros between the radial boundaries (or between the mode rational surfaces).

The variational calculation is often preferred because it has the advantage of providing approximate results with trial functions ξ_r which have a simple z-dependence, but which do not satisfy the equations of motion. A rigorous treatment would expand ξ in terms of a complete set of simple functions, e.g.

$$\xi_r = \mathscr{R} \sum_{n=-\infty}^{\infty} \xi_n(r) \exp i(m\theta + n\pi z/L), \tag{5.146}$$

where the functions ξ_n are constrained by the requirement that ξ_r satisfies the line-tying boundary condition at $z = \pm L$. Einaudi & van Hoven (1983) then show that the energy integral requiring minimization has the form

$$\delta W = \sum_{n=-\infty}^{\infty} \int \left[F_n \left(\frac{\mathrm{d}\xi_n}{\mathrm{d}r} \right)^2 + G_n \xi_n^2 \right.$$

$$\left. + \sum_{n'=-\infty}^{\infty} (\text{cross terms involving } \xi_n, \xi_{n'}) \right] \mathrm{d}V. \tag{5.147}$$

This leads to an infinite set of coupled Euler–Lagrange equations, but they were able to develop an approximate solution for a truncated set. Most analyses, however, simply utilize a single trial function and arrive at an energy integral without cross terms. A single Euler–Lagrange equation results:

$$\frac{\mathrm{d}}{\mathrm{d}r} \left(F \frac{\mathrm{d}\xi}{\mathrm{d}r} \right) - G\xi = 0. \tag{5.148}$$

Such analyses of course investigate the sign of the potential energy change for only a subset of all the possible displacements. They therefore provide necessary but not sufficient criteria for stability.

A third approach which has been espoused recently exploits the ability of large computers to now perform accurate fluid simulations in three dimensions. McClymont & Craig (1987b) and Craig, McClymont & Sneyd (1988) have suggested that the linearized time-dependent equations be solved numerically to follow the development of a random initial velocity distribution. The initial disturbance will either contain a component of each eigenmode or will excite each mode through numerical rounding. Of course, the system may have to be followed through a great number of time steps before the fastest growing mode (in an unstable configuration) or the slowest decaying mode (in a stable configuration) emerges, and the amount of computer time required may be prohibitive. However, by using a technique of cycling through a series of *non-uniform* time steps, these authors showed that several eigenmodes can be detected simultaneously and efficiently.

As a test case, they considered the stability of the constant-twist force-free field, in the low-β limit, confined within a box with no motions allowed at the surface. Since the plasma is neglected in this problem, a fictitious fluid with no inertia has to be introduced in order to generate an artificial equation of motion,

$$\frac{\partial \xi}{\partial t} = \frac{1}{\mu_0}[(\nabla \times \mathbf{B_0}) \times \mathbf{B_1} + (\nabla \times \mathbf{B_1}) \times \mathbf{B_0}], \qquad (5.149)$$

where again $\mathbf{B_1} = \nabla \times (\xi \times \mathbf{B_0})$. By increasing the size of the transverse cross section of the box, they were able to investigate the effects of imposing transverse boundary conditions at varying finite radial distance. The fact that such boundary conditions restrict the analysis to *internal* kink modes, i.e. modes which develop within the tube and vanish beyond it, and eliminate possibility of the *free* kink modes of the infinite system, has been stressed by Gibons & Spicer (1981). The presence of rigid side walls is generally a stabilizing factor.

Each type of stability analysis has its own strengths and limitations. Fortunately there are some benchmark models which have been analysed using a variety of methods, and we may draw with confidence some general conclusions from the results. The equilibrium models fall into three categories and their stability properties, in summary, are as follows.

Force-free fields without plasma

The simplest field structure is a potential field. Since this has no twist, i.e. no azimuthal field component, it is stable against the kink instability and is, indeed, stable to all ideal MHD modes. All other force-free field structures with cylindrical symmetry are unstable, in the absence of line-tying (Anzer, 1968). This includes the linear force-free field (Voslamber & Callebaut, 1962), which is unphysical in so far as the axial field component B_z changes sign in alternate annuli, and the constant-twist field. The fastest growing mode is the long-wavelength (small k) kink $(m = \pm 1)$ mode. The most unstable perturbations are those which are helical in form, like the magnetic field structure. These disturbances produce the least magnetic tension, which is the primary restoring force. The use of non-helical test functions for the perturbations will give only a necessary (but not sufficient) test of stability (An, 1984b). Raadu (1972) was the first to show rigorously that the unstable modes may be stabilized by line-tying, which introduces extra magnetic tension as a result of the curvature of the longitudinal field at the ends.

Hood & Priest (1979b, 1981), Einaudi & van Hoven (1983) and Craig *et al.* (1988) all found that if the tube had a twist less than about 2.5π from end to end, line-tying could stabilize the internal kink modes; otherwise the tube remained unstable. The twist in this case is just $2L/b$ where b is the transverse scale (or 'radius') of the tube. The ratio of total length to radius is known as the *aspect ratio* of the tube. The stability criterion for uniformly twisted tubes can thus be expressed in terms of more easily observed quantities. If the aspect ratio is greater than about 10 – long, thin loops – they are unstable. Short, thick loops are stable.

Force-free fields with plasma

Without line-tying, the models in this category are also generally unstable. The plasma is introduced only to provide inertia; its pressure is neglected, so that the magnetic field configuration remains force-free. However, the structure is assumed to differ between the loop itself, modelled as a tube, and the surrounding medium. Hasan (1980) embedded a constant-twist field in a uniform longitudinal potential field, a cylindrical current sheet separating the two. He found this to be unstable to internal kinks, though Cramer & Donnelly (1985) point out that the surrounding potential field stabilizes the external kink. Rather surprisingly, a twisted potential field of the form of Eqn. (5.122) in a plasma-free external region, which can be matched without a current sheet to a force-free tube field, fails to stabilize the kink (Sillen & Kattenberg, 1980; Cramer & Donnelly, 1985). Adding plasma, i.e. inertia, to the vacuum region reduced the growth rate of the kink instability (Kattenberg & Sillen, 1983) but did not remove it.

A more complicated model involving a modification of the Lundquist field embedded in a potential field was analysed by Einaudi & van Hoven (1983). This is the only case in which line-tying has been introduced, and it resulted in stabilization.

Non-force-free fields

Models which incorporate a gas pressure gradient are generally simple modifications of the force-free structures described in the previous section, and lead to very similar conclusions.

A positive pressure gradient is a stabilizing factor. Models with pressure increasing monotonically away from the axis automatically satisfy the Suydam criterion (or its generalization by An, 1984a) for local stability. Hasan (1980) showed that a constant-twist field embedded in a uniform plasma can also be globally stabilized solely by a sufficiently large positive pressure gradient. Giachetti, van Hoven & Chiuderi (1977) and Chiuderi & Einaudi (1981) found that their more complex model was similarly stabilized. However, the pressures required for such stabilization are higher than those observed in the solar corona, so that line-tying appears to be the more effective stabilizing mechanism. Hood & Priest (1979b) note that line-tying can stabilize constant-twist models even when the pressure gradient is negative. Another model with a negative pressure gradient was studied by Hood & Priest (1979b) and An (1984b). This model possessed variable twist but a constant axial field, B_z. It was found that the largest value of the twist at the 'edge' of the tube $\Phi(b)$ which could be accommodated before instability set in, increased with the aspect ratio. Long, thin tubes are stable at larger twists than short, fat tubes. This result is precisely the opposite to that found for force-free models (An, 1984b).

In summary, it would appear that coronal loop models will be stable to ideal MHD perturbations if their degree of twist, which is of course related to the strength of the axial current compared to the azimuthal current, is not too large. This stability is ensured mainly by the 'tying' of the ends of the loop into the photospheric plasma. The pressure and density structure within the loop is of secondary importance.

However, coronal loops are expected to become unstable to the kink instability if the twist is increased beyond some critical value, which varies from model to model. Loops with twist close to the critical value are marginally stable and any instability will develop with very small growth rates ($n^2 \simeq 0$). Under these circumstances, the argument for adopting the ideal MHD equations, namely that the perturbation develops on the Alfvén time scale, breaks down. Hood & Priest (1979b) and Chiuderi & Einaudi (1981) discuss possible resistive effects in marginally stable ideal MHD situations. The major effect of resistivity is to add a further term to the right-hand side of Eqn. (5.133). This equation now reads

$$n\mathbf{B}_1 = n\nabla \times (\boldsymbol{\xi} \times \mathbf{B}_0) + \eta\nabla^2\mathbf{B}_1. \qquad (5.150)$$

The resistive term will be most important when the first term on the right-hand side vanishes. These loci are just the mode rational surfaces, defined by Eqn. (5.138), when there is no line-tying (Hood & Priest, 1979b). Resistive effects may therefore be incorporated using thin boundary layer analysis. The first such effect is a modification of the kink mode and its growth rate, but Chiuderi & Einaudi (1981) conclude that this is unlikely to be important. The second effect is to introduce a new mode – a local resistive mode confined to the ideal mode rational surface. This is known as a *tearing mode* (cf. Section 5.7.3). Chiuderi & Einaudi estimate the growth rates of this mode and point out that the fastest growth rate occurs for the largest scale perturbations, i.e. those with the smallest k values. These may be unstable when the system is stable to ideal MHD modes. A more detailed analysis by Batistoni, Einaudi & Chiuderi (1985) confirmed the growth rates of the resistive kink mode but reduced that of the tearing mode by over an order of magnitude. These models ignore line-tying. According to Hood & Priest (1979b), line-tying limits the vanishing of the first term on the right-hand side of Eqn. (5.150) to points on the mode rational surfaces at the apex of the tube ($z = L$). Thus tearing modes can develop over surfaces of only limited spatial extent and will possess still slower growth rates than estimated above.

Of more likely significance than resistive effects are non-adiabatic effects. Again, these become important only when the ideal MHD mode is marginally stable, so that the time scales associated with their development become long enough for heat conduction and radiative losses to affect them. This can be the case with high-order ($m \rightarrow \infty$) interchange modes – which are high-order ripples around the tube – along mode rational surfaces.

An (1983) found that, if the radial gradient of the radiative loss function $\mathscr{L}$ is negative ($d\mathscr{L}_0/dr < 0$), incompressible, radiation-driven interchange modes may be unstable when the interchange modes are stable in ideal MHD. An concluded that interchange modes in loops of almost constant density but with a positive radial pressure gradient ($dp_0/dr > 0$) could be unstable. The interchange instability would then mix material across the mode rational surfaces. Such a cross-field transport mechanism is required by some loop scenarios (cf. Section 5.9). Taking account of compressibility allows coupling to waves, which can add to the destabilization of marginally stable ideal states (An, 1984a).

A more complete treatment, including line-tying, was given in a subsequent paper (An, 1984c). Line-tying produces longitudinal curvature of the field and hence brings heat conduction along the tube into play. The results are now more complicated. For models with $dp_0/dr < 0$, line-tying stabilizes the ideal MHD interchange modes which would be unstable without line-tying. Non-adiabaticity has negligible effect in this case. If $dp_0/dr > 0$, radiation can still be destabilizing, although heat conduction tends to counter it. An suggests that cool centres of tubes may possess local interchange instabilities, whereas the hotter outer parts may be stable.

The presence of the magnetic field not only introduces the new set of MHD modes but also modifies the thermal modes described in the previous section. An (1985) has noted that twisting the field reduces heat conduction along the loop. This tends to destabilize the thermal condensation mode discussed in Section 5.5. On the other hand, as a loop is twisted, the pressure on the axis will increase, and this is a stabilizing effect. According to An, the destabilizing effect of heat conduction is the more significant in loops with very small twists, but pressure effects stabilize the condensation mode in loops with larger twists.

So far, we have discussed the stability of coronal loops from a general point of view. Wherever necessary, the heating function has been parametrized. The stability requirements then provide some constraints on the mathematical form of $\mathscr{H}$. These in turn discriminate, albeit rather weakly, between various heating mechanisms. Unfortunately, there have been very few attempts to analyse the stability characteristics of specific heating models. A notable exception is the analysis by Zweibel (1980) of the fast-mode wave heating mechanism (cf. Section 5.7.2). She examined the stability of the condensation mode by introducing a Lorentz force into the momentum equation, to account for the mean effect of the fluctuating currents associated with the assumed wave field, and a heating function $H(T, \rho, \mathbf{B})$ into the energy equation, to account for the ohmic dissipation of the waves in the presence of a uniform background field $\mathbf{B}$. In equilibrium this dissipation balances the radiative losses. When perturbed, the derivatives $\partial H/\partial \mathbf{B}$ must be incorporated in addition to $\partial H/\partial T$ and $\partial H/\partial \rho$. Now that the system is three-dimensional, perturbations transverse to the field may be considered and Zweibel found that modes perpendicular to the field were unstable, the growth time being of the order of 1 hr under typical coronal conditions. Such an instability would presumably lead to the formation of dense plasma sheets or plasma filaments aligned along the field. Clearly, line-tying would change the geometry of the perturbed field and the heat flow, but its effects have not been analysed despite the potential significance of this instability as a means of differentiating plasma loops in the solar corona.

The results of the global stability analyses are fully consistent with our interpretation of coronal loops as plasma delineating a stable magnetic loop structure. However, local interchange instabilities may develop along distinct magnetic surfaces. These may be of limited extent, but may also form an extensive 'boundary' surface of a tube. The possibility of such interchange motions should cause us to question the assumption that the plasma associated with one bundle of field lines is completely isolated from that associated with a neighbouring bundle. The effects resulting from exchange of matter, momentum and energy warrant much further study.

5.6.3 *MHD waves*

In the previous section we considered the response of coronal loop models to a small perturbation and concentrated on the growth of those classes of perturbation which lead to instability. Not all perturbations, though, have a real growth rate. When $n^2 < 0$, the time dependence is imaginary and the displacement may be written as

$$\delta\mathbf{x} = \boldsymbol{\xi} \exp i\omega t \tag{5.151}$$

where the frequency ω is given by $\omega^2 = -n^2$. Such a perturbation describes an oscillation. The modes of oscillation of the loop model are therefore found, in principle, by exactly the same analysis of the equations of motion as described at the beginning of the previous section. These would give the global oscillatory modes. To our knowledge, the global eigenfrequencies of coronal loop models have never been calculated, since stability analyses concentrate exclusively on the marginally stable state ($n^2 = 0$).

Most attention has been devoted not to the large-scale oscillations of coronal loops but to the small-scale. When the wavelength of the oscillation is much less than the typical distances over which the properties of the loop change significantly, the equilibrium model may be treated *locally* as uniform. The analysis can then employ a perturbation of the form

$$\delta\mathbf{x} = \boldsymbol{\xi} \exp i(\mathbf{k} \cdot \mathbf{x} - \omega t). \tag{5.152}$$

Substitution into the linearized adiabatic equations of motion (cf. Eqns. (5.130)–(5.133)) produces a homogeneous set of linear algebraic equations. Non-trivial solutions exist only when the determinant of coefficients vanishes. This requirement produces a relationship between the frequency ω and the wave vector $\mathbf{k}$ known as the dispersion relation for the possible wave modes. This relationship will vary from point to point in a global model, and the nature of the waves will change as they travel from one region to another. However, provided that their wavelengths remain small compared to the scale of the gradients in the equilibrium model, they may be thought of *locally* as waves.

The analysis of ideal wave modes in an *isothermal* stratified plasma permeated by a uniform magnetic field has been summarized in detail in an earlier monograph (Bray & Loughhead, 1974; see also Priest, 1982). There are three independent restoring forces involved, pressure, gravity (through buoyancy), and magnetic (pressure and tension), and these lead to three modes of oscillation. The determinantal equation is a sixth-order polynomial in $\mathbf{k}$ and ω which factorizes to provide two dispersion relations,

$$\omega^2 - v_A^2 k^2 \cos^2 \theta = 0 \qquad (5.153)$$

$$\omega^4 - \omega^2 [(v_a^2 + v_A^2) k^2 - 2i\omega_a v_a k_z] + N^2 v_a^2 (k^2 - k_z^2) + v_a^2 v_A^2 k^2 \cos^2 \theta$$
$$- 2i\omega_a v_a v_A^2 k^3 \cos \theta \cos \zeta = 0. \qquad (5.154)$$

Here k_z is the vertical component of the wave vector $\mathbf{k}$, θ is the angle between the wave vector and the direction of the field and ζ is the angle between the field direction and the vertical. This relation has been written in terms of the acoustic cut-off frequency $\omega_a = \gamma g / 2 v_a$ and the Brunt–Väisälä frequency $N = \sqrt{[(\gamma - 1) g^2 / v_a^2]}$. The former is the frequency at which buoyancy effects suppress acoustic waves, and the latter is the natural frequency of buoyant oscillation (see Bray & Loughhead, 1974).

Since the equilibrium density ρ_0 has an exponential height dependence, the Alfvén speed appearing in these relations is a function of height $v_A(z) = \sqrt{[B_0^2 / \mu_0 \rho_0(z)]}$. Wave solutions of the form (5.152) are thus appropriate only if we can take this quantity to be locally constant. Self-consistent solutions must therefore have a vertical wavelength less than the density scale height.

The first relation (5.153) then describes oblique Alfvén waves, which result purely from magnetic tension. There are no density or gas pressure fluctuations, i.e. the compression $\nabla \cdot \mathbf{v}$ vanishes. This implies that $\mathbf{k} \cdot \mathbf{v} = 0$, so that the fluid displacement is always perpendicular to the wave vector. This property characterizes a transverse wave. However, the *group* velocity $(\partial \omega / \partial \mathbf{k})$ is parallel to the field direction and is equal to the Alfvén speed in magnitude. The mean kinetic energy density $(\frac{1}{2} \rho_0 v_1^2)$ in the wave is equal to the mean magnetic energy density $(B_1^2 / 2\mu_0)$ and energy flux is always in the direction of the magnetic field.

The second relation is much more complicated because it involves all three restoring forces: it thus depends not only on the orientation of the wave vector with respect to the magnetic field direction but also on its orientation with respect to gravity. Only a few special cases have been analysed in the literature (e.g. Bel & Mein, 1971; Leroy & Schwartz, 1982). However, we should notice that we are restricted to wavenumbers $k \gg 2\pi / \Lambda$ where $\Lambda \sim 10^8$ m (Section 5.4.1) and to frequencies $\omega \gg 2\pi / \tau_t$, where the thermal time scale $\tau_t \sim 10^3$ s (Section 5.5.1) in the solar corona. If the dispersion equation is then evaluated for the canonical coronal conditions of Table 5.1, we find that gravity has negligible influence on the solutions in this region of the (k, ω) plane. Furthermore, thermal conduction is effective in the corona when the scales are small, and this tends to smooth out the temperature fluctuations. This situation is mimicked by taking $\gamma = 1$. In this limit, the Brunt–Väisälä frequency vanishes and all buoyancy effects are suppressed. Gravitational effects will therefore be neglected in the further discussion of coronal wave motions.

Dropping the gravitational terms in Eqn. (5.154) leads to the much simpler dispersion relation

$$\omega^4 - (v_a{}^2 + v_A{}^2) k^2 \omega^2 + v_a{}^2 v_A{}^2 k^4 \cos^2\theta = 0, \qquad (5.155)$$

which now involves only the orientation of the wave vector with respect to the magnetic field. This equation factorizes to yield two magneto-acoustic waves,

$$\omega^2 - v_s{}^2 k^2 = 0, \quad \omega^2 - v_f{}^2 k^2 = 0, \qquad (5.156)$$

whose phase velocities are given by

$$v_{f,s}^2 = \tfrac{1}{2}(v_a{}^2 + v_A{}^2) \pm \tfrac{1}{2}[(v_a{}^2 + v_A{}^2)^2 - 4 v_a{}^2 v_A{}^2 \cos^2\theta]^{\frac{1}{2}}. \qquad (5.157)$$

The positive sign selects the fast mode; its phase speed varies between v_A for propagation parallel to the field and $\sqrt{(v_a{}^2 + v_A{}^2)}$ for propagation perpendicular to the field. Since $v_A > v_a$ the characteristics of this mode are almost isotropic. The negative sign selects the slow mode, whose phase speed varies from v_a for propagation parallel to the field to zero when $\mathbf{k}$ is perpendicular to the field. Slow-mode waves thus propagate preferentially in the direction of the magnetic field. The phase velocity curves, presented in a polar diagram of the phase speed as a function of angle of inclination θ, are shown for the three modes in Fig. 5.11. The energy of these waves resides mainly in the magnetic field compression when $v_A > v_a$. Indeed when $v_A \gg v_a$, the gas compression may be neglected completely, so that $v_f \rightarrow v_A$. The slow mode then is eliminated and the dispersion relation (5.155) describes only the fast mode,

$$\omega^2 - v_A{}^2 k^2 = 0. \qquad (5.158)$$

This relation describes *compressional* Alfvén waves, in contrast to (5.153), which refers to shear Alfvén waves. The group velocity of the fast-mode waves is similar to the phase velocity, so that the energy can propagate in all directions. For the slow mode, however, the group velocity is limited to a narrow cone about the magnetic field direction (cf. Fig. 5.11).

The individual wave modes detailed above represent the response of a uniform medium to a steady perturbation with the appropriate wave-number and frequency. However, these modes form a complete set, so that the response to an arbitrary, perhaps transitory, disturbance can be found by taking a suitable superposition of these modes.

In order to apply this wave theory to the analysis of loops in the solar

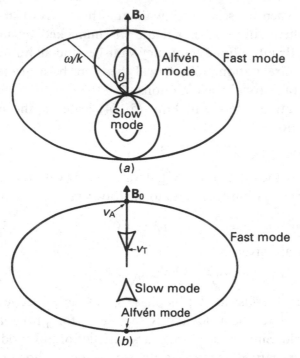

Fig. 5.11. Polar diagrams for MHD waves propagating at an angle θ to the equilibrium magnetic field. Shown here is the case in which $v_A > v_a$. (a) Phase velocity. (b) Group velocity.

corona, we must once again take account of their structure. In our discussion so far the density variation across a coronal loop has played no significant role, because the transverse thermal pressure gradients are so small that they can affect the magnetic structure very little. By the same token, however, very small departures from uniformity in the transverse magnetic field structure can accommodate very large changes in pressure and density across a loop or from one loop to another. The density variation is presumably the major factor in causing the differences in visibility of X-ray loops. As far as wave motions in the corona are concerned, the large variations in Alfvén speed that are possible when the density changes sharply are arguably the most important feature of the coronal system.

In its simplest form, the prototype model of a coronal loop would be an infinite uniform cylinder embedded in a uniform medium possessing a different sound speed and Alfvén speed. The field is assumed to be everywhere longitudinal (the orientation of the loop with respect to gravity is, of course, irrelevant in this context). At the interface between the two media the total pressure $p + B^2/\mu_0$ must balance. This model has been analysed by Wentzel (1979c), Wilson (1980) and Spruit (1982), but the most exhaustive treatments are due to Edwin & Roberts (1983), Cally (1986) and Abdelatif (1988). The following treatment is based on those of Edwin & Roberts and Cally.

The magneto-acoustic waves are governed by the compression, $\Delta = \nabla \cdot \mathbf{v}$. In cylindrical geometry, this has the form

$$\Delta = R(r) \exp i(m\theta + k_z z - \omega t), \tag{5.159}$$

where $R(r)$ is a solution of Bessel's equation

$$\frac{d^2 R}{dr^2} + \frac{1}{r} \frac{dR}{dr} + \left(\alpha^2 - \frac{m^2}{r^2} \right) R = 0, \tag{5.160}$$

and

$$\alpha^2 = \frac{(\omega^2 - k_z^2 v_a^2)(\omega^2 - k_z^2 v_A^2)}{(v_a^2 + v_A^2)(\omega^2 - k_z^2 v_T^2)}. \tag{5.161}$$

Note that in this geometry the 'azimuthal wavenumber' is m/r and the 'radial wavenumber' is α. The quantity v_T occurring in the last expression is the so-called *tube speed*

$$v_T = \frac{v_a v_A}{\sqrt{(v_a^2 + v_A^2)}}. \tag{5.162}$$

It is the minimum group velocity of the slow mode, occurring at the cusp in the polar diagram (Fig. 5.11(b)). Hence it is also known as the *cusp speed*.

Under coronal conditions, $v_T \simeq v_a < v_A$ so that the sign of α^2 changes from positive to negative as the phase speed ω/k_z drops below v_A, changes back to positive in the narrow band $v_T < \omega/k_z < v_a$, and finally reverts to negative for $\omega/k_z < v_T$. In the interior of the cylinder, the appropriate choice of Bessel function depends on the sign of α^2. In order that the fluctuations remain bounded on the axis we must choose

$$R(r) = R_1 \begin{cases} J_m(\alpha r) & \text{if } \alpha^2 > 0 \\ I_m(\beta r) & \text{if } \beta^2 = -\alpha^2 > 0, \end{cases} \tag{5.163}$$

where I_m and J_m are independent Bessel functions of order m. The J solutions oscillate in sign with radius, and are the cylindrical counterpart of the waves in an infinite uniform medium, described above. The I solutions are new; their amplitudes increase monotonically away from the axis, so that they cannot appear in an infinite medium. Their existence is due solely to the presence of a boundary at which they may be truncated. The former are therefore referred to as body waves and the latter as surface waves.

In magneto-acoustic modes the disturbance cannot be made to vanish completely on the cylindrical interface with the external medium, so that the internal disturbance must be matched to an appropriate mode outside.

Two classes of solution are now possible. In the first there is no transfer of energy between the internal and external disturbances – energy can propagate only along the cylinder. In the second, energy may be fed into or extracted from the internal disturbance by inward or outward propagating external waves.

In the first class, the solution in the external medium must be non-propagating and die away as r increases. The appropriate form is therefore

$$R(r) = R_e K_m(\beta r), \tag{5.164}$$

where K_m is the Bessel function which decays monotonically to zero as $r \to \infty$. This solution exists only if $\beta^2 = -\alpha^2 > 0$ in the external region. Otherwise the perturbations must vanish identically there, confining the wave strictly to the interior of the cylinder.

The interior and exterior solutions must be matched at the boundary of the cylinder ($r = b$) in order to preserve continuity of the radial component of velocity and of total pressure. These constraints lead to the dispersion relation for cylindrical waves. For surface waves we obtain

$$\rho_i(\omega^2 - k_z^2 v_{Ai}^2)\beta_e \frac{K'_m(\beta_e b)}{K_m(\beta_e b)} = \rho_e(\omega^2 - k_z^2 v_{Ae}^2)\beta_i \frac{I'_m(\beta_i b)}{I_m(\beta_i b)} \tag{5.165}$$

and for body waves

$$\rho_i(\omega^2 - k^2 v_{Ai}^2)\beta_e \frac{K'_m(\beta_e b)}{K_m(\beta_e b)} = \rho_e(\omega^2 - k^2 v_{Ae}^2)\alpha_i \frac{J'_m(\alpha_i b)}{J_m(\alpha_i b)}. \tag{5.166}$$

The subscripts i and e denote interior and exterior values, and the prime denotes the derivative with respect to r. The ratio of the interior to exterior density appearing in these relations is not arbitrary, but is fixed by the requirement that the total pressure balances across the boundary in the equilibrium state. This leads to the further relation

$$\frac{\rho_e}{\rho_i} = \frac{2v_{ai}^2 + \gamma v_{Ai}^2}{2v_{ae}^2 + \gamma v_{Ae}^2}. \tag{5.167}$$

Thus the wave characteristics are determined entirely by the sound speeds and Alfvén speeds in the two media.

The differences between these dispersion relations and those for an infinite uniform medium (5.153)–(5.154) are important. In the latter case, we may specify k_z and k_y, say, and solve the dispersion relation to find the wavenumber k_x of the modes for *any value* of ω. There is thus a *continuous* spectrum of modes. In a structured system, on the other hand, the dispersion relations (5.165)–(5.166) relate ω to k_z and m. For given k_z and

m, there is only a set of particular ω values which satisfy them. We thus have a *discrete* spectrum of modes – the boundary condition at the interface selects only those that match on either side.

Edwin & Roberts (1983) gave a detailed discussion of the sets of modes for $m = 0$, the sausage modes, and $m = 1$, the kink modes (cf. Section 5.6.2, Fig. 5.10). When the Alfvén speeds are greater than the sound speeds (as is typically the case in the solar corona), the dispersion relations permit no surface waves at all. Wentzel (1979c) concluded otherwise, but he employs a different nomenclature which regards the body waves of Edwin & Roberts as surface waves. The model certainly admits body waves. Indeed, if the Alfvén speed in the external medium exceeds that inside the cylinder, there are two types of body waves, analogous to the fast and slow modes in a uniform medium. A phase–speed diagram of these waves is shown in Fig. 5.12. The phase speed ω/k_z is plotted as a function of the scaled wavenumber $k_z b$. Various sausage and kink modes are possible at each wavenumber, but their phase speeds always lie within the range $v_{Ae} < \omega/k_z < v_{Ai}$ or $v_{ai} < \omega/k_z < v_T$. In the first range we have fast body waves, the kink being similar to the seismic Love waves and the sausage to oceanographic Pekeris waves. In the second range, we have slow body waves.

Slow body waves may have either long or short wavelengths but the only long-wavelength fast body wave is the kink mode. In the long-wavelength limit, this has a phase speed equal to the 'mean' Alfvén speed

$$v_k = \sqrt{\frac{\rho_i v_{Ai}^2 + \rho_e v_{Ae}^2}{\rho_i + \rho_e}}. \tag{5.168}$$

The fundamental kink mode will have $k \simeq 2\pi/2L$, and thus a period of about $2L/v_k$. All other fast body waves have $k_z b > 1.2$. Since $L > b$ in observed loops, these other waves will possess much shorter wavelengths and shorter periods than the fundamental kink mode.

If the Alfvén speed in the cylinder is greater than the external Alfvén speed, only the slow body waves can propagate along the cylinder.

So far the discussion has been based upon the simplest possible model in which the magnetic field is uniform and longitidunal in both the tube and the external medium, differing only in strength. More complicated models are also amenable to analysis – models with uniform plasma but more general force-free magnetic structure within the tube. Cramer & Donnelly (1985) discuss the spectrum of surface modes for such structures. The differences are matters of detail.

Before attempting to relate these characteristics to observations of

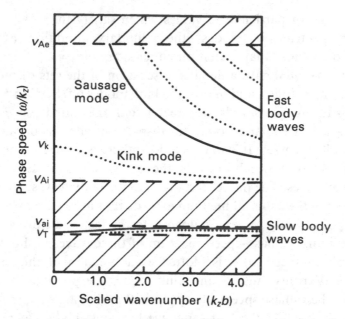

Fig. 5.12. The phase speeds of MHD wave modes in a flux tube with lower temperature and higher density than the external medium. In both the internal and external plasma the Alfvén speed exceeds the sound speed ($v_{ae} = 0.5v_{ai}$, $v_{Ai} = 2v_{ai}$, $v_{Ae} = 5v_{ai}$). Regions where wave motion is forbidden are shaded. The lower propagation band shows just two of the possible slow body waves. (After Edwin & Roberts, 1983.)

coronal loops, we must look briefly at the other class of solution, that in which we have a wave propagating not along the cylinder but across it. The external solution is now an inward or outward propagating wave, described by an appropriate Hankel function. This class of wave has been treated by Spruit (1982) and Cally (1986). If energy is transferred across the tube interface, we can no longer assume that ω is real since this would imply an unphysical source or sink of energy along the axis of the cylinder. Instead, we must allow ω to be complex with the imaginary part describing the growth or decay of the oscillation in time. If the cylinder 'radiates' a wave into the surrounding medium, it will lose energy. This damped mode is called a 'leaky' tube wave by Cally (1986), who provides a systematic classification of them. We can now appreciate the physical reason for the low wavenumber cut-off to the fast body waves in Fig. 5.12. At the cut-off wavenumber, the disturbance in the external medium changes from a non-propagating to a propagating fast mode; the fast body wave switches to being a leaky fast mode.

The response of a flux tube embedded in a magnetized medium to an arbitrary initial disturbance must be expressed in terms of both free (non-leaky or trapped) modes and damped (leaky) modes. The latter will propagate away leaving the cylinder oscillating only in free modes. These are what we should expect to see in the corona.

The analysis of Edwin & Roberts (1983) suggests that high density coronal loops should support free oscillations due to slow body waves, and that low density loops should exhibit also much more rapid oscillations due to the fast body wave. The long wavelength kink mode should be the most visible. These general characteristics are, moreover, not dependent on the particular geometry assumed in this model. Edwin & Roberts (1982) find essentially similar results for a slab geometry.

If the displacement is assumed to vanish at each end of the loop, the wave will be reflected and a standing oscillation will be established in the corona (cf. Section 5.7.2). Roberts, Edwin & Benz (1984) calculate the characteristic frequencies of standing waves in coronal loops with parameters similar to our canonical values. The ordering of the modes is evident from Fig. 5.12. The period of the slow mode is longest, about 1000 s, next comes the kink with a period of about 50 s and the shortest is the fast sausage with a period of about 1 s. The authors then discuss the observational evidence for coronal oscillations with these periods. However, the only observation referring to a definite loop (as opposed to an inferred loop) is of a loop prominence. There is no doubt, though, that the identification and measurement of loop oscillations could provide an important diagnostic tool because knowledge of the Alfvén speed, coupled with spectroscopic estimates of the plasma density, give a direct measure of the magnetic field strength in the loop itself.

Edwin & Roberts (1988) also suggest that the nature of coronal oscillations can provide information about the transverse density gradients in the corona. Using a cartesian *slab* model, they show that shape of the density profile affects the dispersion of the waves. A steep density gradient at the edges of the structure disperses the waves, the duration and time scale of the quasi-periodic oscillations being characteristic of the structure. On the other hand, a diffuse density profile is non-dispersive; then an impulsive disturbance will produce a true periodic oscillation.

Another aspect of coronal disturbances which has attracted theoretical attention recently is their non-linear behaviour. Wave amplitudes may grow into the non-linear regime by propagating down a density gradient or simply by travelling large distances without dissipation (cf. Section 5.6.4). The general non-linear analysis is very difficult, but Molotovshchikov & Ruderman (1987) have looked for soliton solutions for both $m = 0$ (sausage) slow-mode body wave and surface waves. In the long wavelength limit, they derive a third-order differential equation for the former and an integro-differential equation for the latter. Solutions for soliton surface waves on a magnetic cylinder were obtained numerically. The implications of studies of non-linear dynamics are just beginning to be recognized, and this area will undoubtedly assume greater significance in the future.

In the meantime, let us return to our discussion of linear modes. So far we have discussed only the magneto-acoustic modes in the solar corona. Another set of modes is provided by the compressionless waves ($\nabla \cdot \mathbf{v} = 0$), governed by the dispersion relation (5.153). In cylindrical geometry, these waves propagate the longitudinal component of the vorticity $\boldsymbol{\omega} = \nabla \times \mathbf{v}$, which measures the swirl of the motion. The simplest solution is then found when the wave velocity has only an azimuthal component,

$$v_\theta = V_\theta \exp i(k_z z - \omega t), \tag{5.169}$$

where k_z and ω satisfy the dispersion relation

$$\omega^2 - k_z^2 v_A^2 = 0. \tag{5.170}$$

Each cylindrical surface can carry independent torsional waves, so that the radial amplitude variation $V_\theta(r)$ is arbitrary. Other torsional waves with more complicated motions confined within the cylinder are also possible (see, for example, Spruit, 1981).

If we go to scales much smaller than characteristic loop dimensions (i.e. $k \gg 2\pi/2b \sim 10^{-7}\,\mathrm{m}^{-1}$), the exact geometry of the loop will become unimportant and any discontinuities may be treated as plane surfaces. Wave propagation in the neighbourhood of such an interface can then be modelled by considering an interface at $x = 0$ separating two uniform semi-infinite media with the magnetic field in the z-direction. The linearized equations are then satisfied by fluctuations with the form

$$\exp i(\alpha x + k_y y + k_z z - \omega t), \tag{5.171}$$

and the dispersion relation becomes

$$\alpha^2 = \frac{(\omega^2 - k_i^2 v_s^2)(\omega^2 - k_i^2 v_f^2)}{(v_a^2 + v_A^2)(\omega^2 - k_z^2 v_T^2)}. \tag{5.172}$$

Here k_i is the magnitude of the wave vector in the plane parallel to the interface, $k_i^2 = k_y^2 + k_z^2$. The two speeds v_s and v_f are the slow and fast mode speeds given by Eqn. (5.157).

Propagating waves arise if $\alpha^2 > 0$, which occurs in the slow and fast mode domains $k_z^2 v_T^2 < \omega^2 < k_i^2 v_s^2$ and $k_s^2 v_f^2 < \omega^2$, respectively. Otherwise, $\alpha^2 < 0$ and the mode has an exponential dependence in the x-direction away from the interface. These are known as *evanescent* or *tunnelling* modes.

As before, the velocity normal to the interface and the pressure must be continuous across the interface. The dispersion relation analogous to (5.165)–(5.166) is given by Wentzel (1979a). It selects a discrete set of modes. Rae & Roberts (1983) have constructed existence diagrams, which delineate the regions of the phase-speed polar plot where the allowed discrete modes fall. The phase speed is here the phase speed in the plane parallel to the surface ω/k_i and the polar angle is the angle between the wave vector in this plane and the direction of the magnetic field. These authors distinguish four cases. When, for given k_z, k_y and ω, propagating waves occur on both sides of the interface, the wave is simply transmitted across it. If evanescent modes occur on both sides, we have a surface shear wave propagating along the interface. These have been analysed by Wentzel (1979a). For $k_z \ll k_y$, the wave is almost compressionless (like a shear Alfvén wave) and phase propagates at the 'mean' Alfvén speed,

$$\omega^2 - k^2 v_k{}^2 \simeq 0. \tag{5.173}$$

There are equivalent high-order modes in cylindrical geometry with $k_z \ll m/b$, i.e. $m \gg k_z b$.

In reality, changes in the Alfvén speed will not occur discontinuously but smoothly over a short distance. This model was first discussed in the context of models of the solar corona by Ionson (1978). If we take the variation to be in the x-direction as before, the *ideal* magnetoacoustic modes are governed by the homogeneous second-order differential equation

$$\frac{d}{dx}\left\{ \frac{[\omega^2 - k_z{}^2 v_A{}^2(x)][\omega^2 - k_z{}^2 v_T{}^2(x)]}{[\omega^2 - k_i{}^2 v_s{}^2(x)][\omega^2 - k_i{}^2 v_f{}^2(x)]} [v_a{}^2(x) + v_A{}^2(x)] \rho_0(x) \frac{d\xi_x}{dx} \right\}$$

$$= -[\omega^2 - k_z{}^2 v_A{}^2(x)] \rho_0(x) \xi_x, \tag{5.174}$$

where the density ρ_0 and all the velocities are now functions of x. Solutions may be found by integrating across the interface for given ω, k_z and k_y. Surface waves, having an exponential decay in the x-direction on both sides of the interface, will occur for only a discrete set of values of ω — corresponding to those satisfying the dispersion relation (5.165). However, this procedure breaks down when the differential equation becomes singular, i.e. when the coefficient of the highest order derivative vanishes. This occurs when ω/k_z matches the Alfvén speed $v_A(x)$ or the tube speed $v_T(x)$ at some point across the interface. Wentzel (1979b) calls these the Alfvénic and compressive singularities.

If the equation is singular, the boundary conditions are not sufficient to define normal modes in the ideal case, and solutions must be constructed using Green's functions. The characteristics of these solutions can be seen in the artificial but mathematically simpler problem in which the medium is treated as incompressible. The governing equation is then

$$\frac{d}{dx}\left\{[\omega^2 - k_z^2 v_A^2(x)]\frac{\rho_0(x)}{k_i^2}\frac{d\xi_x}{dx}\right\} = [\omega^2 - k_z^2 v_A^2(x)]\rho_0(x)\,\xi_x, \quad (5.175)$$

where $k_i^2 = k_z^2 + k_y^2$. Lee & Roberts (1986) discuss this case in detail for two-dimensional Alfvén waves in the plane perpendicular to the interface ($k_y = 0$). The missing surface mode is then replaced by a *continuum* of Alfvén body waves within the interface, the frequencies lying in the band $\omega = v_A(x)k_z$, together with a 'collective' oscillation extending throughout the interface and into the surrounding media. The latter has a frequency $\omega/k = v_k$ determined by the 'mean' Alfvén speed of the two sides v_k (Eqn. (5.168)). This oscillation transforms into the surface wave in the limit of a discontinuous interface (and is still referred to as a surface wave by Wentzel). However, unlike ideal normal modes, the frequencies are not purely real. The nonzero imaginary component implies a real growth rate, so that the disturbances grow or decay in time.

Lee & Roberts analysed the response of the interface to an impulsive disturbance, solving the governing equations as an initial value problem. They found that the initial perturbation excited the collective disturbance, which then decayed at a rate given by

$$\tau = \frac{8v_k}{\pi k_z^2 a(v_{A1}^2 - v_{A2}^2)}, \quad (5.176)$$

a being the thickness of the interface. As this decayed in amplitude, the Alfvén continuum grew in amplitude. It should be noted that mechanical energy is conserved. As the global disturbance dies away, there is a flow of mechanical energy into the interface, feeding the interface waves. The energy is not dissipated but is transferred from one form of motion to another. The coupling is provided by the singularity in the linear equations; if the linear equations are not singular, non-linear terms in the equations of motion must be retained in order to produce mode coupling. The process by which the energy of the global motion is absorbed by the local interface waves is known as *resonant absorption*.

The compressible case mimics this behaviour when k_y is large enough. The analysis is then restricted to shear Alfvén waves and Eqn. (5.174) reduces to

$$\frac{d}{dx}\left\{[\omega^2 - k_z^2 v_A^2(x)]\frac{\rho_0(x)}{k_y^2}\frac{d\xi_x}{dx}\right\} = [\omega^2 - k_z^2 v_A^2(x)]\rho_0(x)\,\xi_x. \quad (5.177)$$

This is almost identical to Eqn. (5.175), the quantity k_y^2 replacing the $k_i^2 = k_z^2$ assumed by Lee & Roberts. This simply modifies the absorption time scale to read (Ionson, 1978; Wentzel, 1979b)

$$\tau = \frac{8v_k}{\pi k_z k_y \, a(v_{A1}^2 - v_{A2}^2)}. \tag{5.178}$$

Wentzel (1979b) supplies the damping rate for the general case in which k_y is not assumed small.

Wentzel (1979b) also assessed the effects of compressibility, and concluded that it may be neglected whenever the magnetic pressure exceeds the gas pressure. Indeed, absorption at the compressive singularity is dominant only when the gas pressure is at least comparable with the magnetic pressure and $k_y \ll k_z$. This was confirmed by Davila (1987), whose compressible model possessed exactly the same features as the incompressible model of Lee & Roberts.

Moreover, a cylindrical interface shows essentially the same behaviour as a plane interface. Resonant absorption in a loop modelled by an axisymmetric straight cylinder with a force-free Lundquist field has been the subject of a comprehensive study by Grossman & Smith (1988). They employed the ideal equations, but allowed for compressibility, thus admitting not only the Alfvén continuum ($\omega = v_A(x)\,k_z$) but the slow-mode continuum ($\omega = v_T(x)\,k_z$). They concluded, however, that Alfvén absorption is stronger than slow-mode absorption under coronal conditions. In their loop model, the density was highest on the axis where the temperature was least. The factor governing the spatial distribution of energy absorption was the density profile, the absorption increasing away from the axis and peaking where the density gradients began to steepen sharply.

However, it is clear that the energy cannot accumulate in the interface region without some process of dissipation becoming operative. Dissipation can be studied only if we drop the simplifying assumption of ideal MHD behaviour. In the following section we look at such processes as a prelude to our discussion of coronal heating.

5.6.4 Non-ideal effects in waves

We have just discussed small-scale MHD waves in some detail, yet we saw in Section 5.3.1 that the large-scale coronal structures only just satisfied the conditions for fluid behaviour. The validity of the MHD approximation in the case of waves thus calls for some comment. We have seen that the collision frequency in the solar corona is about 1 s^{-1}. In such a gas, waves with higher frequencies could not possibly come to the microscopic equilibrium upon which the hydrodynamic equations are based. However, the solar corona is ionized and behaves as a plasma. The collective motion of charged particles in a disturbance produces an electrostatic field which organizes the motion (even in the absence of collisions) and allows the disturbance to propagate as a wave.

The MHD waves described in the previous section persist in a plasma like the solar corona until the spiral motion of individual particles along the magnetic field becomes important at the ion gyrofrequency. At this frequency, about 10^5 rad s^{-1} in the solar corona, the electrons and ions are no longer coupled closely together and their motions must be treated separately, although each may be treated as a fluid. The physics of plasmas in this regime is the subject of specialist monographs – that by Melrose (1980) is devoted to astrophysical plasmas – and the reader is recommended to consult them for further information. We shall illustrate the results with an illustration from Boyd & Sanderson (1969), which refers to conditions similar to those of the canonical solar corona. Fig. 5.13 traces the changes to the MHD modes in the (k, ω) plane for the case of waves propagating at an angle of 45° to the equilibrium field. Let us look at them in turn.

At frequencies higher than the collision frequency, the slow mode is a longitudinal wave with the electrons and ions coupled by the electrostatic field (not collisions). As the frequency approaches $\Omega_i \cos \theta$, where Ω_i is the ion gyrofrequency and θ is the angle of inclination of the wave vector to the magnetic field, the ions move into circular orbits. However, the longitudinal electron motion forces these orbits to be perpendicular to $\mathbf{k}$ not $\mathbf{B}$. This results in a new wave known as the second ion cyclotron wave.

The fluctuating electric field in an Alfvén wave lies in the plane perpendicular to $\mathbf{B}$. It is elliptically polarized generally, but becomes circularly polarized at the ion gyrofrequency. Since the sense of the rotation is the same as that of the gyrorotations of the ions, the wave field drives a resonance in the ion motion perpendicular to $\mathbf{B}$. However, the build up of plasma pressure adds a longitudinal component to the fluctuating electric field and this allows the wave to continue to propagate, despite the resonance. The first ion cyclotron wave results. Above the ion gyro-frequency, the plasma pressure becomes more important than the magnetic tension, so that the wave takes on the characteristics of a pure acoustic wave. It is now longitudinal, and obeys the dispersion relation $\omega^2 - k^2 v_a^2 = 0$, but it owes its existence to the longitudinal electric field coupling the particles, not collisions. As the frequency increases and wavelength decreases, a point is reached at which the motions of the electrons and ions become decoupled and the temperature of the electrons

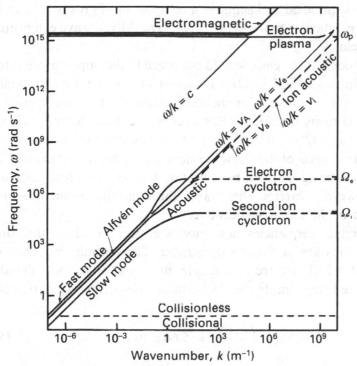

Fig. 5.13. Dispersion curves for waves propagating at 45° to the equilibrium magnetic field in a low-β hydrogen plasma. The parameters are those for the canonical coronal conditions given in Table 5.1. The dashed segments denote regions where damping prevents the propagation of the waves. (After Boyd & Sanderson, 1969.)

can deviate from the temperature of the ions. When $T_e > T_i$, the electron pressure gradient drives the ion motion, producing an acoustic wave that propagates at the 'sound' speed appropriate to the ions alone $v_i = \sqrt{(\gamma k T_i / m_i)}$. The dispersion relation is then

$$\omega^2 - k^2 v_i^2 = 0. \tag{5.179}$$

This ion acoustic wave persists until the frequency reaches $\Omega_e \cos \theta$, at which point the electrons would like to gyrorotate perpendicular to the field but the motion is limited, like that of the ions in the second ion cyclotron wave. The oblique Alfvén wave thus terminates in an electron cyclotron wave.

When the wave vector of the Alfvén wave is almost at right angles to the field, another type of breakdown can occur. According to the MHD equations, the group velocity is always in the direction of the field as $\theta \to \pi/2$. However, when the wavelength $2\pi/k$ becomes comparable with the ion gyroradius (about 1 m for our canonical coronal conditions), the

Alfvén wave is modified and acquires a component of group velocity in the direction perpendicular to the field. This *kinetic* Alfvén wave can propagate across the field.

The fast mode has the electric field polarized in the opposite sense to that of the Alfvén mode and is little modified at the ion gyrofrequency. It develops into the whistler or helicon mode as the ion contribution decreases, due to its inertia. As this occurs, the wave becomes circularly polarized. At $\omega = \Omega_e \cos \theta$, the electric field polarization now matches the sense of electron gyrorotation and the electron cyclotron wave is met again, the resonance being limited by electron inertia. The longitudinal ion motion, however, carries the wave through the resonance into a continuation of the ion acoustic wave.

At still higher frequencies new modes appear in which the ion and electron motions are essentially decoupled. The plasma frequency is the frequency at which electrons oscillate in a stationary ion cloud, the restoring force being simply the electrostatic interaction. This frequency is given by

$$\omega_p = \left(\frac{n_e e^2}{m_e \varepsilon_0}\right)^{\frac{1}{2}} = 5.64 \times 10^7 n_e^{\frac{1}{2}}, \tag{5.180}$$

where ε_0 is the permittivity of free space. It has the value 2×10^{15} rad s^{-1} in the canonical solar corona. Above this frequency, the natural response time of the electrons is insufficient to cancel out electromagnetic disturbances so that transverse electromagnetic waves (radio, light, X-ray, etc.) with dispersion relation

$$\omega^2 - k^2 c^2 = 0, \tag{5.181}$$

where c is the speed of light, can be supported as well as longitudinal electron acoustic waves with dispersion relation

$$\omega^2 - k^2 v_e^2 = 0, \tag{5.182}$$

where $v_e = \sqrt{(\gamma k T_e / m_e)}$ is the 'sound' speed of the electrons. These electron acoustic waves cannot be treated properly in a fluid approximation. The electric field oscillations couple selectively to bunches of electrons whose thermal velocities are suitably resonant. When the electron velocities have a thermal (Maxwellian) distribution, energy is transferred from the wave motion to kinetic energy of these electrons. This process is known as Landau damping; it destroys the thermal velocity distribution of the particles, which is essential for the fluid description to be appropriate. The full plasma kinetic equations must then be solved. These have many more degrees of freedom and allow a complex array of microphysics. Micro-instabilities are thought to be important in flare processes, which give evidence of charged particle beams accelerated to relativistic (i.e. non-thermal) speeds. To pursue this topic would lead us too far afield and the interested reader is referred to texts on the physics of flares (cf. Section 5.8). We here note only that electron acoustic, electron cyclotron as well as ion acoustic and second ion cyclotron waves are all heavily damped and cannot propagate. The end result of the dissipation is a patch of *turbulence*, and the eventual transfer of energy into thermal motions.

Of course, all waves are damped to some extent. The analyses described in the previous section were based on the ideal MHD equations, so that the motion was necessarily adiabatic. As we saw in Section 5.5.1, radiative effects can lead to overstability (Joarder, Gokhale & Venkatakrishnan, 1987). However, the growth rate will be small for frequencies shorter than $2\pi/\tau_r$, where $\tau_r \sim 3000$ s is the radiative time scale in the corona (Section 5.5.1). On the other hand, the inclusion of dissipation due to ohmic losses, viscous losses and plasma loss processes (e.g. Landau damping) causes the amplitude of the motion to decay with time, unless energy is continually fed into the wave. This energy loss can be increased when non-linear terms are reintroduced into the equations of motion.

The loss mechanisms in Alfvén waves are particularly simple, because the waves involve only magnetic forces. The introduction of a finite resistivity, with concomitant ohmic losses, leads to an exponential decay of the wave energy as it propagates away from the source. The characteristic energy loss time τ_0 can be estimated from the energy equation (5.34). The ohmic loss rate on the right-hand side is roughly

$$\frac{j^2}{\sigma} = \frac{B^2}{\mu_0^2 \sigma^2 d^2},\tag{5.183}$$

where d is the characteristic scale. In the case of a wave, B is the fluctuating field and d is the wavelength. Estimating the latter by v_A/ω the loss rate becomes

$$\left(\frac{B_1^2}{\mu_0}\right)\left(\frac{\omega^2}{\mu_0 \sigma v_A^2}\right).$$

The first factor is the energy density of the wave so that the loss time is just

$$\tau_0 = \frac{v_A^2 \mu_0 \sigma}{\omega^2} = \frac{v_A^2}{\omega^2 \eta},\tag{5.184}$$

where η is the magnetic diffusivity. Since the wave is losing energy, the energy density will decay as it propagates. If the decay factor is written as

$\exp(-z/\ell)$ where z is in the direction of energy propagation, the damping length ℓ is the distance travelled during the ohmic loss time, i.e.,

$$\ell \sim v_A \tau_o = \frac{v_A{}^3}{\omega^2 \eta}. \tag{5.185}$$

High-frequency waves are therefore dissipated more rapidly than low frequency waves. Since it is the group velocity of the waves which determines the energy damping length, the kinetic Alfvén waves – short-period Alfvén waves with wave vectors almost perpendicular to the field – have extremely short damping lengths on account of their very low group velocities.

Moreover, there are no non-linear effects in a single wave since the disturbance is always transverse to the motion. A finite amplitude Alfvén wave can propagate without distortion. However, when two Alfvén waves are superposed, the non-linear terms introduce a compression. Under suitable circumstances the compression can propagate away as a slow-mode (acoustic) wave (Chiu & Wentzel, 1972) – cf. Section 5.7.2. The generation of the compressional wave extracts energy from the Alfvén waves and can be considered a loss mechanism.

Magneto-acoustic waves in the collisional regime ($\omega < 0.6$ rad s^{-1} in the corona) experience both ohmic loss and viscous loss, the former being dominant in fast modes with $v_f \sim v_A$ and the latter in slow modes with $v_s \sim v_a$ (Osterbrock, 1961). The ohmic loss time and the damping length of fast modes are the same as those of Alfvén waves, whilst the viscous losses in slow modes may be estimated in a similar manner from (5.34). The loss rate is roughly $\mu v^2/d^2$, which in the case of waves may be written as

$$(\rho_0 v_1{}^2)\left(\frac{\omega^2 \mu}{\rho_0 v_a{}^2}\right).$$

The first factor is the energy density of the wave, so that viscous loss time is

$$\tau_v = \frac{v_a{}^2 \rho_0}{\omega^2 \mu}, \tag{5.186}$$

and the damping length is

$$\ell \sim v_a \tau_v \sim \frac{v_a{}^3 \rho_0}{\omega^2 \mu}. \tag{5.187}$$

When the waves become collisionless, various plasma loss mechanisms – such as Landau damping – become important; these are detailed by

Hung & Barnes (1973) and the papers referenced therein. Zweibel (1980) estimated the damping lengths of fast-mode waves to lie between 5×10^7 and 2×10^8 m for frequences $1 < \omega < 0.05$ rad s^{-1} in the case of a coronal plasma with $B = 10^{-4}$ T and $n_e = 10^{14}$ m^{-3}. For our canonical conditions, these lengths would be about 30 times greater. An approximate form is given by Habbal, Leer & Holzer (1979).

The same processes affect the Alfvénic interface waves that propagate along magnetic structures. Indeed, both viscosity and resistivity can result in much greater dissipation because of the large gradients that are built up across the interface. Einaudi & Mok (1987) and Hollweg (1987) both add viscous or resistive terms to the incompressible problem. The velocity amplitude rises sharply in the vicinity of the resonant surface, where the phase speed matches the local Alfvén speed. In the immediate vicinity of the resonant surface, the gradients become large enough for dissipation to be significant. Einaudi & Mok then match a resistive solution in a narrow resonance layer to ideal solutions throughout the rest of the interface and the surrounding medium. The singularity of the ideal equations is avoided, and the boundary conditions select normal modes which are now *damped* body waves in the interface. They found the damping rate of the modes to be independent of the dissipation coefficient, as the solutions of the singular ideal equations suggested.

Wentzel (1979b) also pointed out that the velocity amplitudes in the vicinity of the resonant layer can easily rise above the Alfvén speed when the amplitudes in the surrounding medium are of the order of 10 km s^{-1}. The linearized wave approximation then breaks down, and various non-linear effects should be taken into account.

Non-linearities are also important in the dissipation of (compressible) magneto-acoustic waves. Since the waves are longitudinal, the disturbance motion combines with the wave motion; the wave crests move faster than the troughs and the wave form steepens as it propagates away from its source, eventually forming a shock wave. The larger the velocity amplitude, the sooner the shock forms. In the absence of dissipation, the amplitude of the slow mode must increase with height. When its group velocity is close to the sound speed, the conservation of the energy flux in the wave requires

$$\tfrac{1}{2}\rho_0 v_1^{\,2} v_a$$

to be constant with height. Since v_a is approximately constant with height whilst the density drops exponentially, the velocity amplitude increases with *twice* the density scale height,

$$v_1 \propto \exp\left(z/2\varLambda\right). \tag{5.188}$$

On the other hand, the fast mode travelling at about the Alfvén speed, which increases with height, requires

$$\tfrac{1}{2}\rho_0 v_1{}^2 v_A = \tfrac{1}{2}\rho_0 v_1{}^2 \frac{B}{\sqrt{(\mu_0 \rho_0)}}$$

to be constant. In this case, the velocity amplitude will increase with *four* times the density scale height. Since the density scale height is comparable to the length of most hot loops, this amplitude growth factor is not large. It can be significant in cool loops, though. Any growth in amplitude of the wave motion accelerates the formation of a shock as the wave crest overtakes the wave trough.

The properties of hydromagnetic shock waves are described in Bray & Loughhead (1974) and Priest (1982). At the shock front there is a sudden change in the pressure, density and temperature, as well as in the magnetic field direction. The change in the field direction is caused by a jump in the tangential field component brought about by a current flowing in the plane of the shock front. The front represents a propagating current sheet. Corresponding to the fast and slow mode waves are two types of shock. In both, the material is compressed on crossing the shock, but the magnetic field increases in strength and is refracted towards the plane of the shock front in a fast shock, and the field decreases in strength and is refracted away from the front in a slow shock. The length scales and associated time scales of the front are both very small and ensure that the dissipation across the shock is both significant and essentially collisionless. Osterbrock (1961) extended the Brinkley–Kirkwood treatment of the dissipation of hydrodynamic shocks (described in Bray & Loughhead, 1974) to estimate the loss rates from hydromagnetic shocks. He obtained a complicated expression for the energy damping length which depends on the effective time for the shock front to pass through a given point t_0 and the relative compression which occurs with the passage of the shock $\bar{\eta} = (\rho_1 - \rho_0)/\rho_0$. This latter quantity is known as the shock strength. When $v_f \simeq v_A \gg v_a$ the energy flux averaged over direction is approximately

$$F = \left(\frac{\rho_0 v_f \bar{\eta}^2}{8\sqrt{3}}\right)\left(\frac{3v_A{}^2}{2}\right), \tag{5.189}$$

and the damping length is

$$\ell = \frac{4v_f t_0}{\bar{\eta}}. \tag{5.190}$$

For a wave of frequency ω developing into a train of shocks $t_0 \simeq 2\pi/8\omega$.

Alfvén waves are not longitudinal and cannot form a shock. However, there are finite-amplitude Alfvén waves in the form of a thin propagating current sheet which simply reverses the sign of the tangential component of the field and velocity across the wave front. There is of course no change in pressure or density, or in the normal component of the field. The wave front simply rotates the field vector. This is known as an intermediate wave – it propagates at the Alfvén speed – or as a *rotational discontinuity*. Since the front is not sustained by non-linear effects, ohmic dissipation will broaden the front and destroy the discontinuity.

This completes our catalogue of dissipation processes occurring in both large-scale, low-frequency MHD motions and small-scale, high-frequency plasma motions. With the exception of the very last (a rotational discontinuity), all can be maintained by mechanical disturbances driven from below the corona. The net result is therefore a steady release of thermal energy into the coronal plasma. Not surprisingly, many attempts have been made to identify the source of coronal heating with one or more of these mechanisms. We are now in a position to address this vexing problem.

5.7 The heating of coronal loops
5.7.1 *Introduction*

In the previous sections we have explored various aspects of the physics of coronal loops, using a series of simplified models of their structure. Actual loops in the solar corona are, however, described only partially by such models, and their appearance and behaviour must be understood in terms of an *interplay* between the physical processes that have emerged so far. When this degree of complexity is reached, firm conclusions are hard to draw from models. It is sometimes difficult to even identify the major physical processes contributing to an observed phenomenon. Nowhere is this dilemma more evident than in the subject that has been postponed until now – the means by which the corona is heated. As Cram (1987) remarks, 'there exist at present a plethora of different and apparently unconnected theoretical concepts, and it is a difficult task to identify those which are likely to remain as key ingredients in the future'. Nevertheless, our discussion would not be complete without an attempt to picture the means by which coronal loops may be heated.

Coronal heating in general has been extensively reviewed in recent years. Excellent discussions are provided by Hollweg (1981b), Kuperus, Ionson & Spicer (1981), Wentzel (1981) and Priest (1982). Many more details may be found in the extensive survey of flare theory by Spicer & Brown (1981). The following précis of the mechanisms is based largely on Wentzel's summary, and is restricted to heating in the specific context of coronal *loops*.

We have seen that coronal loop models must all assume a source of thermal energy to balance the radiative and conductive losses. Most authors look for the immediate source of this energy outside the corona, though van Tend (1980) has suggested that coronal loops surrounding prominences may be heated by waves generated by turbulent motions in the prominence itself. We shall consider here only more ubiquitous heating mechanisms – those which require the energy to be transported to the corona by some means and dissipated there.

It is generally accepted that the energy comes from the subphotospheric regions. It is also generally accepted that the magnetic field plays an important role in the transport and dissipation of this energy, although direct evidence in favour of the link is scanty. Golub *et al.* (1980) have shown that there is a rough relationship between the thermal energy content of an active region, as estimated from the X-ray emission, and the total photospheric magnetic flux of the region. They also showed that such a relation can arise if the active region consists of loops whose energy input is supplied by a twisted (non-potential) component of the magnetic field. The twists were assumed to be generated below the corona at a constant rate, $\dot{B}_\varphi = B_z v_\varphi / L$. This example makes it clear that an atmospheric heating theory must ultimately treat the problem globally: driving (subphotosphere), propagation (photosphere) and dissipation (chromosphere and corona) are eventually to be explained as different aspects of a single phenomenon. This global perspective is pursued in the final chapter. Here, we shall focus attention on the aspects of energy transport and release in individual coronal loops.

Let us consider first what happens to disturbances engendered by footpoint motions having the hydrodynamical time scales of the photosphere. Very low frequency motions with vertical wavelengths $2\pi/k_z$ greater than the coronal scale height $\Lambda \sim 10^8$ m cannot be treated as waves in the context of coronal loops. However, the dissipation lengths of these disturbances are so long that they are unlikely to contribute significantly to the heating of the inner corona, though they might be important in the outer corona and solar wind (Hollweg, 1981a). We shall not consider them further.

High-frequency disturbances, with periods less than about 1 s, can be treated locally as plane waves. Their propagation paths may then be found by the ray-tracing procedure standard in optics. The path is refracted towards regions of low phase speed and away from regions of high phase speed. Since the density decreases and the temperature increases steadily with height above the temperature minimum in the solar atmosphere, both the sound speed and the Alfvén speed increase outwards. There is thus a general tendency for outward propagating waves to be progressively refracted and eventually reflected back downwards. During the course of their propagation through the chromosphere and transition region, the waves may form shocks. Energy is lost in the interactions between shocks and the waves suffer *degradation*, a process which involves a shortening of their wavelengths and periods (Wentzel, 1977).

Disturbances with intermediate frequencies have vertical wavelengths comparable to both the coronal scale height and the typical loop length. The most pronounced *photospheric* dynamical disturbances with periods of the order of 300 s (i.e. 5 min) fall into this range. These waves cannot be treated by ray-tracing, only by the full solution of the governing equations. Schwartz & Leroy (1982) calculated the transmission of these disturbances through a uniform exponential atmosphere permeated by a uniform vertical field. For horizontal wavenumbers in the range 10^{-6}–10^{-9} m^{-1} the fast mode is evanescent in the corona and the chromospheric wave is reflected at the transition region. Only the slow mode and Alfvén mode are able to enter the corona.

However, the model employed by Schwartz & Leroy is not a good representation of the magnetic field structure in the solar photosphere and chromosphere. The coronal loop footpoints are the photospheric flux elements which have a diameter of some 100 km and a magnetic field strength of 0.1 T (cf. Section 6.2.2). If the field strength drops by a factor of 100 as we move into the corona, conservation of magnetic flux requires the diameter of the flux tube to increase tenfold. This expansion occurs over a height of about 1000 km, whereupon the flux merges with that emanating from other footpoints in the neighbourhood. This height is thought to mark the base of the corona. Below it, the magnetic flux breaks up into strands separated by regions of virtually field-free gas (cf. Section 6.2.3). Waves may enter the base of the corona from the field-free regions, whence they impinge on the almost horizontal interface between the non-magnetic ('quiet') chromosphere and the magnetic corona. The geometry of this region is extremely poorly known, so that no meaningful estimates can be made of how much energy is transmitted into coronal wave motions. Waves may also be propagated up the flux tubes into the corona. No analysis along the lines of Schwartz & Leroy has been made for a realistic field geometry. Nevertheless, since these authors found that the fast mode was reflected back from the corona due to the *coronal conditions*, we may surmize that their conclusion – only slow and Alfvén modes reach the corona – would survive changes of geometry.

If the slow mode were responsible for carrying the energy flux required to heat the corona, the amplitude of the motion would have to be large, on account of the low group velocity. So large, in fact, that it would both exceed the observed nonthermal velocity amplitude in the transition region and lead to intense dissipation before reaching the corona. Although Ulmschneider & Bohn (1981) are staunch proponents of coronal heating by slow-mode (acoustic) shocks, most authors favour other modes. The case for Alfvén waves will be described in more detail below.

These arguments do not preclude modes other than the Alfvén mode from being present in the corona, simply because they fail to propagate there from below. The magnetic structure of the Sun is not uniform, so that waves are unable to maintain their identity as the magnetic field flares out and the density falls off steeply. Each initial mode will split into a mixture of modes appropriate to the new local conditions. This phenomenon is known as *mode conversion* or *mode coupling*. Wentzel (1974) shows how a transverse Alfvén wave becomes partially longitudinal as it encounters field curvature. The longitudinal component is described locally as a fast-mode wave. Because the fast mode can dissipate more effectively than the Alfvén wave (cf. Section 5.6.4), mode conversion promotes the transfer of energy from the wave motion to the thermal field, i.e. it produces heating. According to Wentzel, the fractional energy loss from the wave is approximately proportional to the angle through which the field has turned. The damping length is therefore roughly equal to the radius of curvature of the field,

$$\ell \simeq R. \tag{5.191}$$

The process is enhanced if there are density fluctuations in the medium, the mode conversion being greatest when the scale of the inhomogeneity matches the wavelength of the disturbance. Again, a quantitative assessment of the energy reaching the base of the corona demands a model of the structure of chromosphere and transition region which is beyond our present capabilities.

Mode coupling of a different nature was noted by Hollweg (1979), who investigated the propagation of two-dimensional disturbances along a uniform vertical field when the plasma density was stratified with scale height Λ. The Alfvén speed then increases exponentially with height. In the limit where the magnetic pressure outweighs the gas pressure, the governing equations are

$$\frac{d^2v_x}{dz^2} + \left(\frac{\omega^2}{v_A^2} - k_x^2\right) = 0 \tag{5.192}$$

$$\frac{d^2v_z}{dz^2} - \frac{1}{\Lambda}\frac{dv_z}{dz} + \frac{\omega^2}{v_a^2}v_z = -ik_x\left(\frac{dv_x}{dz} - \frac{\gamma-1}{\gamma\Lambda}v_x\right), \tag{5.193}$$

with the velocity having the form $\mathbf{v} = (v_x, 0, v_z)\exp i(k_x x - \omega t)$. The first equation describes transverse compressional Alfvén waves. These then provide a source term in the second equation, which describes driven gravity-modified acoustic waves in the longitudinal direction. Eliminating v_x between these two equations yields a fourth-order for v_z which is singular when

$$\frac{\omega^2}{v_A{}^2} - k_x{}^2 + \frac{(\gamma - 1)^2}{\gamma^2 A^2} = 0. \tag{5.194}$$

Since v_A increases steadily with height a point will always be reached when this condition is fulfilled and the acoustic waves are driven at a resonant frequency. The amplitude of their motion grows sharply and this presumably leads to dissipation, but this aspect remains unexplored.

All these progressive magnetic disturbances have fluctuating currents associated with them, so that they are often referred to as *AC heating mechanisms* when their time scales are less than the time taken for an Alfvén wave to travel along the loop structure. There are also field changes of much longer time scale caused by the development of the active region itself. These changes are not disturbances of the equilibrium structure as discussed throughout this chapter, but represent an irreversible evolution of the loop system. On the time scale of individual loops, the currents associated with the global structure and its changes appear to be steady. These steady currents are also subject to dissipation, giving rise to a whole class of alternative heating mechanisms. These are naturally referred to as *DC mechanisms*. We shall review each class separately, but the fact that all rely ultimately on the dissipation of electric currents should be borne in mind.

The ultimate test of each theory of coronal heating is that it should meet the energy requirements of quiescent coronal loops, both qualitatively and quantitatively. Qualitative constraints are provided by observations such as those of Golub *et al.* (1980), which show that the thermal energy input in an active region is related to the total magnetic flux in the atmosphere below it. A complete theory should explain the thermal structure throughout an active region, as well as providing the reason for its differentiation into loops of varying visibility. Little attention has been given to these questions. Most authors are content to demonstrate that their mechanisms can meet average energy requirements. This requirement is usually translated into the heat flux that must be fed into the base of the loop in order to account for the radiative losses (any heat transferred by conduction into the chromosphere below will only *increase* these figures). Several authors have provided observational estimates of this quantity.

They range from 7×10^2 W m^{-2} (Svestka *et al.*, 1977) to 5×10^3 W m^{-2} (Neupert, Nakagawa & Rust, 1975). Another quantitative constraint is the size of the line-of-sight velocity dispersion as measured by line widths in the corona and transition region. These put limits on the root-mean-square value of v_1 of some 30 km s^{-1} (cf. Tables 2.9 and 3.8). The constraints set by *theory* itself will be considered at the end of the discussion.

5.7.2   *AC heating mechanisms*

The theory of coronal heating by means of wave dissipation was developed in the classic paper by Osterbrock (1961). This work attempted to account for the structure of the chromosphere, corona and active regions in terms of a common theory of mechanical heating. Osterbrock found short-period fast-mode waves to be the most important heat source in a chromospheric model with a *uniform weak magnetic field*. The fast-mode waves form shocks in the chromosphere where dissipation and refraction combine to prevent any significant transmission to the corona. Osterbrock estimated the production of slow-mode shocks from the fast-mode by means of shock interactions and mode coupling. These could propagate along the field lines into the corona, where they would dissipate like essentially hydrodynamic shocks with a damping length

$$\ell = \frac{4v_a t_0}{\eta},$$ (5.195)

(cf. Eqn. 5.190). However, his estimates of the energy available from this source were too small to account for observed coronal losses. He therefore suggested that Alfvén waves, generated in a similar manner, might supply the extra energy, although he recognized the difficulty in transmitting a large Alfvén flux across the transition region. The limitations of assuming a uniform magnetic field in the lower atmosphere have been discussed above, but no quantitative corrections to Osterbrock's picture have yet been made on the basis of the flux-tube concept.

This problem was avoided by Habbal, Leer & Holzer (1979). These authors discussed the fast-mode heating mechanism of coronal loops, ignoring how such waves might get to the corona. They considered a short-period (2 s) wave field and calculated the ray paths through a dipole field geometry containing a high-density (low Alfvén speed) loop. Because of the almost isotropic propagation properties of the fast-mode, waves could approach the loop from all directions. However, they were refracted on entering the loop and ducted along it. The increased path lengths through the loop produced increased heat deposition, the energy being drawn not from a mechanical flux through the footpoints of the loop, but from waves originating over a wide area about the base. According to this picture, differential heating is caused by pre-existing density differences between loops. In contrast, the analysis of condensation modes in Section 5.5.3 suggested that density variations from loop to loop could be the direct result of differential heat input. Most theories of coronal heating adopt the latter viewpoint; enhanced heating in a loop is due to the injection of a greater energy flux in the form of waves or currents into the base of that loop, the energy then being transmitted *along* the field.

Zweibel (1980) also examined a fast-mode heating model. She adopted periods in the range 10–100 s, which yielded damping lengths of 10^8–10^{10} m. With suitable values for the wave amplitudes, it was then found possible to heat the lower corona with short-period waves ($P < 10$ s) and the outer corona with waves of longer period ($P \sim 100$ s). Measurements of root-mean-square field fluctuations in the corona do not, as yet, have the accuracy to provide any constraints on this mechanism. However, the model fails the test of consistency. Zweibel found it to be unstable to filamentation along the field (cf. Section 5.6.2). Filamentation would produce sharp changes in the Alfvén speed from one set of field lines to another. In this case, the plane wave model is no longer appropriate and we should instead look at surface waves, as discussed below.

A stronger case can be made for the penetration of Alfvén waves to the corona. Hollweg (1981a) made a detailed study of the propagation of torsional Alfvén waves along a loop from one photospheric footpoint to the other. For simplicity, the loop was made symmetric and was straightened in order to avoid gravitational effects. A sinusoidal oscillation with a velocity amplitude of 1 km s^{-1} was imposed at one end, and the other end was made to transmit (absorb) all disturbances. He found that long-period waves were strongly reflected by the increasing Alfvén speed before they reached the corona, the transmitted energy flux being too small to account for coronal heating. Shorter-period waves, though, could be transmitted when a standing wave formed in the coronal portion of the loop, i.e. when $\omega = v_A(2\pi m/4L)$ for some integral value of m. The coupling between the photospheric oscillation and the coronal standing wave then allowed the wave to tunnel through the barrier posed by the increase in Alfvén speed at the base of the corona. By this means, the coronal oscillation could acquire sufficient energy to account for observed coronal heating rates (cf. Hollweg & Sterling, 1984). These conclusions were supported by an independent study by Žugžda & Locāns (1982), who solved the one-dimensional Alfvén equation with variation of Alfvén speed. They concluded that the coronal transmission coefficient was about 10^{-4} for waves with periods between 10 and 1000 s, but could rise to about 0.5 when interference took place.

In a subsequent paper, Hollweg (1984) discussed the dissipation of these waves in the corona. So long as the wave dissipation is more important than leakage of waves from the ends of the loop, all the energy will be dissipated eventually, even though the damping length exceeds the length of the loop. However, the damping rate does affect the resonance between the photospheric disturbance and the coronal oscillation. It spreads out the resonance in frequency and allows more energy to be transmitted to the corona. Hollweg estimated the damping length to be $\ell \simeq 5/k$ by requiring the velocity amplitude to be consistent with observations of the nonthermal velocity, namely 30 km s^{-1}. This requires a more effective loss mechanism than simple ohmic heating. Hollweg showed that such a figure could be obtained by postulating turbulent losses from an energy cascade initiated by the Kelvin–Helmholtz instability. This source of turbulence was suggested by Heyvaerts & Priest (1983), and is discussed further below.

Another effective means of dissipating Alfvén waves has been developed by Wentzel. He appeals to the interaction of Alfvén waves propagating up from the different footpoints. Around the top of the loop, a large amplitude wave of frequency ω_0 and wavenumber k_0 travelling in one direction will interact non-linearly with a weaker wave of suitable frequency, $\omega_1 = \omega_0(v_A - v_a)/(v_A + v_a)$, and wavenumber k_1 travelling in the opposite direction. The small compression that then arises can propagate as a slow-mode (acoustic) wave with frequency $\omega_2 = \omega_0 - \omega_1$ and wavenumber $k_2 = k_0 - k_1$, travelling in the original direction. The two weak waves grow at the expense of the original Alfvén wave (Wentzel, 1974). The viscous dissipation of the slow-mode wave provides the immediate source of coronal heating, the damping length being given by

$$\ell \simeq \frac{v_A{}^3}{2\pi\omega v_1{}^2},$$ (5.196)

where v_1 is the original Alfvén wave amplitude. Wentzel (1976) has made a detailed study of the propagation, trapping and dissipation of Alfvén waves in a coronal loop. By requiring the damping length to be comparable with the loop length, he showed that that Alfvén wave flux $\frac{1}{2}\rho_0 v_1{}^2 v_A$ at the base of the corona may be estimated for any given dissipation rate. For our canonical coronal conditions, the observed energy losses can be met only by short-period (10 s) Alfvén waves of large amplitude. Wentzel (1976, 1977) quotes amplitudes in the range 30–50 km s^{-1}, which are uncomfortably close to the upper limits on the velocity dispersion in the corona and transition region.

Each of the models discussed so far has invoked wave modes whose wavelengths are much smaller than the scales over which the coronal properties change. Another class of models focuses on regions where the Alfvén speed may change sharply – for example, from one flux tube to another. Strong absorption of the energy in the disturbance occurs around the surfaces where the phase speed of the disturbance matches the Alfvén speed. This is the phenomenon of resonant absorption, described in Section 5.6.3. The dissipation mechanism depends on whether the plasma behaves in a collisional or collisionless manner, i.e. on whether $\omega\tau_i$ is less than or greater than unity. In the former case, normal viscosity and resistivity become important in the resonant layer and lead to damped normal modes (Einaudi & Mok, 1987; Hollweg, 1987). In the latter case, the large-scale disturbance is damped by mode-conversion to Alfvén waves in the interface (Ionson, 1978; Wentzel, 1979b). These are kinetic Alfvén waves (Section 5.6.4) and have a very small component of group velocity perpendicular to the field. They can travel away from the resonant layer, but so slowly that they are damped within a few wavelengths. Ionson argues that the thickness of the resonant layer is about the wavelength of the kinetic Alfvén wave, of the order of a kilometre in magnitude. Although the dissipation is concentrated in a very thin layer, different components of the large-scale disturbance will dissipate around different resonant layers, so that there will be more or less uniform heating throughout the interface.

The time scale for the loss of energy is one half that for the damping of the velocity amplitude. From Eqn. (5.178) we find this to be

$$\tau = \frac{4v_k}{\pi k_z k_y a(v_{A1}^2 - v_{A2}^2)},\tag{5.197}$$

and the corresponding damping length,

$$\ell_{ra} = v\tau = \frac{4v_k^2}{\pi k_z k_y a(v_{A1}^2 - v_{A2}^2)} = \frac{2(v_{A1}^2 + v_{A2}^2)}{\pi k_z k_y a(v_{A1}^2 - v_{A2}^2)}.\tag{5.198}$$

We can account approximately for cylindrical geometry by estimating the transverse wavenumber as $k_y = 1/b$, where b is the radius of the loop. Then

$$\ell_{ra} = \frac{2b(v_{A1}^2 + v_{A2}^2)}{\pi k_z a(v_{A1}^2 - v_{A2}^2)}.\tag{5.199}$$

Ionson (1978) concluded that disturbances with a velocity amplitude of about 6 km s^{-1} could provide a heating rate comparable to observed radiative loss rates. Einaudi & Mok (1987) reached similar conclusions for a model of resonant heating with viscous dissipation. An energy flux at the base of the corona of 10^4 W m^{-2} required a velocity amplitude of about 30 km s^{-1}.

Another process leading to the dissipation of shear Alfvén waves is *phase-mixing*. Phase-mixing is the production of steep transverse gradients *between* different modes of oscillation, instead of within a single mode. For example, shear waves propagating in the plane of an interface have a different phase speed on each magnetic surface and therefore become rapidly out of phase. Heyvaerts & Priest (1983) demonstrated the effect with an example in which the Alfvén waves had a component only in the *y*-direction. Then the equation of motion in the ideal MHD limit is

$$\frac{\partial^2 v_y}{\partial t^2} = v_A^2(x) \frac{\partial^2 v_y}{\partial z^2}, \tag{5.200}$$

and the dispersion relation simply

$$\omega^2 = v_A^2(x) k_z^2. \tag{5.201}$$

Each magnetic surface has its own frequency and the magnetic structure behaves like a set of independent oscillators. As the oscillations become progressively out of phase, the gradients of the fluctuations in the *x*-direction increase until the ideal MHD approximation is rendered invalid.

Heyvaerts & Priest (1983) estimated the effects of viscosity and resistivity by adopting a constant, isotropic resistivity and viscosity. The modified equation of motion then reads, to first order in η and μ,

$$\frac{\partial^2 v_y}{\partial t^2} = v_A^2(x) \frac{\partial^2 v_y}{\partial z^2} + \left(\eta + \frac{\mu}{\rho_0}\right)\left(\frac{\partial^2}{\partial x^2} + \frac{\partial^2}{\partial y^2}\right)\frac{\partial v_y}{\partial t}. \tag{5.202}$$

The resistivity and viscosity provide a weak coupling between the waves on different magnetic surfaces and allow dissipation of the wave energy in the set of oscillators.

In their study of loop heating by phase-mixing, Heyvaerts & Priest assumed that Alfvén waves on magnetic surfaces are present in the corona and that they are reflected at the footpoints, establishing a standing wave pattern. The damping time (more relevant than a damping length to a standing wave) due to phase-mixing was then found to be

$$\tau_{\text{pm}} = \frac{1}{\omega}\left(\frac{6\omega a^2}{\eta + \mu/\rho_0}\right)^{\frac{1}{3}}, \tag{5.203}$$

where a is the thickness of the region over which the Alfvén speed varies.

This result is based on rather restrictive assumptions, but they are relaxed in Nocera, Leroy & Priest (1984). The mechanism will again supply any amount of heating, granted suitably large amplitudes.

Abdelatif (1987) also appeals to phase-mixing. In this model, surface waves on flux tubes in the photosphere transform into interface Alfvén waves when the flux tube fields merge in the corona. These were assumed to be dissipated by phase-mixing. However, the Heyvaerts & Priest analysis applies to two-dimensional disturbances lying in the plane of each magnetic surface. The incompressible interface modes are two-dimensional disturbances in the plane perpendicular to the magnetic surfaces. The steep gradients which occasion the dissipation arise for different reasons – between different modes in the former case, but within each mode in the latter case. The dissipation of interface modes has been discussed above.

Whilst phase-mixing is an attractive dissipation mechanism, the original authors cast some doubt on its effectiveness. When they assessed the stability of their standing wave pattern, Heyvaerts & Priest (1983) found that a Kelvin–Helmholtz instability can develop where the velocities are greatest (the magnetic nodes), and a tearing-mode instability where the velocities vanish (the magnetic antinodes). Browning & Priest (1984) considered this possibility in more detail and concluded that the instabilities would disrupt the shear wave *before* significant phase-mixing could develop.

One result of this disruption, suggested by Einaudi & Mok (1987), could be the generation of a field of small-scale, short-period Alfvénic waves from the large-scale, long-period disturbances. We noted in the earlier discussion of body waves that short-period waves can be effective heating mechanisms, but are difficult to transmit from the lower atmosphere. *In situ* production would resolve this problem.

Alternatively, the instabilities could lead to a turbulent cascade down to scales in which dissipation can balance the kinetic energy being fed into the instability, as suggested by Browning & Priest (1984). The onset of turbulence can enhance the ohmic and viscous losses. This enhancement is a vital part of the DC heating theories and leads us to a consideration of the competing mechanisms.

5.7.3 *DC heating mechanisms*

We have seen in Section 5.2 that the coronal magnetic field follows predominantly a potential field structure. In such a structure there are no currents and hence no ohmic losses. The magnetic energy in the potential field structure is unavailable for dissipation and can play no role in heating the corona. However, the continual mechanical deformation of the subsurface portion of the loops will generate a current flow throughout the loop. Since the coronal field must be essentially force-free, the current flow will be along the field so that the flux loop may be imagined as an electrical *circuit*. The current produces a non-potential component of magnetic field which threads the circuit, the total flux through the circuit being the product of the total current I and the self-inductance of the circuit L. The latter quantity, which measures the ability of the circuit to store magnetic energy (the ability to store electric and kinetic energy being measured by the capacitance), depends on the shape of the circuit, increasing with its length. The circuit will also possess a resistance R, due to dissipative processes in the plasma. The time scale for current changes in such an LR circuit is $\tau = L/R$. The 'loop' circuit that we are envisaging here will dissipate mainly in the coronal portion. But this is only a small part of the circuit, which will wander through the interior of the Sun. The self-inductance is consequently enormous, and the characteristic time for current change very long. We may therefore treat the current flow along loops as a direct current with very little time-dependence.

The circuit picture may be used to estimate the Joule heating $I^2/R = (j\pi b^2)^2 (2L/\pi b^2 \sigma)$ associated with the current. To produce a heating rate of $1000\pi b^2$ W in a loop of canonical length $2L = 5 \times 10^7$ m and diameter $2b = 2L/10$, requires a uniform current density

$$j \simeq 13 \text{ A m}^{-2},$$

taking the appropriate value (4×10^6) for the classical conductivity σ. This would imply a total current through the loop of $I = j\pi b^2 \simeq 3 \times 10^{15}$ A. This is an enormous current. It would generate an azimuthal magnetic field in the neighbourhood of the loop of $B_\theta = \mu_0 I/2\pi b \simeq 250$ T. This exceeds the assumed longitudinal field component, 10^{-3} T, by several orders of magnitude. Not only does this require the non-potential part of the coronal field to outweigh the potential part, contrary to our conclusions in Section 5.2.3, but would also involve so much twist that the loop would be highly unstable (Section 5.6.2). Therefore, the assumption of a uniform current cannot be correct.

If it is assumed that the currents are concentrated into volumes much smaller than that of the loop itself, two resolutions of the dilemma become possible. At high current densities the MHD approximation breaks down. If we evaluate the drift speed of the electrons relative to the ions v_D, given by $j = n_e e v_D$, for the current density just estimated, we find $v_D \simeq 10^5$ m s^{-1}. This value is comparable to the sound speed in the corona. When this is the case, *plasma waves* – either ion acoustic or ion cyclotron – are generated at the expense of the current. These waves involve relatively strong fluctuating electric fields, which then act back on the drifting electrons carrying the current and scatter them. Under these circumstances, the electrical conductivity is given (Priest, 1982) by

$$\sigma^* = \frac{n_e e^2}{m_e v^*} \simeq \frac{\varepsilon_0 \omega_p{}^2}{v^*},$$
(5.204)

where v^* is the effective scattering frequency. When turbulent scattering completely dominates the Coulomb scattering, this *anomalous* conductivity can fall to several orders of magnitude below the classical value (Section 5.3.2). The heating rate is correspondingly enhanced. These conditions are self-sustaining if the increase in temperature promotes the wave production – such instabilities are known as current-driven instabilities. It should be noted that the electric fields associated with them will accelerate electrons in the direction of the field. Since the scattering cross section *decreases* with speed, a significant fraction will be accelerated to relativistic speeds in a so-called runaway action. These electrons have a characteristic signature in their non-thermal radio emission. They are an integral part of solar flare models.

If the current density is maintained at a level to produce anomalous conditions, the required heating can be generated by a much lower total current. If $\sigma^* \simeq 1$ and $j \simeq 13$, as before, the diameter of the current channel becomes $r = b\sqrt{(\sigma^*/\sigma)} \simeq 2.5 \times 10^3$ m, and the total current $I \simeq 10^8$ A. This reduces the typical azimuthal magnetic field component in the loop to acceptable levels.

A plausible scenario of coronal current heating, first suggested by Tucker (1973) and developed by Rosner, Golub, Coppi & Vaiana (1978a), supposes that the coronal currents are concentrated into filaments occupying only a very small volume compared to the total loop volume. The filamentation could itself be the result of an instability – Heyvaerts (1974) has suggested two possible thermal instabilities. In these current filaments, the current density is assumed to be high enough to trigger plasma instabilities which give rise to anomalous conductivity. The excitation of both ion acoustic waves (Hinata, 1979) and ion cyclotron waves (Hinata, 1980) have been discussed in this context. Benford (1983) investigated the onset of instability in a cylindrical tube as the electron drift speed v_D was increased. He found that the ion acoustic instability develops first if the electrons and ions initially have the same temperature. However, the ion cyclotron instability may be excited if the initial electron temperature exceeds the ion temperature. Coronal conditions are not known with sufficient accuracy to decide which would be the most important mechanism in reality. The distinction is important, though, because the characteristics of the heating process vary greatly between these two mechanisms. In ion acoustic turbulence, the amplitude grows extremely rapidly and energy is released explosively. In ion cyclotron turbulence, the instability is suppressed as the amplitude grows, so that the conditions switch rapidly back and forth between weakly turbulent and classical conditions. It is therefore unclear whether heat is liberated in very short bursts or quasi-continuously in the strong current regions. In either case, the mechanisms by which the heat is distributed more or less uniformly over the whole loop volume have not been elucidated.

The alternative resolution of the current heating dilemma appeals to current sheets. Current sheets are the DC equivalents of MHD shocks. A current flows in a thin layer, causing the tangential component of the magnetic field to rotate from one side to the other. If the direction of the field is reversed, the field strength will vanish on a surface in the layer. Such a sheet is called a neutral sheet. The current in this case flows perpendicular to the field. In other cases, the rotation may be slight (cf. Fig. 5.14) and the current flow almost parallel to the field. The direction of the current flow is used to classify heating mechanisms (cf. Spicer & Brown, 1981). In individual loops, the current flow is generally parallel to the field and we shall concentrate on these heating mechanisms below. Neutral sheets can, however, arise when loops interact or loop systems evolve. Current sheets may indeed be quite generally distributed throughout the corona (cf. Sections 5.1 and 6.3.1), because the magnetic structure is required to readjust to changes at the footpoints enforced by the subphotospheric motions. Reconnection can occur across current sheets, where finite resistivity allows field diffusion and a change of field topology. This process has been called topological dissipation (cf. Parker, 1979). Diffusion and reconnection are, of course, accompanied by heating.

The general equilibrium structure of current sheets and their dissipation mechanisms have been analysed by Chiueh & Zweibel (1987). However, the plausibility of topological heating as a mechanism in loops can be assessed without invoking any particular dissipation mechanism. As discussed in Section 5.2.2, Heyvaerts & Priest (1984) have suggested that the coronal field continually relaxes the stresses imposed on it by subphotospheric motions, whilst maintaining the total helicity constant. The magnetic energy released in this process is assumed to be dissipated on a time scale shorter than those of the stressing motions. This allows the dissipation rate

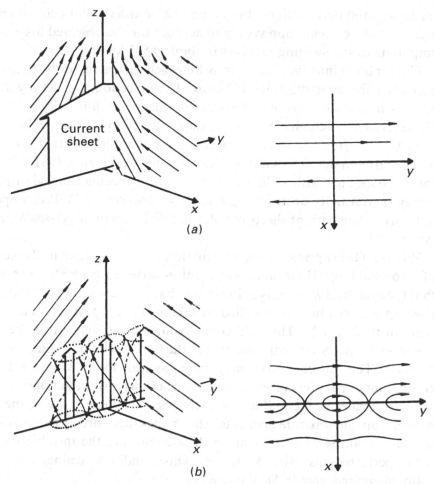

Fig. 5.14. Tearing modes in a twisted magnetic configuration. (*a*) A current sheet in the *yz*-plane parallel to the equilibrium field causes a reversal of the transverse field component on either side of the plane. (*b*) Tearing modes cause the current sheet to break into filaments. The filaments generate an azimuthal field component so that the field interlaces along the *yz*-plane. Reconnection has occurred to produce the X-type neutral points in the *xy*-plane.

to be obtained using a global energy balance argument. In a further discussion of this mechanism, Browning, Sakurai & Priest (1986) considered an ensemble of loops each twisted steadily at the footpoints by a swirling velocity field. They estimated the maximum energy available for dissipation by calculating the 'free' magnetic energy – the difference between the energy of the field which develops in the loop and the energy of the linear force-free field satisfying the same photospheric boundary conditions – and derived an expression for the time-averaged heat flux when the footpoint motions are slow. Vekstein (1987) generalizes the result for faster footpoint motions. Browning *et al.* estimate that enough energy is fed into the coronal loop system to account for the observed losses if the amplitude of the swirling motions is about 0.5–1 km s^{-1}.

If we enquire into the mechanisms by which the energy is dissipated, we encounter the same difficulty as before. Reconnection is generally a slow process in the solar corona, because the magnetic diffusivity is so low. However, it may be enhanced, just as the conductivity is lowered, in regions of instability. In the case of reconnection models, the favoured instability is the tearing mode (e.g. Galeev, Rosner, Serio & Vaiana, 1981). Tearing can be associated with radiative effects or with reconnection via rippling driven by resistivity or density gradients (cf. Section 5.6.2). It can operate not only in thin current sheets but also in thin layers of larger-scale current systems.

Wentzel (1981) gives a clear explanation of how it arises in the context of a coronal loop. He isolates a segment in cartesian geometry, supposing that the axial field $B_z = B_0$ is uniform but that the y-component displays the effect of a twist which is modelled by taking $B_y(x) \propto x/a$ across a current sheet of thickness $2a$. This field component changes sign across the plane $x = 0$, in which the current is parallel to the field, i.e. in the z-direction (Fig. 5.14(a)). The current sheet may be imagined to be composed of a continuum of parallel currents. Since parallel currents attract one another, any disturbance of the uniformity of the sheet will tend to cause the sheet to break up into separate filaments. The sheet is therefore unstable to wave-like disturbances perpendicular to the current, i.e. the instabilities arise from perturbations with $\mathbf{k} \cdot \mathbf{B_0} = 0$. This condition defines the mode rational surfaces met in Section 5.6.2.

In the tearing mode, the finite resistivity allows the magnetic field in the sheet to be torn and broken up into a series of islands, as in Fig. 5.14(b). In the transverse plane, the current filaments produce X-type neutral points in the magnetic field. These are the sites of reconnection.

The simpler the magnetic geometry, i.e. the more uniform the current sheet, the easier it is to excite the tearing-mode instability. It should be noted that the wavelength of the tearing mode is governed by the length of the current sheet, not its thickness. Since the growth rate is proportional to $k^{-\frac{2}{5}}$ (Priest, 1982), the dominant modes have the longest wavelengths possible, i.e. $k < 1/a$. If the geometry is complicated, the wavelengths of the tearing modes are shortened and the current sheet must be correspondingly thinner before instability occurs. Under solar coronal conditions, the current channel thickness would have to be of the order of a kilometre, comparable to that invoked by the current-driven instabilities.

Tearing-mode instabilities grow relatively slowly and quench at a low level. They can therefore provide a quasi-continuous heating of the corona, the energy being extracted from the twist of the magnetic field. However, in order to provide heating at the required rate, Galeev *et al.* (1981) find that the non-potential part of the field must be quite large, its associated magnetic energy being about three times the gas pressure.

The tearing mode in a neutral sheet, where the current is perpendicular to the magnetic field, develops rather differently. As the plasma flows into the sheet, it brings with it new field and the rate of reconnection can accelerate. At relatively low speeds of inflow, the system may evolve into a steady state in which reconnection occurs as fast as the flux is swept into the sheet. The plasma flows out along the sheet through a pair of slow MHD shocks. This is known as the Petschek–Sonnerup mechanism. The outflow speed is the Alfvén speed corresponding to the density in the sheet. Because this is high, most of the magnetic energy is converted into kinetic energy of the flow. If, on the other hand, the inflow is too high for steady-state reconnection, the flux will pile up at the sheet. Long current sheets will eventually break up into a series of regions where bursts of dissipation occur. The non-linear evolution is summarized by Priest (1986). However, current sheets with a perpendicular current are unlikely to be of general occurrence in a loop, and these dissipation mechanisms find more favour in flare models than quiescent loop models (see below).

In summary, it appears to be possible to provide a quasi-continuous supply of energy, sufficient to heat coronal loops, by enhanced current dissipation resulting from either current-driven instabilities or tearing-mode instabilities. Since the latter will operate with a current channel of somewhat greater dimensions than the former, they appear most attractive at present.

The subject of coronal heating is a monument to the ingenuity of theorists. A vast array of possible mechanisms have been invoked with varying degrees of plausibility. However, not one stands out as being able to account comfortably for both the magnitude of the heat input to the corona and the distribution of the heat amongst loops.

The observations of loops and empirical estimates of their properties, listed in Section 5.7.1, do not discriminate between the models. Further constraints may be found, though, by requiring the heating mechanism to be consistent with the properties that our theoretical analysis of loops has suggested.

The overall structure of hot loops is governed by the simple energy balance reflected in the scaling laws derived in Section 5.4.2. These laws are insensitive to the form and distribution of the heat input. Nevertheless, Rosner, Tucker & Vaiana (1978a) in their original paper attempted to utilize the scaling laws (5.74) and (5.76) to decide between rival mechanisms. For a heating function $\mathscr{H}$ uniform along a loop and $\gamma = -\frac{1}{2}$, these predict a relationship between the heating and the length and pressure of the loop, namely,

$$\mathscr{H} \propto p^{\frac{7}{8}} L^{-\frac{5}{8}}. \tag{5.205}$$

They then assumed, firstly, that all mechanisms could be parametrized in terms of their pressure dependence and, secondly, that their associated heating functions could be written

$$\mathscr{H} = H_0 \frac{p}{\ell}. \tag{5.206}$$

They considered slow-mode viscous damping (5.187), slow-mode shock damping (5.195), resistive damping of Alfvén waves (5.185) and mode conversion (5.191) as well as anomalous heating by direct currents, for which they took $\mathscr{H} \propto p$. Only mode conversion and anomalous current heating have an almost linear pressure dependence. However, in order to parametrize the mechanisms in terms of p, several *ad hoc* assumptions were made regarding the dependence of the curvature, magnetic field strength and current generation on the loop pressure. These cannot be justified without a complete model, and it is difficult to exclude any mechanism on this basis alone.

More promising appears to be the evidence of loop stability. As described in Section 5.5.2, McClymont & Craig (1985b) found that stability characteristics provided some constraints on the pressure dependence of the heating mechanism, but none on the temperature dependence. More sensitivity was shown to the *spatial* distribution of heating. They investigated heating functions which depended on position in the loop as measured by the mass column density, calculated down from the top of the loop. If the heating was an *increasing* function of column density, i.e. a *decreasing* function of height, the loop was found to be unstable. They provided a simple physical explanation: extra heating evaporated more material into the loop, increasing the column density and thereby increasing the heat input further. However, McClymont & Craig were unable to analyse heating functions which depend either on a fixed position in space or on non-local parameters. Wave-heating theories fall into the latter category, because the amount of energy available for dissipation at any point depends on the energy losses along the whole path up to that point. The stability arguments are therefore of quite restricted utility.

The only satisfactory test of a coronal heating mechanism is the completeness of the underlying theory, a test stressed by Hollweg (1981b).

5.8 Flare loops

Detailed accounts of solar flare theory have been given in the monographs edited by Sturrock (1980), Priest (1980) and Tandberg-Hanssen & Emslie (1988) and a comprehensive review would be beyond the scope of this book. Instead, we shall conclude our account of coronal loop structure by looking only at how studies of flaring loops relate to those of the normal quiescent loops described in the preceding sections.

5.8.1 *Flare loop dynamics*

Simulations of flare loops differ from the dynamical simulations treated in Section 5.5.3 primarily in the magnitude of the transient heating function applied. Also of significance is the distribution of the assumed source of transient heating along the loop.

There are two classes of flare model, the thermal and the non-thermal. In the former, the flare process is assumed to release heat directly into the coronal portion of the loop, typically around the apex. The heat is then conducted down to the denser layers of the transition region and chromosphere. In the latter, the flare process is taken to release high-energy (non-thermal) electrons into the coronal portion. These electrons beam down into the denser layers where they are eventually thermalized by collisions. In this so-called *thick-target model*, due to Brown (1972), the source of thermal heat is concentrated around the point where the lowest energy electrons are 'stopped'. A large amount of thermal energy is released directly into the upper chromosphere in these models.

Typical of the thermal flare simulations is that of MacNeice (1986). The equilibrium loop was subjected to a weak heating transient of 10^6 W m^{-2} for 100 s, concentrated around the apex of the loop. At first, the temperature rose sharply at the apex, allowing heat to be removed from the deposition region by conduction across the steep temperature gradient that was established between the apex and the rest of the coronal portion of the loop. As heat was thereby fed into the loop on either side of the apex, the temperature rose and the region of steep temperature gradient, the conduction front, migrated away from the top of the loop down each leg. During this phase, the gas remained essentially static. However, when the conduction front eventually reached the transition region, there was rapid heating and a corresponding rise in pressure. The pressure gradient then drove a motion both upward and downward. The upward motion represented ablation or evaporation of erstwhile transition region and chromospheric material into the corona. It remained subsonic throughout. The density rise in the coronal portion of the loop was quite small, because most of the sub-coronal material was driven downwards and compressed as the conduction front moved downwards. In models with larger energy release, Pallavicini *et al.* (1983) found much higher velocities (up to about 100 km s⁻¹) in the material evaporating into the corona. The results of Mariska (1987) described in Section 5.5.3 are in full agreement.

An example of a non-thermal flare simulation is provided by Fisher, Canfield & McClymont (1985a). They modelled thick-target heating over the much shorter interval of 5 s, using heating rates varying from 10^6 W m⁻² up to 10^8 W m⁻². For small flares, i.e. heating rates below 10^7 W m⁻², the coronal temperatures were found to change very little. The chromosphere heated and increased in pressure as a result of direct energy deposition there, but very little material was evaporated into the corona. The maximum upward velocities were of the order of a few tens of kilometres per second. After the end of the transient, the chromosphere reverted rapidly (i.e. on the radiative time scale) to normal.

In larger flares, more of the chromosphere was heated and rose in pressure to values greatly in excess of the coronal pressure. The material was then evaporated explosively into the corona, where it reached speeds of several hundred kilometres per second (Fisher *et al.*, 1985b; Fisher, 1987). The material originally in the coronal portion of the loop increased sharply in temperature throughout, but remained essentially static until the explosive evaporation swept it up. The coronal temperature reached 10^7 K right along the loop, but the coronal density remained at about 10^{15} m⁻³ until the evaporation wave hit it, whereupon the density rose to some 10^{17} m⁻³. These figures are in good agreement with the empirical values for flaring loops quoted in Table 4.17.

However, as in the other models, most of the chromospheric material was driven downwards, compressing the low chromosphere. As the density increased, so did its ability to radiate and the material in the chromosphere *cooled*. Flare models place great emphasis on the treatment of the chromosphere and transition region (e.g., Fisher *et al.*, 1985c), because observations provide much detailed information about the development of flares in these dense regions of the flaring loop. The chromosphere is not our subject, so that chromospheric aspects will not be discussed further. However, the treatment of the transition region does warrant comment. In flares, large amounts of energy are released into regions which are incapable of radiating it away. Conduction is therefore forced to carry very large energy fluxes. At any given temperature, the heat flux can only be increased by increasing the temperature gradient. This naturally imposes severe problems of spatial resolution for the simulations, but it can also lead to a breakdown of the physical basis of the model. Fisher *et al.* note that Eqn. (5.22) allows arbitrarily high conductive fluxes for suitably large temperature gradients. This is not in fact the case. The flux cannot exceed the value set when *all* electrons move together in the same direction with the electron thermal speed. In flare models, this saturation needs to be taken into account. This situation is not simple, though. When the fluid description breaks down, the heat flux is not determined by the local conditions, but by conditions over a volume the size of the mean-free-path. MacNeice (1986) investigated the onset of this breakdown by estimating deviations of the electron velocity distribution from the Maxwellian form. He found that electrons moving downwards through the conduction front developed a high-energy tail which *enhanced* the conductive flux.

Like the quiescent loop simulations in Section 5.5.3, these flare loop simulations are all one-dimensional. The magnetic field in such models is eliminated entirely. Such models may apply to simple loop (or compact) flares, where the flaring loop appears to remain unchanged throughout (cf. Section 4.2). In larger flares, however, any original loop structure appears to be disrupted during the flare process. Dramatic changes of structure could occur if the loop became MHD unstable (Section 5.6.2). Examples are provided by Sakurai (1976), who followed the development of a kink instability in uniform cylindrical geometry into the non-linear regime, showing how the loop twists and forms a buckle. Simple loops appear to reform in the post-flare phase (Section 4.2). Simulations of flare loops that would allow for this degree of complication have yet to be attempted, and will clearly provide a severe test of both computational resources and physical theory for many years to come.

5.8.2 *Flare loop heating*

Theories of flare loop heating are implicit in the flare loop dynamics described above. Flare heating mechanisms differ from quiescent loop heating mechanisms in so far as they have to account for the non-thermal component of flare energy release. So far, however, this extra constraint has not led to any more certain identification of the mechanism in flares. Indeed, almost all the quiescent heating models have also been proposed as flare source models.

According to the exhaustive survey of Spicer & Brown (1981), most of the presently favoured models employ a DC current heating mechanism. They distinguish those involving parallel currents from those employing a perpendicular current. The parallel current models rely for dissipation on current interruption (which leads to a build up of separated electric charge known as a double layer), anomalous Joule heating or tearing modes (or any combinations!). However, only double layer mechanisms produce any substantial amount of non-thermal energy. They are all appropriate to a single loop geometry and are therefore candidates for the source of compact flare heating.

As noted above, larger flares can be more complex, with loops being disrupted and reforming. It is important to note that twisting of a flux tube could excite an MHD instability, such as the kink mentioned above, which would lead to greatly accelerated plasma motion. Although no heating would result so long as the instability remained in the ideal regime, the development of a kink can trigger resistive instabilities – such as tearing modes – within the loop. The *heating* associated with a loop instability can presumably be any one (or more) of the mechanisms listed above.

The heating mechanism in 'post'-flare loops might, however, be different. According to the Kopp & Pneuman (1976) model, these loops reform at the base of a vertical current sheet which was created above the flare site by the flare eruption. This current sheet is a neutral sheet and perpendicular current mechanisms come into play (cf. Section 5.7.3). A more complete model of 'post'-flare loops – which can occur almost from the onset of the flare – has been presented by Cargill & Priest (1982).

The picture of 'post'-flare loops forming by magnetic enclosure *within the corona* is very different to that envisaged for the quiescent loops, which are formed below the surface. Yet after the flare has died away, the two types of loop merge and become indistinguishable within the active region loop system. Flare loops are clearly very closely related to quiescent coronal loops; they are presumably governed by the same physical processes and should be described by essentially the same models. Our understanding of flare loops can be no better than our understanding of coronal loops in general. We close this chapter with a brief assessment of the extent to which today's models both succeed and fail to provide this understanding.

5.9   Summary and conclusion

We conclude this chapter with a brief review of the extent to which theory can provide a physical framework with which to interpret the observations of coronal loops described in Chapters 2 and 3.

Especially successful is the description of hot loops. In hot, diffuse plasma the transport of heat and momentum in the direction of the magnetic field is much more effective than that perpendicular to the field, so that one-dimensional models are a good first approximation. The effectiveness of heat conduction further ensures that the coronal loop is almost isothermal along its length, and the large thermal scale height ensures that it is almost isobaric. A loop of given length then adjusts its temperature and density (by chromospheric evaporation) until the total radiative losses balance the heat input and the pressure matches that of the upper chromosphere, where the loop is rooted. The latter condition ensures that pressures are in the range 0.03–0.3 Pa, as generally observed (cf. Table 3.8). The length of the loop is fixed by the geometry of the coronal magnetic field; hence the only free parameter is the heating rate. Whilst the source of this heating remains at present equivocal, several mechanisms, drawing their energy from subcoronal forcing, would be able to meet the energetic requirements of hot loop systems.

The processes of evaporation and condensation at the chromospheric interface, when correctly modelled, ensure that hot loops are thermally stable, although an *isolated* hot loop would be unstable. Loop models subjected to even large thermal perturbations relax to a new equilibrium over the thermal time scale of about an hour. The lifetime of a loop is therefore governed by the processes governing changes in the coronal field geometry and the coronal heating rates. The observation that lifetimes of hot loops range from hours to days is consistent with the assumption that these processes take place beneath the solar surface.

Numerical simulations of loops provide also much dynamical information. The flows induced in the corona during the relaxation to equilibrium tend to be quite gentle, the velocities remaining well subsonic. Subsonic flows will also arise if thermal and hydrostatic imbalances between the two legs are caused by asymmetries in the loop. In the absence of direct measurements of flows in hot loops, this feature of the models cannot be tested, but the fact that the non-thermal velocities are estimated to be less than 20 km s^{-1} (Table 3.8) indicates that there is little variation of flow speed either along or across a hot loop.

Of course, one-dimensional models, in which the magnetic field is just the passive determinant of the geometry of the loop axis, make no allowance for transverse structure in a loop and therefore throw no light on the diameters or aspect ratios of loops. In order to progress beyond one-dimensional models, the dynamical and thermal effects associated with the magnetic field must also be considered, at the cost of an enormous increase in complexity. Some authors suggest that the diameter of a loop is determined by the scale of the twist in the magnetic structure, the 'diameter' being limited by the requirement of MHD stability. But a twisted field can also produce thermal variations across the loop as a result of the systematic change in effective length of the loop along the field lines. The thermal variations can themselves affect the stability of the structure by introducing a radial pressure gradient. It is also possible that variations in heat input and of footpoint boundary conditions could be significant in determining the cross-sectional extent of individual loops. Multi-dimensional modelling has not yet reached the point at which firm conclusions can be drawn.

Turning now to cool loops, the situation appears less satisfactory. Models suggest that 'cool' loops with temperatures in the range 10^5–10^6 K are governed by exactly the same processes as 'hot' loops with temperatures in excess of 10^6 K. Conduction still plays a major role in removing heat from the coronal portion of the loop, but radiation losses are more effective than in hotter loops at the same pressure. The pressure must again match that in the upper chromosphere (cf. Table 2.8). According to the scaling laws, cool loops require less heating and are much shorter. Although the average figures quoted in Tables 2.9 and 3.7 show little evidence that the lengths and heights of loops vary with temperature, observations of specific regions suggest that these trends can be found (cf. Section 2.4.3).

Loop models with temperatures less than 10^5 K are found to be very different. Conduction becomes ineffective at carrying the heat flux in the corona and radiation is required to remove the heat input. This is only possible with high densities. These loops are short and low. Such loops would be difficult to observe at the limb and difficult to distinguish on the disk. Indeed, whilst plasma at these temperatures makes a substantial contribution to the total emission from active regions (cf. Antiochus & Noci, 1986), it is not known whether the emission stems mainly from arcs at the base of hotter loops or from complete but unresolved loops. Indirect evidence from the dynamical behaviour would support the latter interpretation. The only models which can reproduce the high velocities of some tens of kilometre per second observed in cool coronal plasma are those of complete cool loops. Transition zone material at the base of hot loops can very rarely be made to move substantially faster than the hot coronal plasma, a few kilometres per second at most. Models of large-scale cool loops cannot be constructed, which suggests that the loops shown in Fig. 2.13, for example, are not stable structures and will survive little longer than the thermal time scale. This is certainly true of $H\alpha$ loops (cf. Table 2.8).

As in the case of hot loops, the cool loop models take no account of the transverse structure of the loop. This failing is in fact more critical for cool loops. The form of the radiative loss function depends strongly on the treatment of the Lα line. Some authors adopt the optically thick treatment of McClymont & Canfield (1983). This is appropriate to the almost plane geometry of the chromosphere, but has not been tested for the thin cylindrical geometry of a loop. Not only do the structural equations require a full multi-dimensional treatment, but the radiation field does as well. Such models are still a long way off, but they seem to offer the only means of resolving the outstanding problems of loop structure.

Foremost amongst these problems is the relationship of cool to hot loops in loop systems. Any cool loop residing within a system of predominantly hot loops will of course appear as a cool structure surrounded by hotter material. There need be no implication that the multi-temperature plasma forms part of a single loop volume. The observations that inspired models of loops with cool axial cores surrounded by hot sheaths of plasma were critically evaluated in Section 3.7.3. There it was concluded that hot and cool loops must be regarded as separate physical structures. Nevertheless, many *ad hoc* models of axisymmetric loops with cool centres have been discussed. If the gas pressure also increases outwards, these models have the theoretical attraction of being stable to interchange modes. However, this simplification cannot be maintained beyond the point where the loop merges with its neighbours. As noted above, we do not know what process causes one loop volume to be distinguished from its neighbours in the corona.

We have already quoted Hollweg (1981) as saying that the only satisfactory test of a coronal heating mechanism is the completeness of the underlying theory. The same is true of loop modelling in general. Analysis and computation have focused on many aspects of coronal loops and have taken account of as many factors as present-day techniques and capacity will allow. This still falls short of what is required. Nevertheless, a few authors have attempted to fill in the picture and provide a self-consistent, if yet qualitative, sketch of coronal loop physics. The outstanding example is that of Ionson (1978), who took up and developed many of the ideas put forward earlier by Foukal (1976).

In his classic paper on resonant heating, Ionson took into account not only the generation of the energy by photospheric motions, the propagation of the surface waves and their dissipation via kinetic Alfvén waves, but also the dynamical consequences of the intense energy release along the resonant surfaces. In Ionson's scenario, the resonant heat release is confined to very thin layers about the resonant surface – layers too thin to be observable. Very large cross-field temperature gradients then arise which can drive a heat flow out from the layer, despite the low values of the cross-field conductivity. The plasma in the neighbourhood of the sheet is thereby heated and becomes buoyant. It flows up along the loop. At the apex, the hot plasma has to flow between regions of cooler, higher density plasma both above and below. The situation in which higher density material lies above less dense is unstable and a Rayleigh–Taylor instability can develop. Priest (1978) suggests that a Kelvin–Helmholtz instability might also develop along the velocity shear layer between the flow of hot plasma and the cooler surrounding material. Both instabilities are a form of interchange, and material is exchanged between the heated layer and the surrounding volume of the loop.

Ionson then suggests that the plasma cools in the *interior* of the loop, increases in density and sinks back downwards. A dynamical balance exists between the upflow in the sheath and the downflow in the cool interior. The observed loop might therefore be maintained out of hydrostatic equilibrium (cf. Noci & Zuccarello, 1983).

Hinata (1982) pointed out another means of generating steady flows in the context of DC heating. If the ohmic losses are continually balanced by the transport of electromagnetic energy from below, the electromagnetic Poynting flux will produce a cross-field plasma drift into the loop from the surrounding regions. He also suggests that mass balance is maintained by a downflow of material along the loop.

Both of these examples point to the need of a *complete* theory of loop structure that looks beyond the individual loop to the coronal loop *system* in which it is embedded and to its continuation within the body of the Sun. The resolution of many problems of coronal loop physics may well lie in the proper recognition of their global properties. The global approach to modelling coronal loops in both the Sun and other stars forms the subject matter of Chapter 6.

References

Abdelatif, T.E. (1987). Heating of coronal loops by phase-mixed shear Alfvén waves. *Astrophysical Journal*, **322**, 494–502.

Abdelatif, T.E. (1988). Surface and body waves in magnetic flux tubes. *Astrophysical Journal*, **333**, 395–406.

Adams, J. & Pneuman, G.W. (1976). A new technique for the determination of coronal magnetic fields: a fixed mesh solution of Laplace's equation using line-of-sight boundary conditions. *Solar Physics*, **46**, 185–203.

Alissandrakis, C.E. (1981). On the computation of constant α force-free magnetic field. *Astronomy and Astrophysics*, **100**, 197–200.

Altschuler, M.D. & Newkirk, G. (1969). Magnetic fields and the structure of the solar corona I: methods of calculating coronal fields. *Solar Physics*, **9**, 131–49.

Altschuler, M.D., Levine, R.H., Stix, M. & Harvey, J. (1977). High resolution mapping of the magnetic field of the solar corona. *Solar Physics*, **51**, 345–75.

Aly, J.J. (1987). On the uniqueness of the determination of the coronal potential field from line-of-sight boundary conditions. *Solar Physics*, **111**, 287–96.

Aly, J.J. (1989). On the reconstruction of the nonlinear force-free coronal magnetic field from boundary data. *Solar Physics*, **120**, 19–48.

An, C.-H. (1982). The effects on the MHD stability of field line-tying to the end faces of a cylindrical magnetic loop. *Solar Physics*, **75**, 19–34.

An, C.-H. (1983). MHD stability of incompressible coronal loops with radiative energy loss. *Astrophysical Journal*, **264**, 302–8.

An, C.-H. (1984a). MHD stability of compressible coronal loops. *Astrophysical Journal*, **276**, 352–6.

An, C.-H. (1984b). Comments on the MHD stability of coronal plasmas with line-tying. *Astrophysical Journal*, **281**, 419–25.

An, C.-H. (1984c). The effect of line-tying on the radiative MHD stability of coronal plasmas with radial pressure gradient. *Astrophysical Journal*, **284**, 422–8.

An, C.-H. (1985). Formation of prominences by condensation modes in magnetized cylindrical plasma. *Astrophysical Journal*, **298**, 409–13.

An, C.-H., Canfield, R.C., Fisher, G.H. & McClymont, A.N. (1983). Flare loop radiative hydrodynamics. IV. Dynamic evolution of unstable semiempirical loop models. *Astrophysical Journal*, **267**, 421–32.

Antiochus, S.K. (1979). The stability of solar coronal loops. *Astrophysical Journal*, **232**, L125–9.

Antiochus, S.K. & Noci, G. (1986). The structure of the static corona and transition region. *Astrophysical Journal*, **301**, 440–7.

Antiochus, S.K., Shoub, E.C., An, C.-H. & Emslie, A.G. (1985). Thermal stability of static coronal loops. I. Effects of boundary conditions. *Astrophysical Journal*, **298**, 876–86.

Anzer, U. (1968). The stability of force-free magnetic fields with cylindrical symmetry in the context of solar flares. *Solar Physics*, **3**, 298–315.

Basri, G.S., Linsky, J.L., Bartoe, J.-D.F., Brueckner, G. & Van Hoosier, M.E. (1979). Lyman-alpha rocket spectra and models of the quiet and active solar chromosphere based on partial redistribution diagnostics. *Astrophysical Journal*, **230**, 924–49.

Batistoni, P., Einaudi, G. & Chiuderi, C. (1985). Resistive instabilities in coronal conditions. *Solar Physics*, **97**, 309–20.

Bel, N. & Mein, P. (1971). Propagation of magneto-acoustic waves along the gravitational field in an isothermal atmosphere. *Astronomy and Astrophysics*, **11**, 234–40.

Benford, G. (1983). Turbulent resistive heating of solar coronal arches. *Astrophysical Journal*, **269**, 690–7.

Berger, M.A. (1985). Structure and stability of constant-α force-free fields. *Astrophysical Journal Supplement Series*, **59**, 433–44.

Berstein, I.B., Frieman, E.A., Kruskal, M.D. & Karlsrud, R.M. (1958). An energy principle for hydromagnetic stability problems. *Proceedings of the Royal Society of London, Series A*, **244**, 17–40.

Bogdan, T.J. (1986). The determination of coronal potential magnetic fields using line-of-sight boundary conditions. *Solar Physics*, **103**, 311–15.

Bogdan, T.J. & Low, B.C. (1986). The three-dimensional structure of magnetostatic atmospheres. II. Modelling the large-scale corona. *Astrophysical Journal*, **306**, 271–82.

Borrini, G. & Noci, G. (1982). Non-equilibrium ionization in coronal loops. *Solar Physics*, **77**, 153–66.

Boyd, T.J.M. & Sanderson, J.J. (1969). *Plasma Dynamics*. London: Nelson.

Bray, R.J. & Loughhead, R.E. (1974). *The Solar Chromosphere*. London: Chapman and Hall.

Bray, R.J., Loughhead, R.E. & Durrant, C.J. (1984). *The Solar Granulation*. Second edition. Cambridge: Cambridge University Press.

Brown, J.C. (1972). The directivity and polarization of thick target X-ray bremsstrahlung from solar flares. *Solar Physics*, **26**, 441–59.

Browning, P.K. & Priest, E.R. (1984). Kelvin-Helmholtz instability of a phase-mixed Alfvén wave. *Astronomy and Astrophysics*, **131**, 283–90.

Browning, P.K. & Priest, E.R. (1986a). Heating of coronal arcades by magnetic tearing turbulence, using the Taylor-Heyvaerts hypothesis. *Astronomy and Astrophysics*, **159**, 129–41.

Browning, P.K. & Priest, E.R. (1986b). The shape of buoyant coronal loops in a magnetic field and the eruption of coronal transients and prominences. *Solar Physics*, **106**, 335–51.

Browning, P.K., Sakurai, T. & Priest, E.R. (1986). Coronal heating in closely-packed flux tubes: a Taylor-Heyvaerts relaxation theory. *Astronomy and Astrophysics*, **158**, 217–27.

Cally, P.S. (1986). Leaky and non-leaky oscillations in magnetic flux tubes. *Solar Physics*, **103**, 277–98.

Cargill, P.J. & Priest, E.R. (1980). Siphon flows in coronal loops: I. Adiabatic flow. *Solar Physics*, **65**, 251–69.

Cargill, P.J. & Priest, E.R. (1982). Slow-shock heating and the Kopp-Pneuman model for 'post'-flare loops. *Solar Physics*, **76**, 357–75.

Chandrasekhar, S. & Kendall, P.C. (1957). On force-free magnetic fields. *Astrophysical Journal*, **126**, 457–60.

Chapman, G.A. (1981). Active regions from the photosphere to the chromosphere. In *Solar Active Regions*, ed. F.Q. Orrall, pp. 43–82. Boulder: Colorado Associated University Press.

Chen, Z.-C. & Chen, Y. (1989). Solar force-free magnetic fields on and above the photosphere. *Solar Physics*, **119**, 279–99.

Chiu, Y.-C. & Wentzel, D.G. (1972). Nonlinear dissipation of Alfvén waves. *Astrophysics and Space Science*, **16**, 465–77.

Chiu, Y.T. & Hilton, H.H. (1977). Exact Green's function method of solar force-free magnetic-field computations with constant α. I. Theory and basic test cases. *Astrophysical Journal*, **212**, 873–85.

Chiuderi, C. & Einaudi, G. (1981). Current confinement in solar coronal loops. *Solar Physics*, **73**, 89–103.

Chiuderi, C., Einaudi, G. & Torricelli-Ciamponi, G. (1981). What can we learn from static models of coronal loops? *Astronomy and Astrophysics*, **97**, 27–32.

Chiuderi, C., Giachetti, R. & van Hoven, G. (1977). The structure of coronal magnetic loops I: equilibrium theory. *Solar Physics*, **54**, 107–22.

Chiueh, T. & Zweibel, E.G. (1987). The structure and dissipation of forced current sheets. *Astrophysical Journal*, **317**, 900–17.

Clemmow, P.C. & Dougherty, J.P. (1969). *Electrodynamics of Particles and Plasmas*. Reading: Addison-Wesley.

Craig, I.J.D. & McClymont, A.N. (1981). The dynamic formation of quasi-static active region loops. *Solar Physics*, **70**, 97–113.

Craig, I.J.D. & McClymont, A.N. (1986). Quasi-steady mass flows in coronal loops. *Astrophysical Journal*, **307**, 367–80.

Craig, I.J.D. & McClymont, A.N. (1987). Why coronal loops cannot exist without solar gravity. *Astrophysical Journal*, **318**, 421–7.

Craig, I.J.D. & Schulkes, R.M.S.M. (1985). Limit cycle behavior in solar and stellar coronal loops. *Astrophysical Journal*, **296**, 710–18.

Craig, I.J.D. & Sneyd, A.D. (1986). A dynamic relaxation technique for determining the linear stability of coronal magnetic fields. *Astrophysical Journal*, **311**, 451–9.

Craig, I.J.D., McClymont, A.N. & Sneyd, A.D. (1988). General methods for determining the linear stability of coronal magnetic fields. *Astrophysical Journal*, **335**, 441–55.

Craig, I.J.D., McClymont, A.N. & Underwood, J.H. (1978). The temperature and density structure of active region coronal loops. *Astronomy and Astrophysics*, **70**, 1–11.

Craig, I.J.D., Robb, T.D. & Rollo, M.D. (1982). The stability and uniqueness of coronal loops. *Solar Physics*, **76**, 331–55.

Cram, L.E. (1987). Heating of chromospheres and coronae: present status of theory. In *Cool Stars, Stellar Systems, and the Sun*, eds. J.L. Linsky and R.R. Stencel, pp. 123–134. Lecture Notes in Physics, Vol. 291. Berlin: Springer-Verlag.

Cramer, N.F. & Donnelly, I.J. (1985). Perturbations of a twisted solar coronal loop: the relation between surface waves and instabilities. *Solar Physics*, **99**, 119–32.

Davila, J.M. (1987). Heating of the solar corona by the resonant absorption of Alfvén waves. *Astrophysical Journal*, **317**, 514–21.

Dowdy, J.F., Moore, R.L. & Wu, S.T. (1985). Inhibition of conductive heat flux by magnetic constriction in the corona and transition region: dependence on the shape of the constriction. *Solar Physics*, **99**, 79–99.

Durney, B.R. & Pneuman, G.W. (1975). Solar-interplanetary modeling: 3-D solar wind solutions in prescribed non-radial magnetic field geometries. *Solar Physics*, **40**, 461–86.

Durrant, C.J. (1989). Linear force-free fields and coronal models. *Australian Journal of Physics*, **42**, 317–29.

Edwin, P.M. & Roberts, B. (1982). Wave propagation in a magnetically structured atmosphere III: the slab in a magnetic environment. *Solar Physics*, **76**, 239–59.

Edwin, P.M. & Roberts, B. (1983). Wave propagation in a magnetic cylinder. *Solar Physics*, **88**, 179–91.

Edwin, P.M. & Roberts, B. (1988). Employing analogies for ducted MHD waves in dense coronal structures. *Astronomy and Astrophysics*, **192**, 343–7.

Einaudi, G. & Mok, Y. (1987). Alfvén wave dissipation in the solar atmosphere. *Astrophysical Journal*, **319**, 520–30.

Einaudi, G. & van Hoven, G. (1983). The stability of coronal loops: finite length and pressure profile limits. *Solar Physics*, **88**, 163–77.

Einaudi, G., Torricelli-Ciamponi, G. & Chiuderi, C. (1984). Magnetic fields and thermal structure of solar plasmas. *Solar Physics*, **92**, 99–107.

Fisher, G.H. (1987). Explosive evaporation in solar flares. *Astrophysical Journal*, **317**, 502–13.

Fisher, G.H., Canfield, R.C. & McClymont, A.N. (1985a). Flare loop radiative hydrodynamics. V. Response to thick-target heating. *Astrophysical Journal*, **289**, 414–24.

Fisher, G.H., Canfield, R.C. & McClymont, A.N. (1985b). Flare loop radiative hydrodynamics. VI. Chromospheric evaporation due to heating by nonthermal electrons. *Astrophysical Journal*, **289**, 425–33.

Fisher, G.H., Canfield, R.C. & McClymont, A.N. (1985c). Flare loop radiative hydrodynamics. VII. Dynamics of the thick-target heated chromosphere. *Astrophysical Journal*, **289**, 434–41.

Foukal, P.V. (1976). The pressure and energy balance of the cool corona over sunspots. *Astrophysical Journal*, **210**, 575–81.

Galeev, A.A., Rosner, R., Serio, S.V. & Vaiana, G.S. (1981). Dynamics of coronal structures: magnetic field-related heating and loop energy balance. *Astrophysical Journal*, **243**, 301–8.

Gary, G.A. (1989). Linear force-free fields for solar extrapolation and interpretation. *Astrophysical Journal Supplement Series*, **69**, 323–48.

Giachetti, G., van Hoven, G. & Chiuderi, C. (1977). The structure of coronal magnetic fields II: MHD stability theory. *Solar Physics*, **55**, 371–86.

Gibons, M. & Spicer, D.S. (1981). On line-tying. *Solar Physics*, **69**, 57–61.

Glencross, W.M. (1980). Plasma flow along sheared magnetic arches within the solar corona. *Astronomy and Astrophysics*, **83**, 65–72.

Golub, L., Maxson, C., Rosner, R., Serio, S. & Vaiana, G.S. (1980). Magnetic fields and coronal heating. *Astrophysical Journal*, **238**, 343–8.

Grad, H. & Rubin, H. (1958). Hydromagnetic equilibria and force-free fields. *Proceedings of the 2nd United Nations International Conference on the Peaceful Uses of Atomic Energy*, **31**, 190–7.

Grossman, W. & Smith, R.A. (1988). Heating of solar coronal loops by resonant absorption of Alfvén waves. *Astrophysical Journal*, **332**, 476–98.

Habbal, S.R., Leer, E. & Holzer, T.E. (1979). Heating of coronal loops by fast-mode MHD waves. *Solar Physics*, **64**, 287–301.

Habbal, S.R. & Rosner, R. (1979). Thermal instabilities in magnetically confined plasmas: solar coronal loops. *Astrophysical Journal*, **234**, 1113–21.

Hannakam, L., Gary, G.A. & Teuber, D.L. (1984). Computation of solar magnetic fields from photospheric observations. *Solar Physics*, **94**, 219–34.

Hasan, S.S. (1980). Magnetohydrodynamic equilibrum and stability of pre-flare loops. Constant pitch field. *Solar Physics*, **67**, 267–83.

Heyvaerts, J. (1974). The thermal instability in a magnetohydrodynamic medium. *Astronomy and Astrophysics*, **37**, 65–73.

Heyvaerts, J. & Priest, E.R. (1983). Coronal heating by phase-mixed shear Alfvén waves. *Astronomy and Astrophysics*, **117**, 220–34.

Heyvaerts, J. & Priest, E.R. (1984). Coronal heating by reconnection in dc current systems. A theory based on Taylor's hypothesis. *Astronomy and Astrophysics*, **137**, 63–78.

Hinata, S. (1979). The role of turbulent heating in the solar atmosphere. *Astrophysical Journal*, **232**, 915–22.

Hinata, S. (1980). Electrostatic ion-cyclotron heating of solar atmosphere. *Astrophysical Journal*, **235**, 258–67.

Hinata, S. (1981). Stability of coronal loops heated by direct current dissipation. *Astrophysical Journal*, **246**, 532–7.

Hinata, S. (1982). Plasma flow around coronal loops. *Solar Physics*, **80**, 173–83.

Hollweg, J.V. (1979). A new resonance in the solar atmosphere I. Theory. *Solar Physics*, **62**, 227–40.

Hollweg, J.V. (1981a). Alfvén Waves in the solar atmosphere II: open and closed magnetic flux tubes. *Solar Physics*, **70**, 25–66.

Hollweg, J.V. (1981b). Mechanisms of energy supply. In *Solar Active Regions*, ed. F.Q. Orral, pp. 277–318. Boulder: Colorado Associated University Press.

Hollweg, J.V. (1984). Resonances in coronal loops. *Astrophysical Journal*, **277**, 392–403.

Hollweg, J.V. (1986). Viscosity and the Chew-Goldberger-Low equations in the solar corona. *Astrophysical Journal*, **306**, 730–9.

Hollweg, J.V. (1987). Incompressible magnetohydrodynamic surfaces waves: nonlinear aspects. *Astrophysical Journal*, **317**, 918–25.

Hollweg, J.V. & Sterling, A.C. (1984). Resonant heating: an interpretation of coronal loop data. *Astrophysical Journal*, **282**, L31–3.

Hood, A.W. (1986a). Photospheric line-tying conditions for the MHD stability of coronal magnetic fields. *Solar Physics*, **105**, 307–12.

Hood, A.W. (1986b). Ballooning instabilities in the solar corona: conditions for stability. *Solar Physics*, **103**, 329–45.

Hood, A.W. & Anzer, U. (1988). Thermal condensations in coronal magnetic fields. *Solar Physics*, **115**, 61–80.

Hood, A.W. & Priest, E.R. (1979a). The equilibrium of solar coronal magnetic loops. *Astronomy and Astrophysics*, **77**, 233–51.

Hood, A.W. & Priest, E.R. (1979b). Kink instability of solar coronal loops as the cause of solar flares. *Solar Physics*, **64**, 303–21.

Hood, A.W. & Priest, E.R. (1980). Are solar coronal loops in thermal equilibrium? *Astronomy and Astrophysics*, **87**, 126–31.

Hood, A.W. & Priest, E.R. (1981). Critical conditions for magnetic instabilities in force-free coronal loops. *Geophysical and Astrophysical Fluid Dynamics*, **17**, 297–318.

Hung, R.J. & Barnes, A. (1973). Dissipation of hydromagnetic waves with application to the outer corona. III. Transition from collisional to collisionless protons. *Astrophysical Journal*, **181**, 183–208.

Ionson, J.A. (1978). Resonant absorption of Alfvénic surface waves and the heating of solar coronal loops. *Astrophysical Journal*, **226**, 650–73.

Ionson, J.A. (1983). Electrodynamic coupling in magnetically confined X-ray plasmas of astrophysical origin. *Astrophysical Journal*, **271**, 778–92.

Joarder, P.S., Gokhale, M.H. & Venkatakrishnan, P. (1987). Thermal overstability of hydromagnetic surface waves. *Solar Physics*, **110**, 255–70.

Jordan, C. (1976). The structure and energy balance of solar active regions. *Philosophical Transactions of the Royal Society*, **281**, 391–404.

Jordan, C. (1980). The energy balance of the solar transition region. *Astronomy and Astrophysics*, **86**, 355–63.

Kattenberg, A. & Sillen, R. (1983). A low-β coronal loop model II: kink instabilities in loops surrounded by plasma. *Solar Physics*, **79**, 343–51.

Klimchuk, J.A. & Mariska, J.T. (1988). Heating related flows in cool solar loops. *Astrophysical Journal*, **328**, 334–43.

Klimchuk, J.A., Antiochus, S.K. & Mariska, J.T. (1987). A numerical study of non-linear thermal stability of solar loops. *Astrophysical Journal*, **320**, 409–17.

Kopp, R.A. & Pneuman, G.W. (1976). Magnetic reconnection in the corona and the loop prominence phenomenon. *Solar Physics*, **50**, 85–98.

Kopp, R.A., Poletto, G., Noci, G. & Bruner, M. (1985). Analysis of loop flows observed on 27 March, 1980 by the UVSP instrument during the Solar Maximum Mission. *Solar Physics*, **98**, 91–118.

Koshlyakov, N.S., Smirnov, M.M. & Gliner, E.B. (1964). *Differential Equations of Mathematical Physics*. Amsterdam: North Holland Publishing Company.

Kovitya, P. & Cram, L.E. (1983). Electrical conductivity in sunspots and the quiet photosphere. *Solar Physics*, **84**, 45–8.

Krall, K.R. & Antiochus, S.K. (1980). The evolution of active region loop plasma. *Astrophysical Journal*, **242**, 374–82.

Krieger, A.S., de Feiter, L.D. & Vaiana, G.S. (1976). Evidence for magnetic energy storage in coronal active regions. *Solar Physics*, **47**, 117–26.

Krishan, V. (1983). Spatial profiles of lines in active region loops. *Solar Physics*, **88**, 155–61.

Krishan, V. (1985). Two-dimensional pressure structure of a coronal loop. *Solar Physics*, **97**, 183–9.

Kuin, N.P.M. & Martens, P.C.H. (1982). On the thermal stability of hot coronal loops: the coupling between chromosphere and corona. *Astronomy and Astrophysics*, **108**, L1–4.

Kuperus, M., Ionson, J.A. & Spicer, D.S. (1981). On the theory of coronal heating mechanisms. *Annual Reviews of Astronomy and Astrophysics*, **19**, 7–40.

Lakshmi, P. & Gokhale, M.H. (1987). Equilibrium of a thin, isolated, axisymmetric force-free magnetic flux tube in a stratified atmosphere. *Solar Physics*, **114**, 75–80.

Landini, M. & Monsignori Fossi, B.C. (1975). A loop model of active coronal regions. *Astronomy and Astrophysics*, **42**, 213–20.

Landini, M. & Monsignori Fossi, B.C. (1981). Coronal loops in the Sun and in the stars. *Astronomy and Astrophysics*, **102**, 391–400.

Leroy, B. & Schwartz, St.J. (1982). Propagation of waves in an atmosphere in the presence of a magnetic field V. The theory of magneto-acoustic-gravity oscillations. *Astronomy and Astrophysics*, **112**, 84–92.

Lee, A. & Roberts, B. (1986). On the behaviour of hydromagnetic surface waves. *Astrophysical Journal*, **301**, 430–9.

Levine, R.H. (1976). Evidence for opposed currents in active region loops. *Solar Physics*, **46**, 159–70.

Levine, R.H. & Altschuler, M.D. (1974). Representations of coronal magnetic fields including currents. *Solar Physics*, **36**, 345–50.

Levine, R.H. & Pye, J.P. (1980). The coronal and transition region temperature structure of a solar active region. *Solar Physics*, **66**, 39–60.

Low, B.C. (1982). Magnetostatic atmospheres with variations in three dimensions. *Astrophysical Journal*, **263**, 952–69.

McClymont, A.N. & Canfield, R.C. (1983). Flare loop radiative hydrodynamics. II. Thermal stability of empirical models. *Astrophysical Journal*, **265**, 497–506.

McClymont, A.N. & Craig, I.J.D. (1985a). Thermal stability of coronal loops. I. The equilibrium structure and the stability equation. *Astrophysical Journal*, **289**, 820–33.

McClymont, A.N. & Craig, I.J.D. (1985b). Thermal stability of coronal loops. II. Symmetric modes and constraints on the heating mechanism. *Astrophysical Journal*, **289**, 834–43.

McClymont, A.N. & Craig, I.J.D. (1987a). Fast downflows in the cool transition region explained. *Astrophysical Journal*, **312**, 402–11.

McClymont, A.N. & Craig, I.J.D. (1987b). The structure and stability of coronal magnetic fields. *Solar Physics*, **113**, 131–6.

MacNeice, P. (1986). A numerical hydrodynamic model of a heated coronal loop. *Solar Physics*, **103**, 47–66.

McWhirter, R.W.P., Thonemann, P.C. & Wilson, R. (1975). The heating of the solar corona II. A model based on energy balance. *Astronomy and Astrophysics*, **40**, 63–73.

Mariska, J.T. (1987). Solar transition region and coronal response to heating rate perturbations. *Astrophysical Journal*, **319**, 465–80.

Mariska, J.T. & Boris, J.P. (1983). Dynamics and spectroscopy of asymmetrically heated coronal loops. *Astrophysical Journal*, **267**, 409–20.

Martens, P.C.H. & Kuin, N.P.M. (1982). On cool coronal loops. *Astronomy and Astrophysics*, **112**, 366–8.

Melrose, D.B. (1980). *Plasma Astrophysics: Nonthermal Processes in Diffuse Magnetized Plasmas*. New York: Gordon and Breach Science Publishers.

Meyer, F. & Schmidt, H.U. (1968). Magnetisch ausgerichtete Strömungen zwischen Sonnenflecken. *Zeitschrift für Angewandte Mathematik und Mechanik*, **48**, T218–21.

Molotovshchikov, A.L. & Ruderman, M.S. (1987). Long nonlinear waves in a compressible magnetically structured atmosphere IV. Slow sausage waves in a magnetic tube. *Solar Physics*, **109**, 247–63.

Neupert, W.M., Nakagawa, Y. & Rust, D.M. (1975). Energy balance in a magnetically confined coronal structure observed by OSO-7. *Solar Physics*, **43**, 359–76.

Newcomb, W.A. (1960). Hydromagnetic stability of a diffuse linear pinch. *Annals of Physics*, **10**, 232–67.

Newkirk, G. & Altschuler, M.D. (1970). Magnetic fields and the solar corona III: the observed connection between magnetic fields and the density structure of the corona. *Solar Physics*, **13**, 131–52.

Nocera, L., Leroy, B. & Priest, E.R. (1984). Phase mixing of propagating Alfvén waves. *Astronomy and Astrophysics*, **133**, 387–94.

Noci, G. (1981). Siphon flows in the solar corona. *Solar Physics*, **69**, 63–76.

Noci, G. & Zuccarello, F. (1983). Flow in coronal loops with a mass source. *Solar Physics*, **88**, 193–209.

Oran, E.S., Mariska, J.T. & Boris, J.P. (1982). The condensational instability in the solar transition region and corona. *Astrophysical Journal*, **254**, 349–60.

Osterbrock, D.E. (1961). The heating of the solar chromosphere, plages, and corona by magnetohydrodynamic waves. *Astrophysical Journal*, **134**, 347–88.

Pakkert, J.W., Martens, P.C.H. & Verhulst, F. (1987). The thermal stability of coronal loops by nonlinear diffusion asymptotics. *Astronomy and Astrophysics*, **179**, 285–93.

Pallavicini, R., Peres, G., Serio, S., Vaiana, G.S., Acton, L., Leibacher, J. & Rosner, R. (1983). Closed coronal structures. V. Gasdynamic models of flaring loops and comparison with *SMM* observations. *Astrophysical Journal*, **270**, 270–87.

Pallavicini, R., Peres, G., Serio, S., Vaiana, G.S., Golub, L. & Rosner, R. (1981). Closed coronal structures. III. Comparison of static models with X-ray, EUV and radio observations. *Astrophysical Journal*, **247**, 692–706.

Parker, E.N. (1953). Instability of thermal fields. *Astrophysical Journal*, **117**, 431–6.

Parker, E.N. (1977). The origin of solar activity. *Annual Review of Astronomy and Astrophysics*, **15**, 45–68.

Parker, E.N. (1979). *Cosmical Magnetic Fields*. Oxford: Oxford University Press.

Peres, G., Rosner, R., Serio, S. & Vaiana, G.S. (1982). Coronal closed structures. IV. Hydrodynamical stability and response to heating perturbations. *Astrophysical Journal*, **252**, 791–9.

Poletto, G., Vaiana, G.S., Zombeck, M.V., Krieger, A.S. & Timothy, A.F. (1975). A comparison of coronal X-ray structures of active regions with magnetic fields computed from photospheric observations. *Solar Physics*, **44**, 83–99.

Priest, E.R. (1978). The structure of coronal loops. *Solar Physics*, **58**, 57–87.

Priest, E.R. (ed.) (1980). *Solar Flare Magnetohydrodynamics*. New York: Gordon and Breach Science Publishers.

Priest, E.R. (1982). *Solar Magnetohydrodynamics*. Dordrecht: D. Reidel Publishing Company.

Priest, E.R. (1986). Magnetohydrodynamic theories of solar flares. *Solar Physics*, **104**, 1–18.

Pye, J.P., Hutcheon, R.J., Gerassimenko, M., Davis, J.M., Krieger, A.S. & Vesecky, J.F. (1978). The structure of the X-ray bright corona above active region McMath 12628 and derived implications for the description of equilibria in the solar atmosphere. *Astronomy and Astrophysics*, **65**, 123–38.

Raadu, M.A. (1972). Suppression of the kink instability for magnetic flux ropes in the chromosphere. *Solar Physics*, **22**, 425–33.

Rae, I.C. & Roberts, B. (1983). Wave diagrams for MHD modes in a magnetically structured atmosphere. *Solar Physics*, **84**, 99–103.

Raymond, J.C. & Smith, B.W. (1977). Soft X-ray spectrum of a hot plasma. *Astrophysical Journal Supplement Series*, **35**, 419–39.

Roberts, B. & Frankenthal, S. (1980). The thermal statics of coronal loops. *Solar Physics*, **68**, 103–9.

Roberts, B., Edwin, P.M. & Benz, A.O. (1984). On coronal oscillations. *Astrophysical Journal*, **279**, 857–65.

Rosner, R., Golub, L., Coppi, B. & Vaiana, G.S. (1978a). Heating of coronal plasma by anomalous current dissipation. *Astrophysical Journal*, **222**, 317–32.

Rosner, R., Tucker, W.H. & Vaiana, G.S. (1978b). Dynamics of the quiescent solar corona. *Astrophysical Journal*, **220**, 643–65.

Sakurai, T. (1976). Magnetohydrodynamic interpretation of the motion of prominences. *Publications of the Astronomical Society of Japan*, **28**, 177–98.

Sakurai, T. (1979). A new approach to the force-free field and its application to the magnetic field of solar active regions. *Publications of the Astronomical Society of Japan*, **31**, 209–30.

Schmidt, H.U. (1964). On the observable effects of magnetic energy storage and release connected with solar flares. In *AAS-NASA Symposium on the Physics of Solar Flares*, ed. W.N. Hess, pp. 107–114. Washington: NASA.

Schatten, K.H., Wilcox, J.M. & Ness, N.F. (1969). A model of interplanetary and coronal magnetic fields. *Solar Physics*, **6**, 442–55.

Schulz, M., Frazier, E.N. & Boucher, D.J. (1978). Coronal magnetic-field model with non-spherical source surface. *Solar Physics*, **60**, 83–104.

Schwartz, St.J. & Leroy, B. (1982). Propagation of waves in an atmosphere in the presence of a magnetic field. VI. Application of magneto-acoustic-gravity mode theory to the solar atmosphere. *Astronomy and Astrophysics*, **112**, 93–102.

Seehafer, N. (1978). Determination of constant-α force-free solar magnetic fields from magnetograph data. *Solar Physics*, **58**, 215–23.

Seehafer, N. (1982). A comparison of different solar magnetic field extrapolation procedures. *Solar Physics*, **81**, 69–80.

Semel, M. (1988). Extrapolation functions for constant-α force-free fields. *Astronomy and Astrophysics*, **198**, 293–9.

Serio, S., Peres, G., Vaiana, G.S., Golub, L. & Rosner, R. (1981). Closed coronal structures. II. Generalized hydrostatic model. *Astrophysical Journal*, **243**, 288–300.

She, Z.S., Malherbe, J.M. & Raadu, M.A. (1986). Transition zone effects on thermal non-equilibrium and plasma condensation in solar coronal loops. *Astronomy and Astrophysics*, **164**, 364–72.

Sillen, R. & Kattenberg, A. (1980). A low-β coronal loop model I: kink instabilities in the β = 0 limit. *Solar Physics*, **67**, 47–56.

Spicer, D.S. & Brown, J.C. (1981). Solar flare theory. In *The Sun as a Star*, ed. S. Jordan, pp. 413–70. NASA SP-450. Paris/Washington: CNRS/NASA.

Spruit, H.C. (1981). Magnetic flux tubes. In *The Sun as a Star*, ed. S. Jordan, pp. 385–412. NASA SP-450. Paris/Washington: CNRS/NASA.

Spruit, H.C. (1982). Propagation speeds and acoustic damping of waves in magnetic flux tubes. *Solar Physics*, **75**, 3–17.

Sturrock, P.A. (1980). Flare models. In *Solar Flares*, ed. P.A. Sturrock, pp. 411–49. Boulder: Colorado Associated University Press.

Svestka, Z., Solodyna, C.V., Howard, R. & Levine, R.H. (1977). Open magnetic fields in active regions. *Solar Physics*, **55**, 359–69.

Tandberg-Hanssen, E. & Emslie, A.G. (1988). *The Physics of Solar Flares*. Cambridge: Cambridge University Press.

Torricelli-Ciamponi, G., Einaudi, G. & Chiuderi, C. (1982). Diagnostic of coronal heating processes based on the emission measure of UV lines. *Astronomy and Astrophysics*, **105**, L1–4.

Tucker, W.H. (1973). Heating of solar active regions by magnetic energy dissipation: the steady-state case. *Astrophysical Journal*, **186**, 285–9.

Ulmschneider, P. & Bohn, H.U. (1981). Comments on the acoustic heating of stellar coronae. *Astronomy and Astrophysics*, **99**, 173–6.

Van Ballegooijan, A.A. (1985). Electric currents in the solar corona and the existence of magnetostatic equilibrium. *Astrophysical Journal*, **298**, 421–30.

Van der Linden, R., Goosens, M. & Hood, A.W. (1988). The effects of parallel and perpendicular viscosity on resistive ballooning modes in line-tied coronal fields. *Solar Physics*, **115**, 235–49.

Van Hoven, G., Ma, S.S. & Einaudi, G. (1981). The stability of coronal loops with realistic photospheric boundary conditions. *Astronomy and Astrophysics*, **97**, 232–4.

Van Tend, W. (1980). Coronal heating by prominence turbulence. *Solar Physics*, **66**, 29–37.

Vekstein, G.E. (1987). The theory of magnetic coronal heating. *Astronomy and Astrophysics*, **182**, 324–8.

Velli, M. & Hood, A.W. (1987). Resistive ballooning modes in line-tied coronal fields. II. Loops. *Solar Physics*, **109**, 351–4.

Vernazza, J., Avrett, E.H. & Loeser, R. (1981). Structure of the solar chromosphere. III. Models of the EUV brightness components of the quiet sun. *Astrophysical Journal Supplement Series*, **45**, 635–725.

Vesecky, J.F., Antiochus, S.K. & Underwood, J.H. (1979). Numerical modeling of quasi-static coronal loops. I. Uniform energy input. *Astrophysical Journal*, **233**, 987–97.

Voslamber, D. & Callebaut, D.K. (1962). Stability of force-free magnetic fields. *Physical Review, Second Series*, **128**, 2016–21.

Wallenhorst, S.G. (1982). Enthalpy flux cooling of the solar corona II. Applications to closed coronal structures. *Solar Physics*, **79**, 333–41.

Wentzel, D.G. (1974). Coronal heating by Alfvén waves. *Solar Physics*, **39**, 129–40.

Wentzel, D.G. (1976). Coronal heating by Alfvén waves, II. *Solar Physics*, **50**, 343–66.

Wentzel, D.G. (1977). On the role of hydromagnetic waves in the corona and the base of the solar wind. *Solar Physics*, **52**, 163–77.

Wentzel, D.G. (1979a). Hydromagnetic surface waves. *Astrophysical Journal*, **227**, 319–22.

Wentzel, D.G. (1979b). Dissipation of hydromagnetic surface waves. *Astrophysical Journal*, **233**, 756–64.

Wentzel, D.G. (1979c). Hydromagnetic surface waves on cylindrical flux tubes. *Astronomy and Astrophysics*, **76**, 20–23.

Wentzel, D.G. (1981). Coronal Heating. In *The Sun as a Star*, ed. S. Jordan, pp. 331–53. NASA SP-450. Paris/Washington: CNRS/NASA.

Wilson, P.R. (1980). The general dispersion relation for the vibration modes of magnetic flux tubes. *Astronomy and Astrophysics*, **87**, 121–5.

Woltjer, L. (1958). A theorem on force-free magnetic fields. *Proceedings of the National Academy of Science USA*, **44**, 489–91.

Wragg, M.A. & Priest, E.R. (1981). The temperature–density structure of coronal loops in hydrostatic equilibrium. *Solar Physics*, **70**, 293–313.

Wragg, M.A. & Priest, E.R. (1982a). The thermal stability of solar coronal loops in hydrostatic equilibrium. *Astronomy and Astrophysics*, **113**, 269–76.

Wragg, M.A. & Priest, E.R. (1982b). Thermally isolated coronal loops in hydrostatic equilibrium. *Solar Physics*, **80**, 309–12.

Xue, M.L. & Chen, J. (1983). MHD equilibrium and stability properties of a bipolar current loop. *Solar Physics*, **84**, 119–24.

Yeh, T. (1987). Polarity neutral lines on the solar surface and magnetic structures in the corona. *Solar Physics*, **107**, 247–62.

Yeh, T. & Pneuman, G.W. (1977). A sheet current approach to coronal-interplanetary modeling. *Solar Physics*, **54**, 419–30.

Žugžda, Y.D. & Locāns, V. (1982). Tunneling and interference of Alfvén waves. *Solar Physics*, **76**, 77–108.

Zweibel, E. (1980). Thermal stability of a corona heated by fast-mode waves. *Solar Physics*, **66**, 305–22.

Zweibel, E.G. & Boozer, A.H. (1985). Evolution of twisted fields. *Astrophysical Journal*, **295**, 642–7.

6

The plasma loop model of the coronae of the Sun and stars

Introduction

The preceding chapter has discussed the structure and heating of *single* plasma loops in the solar corona. We turn now to consider *systems* of loops, exploring the questions of the origin, evolution and global topology of plasma loops and their associated magnetic fields in the corona of the Sun and in the coronae of other stars.

Two major physical processes determine the structure and development of the plasma loops that are so frequently delineated in the solar corona. First, the coronal magnetic field in regions occupied by loops must conform to an elongated or tubular topology compatible with the presence of potential or force-free conditions in the corona (Sections 5.2.2, 5.2.3) and with the existence of spatially isolated sources with complex polarities intermingled in the photosphere. Secondly, the injection, transport and dissipation of matter and energy fluxes in the corona must be localized in and guided by this tubular magnetic topology. The anisotropic character of energy and mass transport coefficients in the presence of a magnetic field helps to explain this behaviour. However, anisotropic transport coefficients by themselves do not explain why only some magnetic tubes are delineated at any time. Spatially localized conversion and concentration of the mass and energy fluxes must also occur and, as we shall see, this aspect of the phenomenon is mediated by processes occurring beneath the visible layers of the Sun. Moreover, the simple tubular topology which appears on spatial scales of a few megametres or more can only arise if the coronal field undergoes significant reorganization (reconnection) as time passes, since the photospheric sources of the coronal magnetic field are observed to be changing persistently and inexorably in response to phenomena occurring inside the Sun. There is evidently a close link between the behaviour of the corona and the structure and evolution of the underlying chromosphere, photosphere and convection zone.

Further steps in our investigation of coronal plasma loops thus take us downwards from the corona into the deeper parts of the atmosphere and into the invisible layers of the solar interior. The next two sections of this chapter deal with these matters: Section 6.2 explores issues related to the generation, structure and evolution of magnetic fields beneath the corona, while Section 6.3 discusses the coupling of these fields into the corona.

In the final section we turn to the examination of phenomena related to the presence of plasma loops in the coronae of stars. Here, we do not have the abundance of detailed observations available for the Sun, as described in Chapters 2–4. However, tantalizing observations now exist of coronal phenomena in a large number of stars in diverse structural and evolutionary states, and the plasma loop model has been used widely to interpret these data. As we show, the stellar data reveal strong, systematic relations between the characteristics of the coronae and other stellar properties, providing important clues to the basic physical mechanisms responsible for coronal activity in the Sun and in other stars.

6.1   Solar magnetic fields

The observations and theories described in previous chapters have presented a compelling case for the central importance of coronal magnetic fields in the formation and maintenance of coronal plasma loops. But these fields do not exist in the corona in isolation from the rest of the Sun. On the contrary, it is believed that almost all magnetic field lines present in the corona at any time enter there by passing through the underlying parts of the atmosphere from the invisible solar interior. Moreover, the high electrical conductivity and relatively low density (inertia) of the coronal plasma imply that the behaviour of the coronal parts of the magnetic field is often likely to be a *response* to processes taking place in the deeper and denser layers. We thus turn now to consider the behaviour of magnetic fields and related phenomena in regions beneath the corona, in a search for the *origins* of many of the phenomena described in earlier parts of this volume.

A fairly comprehensive picture of photospheric magnetic fields (and related phenomena) has been built up, since we are able to make useful observations of many physical parameters in this region. Not surprisingly, many aspects of the theory of the photospheric field seem to be well understood. Direct observations of the behaviour of magnetic fields and related phenomena in the chromosphere and in the transition region between the chromosphere and the corona are not so straightforward, but nevertheless it is possible to explore many of the important physical mechanisms by judiciously linking observations and theory.

However, the structure and evolution of the chromospheric and photospheric fields are controlled in large measure by phenomena occurring in the deep zone of vigorous convection located beneath the visible layers, and we should move to the solar interior to begin our discussion. Here, the picture is much less clear since observational inferences are necessarily indirect and the physical conditions are difficult to study using currently available theoretical methods. Nevertheless, important aspects of the formation and evolution of coronal loops and loop systems can only be appreciated in the light of an understanding of the *coupling* provided by the magnetic field between the interior and the corona.

6.1.1 *The origin of magnetic fields*

Three key concepts underlie our understanding of the connections between magnetic fields in the solar interior and the behaviour of plasma loops in the corona. The first is the possibility of regenerative amplification of magnetic fields through the interaction of two (or more) flow patterns having different topologies (such as rotation and cyclonic convection). This process is embodied in theories that ascribe the existence of solar and stellar magnetic fields to dynamo action. The second is the tendency for convection to advect any magnetic flux frozen into the flow towards regions of streamline convergence, and thence to maintain the resulting magnetic topology in its intermittent state by adjusting the flow pattern and the field to be almost mutually exclusive. This process may be responsible for much of the highly inhomogeneous magnetic structure seen in the solar atmosphere and inferred in other stars. The third is the tendency for magnetic concentrations in the solar interior to rise to the surface (perhaps as a result of buoyancy) and to 'balloon out' into the atmosphere where they may participate in the fascinating phenomena associated with magnetic activity.

The combination of these three processes appears to be responsible for many aspects of the behaviour of magnetic fields – and hence plasma loops – in the coronae of the Sun and stars, and thus it is appropriate to present here an overview of the basic physics involved. The ideal MHD approximation, leading to the condition that the magnetic field is frozen into the flow pattern (Section 5.3.2), underlies much of the discussion, although non-zero resistivity is an essential feature of some aspects of the theory.

(1)   *Dynamo theory*

An extensive review of the theory of solar and stellar dynamos would take us too far from the subject matter of this monograph. However, we are concerned with several connections that exist between the physics of dynamos and the behaviour of coronal magnetic fields, including (i) the rate of formation of 'new' magnetic flux, its availability for injection into the corona, and its ultimate fate; (ii) the temporal development of coronal magnetic structure over the time scale of the solar (or stellar) cycle; and (iii) the connections between coronal behaviour and fundamental stellar properties such as mass, age and angular momentum. These issues are the focus of our discussion of dynamo theory.

It is generally believed that the physical conditions in the solar interior are such that the MHD induction equation provides an adequate description of the relation between the macroscopic velocity field $\mathbf{v}$ and the magnetic induction $\mathbf{B}$ (cf. Equation 5.30):

$$\frac{\partial \mathbf{B}}{\partial t} = \nabla \times (\mathbf{v} \times \mathbf{B}) + \eta \nabla^2 \mathbf{B}. \tag{6.1}$$

Here, $\eta = 1/\sigma\mu_0$ is the magnetic diffusivity, taken to be a scalar quantity in the present discussion. Regenerative hydromagnetic dynamo action arises when the flow pattern $\mathbf{v}$, established and maintained by some external force, generates by its inductive interaction with the magnetic field $\mathbf{B}$ just the current system $\mathbf{j} = \nabla \times \mathbf{B}/\mu_0$ required to provide the self-same field $\mathbf{B}$ (Moffatt, 1978).

Although it is known that a large class of velocity fields is capable of sustaining dynamo action in this way, it remains to identify the important flows involved in the putative solar dynamo. Many models have been based on a flow pattern first investigated in depth by Parker (1955b), who argued that non-uniform (differential) rotation will generate new toroidal field components from an initial poloidal field, while cyclonic upwellings and downdrafts with associated Ohmic decay (reconnection) will regenerate the poloidal components from the toroidal field. Parker (1955b) found simple solutions of the induction equation that arises when such motions are present, and showed that some of these correspond to migratory dynamo waves. His suggestion that these 'may be of interest in stellar magnetic activity' was certainly accurate, and many models of the solar dynamo have been constructed following this basic recipe. Overviews of the theory have been prepared by Moffatt (1978), Parker (1979) and Zel'dovich & Ruzmaikin (1983).

Partly appealing to the considerations outlined above, Babcock (1961) advanced an influential and remarkably complete account of the evolving

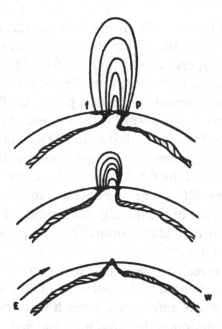

Fig. 6.1. The eruption of magnetic field lines through the solar photosphere and their penetration into the corona, according to Babcock (1961).

topology of the Sun's magnetic field throughout the 22-year activity cycle. Babcock's model describes the configuration and evolution of the magnetic field inside the Sun and in the solar atmosphere in five distinct stages, beginning about three years before the birth of an obvious new sunspot cycle. The stages are: (1) an initial dipolar (poloidal) field principally located inside the Sun, but visible as polar cap streamers; (2) amplification of the submerged parts of this field by differential rotation, with the formation of azimuthal (toroidal) field having local irregularities ('ropes') as a consequence of convective distortions; (3) emergence of 'stitches' in the toroidal flux ropes as a result of magnetic buoyancy, to form bipolar magnetic regions (BMRs) having many of the observed properties (polarity laws, etc.) of solar activity; (4) neutralization of the dipolar field by the dispersal of BMR flux in the sense that preceding (p) flux passes towards the equator (to merge with flux from the opposite hemisphere) while following (f) flux migrates to the poles (to form a dipolar field of opposite sense to the initial field); and (5) repetition of the cycle with the dipolar field of opposite sense.

Stage (3), involving the appearance of magnetic 'stitches' in the visible layers, provides (according to Babcock) the link between the behaviour of the internal field and the observed manifestations in the visible layers. The emergence of the stitch corresponds to the formation of contiguous patches of p and f magnetic flux in the photosphere. These patches are the 'footprints' of the magnetic lines of force in a flux rope which emerges from and returns to the solar interior. Above the photospheric BMR (see Fig. 6.1, reproduced from Babcock's paper), the '...arching field lines are related to condensations in the corona from which the disposition of the magnetic field lines can sometimes be inferred...'. As time passes, the BMR expands and the arching lines of force loop outward with increasing height. As Stage (4) passes, the 'full-blown flux loops' above old BMRs continue to expand towards each other, and large segments of antiparallel field are gradually pushed together. The ends of long flux loops may be severed and reconnected, and low-intensity loops of field lines are liberated to drift into the interplanetary region. A major part of the initially submerged toroidal field is thus gradually converted to individual coronal flux loops which reconnect and/or ultimately leave the Sun.

Although subsequent work has revealed problems with aspects of both the observational material and the theoretical foundations of Babcock's model, it remains even today a point of reference for much work on the solar cycle and the structure and evolution of the gross solar magnetic field. Its success as a *phenomenological* model can be judged by comparing the summary in the previous paragraph with the distillation of the observed behaviour of X-ray loops presented in Section 3.4.5. Almost all aspects of the observations can be woven into the model – and the model was developed more than a decade before the X-ray data became available!

Dynamo theory has been developed in many directions since the seminal work of Parker and Babcock. One of the most challenging problems has been the search for a satisfactory account of the contribution made by turbulent convection both in the dynamo itself and in promoting the differential rotation and other large-scale flows that are regarded as essential parts of virtually all theories of solar magnetism. The theory of mean-field electrodynamics (Krause & Radler, 1980) has been widely used to explore the role of turbulence in stellar dynamos, although the highly inhomogeneous nature of the observed fields and the need to rely on very crude models of convection point to the need for the development of more powerful analytical techniques than this approach exploits. In particular, it is difficult to see how a mean-field approach can make fruitful contact with the details of the time-dependent, spatially resolved coronal observations described in Chapters 2, 3 and 4 – and this is, of course, what is ultimately needed.

The dynamo problem is closely connected to the problems of the maintenance of differential rotation and meridional circulation (e.g. Durney & Sofia, 1987) and the turbulent diffusion of magnetic flux concentrations (e.g. Sheeley *et al.*, 1987). These are challenging problems that remain subjects of active research in magnetohydrodynamics. The observational data are being steadily improved and expanded, and new insights regarding the interactions between the gross patterns of surface magnetic fields and the global circulation of the Sun are emerging (e.g. Wilson, 1987). Of special relevance to our work is the possibility of developments stemming from the widening acceptance that the 'flux rope' character of the internal solar magnetic field may be central to the operation of the dynamo (e.g. Zel'dovich & Ruzmaikin, 1983) and from progress on the controversial question of the ultimate fate of magnetic fields that do emerge to become visible in the atmosphere. It is not clear whether these fields disperse by diffusion, become lost from the atmosphere by expulsion from the Sun, or retract into the interior (see Zwaan, 1987; Parker, 1988b). The answer to this question would help improve our understanding of the regeneration of the general field and also help explain the final stages of the evolution of magnetic loop systems.

(2)   *Magnetoconvection*
        An important but somewhat obscure aspect of Babcock's model concerns the formation of 'flux strands or ropes', which are required to account for the appearance of magnetic stitches in *localized* areas on the solar surface. The process of formation and maintenance of flux strands has since been investigated several times, and there is now a much sounder theoretical understanding of the tendency for flows containing eddies to concentrate magnetic fields into regions of convergence or divergence (hereinafter termed vertices) in the flow. The phenomenon can be illustrated by considering the distribution of magnetic field that is maintained by specified ('kinematic') flow patterns, according to the MHD magnetic induction Eqn. (6.1). As shown by Procter & Weiss (1982), one simple example involves a velocity field $\mathbf{v}$ in a cylindrically symmetric geometry with a solenoidal stagnation point flow satisfying

$$\mathbf{v} = (V/L)(-r\hat{\mathbf{r}}, 0, 2z\hat{\mathbf{z}}). \tag{6.2}$$

If the initial field $\mathbf{B} = B\hat{\mathbf{z}}$ is uniform and parallel to the $\hat{\mathbf{z}}$ axis, the induction equation has a steady-state solution in which the axial direction of the magnetic field is preserved, but the intensity is concentrated in a Gaussian distribution of the form

$$B(r) = \tfrac{1}{2}R_{\mathrm{m}} B_0 \exp\left(-\tfrac{1}{2}R_{\mathrm{m}} r^2/L^2\right). \tag{6.3}$$

Inward advection due to the flow is balanced everywhere by the outward diffusion of the field, Ohmic losses being sustained by the continual addition of flux from large radii. The field intensity on the axis is enhanced by a factor approximately equal to the magnetic Reynolds number ($R_m = LV/\eta$ – see Section 5.3.2) in this kind of model. Such enhancement would correspond to an extremely intense concentration given the conditions in the solar interior.

However, in many circumstances the field gradients cannot rise to the level required for efficient outward diffusion to occur, since the Lorenz force in the fluid momentum balance equation (Eqn. 5.33) becomes large enough to modify the flow pattern. This regime has been explored by Galloway & Moore (1979) and others, who have shown that, depending on the vigour of the flow, the total flux present and the field intensity, there are three domains of interaction. If the total flux is small, the maximum concentration is determined by the balance between advection and diffusion as described in the simple model above. If, on the other hand, the total flux and the field intensity is large the flow cannot distort the field grossly, and the phenomenon of oscillatory (overstable) convection will set in. For intermediate magnetic fluxes, the field may be concentrated to values lying somewhere between the equipartition value, B_{eq}, and the magnetostatic limit, B_P, defined by

$$B_{eq}^2 = 2\mu_0 \cdot \tfrac{1}{2}\rho v^2; \quad B_P^2 = 2\mu_0 \cdot P. \tag{6.4}$$

Here, ρ is the fluid density, v the fluid velocity and P the gas pressure in the fluid outside the flux concentration. Estimates of these limiting field strengths in the solar convection zone are illustrated in Table 6.1.

Should part of the magnetic field become concentrated at a vertex of the flow, the back-reaction of the Lorenz force on the flow will tend to maintain the separation of the field from the flow. Even when the flow is highly turbulent, new eddies will respect the topology of the field (if it is strong enough) as they form, thereby tending to maintain the separation of field and flow (Galloway & Weiss, 1981). Thus, the net outcome of turbulent magnetoconvection is an environment in which diffuse fields cannot persist: the preferred state is one in which the magnetic flux occupies a small fraction of the total volume, and adopts a topology which tends to remain relatively undisturbed.

The observed relations between magnetic field and flow patterns in the solar photosphere provide strong evidence in favour of the relevance of these physical concepts. For example, the tendency for magnetic regions and convective flows to be mutually exclusive is thought to account for the

Table 6.1. *Equipartition and magnetostatic magnetic fields*

Place	T (K)	P (Pa)	ρ (kg m^{-3})	$v^{(a)}$ (km s^{-1})	B_P (mT)	B_{eq} (mT)
Corona	10^6	0.014	1.2×10^{-12}	100	0.2	0.12
Chromosphere	10^4	0.017	1.4×10^{-10}	10	0.2	0.13
Photosphere	7×10^3	1.5×10^4	1.8	3	200	45
0.9 R$_\odot$	6×10^5	2.6×10^{11}	3.6×10	0.1	8×10^5	670
0.8 R$_\odot$	1.4×10^6	2×10^{12}	1.2×10^2	0.07	2×10^6	860
Base of convection zone	2.2×10^6	6.6×10^{12}	2.5×10^2	0.0	4×10^6	0

(a) Speed of sound in atmosphere; convective speed in interior.

appearance of discrete flux concentrations (or 'tubes') in the solar photosphere. Indeed, observations of the interactions between small flux tubes and the photospheric granulation and supergranulation provide direct evidence of the tendency for field to occupy regions of convergence at the boundaries of the turbulent eddies (Simon *et al.*, 1988; Hurlburt & Toomre, 1988). Sunspots, too, are associated with patterns of flow apparently modified by the presence of the field (e.g. the Evershed effect and the moat – see Moore & Rabin (1985)) although it seems probable that in this case the interaction between field and flow is due not only to induction and diffusion effects, but also to lateral pressure gradients arising from temperature differences.

It is important to recognize that the flux expulsion and concentration mechanism described above does not lead to a *steady* pattern of magnetic flux and fluid flow. Convection in the solar interior is highly turbulent, because the viscous drag on the flowing gas is extremely small compared with the inertia (i.e. the Reynolds number is very large), and the structure of the convection is ceaselessly changing (see Bray, Loughhead & Durrant, 1984). Consequently, the advective term in the magnetic induction equation is never steady, and the magnetic field continues to change as time passes. These changes are responsible (in part) for inducing variations in the electric currents flowing through the gross magnetic system* and, as we shall see, these currents may play a central role in coronal physics. The unsteady (turbulent) nature of the convection may also promote the dispersal of magnetic flux from the initially concentrated form that makes

---

* As described by Ampère's law, Eqn. (5.4).

up active regions and complexes, through a process of diffusion (e.g. Sheeley *et al.*, 1987).

(3)   *Magnetic flux emergence*
In addition to the action of Lorenz and inertial forces, magnetic flux concentrations in the solar interior may be subject to significant buoyancy forces (Parker, 1955a). These arise because the lateral balance of total pressure implies that

$$P_e = P_i + \frac{B^2}{2\mu_0},$$
(6.5)

where P_e and P_i are the external and internal gas pressures. If the temperature is uniform in the horizontal direction, the internal and external densities, ρ_e and ρ_i, then satisfy

$$\rho_e = \rho_i + \frac{B^2}{2\mu_0} \frac{m}{kT},$$
(6.6)

where m is the mean particle mass. The material in the magnetic region is less dense and thus it experiences an upward force due to buoyancy.

The buoyancy force may be countered by the Lorenz force associated with curvature of the magnetic field (Parker, 1984), by advection in certain patterns of flow (Arter, 1983) or by changes in the temperature of the surrounding plasma (Parker, 1987b, 1988a). These other forces may be sufficient to ensure that a magnetic concentration deep in the solar interior will remain 'floating' there for some considerable time: this tendency is required in many dynamo theories to permit sufficient time for flux amplification to occur (Parker, 1975; Galloway & Weiss, 1981). However, in the outer parts of the convection zone the gas pressure is relatively low and the magnetic pressure is relatively high, and there will be a strong – perhaps irresistible – tendency for magnetic flux concentrations to rise to the visible layers and beyond. The buoyancy of sub-surface flux concentrations may thus play an important role in the apparently inexorable rise of magnetic structures which takes place during the birth of an active region (Bray & Loughhead, 1964; Brants & Steenbeek, 1985), and it may contribute to the 'unity' exhibited by the magnetic elements that comprise active regions (Gaizauskas *et al.*, 1983). These processes in turn may provide a source of energy for the corona, either through the expansion work that must be done as new flux pushes into the corona (cf. Rosner, Low & Holtzer, 1986) or through electrodynamic stimulation of the often violent coronal phenomena that occur as new magnetic flux enters the atmosphere (see Section 4.2).

The simple picture of magnetic buoyancy has been criticized and modified by Parker (1988a), who has pointed out that the temperature distribution in the neighbourhood of a submerged magnetic concentration must be disturbed by the interaction between the field and the energy flux in the convective flow (see above). This interaction may tend to make the gas beneath the field somewhat hotter, and the gas above the field somewhat cooler, relative to the mean stratification, leading to reduced (or even reversed) buoyancy forces and Rayleigh–Taylor instabilities. The implications of this more refined picture include thermal relaxation oscillations and possible retraction of flux concentrations from the atmosphere to the interior. Both of these processes could have important consequences for the behaviour of coronal plasma loops. Parker's work exposes the probable complexity of the magneto-convective 'meteoro-logical system' in the solar interior, which may behave '...with all the statistical regularity and annoying local irregularity of the terrestrial atmosphere...'. The work also highlights the rudimentary level of knowledge we possess at this time regarding the behaviour of magnetic fields inside the Sun.

6.1.2 *Photospheric magnetic fields*

In the previous section we have reviewed the main ideas involved in the generation and structuring of magnetic flux within the solar convection zone, and discussed the processes that might be responsible for its eventual emergence into the solar atmosphere. We turn now to consider the behaviour of magnetic fields when they become visible in the photosphere.

Our most direct knowledge of the behaviour of solar magnetic fields comes from instruments designed to measure circular polarization in spectral lines formed in the photosphere. Interpreted in terms of the Zeeman effect, the observations made by these instruments provide fairly accurate and comprehensive information on the distribution in space and time of the longitudinal (i.e. parallel to the line-of-sight) component of the magnetic field, $B_\parallel$. Measurements of linear polarization, which would give insight into the transverse components of the magnetic field, are more difficult to make and thus our understanding of the three-dimensional structure of the field in the photosphere is far from complete. Stenflo (1984) has reviewed the techniques used to obtain and interpret Zeeman observations of magnetic fields in the solar photosphere.

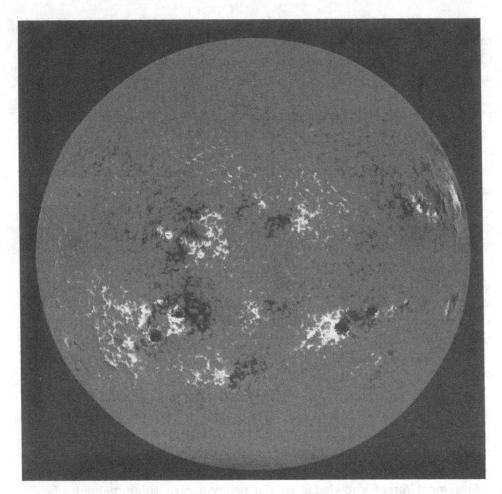

Fig. 6.2. An image of the line-of-sight component of the Sun's magnetic field (a magnetogram). Black areas indicate a field pointing away from the Earth, and white toward. North is at the top. Note the widespread existence of fine-scale magnetic fields. The structure of the coronal magnetic field is controlled to a large degree by the patterns of the magnetic field in the photosphere. Magnetogram courtesy of Dr J. Harvey, NSO/Kitt Peak, produced cooperatively by NSF/NOAO, NASA/GSFC and NOAA/SEL.

(a)   *Structure of the longitudinal field*

The result of such measurements at any instant of time may be displayed in a magnetogram, as illustrated in Fig. 6.2. This figure shows clearly the highly inhomogeneous or 'intermittent' character of the magnetic field, a fundamental property that is apparently a consequence of the turbulent convective flows beneath the photosphere (Section 6.2.1). It can be seen that the longitudinal component of the photospheric field is strong only in discrete concentrations that are surrounded by regions of

Table 6.2. *The hierarchy of magnetic elements*

Property[a]	Sunspot with penumbra		Pore	Magnetic knot (micropore)	Faculae, network clusters	Flux fibre
	Large	Small				
Φ (10^{10} Wb)	3×10^4	500	250–50	≈ 10	$\lesssim 20$	$\lesssim 0.5?$
R (Mm)	28	4	—	—	—	—
R_u (Mm)	11.5	2.0	1.8–0.7	≈ 0.5	—	$\lesssim 0.01$
B (mT)	290 ± 40	240 ± 200	220 ± 20	≈ 150–200	—	≈ 150
Overall contrast:	darker					brighter
Cohesion:	single compact structure					cluster
Behaviour in time:	remain sharp during decay, shrinking		—			modulated by granulation
Occurrence:	exclusively in active regions					both inside and outside active regions

[a] Φ – magnetic flux.
R – average radius of sunspot including the penumbra.
R_u – radius of sunspot umbrae and smaller elements.
B – magnetic field strength on axis of element.
Table after Zwaan (1987).

very low (or zero) longitudinal field strength. The concentrations themselves occur in a *spectrum* or *hierarchy* of scales reflecting a distribution of total magnetic flux, ranging down in size from the largest sunspots having more than 10^{14} Wb of flux, to very small (sub-telescopic) knots having no more than 10^9 Wb (Zwaan, 1978, 1987). Within the larger (spatially resolvable) concentrations there is abundant evidence for the existence of magnetic fine structure (e.g. Moore & Rabin, 1985). Although the peak field intensity is generally larger in structures containing a larger total flux, the variation in peak intensity is far smaller than that in total flux or, equivalently, in area. Table 6.2, after Zwaan (1987), summarizes the major systematic variations seen in the hierarchy of magnetic concentrations in the photosphere.

Although they appear to be formed by the accumulation of small-scale elements (Sheeley, 1981; Parker, 1979), mature large concentrations do not behave simply as a superposition of the component parts. On the contrary, different scales in the hierarchy of magnetic elements exhibit marked differences in a range of physical properties, including their temperature, lifetime and optical appearance. For example, Dunn & Zirker (1973) have shown that the smallest magnetic concentrations, known as filigree elements, facular granules or magnetic knots, are associated with weakenings in photospheric absorption lines and thus appear as bright dots in filtergrams and spectroheliograms – at fixed optical depth these features are apparently hotter than the ambient photosphere. On the other hand the largest concentrations, sunspots, develop a characteristic umbra–penumbra structure and are seen as very dark optical features (Bray & Loughhead, 1964; Moore & Rabin, 1985) – they are clearly colder than their surroundings. The size of a magnetic concentration clearly plays an important role in determining the outcome of its interaction with the surrounding gas.

The fact that magnetic elements of different sizes display characteristic differences in many spectral features provides a means to study magnetic concentrations by 'proxy'. This association between magnetic elements and other spectral features is particularly useful in guiding stellar work, since direct observations of stellar photospheric fields are extremely difficult (Section 6.3.4).

Not only do the magnetic concentrations occur in a spectrum of sizes, but they also exhibit a strong tendency to be organized into various patterns; i.e. the various elements exhibit a very non-uniform spatial correlation function. In the 'quiet' Sun (i.e. those extensive parts of the solar surface where the mean magnetic field is less than about 1 mT), the smallest observable features – the filigree elements – form chains or irregular rings (the origin of the term 'filigree') with scales of 1–2 Mm (Dunn & Zirker, 1973), while the filigree chains are themselves arranged into a diffuse web or network ('supergranular network') with a cell size of about 30 Mm (Leighton, Noyes & Simon, 1962). In regions of greater magnetic activity (i.e. mean fields greater than about 1 mT), the smallest observable features – facular points or granules – are often arranged in structures called enhanced network, and in regions with even higher mean field the facular granules become closely packed to form faculae. Faculae are in turn often associated with the other members of the hierarchy of magnetic elements – sunspots and pores – in even larger conglomerates called bipolar magnetic regions (BMRs), spot groups, active regions, or active complexes. The arrangement of the individual magnetic concentrations in such large active regions tends to follow certain regular patterns of size, number density and polarity that are detailed by Bray &

Loughhead (1964). These patterns are strongly imprinted onto the structure of the plasma loops located in the overlying corona (Sections 2.3.2; 2.4.2; 3.4.2), and they play an important part in determining the general character of the flare activity in the region.

Most of the characteristic structures assumed by the photospheric magnetic concentrations are seen to be associated with particular patterns of flow in the photospheric plasma lying in and adjacent to the magnetic fields. For example, the filigree tends to appear in the dark lanes between the photospheric granulation (Dunn & Zirker, 1973; Bray *et al.*, 1984) where there are downflows with velocities often exceeding 1 km s^{-1}. Strong vortices have been seen in such locations when observing conditions are propitious (Simon *et al.*, 1988). On a larger scale, the magnetic network is closely associated with the supergranulation, a pattern of velocity cells characterized by horizontal flows converging at the points and lines occupied by magnetic concentrations (Leighton *et al.*, 1962). The supergranulation pattern is significantly disturbed in the neighbourhood of active regions. Mature sunspots are often associated with radial outflows such as the photospheric Evershed effect (Bray & Loughhead, 1964) and the moat (Vrabec, 1974). The observed associations between plasma flow and field structure can be seen as strong evidence for the importance of the magnetoconvective phenomena discussed in the previous section.

(b)   *Time development*

A snapshot of the photospheric magnetic field at any instant reveals only an incomplete picture of the behaviour of the field. In fact, it is only through the study of the *evolution* of the photospheric field that the significance of many aspects of the magnetic structure becomes clear. We first consider the evolution of the magnetic structure in active regions.

New magnetic flux first appears in the photosphere as a bipolar field which significantly disturbs the normal pattern of photospheric granulation (Brants & Steenbeek, 1985). The bipoles seen in the photosphere must represent a horizontal cross-section through a magnetic arch, loop or stitch which often becomes outlined as a three-dimensional structure in the chromosphere through the formation of an arch filament system (Gaizauskas *et al.*, 1983). When it first appears in the photosphere, the new flux has a field strength of about 50 mT and is often associated with relatively dark material. A significant portion, but not necessarily all, of this dark material aggregates to form larger dark areas known as pores. The field strength in pores exceeds 200 mT. After a few hours the total flux in some aggregating sites may have increased to about 10^{12} Wb and well-developed pores will be present. Many instances of flux emergence go no further than this: the structure lasts for a day or so as a so-called *ephemeral active region*, and then its magnetic flux merges into the background (network) field or disappears by 'cancellation' (Livi *et al.*, 1985).

If flux continues to emerge, some of the pores will develop further to become (after a day or so) sunspots with growing penumbrae. Generally, spots of both magnetic polarities will appear, and they will move apart at about 1 km s^{-1} as the total flux in the newly-formed *active region* continues to increase. After a period of 2–3 days, more than 10^{14} Wb may have emerged and the active region will have well developed spots and faculae. Violent flares may take place in the chromosphere and corona above the photospheric active region, particularly when new magnetic flux emerges. The topological pattern taken up by the magnetic flux as it emerges (as defined by the distribution of the longitudinal component of the field in the photosphere) dominates the structure of the active region, and appears to play a major part in determining the behaviour of the plasma within and near the active region. In particular, the patterns taken up by coronal plasma loops are strongly conditioned by this topology (Section 5.2).

The flux forming an active region displays remarkable cohesion, for example by rotating in bulk at a rate faster than the neighbouring quiet photosphere and by displaying large-scale internal drifts (Vrabec, 1974; Zwaan, 1987). This behaviour points to the existence of a cohesive 'source' of the visible field located inside the Sun, and as noted above (Section 6.1.1, Fig. 6.1) the observations are remarkably consistent with many aspects of the phenomenological description advanced by Babcock (1961).

Observations reported by Gaizauskas *et al.* (1983) suggest that, although the total magnetic flux in an active site appears to be relatively constant over much of its lifetime, there is actually a persistent emergence and removal of flux within the region. The mode of removal has not yet been clearly identified, but it does not appear to be the diffusive process described by Babcock (1961). However, when an active site reaches the end of its life, sunspots are seen to break up by shedding their flux in small concentrations that flow from the spot to surrounding faculae (Vrabec, 1974). Faculae themselves dissolve slowly as a result of the incursion of supergranulation into the active region: the facular magnetic field then becomes a part of the enhanced network, and eventually a part of the normal network in the quiet Sun. Because the two magnetic polarities of the active region remain relatively well separated, the dispersal of the magnetic flux leads to the formation of rather large areas dominated by magnetic flux of one polarity: these are called unipolar regions. It has been suggested by Babcock (1961), Sheeley *et al.* (1987) and others that this dispersal of active region flux reflects the diffusion or random walk of the component flux concentrations in response to the ever-changing pattern of turbulent convection.

The life cycle of individual active regions takes place within the ambit of the longer 22-year solar cycle which is characterized by its own idiosyncratic patterns of change. For example, although new magnetic flux may appear anywhere in the photosphere in ephemeral regions, the large flux conglomerations that become active regions appear preferentially at certain latitudes, obeying the Hale polarity 'laws' and the phenomenology of the 'butterfly diagram' (see Bray & Loughhead, 1964). Active regions often tend to reappear at particular localities in the photosphere that are known as activity complexes (Gaizauskas *et al.*, 1983). As each active complex disperses, the old flux pattern is distorted by differential rotation and part of it is carried towards the poles, where its presence is marked by polar filaments and polar coronal holes. This pattern of magnetic evolution through the cycle is associated with remarkable changes in the form of the solar differential rotation, including a torsional oscillation and possible long-term changes in the rotation rate of each hemisphere (Howard & LaBonte, 1980).

The phenomenology of the solar cycle has been the subject of intense study, and important new insights into the evolution of magnetic and velocity patterns have emerged in recent years (e.g. Wilson, 1987). As we have indicated above, explanations for many aspects of these observations have been developed in terms of the theories described in Section 6.1.1. However, there is still no satisfactory answer to many of the most important problems, including the nature of the mechanisms controlling (i) the spectrum of sizes in the emerged field, (ii) the striking spatial correlations between magnetic elements and (iii) the dispersal and cancellation of the visible field. These processes are, of course, of great importance in the study of the form and development of coronal plasma loops.

(c)   *Theory of photospheric magnetic fields*
Whilst the spectrum of sizes and the form and evolution of the spatial correlation function of the photospheric field is likely to be controlled from below, many aspects of the behaviour of flux concentrations *in the photosphere itself* can be explained in terms of the simultaneous action of two basic physical processes: (i) magnetostatic confinement of a magnetic flux tube by the pressure of a surrounding field-

free plasma and (ii) interference by the magnetic tube with the normal radiative and convective modes of energy transfer in and near the tube.

A steady-state model of a flux tube controlled by these two factors may be constructed by satisfying simultaneously equations describing force balance and the conservation of energy, in the form (cf. Eqns. (5.33)–(5.34))

$$\nabla P = \rho \mathbf{g} + \mathbf{j} \times \mathbf{B} \tag{6.7}$$

$$\nabla \cdot (F_{\mathrm{R}} + F_{\mathrm{C}} + F_{\mathrm{M}}) = 0. \tag{6.8}$$

Here, we have expressed the principle of energy conservation as a symbolic balance between the divergences of the radiative energy flux, F_{R}, thermal convection, F_{C}, and any mechanical or electrodynamic energy flux that may be present, F_{M}.

Although satisfactory models of magnetic flux concentrations in the photosphere and chromosphere must ultimately consider the simultaneous balance of force and energy flux symbolized by these equations, the problem is fiendishly complicated and no general solutions have yet been constructed. However, simplified models have been used with some success to illustrate much of the important physics. In particular, the *thin flux tube* approximation has been used by several authors (Spruit, 1976; Spruit & Roberts, 1983; Kalkofen *et al.*, 1986) to demonstrate the interaction between the following processes: (i) the reduction in gas pressure and density within the tube consistent with the lateral pressure balance embodied in Eqn. (6.5), (ii) the inhibition of convective heat flow by the field, tending therefore to cool the tube and reduce the pressure scale height within it, (iii) the formation of a 'Wilson depression' as a result of the reduction of opacity within the tube, (iv) the flow of radiant energy into the tube as a consequence of its relative transparency, changing not only the energy balance but also the optical appearance of the tube, and (v) the initial intensification of the tube through the collapse of the interior when it is first exposed to freely radiate at the photosphere. The combination of these effects leads to the formation of *intense* flux concentrations, with axial field strengths approaching the magnetostatic limit B_{P} defined in Eqn. (6.4). This corresponds to about 100 mT in a small structure with a shallow Wilson depression, and more than 300 mT in large spots having a deep Wilson depression. In addition, studies of *thick flux tubes* have shown that curvature and tension forces, omitted from thin models, play a central role in establishing the equilibrium shape of the magnetic structure (Low, 1982; Pizzo, 1986; Steiner *et al.*, 1986). Figure 6.3 illustrates the form of the magnetic configuration that might occur in topologically simple flux tubes: note in particular the 'necking' right at the photosphere and the rapid flaring of the field lines immediately above the photosphere.

Hypothetical magnetic configurations

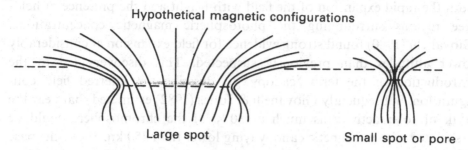

Large spot   Small spot or pore

Fig. 6.3. Hypothetical magnetic field configurations in a vertical plane passing through a magnetic cylinder that penetrates the photosphere (shown as a dashed line). Note how magnetostatic pressure balance leads to a marked constriction in the deep photosphere and to a rapid flaring-out of the field in the chromosphere. The diagram is from Pizzo (1986).

No self-consistent models of photospheric flux tubes containing all of the above physical processes have yet been constructed. There is a pressing need for a deep and thorough confrontation of theory by observation, to demonstrate that the combination of the listed processes is in fact able to account for the behaviour of individual tubes and for the systematic changes associated with the hierarchy described in Table 6.2.

6.1.3 *Chromospheric magnetic fields*

In the previous section we have seen that the photospheric magnetic field is concentrated into patches having field strengths of order 150 mT or larger, occupying only a small fraction of the total surface area. On the other hand, as explained in Section 5.2.1 the coronal magnetic field must be widespread and relatively uniform, for there is no way for the coronal gas pressure to confine it. By implication, the photospheric field must spread out with height through the chromosphere and transition region to connect the different spatial scales of the field in the photosphere and the corona.

Although it is more difficult to directly observe the chromospheric magnetic field than it is to study the photospheric field, several authors have offered direct observational evidence for the rapid spreading of photospheric fields as they enter the chromosphere. In particular, Pope & Mosher (1975) noted that magnetograms of active regions exhibit a more diffuse appearance in chromospheric lines (such as Ca II $\lambda 854$ nm) than in photospheric lines, and that the appearance of chromospheric lines near the limb is strongly indicative of inclined field lines in an expanding geometry.

The structure of the expanding field in the upper photosphere and chromosphere has been studied extensively by Giovanelli (1980) and Giovanelli & Jones (1982). By applying a diagnostic model that recognized both the rapid expansion of the field with height and the presence of field-free regions surrounding the photospheric magnetic concentrations, Giovanelli (1980) found strong evidence for field expansion at considerably lower altitudes than previously suspected. This discovery led to the introduction of the term 'canopy' to describe the inferred field configuration. Subsequently Giovanelli & Jones (1982) estimated that near the peak of solar activity as much as 30% of the chromosphere could be associated with a magnetic canopy lying lower than 750 km above the base of the photosphere; a much larger area would be associated with a magnetic canopy at higher altitudes.

An expansion of the magnetic field with increasing height in the chromosphere is expected on theoretical grounds, since the scale height of the gas pressure in the non-magnetic regions of the photosphere and chromosphere is only a few hundred kilometres. The equation of horizontal magnetostatic equilibrium (Eqn. (6.5)) then implies that the field strength must decrease rapidly with height, and flux conservation implies that the area occupied by the field must increase rapidly. Throughout the volume into which rapid spreading occurs much of the field will have a significant horizontal component. It is believed that the spreading field in the chromosphere is outlined in many places by prominent fine structure in the Hα line, as superpenumbrae, fibrils and dark mottles (see Bray & Loughhead, 1974). The low-lying Hα structures may be regarded as strong evidence for the existence of plasma loops in the chromosphere: in some regions these loops presumably meld into cool loops that extend to coronal heights (see Section 5.4.3 and Antiochos & Noci, 1986).

An estimate of the merging height of photospheric flux tubes has been offered by Spruit (1981), who argues that transverse magnetostatic equilibrium in an isothermal atmosphere leads to a merging height z_{m} given by

$$z_{\mathrm{m}} = H_{\mathrm{P}} \ln\left[\left(\frac{\pi B_0}{4\bar{B}}\right)^2\right]. \tag{6.9}$$

Here H_{P} is the pressure scale height and $\bar{B}$ and B_0 are respectively the mean field and the peak field in the photosphere. With $H_{\mathrm{P}} = 150$ km and $\bar{B}/B_0 = 15$ (active region) or 300 (quiet region), this model predicts merging heights of about 750 km and 1500 km respectively. These are compatible with the canopy heights inferred from the measurements and indicate in particular that the merging heights are rather insensitive to the mean field strength. We may conclude that the magnetic field at the *base* of the corona is much more uniform than the field in the photosphere. However, the highly inhomogeneous photospheric field still imposes significant variations on the field at the base of the corona (Sections 5.2.2, 5.2.3) and, in particular, the upper regions of large sunspots are thought to possess relatively intense fields (Bray & Loughhead, 1964).

As discussed in the previous section, a thorough investigation of the structure of magnetostatic flux tubes requires a numerical solution of the force balance equation (Eqn. (5.33)) combined with a detailed description of the energy balance. Solutions of this problem spanning the photosphere and chromosphere have been reported by Pizzo (1986), Pneuman, Solanki & Stenflo (1986) and others. They reveal a complex picture in which gas pressure gradients, curvature forces and the Lorentz force all play a part in determining the configuration of the expanding field.

The picture of canopies and magnetic field merging described above fails to include the wealth of dynamical processes observed in the chromosphere. In particular, energetic mass transfer appears to occur in spicules and fibrils (Bray & Loughhead, 1974) and in the high-velocity 'jets' seen in EUV observations of the transition region (Dere, Bartoe & Brueckner, 1983). It has been suggested by several authors that this mass transfer may be associated with energy (or enthalpy) flows that play a key part in the energy budget of the chromosphere and corona (e.g. Pneuman & Kopp, 1977; Athay, 1982). We may speculate that these phenomena are associated with cool, low-altitude plasma loops in the chromosphere, but as pointed out in Section 2.4.1 the data available now present only a tantalizing glimpse of a possible loop phenomenon that remains to be explored.

6.1.4 *Summary*

The Sun possesses a deep and vigorous convection zone in which turbulent flows carry much of the solar luminosity. Solar magnetic fields may be formed by regenerative dynamo action deep within this region. If this is so the dynamo presumably controls the overall form of the solar cycle. The rate of generation and the large scale patterns of the resulting fields are thought to depend on the interaction between convection and rotation, although the detailed physical mechanisms are not yet fully understood. The turbulent flows occurring within the solar convection zone promote and maintain localized concentrations in the magnetic field as a consequence of the interplay between the advection of frozen-in flux to vertices in the flow, and the subsequent adjustment of the flow pattern through Lorenz forces.

The magnetic concentrations produced in this way will be buoyant if they are in thermal equilibrium with their surroundings, and unless the buoyancy is counteracted the flux concentration will rise towards the visible layers. Magnetoconvection and buoyancy, in combination, may thus account for the appearance in the photosphere of magnetic flux in localized concentrations. The observed relations between convective flows and the photospheric field are compatible with this general picture. However, we have little insight at present into the processes responsible for determining the spectrum of sizes present in the convection zone or visible in the photosphere, nor do we have a satisfactory understanding of the remarkable cohesiveness of magnetic elements in the photosphere.

The pattern of confined spots, pores, facular granules and filigree elements that occupies the photosphere at any time represents a *boundary condition* on the coronal magnetic field. The fact that the subatmospheric convection is highly turbulent ensures that the structure of the magnetic field in the interior and in the photosphere is ceaselessly changing. Most importantly, the evolution of the photospheric pattern imposes a need for adjustment on the coronal field (Section 6.2) and must thus be responsible for time-dependent processes in the corona. The time scales involved range from a few tens of minutes in regions of vigorous flux emergence to years on the scale of the largest patterns of the photospheric field. The eventual dispersal of flux concentrations may be due to a combination of retraction, expulsion and diffusion by the turbulent convection.

At first sight it might appear that the existence of highly inhomogeneous magnetic fields in the photosphere – i.e. *photospheric flux tubes* – provides a ready explanation for the existence of *coronal plasma loops*. One might simply envisage that the photospheric footpoints preserve their identity as they project into the corona. Each coronal plasma loop could then be regarded as an identifiable magnetic structure connected to a pair of photospheric footpoints. Were this true, the behaviour of an individual photospheric flux tube would be expected to play a large role in controlling the behaviour of the coronal loop attached to it.

It is not clear, however, that this simple picture is correct. Observation and theory both show that the photospheric magnetic concentrations expand rapidly with increasing altitude, implying that the field lines of individual photospheric structures tend to merge in the chromosphere. Some of the magnetic field lines may even close back to the photosphere without extending above the chromosphere. Although observations do reveal continuity in what are believed to be magnetically controlled phenomena linking the photosphere, chromosphere and transition region, no observations have established as yet the existence of a simple one-to-one connection between photospheric magnetic elements and coronal flux tubes. Indeed, one of the most challenging and significant problems in solar physics is the pressing need for a clearer picture of the physical linkages due to magnetic fields threading the solar atmosphere.

6.2   Magnetic fields and coronal plasma loops

In the previous sections we have discussed the physical character of some of the main magnetic processes thought to occur in the interior of the Sun. We have also explained how these processes lead to observable phenomena in the photosphere and chromosphere. These processes and phenomena underpin the behaviour of the coronal magnetic field, and we are now in a position to trace the connections in detail.

The first problem to be addressed concerns the topology of the coronal magnetic field, as controlled by the boundary conditions imposed at the photosphere. This problem was introduced in Sections 5.2.2 and 5.2.3, where it was shown how potential field solutions can be constructed outside any given photospheric field distribution. It was also noted that several studies have established a 'general correspondence' between field distributions calculated in this way and the observed coronal structure. However, a fundamental difficulty to be confronted by potential field models arises when it is recognized that the presence of *highly conductive coronal plasma* restricts the ability of the coronal magnetic field to respond to *changes* in its boundary conditions. This fact has led to a heated and still unresolved debate regarding the evolution of the topology of the corona, which is explored in detail in the next section.

A more general problem emerges from a study of the structure of the coronal magnetic field, namely the precise nature of the coupling between the sub-photospheric convection zone and loops in the solar corona. It has long been recognized that a satisfactory theory of coronal magnetic phenomena requires the establishment of concepts to account for the *links* between the corona and the deeper layers, but these have proved to be quite elusive. There are prospects for improving our description of these important but difficult problems through the investigation of the global electrodynamics of the entire magnetic system that threads the corona and the solar interior: the matter is discussed in detail in Section 6.2.2.

6.2.1   *Magnetic topology in the corona*

The complex character of the evolution of coronal magnetic fields in response to changes taking place beneath the photosphere has been recognized for many years. Gold (1963) provided an early but lucid picture of some of the important physical processes, when he described in qualitative terms the structure of the magnetic field in a (hypothetical)

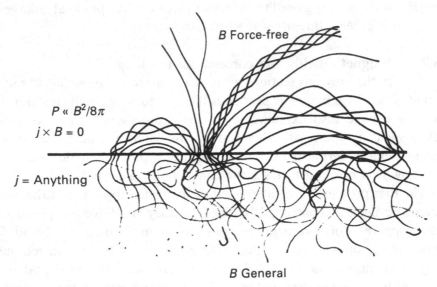

Fig. 6.4. The magnetic topology in a region consisting of a high density conductor (below) and a low density conductor (above). In the low density region the magnetic configuration must be force-free almost everywhere, since the gas pressure is unable to support magnetic pressure gradients. The twisted character of the field is due to the presence of turbulence in the high density region. The figure is from Gold (1963).

medium consisting of two parts bounded by a horizontal dividing plane (see Fig. 6.4). The medium beneath the plane is assumed to be a perfect conductor with a gas pressure larger than the energy density of the magnetic field, and with turbulent motions that can carry magnetic fields freely across the dividing plane. In this region the magnetic field and the electric current may be of any form, free from any particular restrictions on the topology since the gas pressure and the fluid accelerations can account for any imbalance in the electromagnetic stresses. The medium above the plane is also a conductor, but in it the gas pressure is supposed to be negligibly small compared with the magnetic energy density. As a result of the low pressure, the fields above the plane must be *force-free* (see Section 5.2.2) and must also match to the normal components of the field at the dividing plane. Electric currents will inevitably flow in the upper plane, although they will be restricted to lie parallel to the magnetic field vector.

Gold (1963) pointed out that as the motion of the system proceeds, destruction of magnetic energy must take place throughout the upper region. This may occur through the production of very high speeds (i.e. significant kinetic energy) in the upper regions which will then in practice dissipate through viscosity and heat conduction, or through the formation of magnetohydrodynamic instabilities in which the gas will be accelerated or compressed as the magnetic system adjusts to a stable configuration of lower energy content. Gold also noted that the magnetic field in the upper space is *not* uniquely defined by the distribution of the field at the dividing plane. It is '. . . not only the instantaneous distribution of the feet which determines the field above the plane but also the entire history of the fields and the motion . . .'. Gold continued by discussing the implications of the destruction of magnetic energy in the context of solar flares.

The qualitative picture sketched by Gold has been remarkably resistant to quantitative theoretical analysis. In particular, the problem of the structure of a force-free magnetic field evolving in response to changes in the boundary conditions has been difficult to solve. One approach to the problem (e.g. Low, 1981) involves the search for a sequence of stable solutions of the equations of magnetostatic equilibrium as *ad hoc* simple magnetic boundary conditions are systematically and continuously changed. Provided that the changes take place slowly enough (e.g. compared with the Alfvén transit time in the magnetic configuration), the problem can be regarded as a stability analysis of each stage of the sequence of configurations adopted by the field. Several models have been found in which the magnetostatic equilibrium (usually containing singular current sheets or lines) evolves continuously until a maximum distortion is attained beyond which there is no stable neighbouring configuration. Such models may be particularly pertinent in solar flares and coronal mass ejections, but they often appear to predict catastrophic eruption of the magnetic field and are thus unlikely to be relevant in explaining the formation and continuing existence of coronal plasma loops and loop systems.

A rather different picture, developed by Parker (e.g. 1983a, b; 1988b) in particular, suggests that the magnetic field in a low pressure, highly conducting plasma with boundary conditions that are ceaselessly readjusting will *never* exist in an equilibrium configuration. Instead, the close-packed magnetic tubes will be subject to dynamical nonequilibrium and what Parker terms *topological dissipation*. The dissipation takes the form of neutral point reconnection at those sites where there is a topologically unavoidable reversal of sign of the transverse component of the magnetic field, for example across the boundaries of contiguous flux tubes having the same sense of twist. By considering idealized topologies and specific boundary motions, Parker was able to explore quantitatively the implications of Gold's (1963) claim that magnetic dissipation must occur throughout the low pressure 'coronal' region. Parker forcefully supports Gold's conjecture with the conclusion that '. . . all transverse field components are subject to rapid reconnection and dissipation, so that the work done by the stochastic rotation and shuffling of the footpoints of the flux tubes proceeds directly into heat . . . no matter how large the electrical conductivity of the gas in which the tubes are embedded . . .'.

Parker's conclusions have been disputed. For example, van Ballegooijen (1985) exhibited a class of three-dimensional magnetostatic equilibria in which footpoint motion does not inevitably lead to the formation of current sheets in the model corona. Zweibel & Li (1987) and Antiochos (1987) also presented calculations of sequences of magnetostatic equilibria in which the field topology is determined by the boundary displacements without essential nonequilibrium or the spontaneous formation of singularities. These apparent counter examples to Parker's work have been criticized by Parker (1987a) and by Low (1987). In particular, Low (1987) showed by means of a simple example how the interplay between the need for force balance and the need to satisfy the frozen-in condition in a perfect conductor can lead to the formation of a current sheet as the result of the continuous deformation of the footpoints. The general validity of the Gold–Parker picture was further confirmed by non-linear numerical simulations reported by Strauss & Otani (1988), who showed how kink-ballooning instabilities arise as one- or two-dimensional initial configurations are twisted by motions imposed at anchored endpoints. These instabilities evolve (non-linearly) into three-dimensional, near-to-equilibrium states containing current sheets. As Strauss & Otani pointed out, once a current sheet has formed ohmic heating becomes very efficient: further twisting will not drive any more current, but will only produce more ohmic dissipation.

This work on the dynamical and topological response of the coronal magnetic field to motions imposed by the underlying convection zone provides an important link between the theory of magnetoconvection and photospheric magnetostatic flux tubes on the one hand (Section 6.1.2), and the structure and energy balance of coronal plasma loops on the other (Section 5.4). Perhaps the most accurate picture that can be drawn at present shows the gross pattern of the coronal field closely conforming to a force-free configuration compatible with the instantaneous magnetic (and dynamical) disposition of the footpoints. The coronal configuration almost certainly contains many singular surfaces and lines on which the current density is large enough to generate self-magnetic fields having strengths comparable with the 'ambient' potential field, and in which intense (anomalous) ohmic heating may be occurring.

As the footpoints move in response to changes in the convection pattern, the system of magnetic tubes and current singularities will evolve continuously. Dissipation will tend to destroy tangential components in the field while the longitudinal pattern will tend to be more stable. This process will be seen as excitation and re-excitation of particular sites in the corona: the topology of the field lines passing through these sites will be predominantly longitudinal (since the non-force-free components are persistently dissipating) and the fact that mass and energy transport along the field vector is relatively efficient will help to ensure that loop-like structures are seen. On occasion, however, the evolving field may encounter a state where smooth change is no longer possible, and a sudden and major change in the topology of the field and the current systems may take place. Should this occur, the associated coronal loop system will evolve rapidly, and if sufficient energy is dissipated a flare will take place.

Although this qualitative description may provide a plausible outline of the physics of coronal plasma loops, it is far from satisfactory since so few of the suggested ingredients have yet been thoroughly explored. Many important questions must be answered before the final picture is clear, and the concepts discussed in the next section promise to assist this process.

6.2.2 *Global electrodynamics*

In the discussion above we have been primarily concerned with questions regarding the topology and evolution of the magnetic field in the solar corona. In these discussions the presence of material plasma has been important only insofar as the systems of electric currents induced in the conducting matter may exert a strong influence on the magnetic structure and topology. However, coronal plasma loops are in the final analysis composed of matter as well as field, and we must now turn to consider the processes that condition the detailed physical state of this matter.

As emphasized in Chapter 5, '... coronal loops cannot be understood in isolation. They form systems which are essentially governed by the high density regions of the Sun in which they are rooted ...' It has proved to be a difficult task to characterize the physical nature of these systems, but substantial progress has been made in recent years through the development of the concept of electrodynamic coupling.

Early recognition of the possible importance of electrodynamic interactions between the solar interior and atmosphere appears in the qualitative model of heating in flares and the active corona advanced by Gold (1963 – see above). Tucker (1973) developed the model further, first by providing an order-of-magnitude estimate of the rate at which magnetic energy can be generated in the convection zone and then by showing how this energy flow might be responsible for coronal heating. Specifically, Tucker argued that the 'rate of generation of magnetic energy' is given by

$$\dot{W}_m \approx B^2 v_\phi A / \mu_0, \tag{6.10}$$

where v_ϕ is the effective speed at which the photospheric magnetic field is twisted and A is the area of the photospheric region in question. Tucker showed that the dissipation of this magnetic energy in the corona cannot proceed by classical (Coulomb) ohmic losses, since the current density required to provide sufficient heating would definitely lead to (twisted) magnetic structures that are not observed. He thus suggested that the dissipation would occur in small volumes where the current density is large and the plasma is in a turbulent state (i.e. in current sheets). Tucker showed that the observed coronal power loss of about 3 kW m^{-2} could be supplied by current sheets dissipating the magnetic power that would be generated by photospheric conditions corresponding to the quantity $B^2 v_\phi$ being of the order of $(20 \text{ mT})^2 \times 10 \text{ km s}^{-1}$. This result represents a plausible balance between generation and loss and lends some support to the model. Moreover, Tucker pointed out that the expected alignment between the magnetic field and the X-rays in this model (arising because the energy release is localized and the transfer is oriented preferentially along the field lines) is also consistent with observations, as is the fact that the X-ray emission measure is a swiftly falling function of temperature.

Although Tucker (1973) was obliged to make a number of coarse approximations in this pioneering study, the possible importance of the electrodynamic linkage between photospheric motions and coronal structure and heating had been quite firmly established. Several authors have subsequently refined aspects of the theory. For example, Golub *et al.* (1980) combined an estimate of the photospheric heating rate similar to that given in Eqn. (6.10) with the scaling relations of Rosner, Tucker & Vaiana (1978) for a single coronal loop (see Section 5.4.2) to arrive at a relation between the coronal base pressure P_c and the average longitudinal photospheric magnetic field strength $\langle B \rangle$ *in an entire active region* of the form

$$P_c \sim (\alpha v_\phi)^{0.86} \langle B \rangle^{1.71} L^{-0.14}. \tag{6.11}$$

Here, αv_ϕ is the product of the twist $\alpha = B_\phi / B_z$ and the azimuthal velocity v_ϕ, while L is the size of the active region. The connection between the scaling law for a single loop and that for an entire active region emerged from the assumption that the region consists of loops having identical aspect ratios (i.e. the length-to-diameter ratio is constant), combined with a length distribution determined by an *ad hoc* field distribution in the photosphere (and not on the erection of a self-consistent, approximately force-free field). Golub *et al.* (1980) showed that there is a fairly tight correlation between the observed thermal energy content and the value predicted using the above scaling law in active regions of very different sizes, suggesting that the quantity $\alpha v_\varphi \approx 0.5 \text{ km s}^{-1}$ over many active regions. Why this product should be an invariant is not explained, but the consequent agreement between observation and theory when this relation is adopted is (rather curiously) advanced as evidence in favour of the model.

Sturrock & Uchida (1981) have presented a similar calculation with a rather more detailed investigation of the relation between the azimuthal components of the photospheric velocity field and what they regard as the *consequent* toroidal component of the photospheric magnetic field. Combined with a simplified analysis of the stochastic properties of the photospheric velocity field, this work leads to an estimate of the mean rate of increase of energy in the flux tube of the form

$$\dot{W}_m = \frac{1}{\mu_0} (B_* B_c) \frac{\langle v^2 \rangle}{(L/\tau_c)} \left(\frac{V}{L}\right), \tag{6.12}$$

where $V = \pi R_c{}^2 L$ is the volume of the flux tube *in the corona*, B_c is the mean coronal magnetic field strength, B_* is the photospheric magnetic field strength and $\langle v^2 \rangle \tau_c$ is the product of the mean square amplitude and the correlation time of the photospheric velocity field. This result is combined with the loop scaling relation due to Rosner *et al.* (1978) to predict relations between density, temperature, pressure and magnetic field in coronal flux tubes. Although accurate observational tests of the theory are not possible, Sturrock & Uchida (1981) concluded '. . . that the heating mechanism proposed in this article is not obviously ruled out by observational data . . .'

The work described above has tentatively demonstrated the viability of coronal heating models based on the basic idea of (non-classical) ohmic dissipation of coronal electric currents generated by the inductive effects of photospheric motions. However, it has proved to be difficult to develop the theory further, and even today there remain unsatisfying weaknesses regarding both the rigour of published analyses and the indistinct character of predictions flowing from the theory.

One development that promises to help elucidate important aspects of the problem is the growing recognition of the importance of global connections and processes. In a remarkable period of conceptual development in the early 1980s, Ionson (1982, 1983), Kan, Akasoufu & Lee (1983), Kuperus (1983), Spicer (1983) and others strove to clarify the

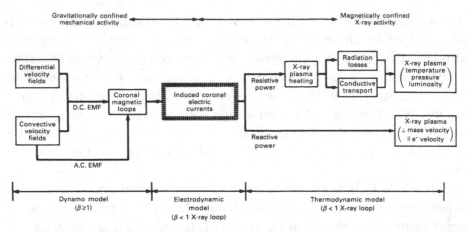

Fig. 6.5. A schematic of the electrodynamic coupling model of coronal structure due to Ionson (1983). According to this model, convective flows in the high-density sub-photospheric regions induce electric currents in the corona. The dissipation of these currents controls the thermodynamic state of the corona. In many cases the coupling takes place through magnetic systems with a loop-like topology.

nature of the *global electrodynamic coupling* between the photosphere and the corona. Figure 6.5, from Ionson (1983), illustrates the general principles underlying these studies of global electrodynamics. According to these principles, the thermodynamic state of the corona and its constituent loops is due to the dissipation of a Poynting flux of electromagnetic energy. This energy is carried to the corona as a system of electric currents along the common magnetic field that threads both the corona and the photosphere. The electromagnetic energy flux is produced by conversion of mechanical energy in the photospheric and sub-photospheric parts of the magnetic field.

Many aspects of this global phenomenon are the subject of active research and to controversy. For example, Spicer (1983) – see also Spicer, Mariska & Boris (1986) – has addressed the question of the generation of electric currents flowing parallel to the magnetic field direction: these currents directly connect the photosphere and the corona. According to Spicer *et al.* (1986) the field-aligned current, $j_{\parallel}$, can be regarded as the consequence of a cross-field current, $j_{\perp}$, obtained through the condition $\nabla \cdot j_{\parallel} = -\nabla \cdot j_{\perp}$. The cross-field current in turn is produced by mechanical effects such as pressure gradients or accelerations, according to

$$\mathbf{j}_{\perp} = \frac{\mathbf{B} \times \nabla P}{B^2} - \frac{\rho}{B^2} \frac{d\mathbf{v}}{dt} \times \mathbf{B}. \qquad (6.13)$$

A similar model underlies the work of Ionson (1983) and Kuperus (1983).

The electrical conductivity of the photosphere (and subphotosphere) does not appear explicitly in this equation, despite the fact that Ionson (1983), Kuperus (1983) and Spicer (1983) all emphasize the crucial importance of non-zero resistivity in the mechanical driving regions.

Kan *et al.* (1983), on the other hand, have developed a theory of the generation of field-aligned currents in the photosphere by the action of a 'neutral wind' involving differential flow of the ionized and neutral components across the magnetic field lines. This approach explicitly refers to the resistivity of the high density photospheric plasma. The study by Hénoux & Somov (1987) of a system of photospheric and coronal currents in a flaring activity complex follows a similar line in its discussion of the generation of the coronal current systems. Since the electrical resistivity of the photosphere and subphotosphere is very low, it is not clear that a significant neutral wind can exist, and there is a pressing need for a deeper analysis of the 'microscopic' physical process responsible for the generation of cross-field and field-aligned currents in the deeper parts of the solar atmosphere and in the solar interior.

It must be stressed that the generation of coronal current systems cannot take place without the presence of a region of non-zero resistivity (since no closed system of entirely force-free currents can exist). As pointed out by Ionson (1983), there is an intimate connection between this problem and the more general problem of the behaviour of the dynamo responsible for generating and structuring the slowly varying and large-scale magnetic field of the Sun. It is, however, not clear that Ionson's separation of the current generation process into a DC part (the conventional dynamo problem) and an AC part (directly relevant to coronal heating) is certain to be fruitful, since the poorly understood phenomenon of 'turbulent' diffusion under-lying many dynamo theories presumably bridges the two aspects of the process.

Global coupling between the photospheric 'generator' and the coronal 'load' can be instructively illustrated by means of an equivalent circuit. The circuit equation can be constructed (Ionson, 1982, 1983) from the time-dependent electrodynamic induction equation (see Eqn. (5.30)) by a sequence of simplifications which basically rely on the assumption that plasma parameters vary slowly in space and time, and that the magnetic field guides the energy flow. The circuit equation is

$$\mathscr{L}\frac{d^2 I}{dt^2} + \mathscr{R}\frac{dI}{dt} + \frac{I}{\mathscr{C}} = \frac{d\mathscr{E}}{dt}, \tag{6.14}$$

where I is the current, $\mathscr{L} = \mu_0 l_\parallel / \pi^2$ is the inductance ($l_\parallel$ is the length of the dissipation region), $\mathscr{C} = l_\parallel / \mu_0 v_A{}^2$ is the capacitance and $\mathscr{E} = v_{phot} B_{phot} l_{B_{phot}} / \mu_0$ is the EMF due to the interaction between photospheric motions and magnetic fields having a length scale l_B. The resistance $\mathscr{R}$ is the sum of several parts (Ionson, 1984), one corresponding to dissipation in the photospheric regions, and others corresponding to various 'modes' of dissipation in the outer atmosphere. Dissipative modes included in the model to date encompass viscous and Joule heating, phase mixing of wave modes as a means of limiting the amplitudes of coronal waves and the leakage of energy from the circuit by the release of magnetic stresses.

The electrodynamic coupling model has the interesting implication that in loops with slow dissipation rates (compared with the Alfvén transit time), the dissipation does not depend explicitly on the resistance. Rather, the model implies that the *size* of the heated region will vary so as to absorb the energy flow that is available for conversion. This phenomenon is a close analogue of the fact that the electrical loss in a high-Q 'resonant' circuit* is controlled not only by the resistance but also by the properties of the circuit *in toto*. It might be noted that previous models of coronal electrodynamic heating (e.g. Tucker, 1973) also assume that the corona is capable of dissipating the available power from the photosphere, but do not include the additional constraints imposed by the need to satisfy a global circuit equation.

The possibility of resonance in loops with high quality Q has important consequences for attempts to test theories of coronal heating. In particular, tests based on energy balance considerations are likely to be of reduced importance in coronal structures (such as loops) that are able to adjust their dissipation to accommodate to the available power. Ionson (1983, 1985) has argued that the behaviour of young active-region loops suggests that they are low-Q structures. This implies that such loops will respond readily to the rapid dynamical readjustments required as they emerge into the corona. The heating mechanism in these loops might be identified by energy balance studies in association with observations of the changes in loop configuration. On the other hand, Ionson suggests that old active region loops and large scale coronal loops are most likely to be high-Q regions. In this case it is not possible to identify the heating mechanism by observing the X-ray heating rate as a function of loop length or magnetic field strength. Instead, a close comparison between the heating process and the associated coronal velocity fields may offer the best chance of identifying the energy conversion process. Ionson (1984) suggested that the

---

* Recall that $Q = \sqrt{(\mathscr{L}/\mathscr{C})}/\mathscr{R}$ is a measure of the sharpness of a resonance in a series LRC circuit.

older active region loops are heated by electrodynamic coupling to the five-minute oscillations while the large-scale old loops are coupled to the granulation.

According to the electrodynamic coupling picture, the appearance of the plasma loops in the solar corona at any time would be principally a consequence of (i) the length of the field lines between corresponding photospheric footpoints and (ii) the spectrum of velocity fields in the photosphere near these footpoints. While this suggestion is interesting, it does seem that further work should be done on the global electrodynamic model to establish a more convincing argument connecting photospheric velocity fields with specific coronal properties.

Heyvaerts & Priest (1984) have made significant progress in this direction by developing a novel (albeit approximate) global heating theory applicable in conditions where the time scales of the photospheric velocity fields are 'slow' compared with the Alfvén transit time in the loop: the so-called DC case.* These authors made two important points regarding the assumption that the electrodynamic energy generated in the photosphere is entirely dissipated in the corona. First, they noted that a non-potential coronal magnetic configuration may be 'de-stressed' by certain footpoint motions, leading to a downward flow of energy. Second, they noted that part of the magnetic energy fed into the corona may not be converted to heat on a short time scale, since ohmic losses may proceed very slowly. Thus, there is a need to develop a theory containing a description of the lowest energy configuration accessible from a given initial state: this permits the division of the generated energy into the part available for coronal heating and the part stored in non-potential coronal magnetic fields.

To illustrate this division, Heyvaerts & Priest (1984) considered the evolution of a magnetic arcade subject to shearing motion of the footpoints. By appealing to a variant of Taylor's hypothesis (cf. Section 5.2.2), they suggest that a DC current flow provides a power $\mathscr{F}$ to the corona given by

$$\mathscr{F} = \mathscr{F}_{\max} \left(\frac{l_B}{l_B + l_V} \right)^2 \left(\frac{\tau_{\mathrm{rec}} v_A}{l_B} \right), \qquad (6.15)$$

where

$$\mathscr{F}_{\max} = \frac{B_o^{\,2}}{\mu_0} v \qquad (6.16)$$

is the maximum Poynting flux available (cf. Eqn. (6.10)). The power available for heating is reduced relative to the maximum by two factors, one, $(l_B/(l_B + l_V))^2$, related to the geometrical match between the length

* AC and DC mechanisms are discussed in Section 5.6.

scales of the magnetic and velocity fields in the photosphere and the other, $(\tau_{rec} v_A / l_B)$, to the ratio of the reconnection time τ_{rec} to the Alfvén transit time. The implication of the appearance of the factor τ_{rec} is that the actual heating rate will depend on the detailed physics of the dissipation process. It is interesting to note that although the form of Eqn. (6.15) is identical to that derived by Ionson (1983), there is an inconsistency between Ionson's model and the work of Heyvaerts & Priest since the former formally predicts zero heating in the limit of slow motions. This result has been discussed and satisfactorily explained by Heyvaerts & Priest (1984) and Ionson (1985). Browning, Sakurai & Priest (1986) have applied the approach of Heyvaerts & Priest to investigate the heating of closely-packed flux tubes excited by various cellular flow patterns, showing that the observed heating rates are consistent with plausible photospheric flow.

In an entirely different approach to the problem of electrodynamic coupling, Steinolfson & Tajima (1987) have used a numerical method to solve the momentum balance and induction equations for a constant density plasma in which forces due to gas pressure gradients and gravity are ignored. The simulation, applied to a cylinder twisted in opposite senses at either end, reveals the evolution of the system through initial force-free states, to states in which the kinetic energy of the fluid is sufficient to influence the system, and ultimately to configurations which appear to be unstable. Despite the crudeness of the model, the computations provide considerable insight into the processes that might occur in a coronal magnetic configuration subject to stresses at the photospheric footpoints. The indications are that numerical studies of this kind have much to offer including, perhaps, close contact with observations once more complex models can be treated.

The value of circuit theories of global electrodynamic coupling has been frequently debated (see, for example, Chiuderi (1983) and the subsequent discussion). While it might be argued that they are excessively crude in the face of quite detailed observations of coronal physical conditions, the approach does provide a new and apparently productive framework for discussing the challenging problems that have arisen as a result of the evidence of important electrodynamic processes in the corona. The way forward will presumably encompass further studies of abstract models which can be used to sharpen our ideas in this complex and incompletely understood field, as well as numerical studies of models that conform more closely to the observed behaviour of plasma loops in the corona.

6.3 Plasma loops in stellar coronae

The promise of a fruitful synergy between solar and stellar physics has been recognized since the birth of astrophysics, late in the nineteenth century. The essence of the promise has been enunciated often: The Sun lets us study phenomena *in detail* on account of its proximity, but only for a single set of stellar parameters since the mass, age, angular momentum, element abundances and many other factors are fixed. The stars, on the other hand, are too remote to permit studies in detail, but they do offer a *wide range* of physical conditions which allow conjectures and speculations to be tested more thoroughly than is possible on the Sun alone.

There are many examples of the successful exploitation of this synergy. Techniques used in the observation and analysis of photospheric spectra, the application of non-local equilibrium thermodynamics in astronomical spectroscopy, the extension of the chromosphere-corona-wind paradigm to stars and to entire galaxies, and increasing interest in the similarities and differences between solar, stellar and galactic magnetic activity are all cases where cross-fertilization has been fruitful. In the present section we explore the properties of plasma loops in the coronae of stars, and offer a tentative synthesis of the solar and stellar data based on the concept of electrodynamic coupling that has been developed in the previous section.

Beginning with a discussion of the observational material, we are immediately faced with two significant and unavoidable problems posed by the vast distances to the stars: the lack of instrumental sensitivity leading to data that are limited in signal-to-noise ratio, spectral resolution and wavelength coverage, and the lack of capacity to spatially resolve any structure that might exist in the stellar atmospheres. As a result, no direct observations of stellar coronal loops exist at the present time, and we are obliged to consider chains of inference regarding *possible* loops. These chains possess many weak links, and it is thus not surprising that it is difficult to make reliable estimates regarding the physical conditions in such stellar coronal loops as might exist. As a consequence, little has yet been learned about the basic physics of coronal plasma loops from comparisons between stellar and solar data, beyond the demonstration of consistency between simple theory and rudimentary observations and the recognition that stellar plasma loops may possess a wider range of physical properties than encountered on the Sun. However, by accepting the hypothesis that coronal radiative emission from stars originates in the analogues of solar loops we find in the stellar data a number of intriguing connections between the prominence of the loops and other properties of the star, such as the age, mass, angular momentum and the nature of any companion(s) that might be present. These connections provide insights into the physics of coronae that are simply not available on the Sun.

In Chapters 2, 3 and 4 we considered observations and inferred properties of, respectively, cool loops, hot loops and flare loops in the solar corona. It is convenient to present the stellar data in a different sequence, since our picture of stellar coronal loops has emerged almost exclusively from studies of hot loops and flares. More specifically, all of the available data on hot loops in stellar coronae (i.e. structures with temperatures in excess of 1 MK – see Section 2.1) have been derived from observations made using X-rays or radio wavelengths: these data are described in Section 6.3.1. *None* of the EUV lines used to study hot loops in the solar corona (Table 3.2) have yet been observed in stellar spectra. As we explain in Section 6.3.2, there are very few stellar data that have been interpreted in terms of cool coronal loops of the kind discussed in Chapter 2. The interpretation of observations of stellar coronal emission to infer the physical conditions existing in possible coronal loops is discussed in Section 6.3.3.

The connections between indicators of stellar magnetic activity (such as X-ray emission from putative coronal loops) and other stellar properties (such as mass, rotation rate and degree of evolution) are described in Section 6.3.4. These connections are important because they allow us to test models of dynamo mechanisms in a variety of circumstances. In Section 6.3.5 we discuss theoretical aspects of plasma loops in stellar coronae, in an attempt to unify the observational material described in Sections 6.3.1–6.3.4. The discussion is developed in two stages, the first dealing with stellar dynamos and related phenomena and the second the character of electrodynamic coupling in the stars.

Section 6.3 concludes with a brief account of the prospects for applying our understanding of the behaviour of plasma loops in the coronae of the Sun and stars to more remote astronomical objects, such as the enormous haloes that surround some spiral galaxies, and the mysterious regions that lie near the centre of our Galaxy and in other galaxies.

6.3.1   *Hot loops in stellar coronae*

In Chapter 3 we noted that the crucial importance of hot loops as a fundamental part of solar coronal structure was recognized only after X-ray images were obtained by using sounding rockets and satellites, even though prior observations in visible light (such as the green line – Fe XIV λ530.3 nm – and the electron scattering continuum) had provided glimpses of the loop structure of the hot solar corona. Since there is little evidence of hot coronal material in the visible spectrum of any stable star* the opportunity to study hot material in stellar coronae and any associated plasma loops arose only after sensitive X-ray and radio observations became possible in the 1970s. In this section we will first discuss the X-ray data and then the radio observations: the analysis of these data to infer physical conditions is presented in Section 6.3.3.

(a)   *X-ray observations*

Early observations of hot material in stars were made at X-ray wavelengths using non-focusing telescopes carried by rockets. The first detection of a stellar X-ray source (Sco X-1) was made in June 1962, and a series of rocket flights through the 1960s subsequently identified about 30 galactic X-ray sources. The next major step in instrumentation was made in 1970 with the flight of the *UHURU* satellite, still equipped with non-focusing optics but able to obtain much longer integrations of the X-ray fluxes from cosmic objects. *UHURU* confirmed that many, and perhaps all, of the brightest galactic X-ray sources are mass-exchange binary stars which we do not discuss in this volume.

The first detections of quiescent X-ray emission from 'normal' stars were reported in 1975. Catura, Acton & Johnson (1975) detected the binary system Capella (α Aur) serendipitously, during the calibration of an attitude control system of a rocket-borne X-ray reflector. Using the Astronomical Netherlands Satellite (*ANS*), Mewe *et al.* (1975) detected Capella and Sirius (α CMa), and placed an upper limit on the integrated X-ray flux from α Cen only three times that of the active Sun. *ANS* also provided the first detection of X-ray emission from flares on the flare stars YZ CMi and UV Ceti, together with a strong indication that the ratios of radio, optical and X-ray flare luminosities are neither fixed nor always comparable to solar values (Heise *et al.*, 1975). The flare luminosities in soft (0.2–0.28 keV) X-rays were of order 10^{22} W, and the corresponding powers about 10^{24-25} J. The *Soyuz-Apollo* observation of an XUV (44–190 Å) flare on Proxima Cen (Haisch *et al.*, 1977) involved a peak luminosity of about 10^{23} W. These luminosities and energies are at least an order of magnitude larger than those of the largest solar flares. None of these data were seen as an opportunity to address the question of the role of plasma loops in stellar coronae or flares.

The low-energy detectors (0.15–2.8 keV) of the A-2 experiment on the

---

* Lines of coronal ions such as Fe XIV are not detected in the visible spectra of stars. However, Zirin (1975) has suggested that the presence of He I absorption features in optical stellar spectra may indicate indirectly the existence of coronal X-rays that could be responsible for exciting the upper bound levels of the He atoms.

satellite *HEAO 1* provided evidence of X-ray emission from a sufficient number of 'normal' stars to sustain the view that stellar coronae are common. In particular, the detection of α Cen represented the first observation of coronal X-rays from a main-sequence star similar to the Sun: the measured flux was comparable with that of the Sun in its active state (Nugent & Garmire, 1978). *HEAO 1* also detected several X-ray flares from nearby flare stars (Kahn *et al.*, 1979), with peak X-ray luminosities in excess of 10^{23} W, temperatures as high as 30 MK and emission measures as much as three to four orders of magnitude larger than that of a very large solar flare. For these flares the ratio of X-ray to flare optical luminosity, L_x/L_{opt}, was found to be generally of order unity: as we shall see, this ratio is an important point of contact between theory and observation.

A significant achievement of the flight of *HEAO 1* was the demonstration that RS Canum Venaticorum (RS CVn) stars* considered as a *class* are X-ray sources. This possibility was suggested by Walter, Charles & Bowyer (1978) on the basis of detections of RS CVn itself, HR1099, and UX Ari, together with the prior detection of the 'long-period RS CVn' system Capella (see above). Confirmation was forthcoming when Walter *et al.* (1980) reported that a *HEAO 1* survey of 59 known or suspected RS CVn systems had detected a total of 15 systems, with X-ray luminosities in the range 3×10^{23}–4×10^{24} W and temperatures of order 10 MK. The non-detections could be explained as a result of either too large a distance or reduced sensitivity in the X-ray detector. In the same report, Walter *et al.* (1980a) noted that three types of variability are seen in the X-ray emission from RS CVn stars: (i) flaring, with flare luminosities in excess of 10^{25} W and time scales of many hours, (ii) short-term fluctuations of as much as a factor of 2 over time scales of an hour or so and (iii) rotational modulation produced as bright X-ray emitting regions are carried behind the stellar photosphere. This last observation shows that the X-ray sources are spatially localized on the stellar surface and suggests an analogue of the active coronal regions seen on the Sun.

An important conclusion drawn by Walter *et al.* (1980a) from their studies of *HEAO 1* observations of RS CVn stars is that '...there appears to be no fundamental distinction between the activity in RS CVn systems and that in the Sun...RS CVn systems may just be an extreme case of

---

* RS CVn stars form a rather inhomogeneous taxon characterized by the following properties: (i) binarity, with a period exceeding about 1 day implying that the components are detached, (ii) late-type components, with at least one being 'evolved' to a position above the main sequence and (iii) optical photometric variability attributable to photospheric starspots and to flares. The properties of the class are reviewed by Hall (1975).

normal late-type stellar activity . . .'. A similar point of view was advanced by Topka *et al.* (1979) in their report of the detection of X-ray emission (in a pre-*HEAO 1* flight) from Vega (α Lyrae) and η Bootis. Topka *et al.* (1979) noted that the detection of X-rays from Vega, in particular, was quite unexpected in terms of the acoustic shock heating theory current at the time. Appealing to then-recent advances in solar physics, these authors argued that any star having a magnetic field with a strength and geometry similar to that of the Sun's magnetic field would possess a corona. Further, they had the foresight to note that '. . . if one accepts the hypothesis that stellar surface activity is the consequence of dynamo processes in convectively unstable surface layers . . . then the vigour of coronal activity and hence the X-ray flux should be correlated with the stellar rotation rate . . .'. We return to this point in Section 6.3.4.

This picture of stellar X-ray emission has been strongly reinforced by the flight of the *Einstein* satellite (*HEAO 2*), which expanded enormously the quantity of high quality data available on X-rays from normal stars. The improved sensitivity of the satellite was due principally to the successful application of an 0.6 m grazing-incidence telescope which could be used to image X-rays onto any one of a number of focal-plane instruments (Giacconi *et al.*, 1979). In combination with the Imaging Proportional Counter (IPC) detector, the telescope offered a sensitivity better than 10^{-15} W m^{-2} in the 0.2–0.4 keV band. This corresponds to the detection of an X-ray luminosity of 10^{21} W at a distance of 100 parsec, an improvement in sensitivity over previous instruments by about three orders of magnitude. This step was rewarded by the detection of many hundreds of 'normal' stars as X-ray sources (and, of course, many other important astrophysical discoveries). The major advances stemming from the flight of the *Einstein* satellite are presently being strengthened by reports from the subsequent *EXOSAT* mission which, despite having lower sensitivity, covered a broader spectral range (0.04–10 keV) than *Einstein*.

The *Einstein* stellar surveys (Vaiana *et al.*, 1981) demonstrated that X-ray emission at intrinsic levels of about 10^{19}–10^{26} W is present in the spectra of many kinds of stars. Among the early-type stars, with spectral types from early O to about B5, the X-ray flux is correlated with bolometric luminosity ($L_x/L_{bol} \approx 10^{-7}$) although there are significant deviations from this trend, and variability of the X-ray flux is common (Pallavicini *et al.*, 1981). The mechanisms responsible for X-ray emission from early-type stars are poorly understood, but it is thought that instabilities in the stellar winds or interactions between the wind and interstellar matter may be important. The subject is explored extensively in the volume edited by

Underhill & Michalitsianos (1985). There has been little discussion of the possible role of coronal plasma loops in the production of X-rays in these objects. However, Landini & Monsignori-Fossi (1981) and Mangeney & Praderie (1984) have advanced simple but unified models of X-ray emission, based on the proposed existence of magnetic activity in all normal stars, that are claimed to account for the broad range of observed X-ray characteristics.

The results of *Einstein* observations of the cooler main-sequence (i.e. dwarf) stars of spectral types F, G, K and M are consistent with the conclusion that *all* such stars emit X-rays, with quiescent luminosities ranging between roughly 10^{19} and 10^{24} W (Vaiana *et al.*, 1981; Ayres *et al.*, 1981; Rosner, Golub & Vaiana, 1985). The X-ray spectra are apparently thermal, indicating temperatures ranging from about 1 MK to well above 10 MK. A remarkable feature of the X-ray emission from dwarf stars is the wide range of X-ray luminosities seen at any given spectral type. The statistics of X-ray detections and upper limits permit the construction of X-ray luminosity functions as exhibited in Fig. 6.6 (after Rosner *et al.*, 1985). These show that the X-ray luminosities of samples of late-type dwarf stars having almost identical photospheres are observed to span more than two orders of magnitude. Even the least prominent F stars are quite strong X-ray sources. The strongest X-ray M stars are, however, strong relative to all other 'normal' late-type stars. Equally remarkable is the result that the observed total X-ray luminosity, L_x, does not decline towards later spectral types, although the decrease in the apparent photospheric area is about a factor of 100 from the early F to the late M dwarfs. Clearly, the average X-ray surface brightness of the most active main sequence stars increases markedly as the stellar radius decreases. Since the bolometric luminosity, L_{bol}, also falls sharply in cooler stars, the ratio of L_x to L_{bol} also tends to increase towards later types.

The systematics of the X-ray luminosity observed in late-type giant and supergiant stars have excited considerable interest. In particular, the data bear on the proposal by Linsky & Haisch (1979), initially based on EUV spectra made using the *IUE* satellite, that the Hertzsprung–Russell (HR) diagram is divided into two parts, the one containing F–M dwarfs, giants earlier than K2 III and supergiants earlier than G5 Ib, all of which have detectable hot outer atmospheres (temperatures in excess of about 0.1 MK), and the other comprising the later giants and supergiants, few of which appear to have detectable hot material in their outer atmospheres.

The *Einstein* satellite data reinforce the significance of this division, although Hartmann, Dupree & Raymond (1982) have argued that the

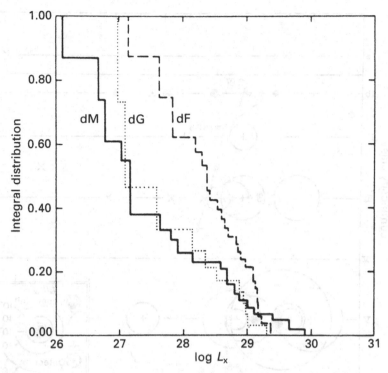

Fig. 6.6. The X-ray luminosity function of late-type main sequence stars, published by Rosner *et al.* (1985). The histograms illustrate the fraction of stars of the designated spectral type having an intrinsic X-ray luminosity of L_x (erg cm^{-2} s^{-1}). Note that even the weakest F-dwarfs are relatively strong X-ray emitters, while the strongest X-ray M-dwarfs are more luminous than all other stars in these categories.

apparent sharpness of the temperature transition is accentuated by the sensitivity of the instrument to changes in source temperature. An indication of the significance of the trend is found in the work of Vaiana *et al.* (1981) and Ayres *et al.* (1981), who report that X-ray luminosities ranging up to 10^{23} W (approaching 10^{-5} of the bolometric luminosity) have been detected in several G giants. However, virtually no giants later than spectral type K3 have detectable X-ray fluxes, at a detection limit of about 3×10^{-8} of the bolometric luminosity for the closest stars. This is a surface flux limit about three orders of magnitude smaller than the X-ray surface flux from solar coronal holes. As shown in Fig. 6.7, tight upper limits have also been placed on the X-ray luminosities of supergiants later than about G5. The 'cut-off' in X-ray emission (Ayres *et al.*, 1981; Gondoin, Manganey & Praderie, 1987) is associated with significant changes in several other spectroscopic features (such as EUV lines, the Ca II and Mg II resonance lines, circumstellar lines and He I λ1083 nm) in a complex

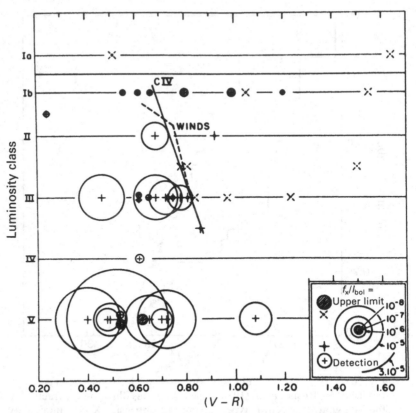

Fig. 6.7. A Hertzsprung–Russell diagram showing the systematic variation of X-ray luminosity among late-type stars. The circles indicate the ratio of X-ray to bolometric luminosity, as indicated in the inset. Note that no X-rays are observed from the cooler red giants: there appears to be a 'dividing line' near $(V$-$R) = 0.8$ that separates the hotter stars which are X-ray emitters from the cooler stars having very small (or zero) X-ray fluxes. The figure is from Ayres *et al.* (1981).

phenomenology that has become known as the 'dividing line' problem (Haisch, 1987).

Pre-main-sequence (PMS) stars (T Tauri stars and related types)* display a complex pattern of behaviour at X-ray wavelengths. As reported by Gahm (1980), Feigelson & DiCampli (1981) and others there is an anti-correlation between the presence of strong Hα emission lines in the optical spectrum (an important signature of the 'strength' of anomaly in the spectrum of T Tauri stars) and strong X-ray emission: most T Tauri stars

---

* T Tauri and other PMS stars exhibit a number of signs of youthfulness, including their proximity to young massive stars, their association with nebulosity and molecular clouds, and the high abundance of lithium in their photosphere. They are located in the H–R diagram in a region corresponding to models of contracting stars. The spectra of these objects are often complex, providing evidence of non-thermal processes in the stellar atmospheres and complex interactions between the stars and their environments. Cram & Kuhi (1989) have reviewed these objects, and published an extensive survey of the relevant literature.

having strong Hα and other optical emission line spectra are not detected at all as X-ray sources, whereas several other stars having weak emission line spectra exhibit high X-ray luminosities, occasionally exceeding 10^4–10^5 times the solar value.

Initially it was suggested that the anti-correlation might be due to a 'smothered' corona (Walter & Kuhi, 1981), in which the Hα emitting region blanketed an active stellar corona. It is now believed, however, that stars having strong Hα possess a massive, rather cool envelope or disk in which few if any X-rays are produced. On the other hand, it is believed that stars having strong X-ray emission do not have such an envelope or disk (i.e. they are 'naked' – Walter (1986)). The strength of the X-ray emission in the so-called 'naked T Tauri stars' has led Walter (1986) to suggest that they possess extremely active coronae (presumably with abundant hot plasma loops) similar in many respects to very active main sequence stars. According to this picture, the difference between the strong Hα and strong X-ray stars reflects a developmental sequence associated with the clearing away of placental interstellar material. Since many X-ray stars appear to be as young as the neighbouring Hα stars (e.g. Feigelson & Kriss, 1989), the time scale for clearing away the shroud of interstellar (or prestellar) material appears to vary from star to star. These results raise the important problem of the initial development of magnetic activity in stars: is it true that a corona only appears after a star becomes well separated from its placental material and, if so, why? The question is important in relation not only to the development of stellar activity, but also to the formation of planetary systems and, perhaps, the development of life on some of those planets.

In some star-forming regions the putative placental envelopes have been cleared from many objects and the consequence is a spectacular grouping of very intense X-ray stars characterized by the following properties (e.g. Montmerle *et al.*, 1983; Feigelson & Kriss, 1989): (i) X-ray fluxes up to 10^3 times larger than fluxes in main sequence stars of similar spectral type; (ii) extreme variability, amounting to as much as an order of magnitude in one day – Montmerle *et al.* (1983) describe the ρ Ophiuchi T-association as an 'X-ray Christmas tree' on account of its variable PMS X-ray stars; and (iii) an unusually hard spectrum indicating either very hot coronal material or non-thermal excitation.

These observations have been interpreted as evidence for the occurrence of frequent X-ray flares on these young stars. Montmerle *et al.* (1983) suggested that the flares obey a power-law relation between the normalized amplitude distribution and the flare frequency, in the sense that large flares

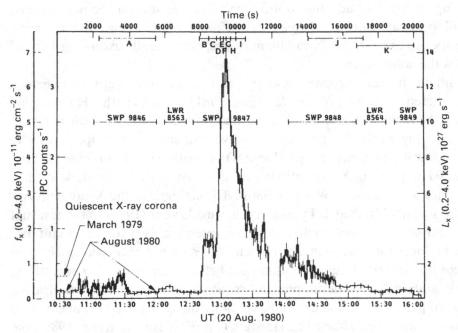

Fig. 6.8. X-ray light curve of a flare observed on the nearby flare star Proxima Centauri by Haisch *et al.* (1983). Over an interval of about 30 minutes the X-ray luminosity rises to about 20 times the quiescent level. The bands designated SWP or LWR refer to simultaneous *IUE* spectra (see Haisch *et al.*, 1983 for details).

are rare, while small flares are very frequent. The mean X-ray energy loss from individual flares – if this is the correct interpretation – is observed to be about 10^{27} J; the linear dimension of the source can be estimated from the cooling time (see below) to be about 100 Mm, and the typical duration is several hours. These flares could occur in magnetic loops, although there is no convincing evidence for this at the present time.

Although X-ray flaring is a prominent characteristic of pre-main sequence stars, it is by no means confined to that class. As noted above, RS CVn systems exhibit flare-like variability, and several flares have been observed on classical (i.e. optical) flare stars by the *Einstein* and *EXOSAT* satellites. The two most prominent X-ray flares observed by *Einstein* involved (i) a brightening of the X-ray flux of the Hyades binary HD 27130 (possibly an RS CVn system), lasting many hours and attaining a peak at least a factor of 20 larger than the quiescent level of the star, and (ii) a brightening of Proxima Cen over an interval of about 2 hours which attained a peak of more than 30 times the quiescent flux level (Haisch *et al.* (1983) – Fig. 6.8 shows the X-ray light curve). The dozen or so flares observed by *Einstein* have been discussed by Haisch (1983), who also summarized the theoretical interpretation of the X-ray light curves of these events.

The nature of stellar X-ray flares has been further clarified by the *EXOSAT* observation of a flare on the G0 V star π^1UMa, reported by Landini *et al.* (1986). Specifically, the broad spectral coverage of the *EXOSAT* instrumentation permitted the investigation of the time-development of the spectrum during the flare. Although the measurements were conducted at the limits of sensitivity, there is evidence that the hard X-ray flux peaked before the soft X-ray flux, in accord with the pattern encountered in solar flares (Section 4.5.4). The observations have been used to estimate temperatures and emission measures (see below).

Many, but not all, of the stellar X-ray flares observed to date have been satisfactorily interpreted in terms of energy balance models that are crude but consistent with our understanding of large solar flares. The main differences are encountered in the very energetic events seen in RS CVn and 'naked' T Tauri stars, where the inferred spatial scale may exceed the stellar radius and the flare duration is very long compared with solar flares. But even in these cases, the observations available at present are compatible with the conclusion drawn by Landini *et al.* (1986), that '...in spite of the widely different classes of objects (on which flares are found), there are enough similarities suggesting an underlying common physical mechanism, i.e. sudden release of magnetic energy in one or more arch-like coronal structures...'. On the other hand it must be noted that the observations are very fragmentary, and far from adequate to reveal analogues of the great number of intriguing and unexplained phenomena that exist in solar flares.

An observation of particular significance in the study of hot loops in stellar coronae was reported by Walter, Gibson & Basri (1983). These authors obtained an extensive and unique time series of observations of X-ray and EUV emission from the eclipsing RS CVn system AR Lac which could be used to infer the three-dimensional distribution of the source regions. Despite having incomplete phase coverage Walter *et al.* (1983) were able to advance a credible case for the geometry depicted in Fig. 6.9, in which both stellar components have compact coronal regions bounded in latitude and longitude. Using the diagnostic techniques described below, these regions are found to consist of large numbers ($\approx 10^6$) of compact loops, with lengths and outward extensions no more than 2 % of the stellar radii, and having (assumed) temperatures of about 8 MK, pressures of about 7.5 Pa and electron densities of 6×10^{14} m^{-3}. The individual loops are not unlike normal solar *flare* loops; the totality of the RS CVn compact emission would correspond to about one-half of the Sun being covered by

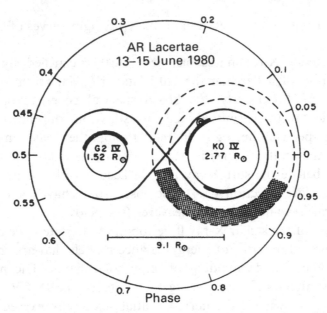

Fig. 6.9. Diagram of the X-ray emitting regions (and possible coronal plasma loop systems) on the RS CVn system AR Lac, deduced by Walter *et al.* (1983) from time-series X-ray observations. Low-lying loop systems are inferred to exist on both stars (heavy lines), and a system of high loops is identified above the cooler component. The marked circle indicates viewing directions for various phases for the system.

such loops. In addition, the more active (K star) component possesses a small number (≈ 10) of hot (≈ 30 MK) loops which the X-ray light curve shows must extend outward over distances larger than the stellar radius: such enormous structures have no obvious solar counterpart.

(b)   *Radio observations*
       The picture of X-ray emission sketched above can be enhanced by adding information gained from radio observations of active stars. As with X-rays, the radio data often indicate emission levels many orders of magnitude stronger than the probable solar analogues of the stellar phenomena. For example, radio 'flares' have been reported in flare stars at metre wavelengths with brightness temperatures exceeding 10^{15} K, and even the weakest reported events are 10^4 times more energetic than a large solar flare in the same band (Gibson, 1983). Several metre-wavelength studies of stellar flares on the 'classical' flare stars (M dwarfs) have been undertaken, but the link between stellar and solar radio emission processes at metric wavelengths (in both cases, presumably plasma emission or some other coherent mechanism) has yet to be explored in depth.
       The detection of *microwave* radio emission from stars perhaps provides

a closer link between stellar source structure and the solar loop phenomena discussed in Sections 3.5, 3.6.5 and 4.6. Quiescent (i.e. relatively steady) microwave emission has been detected from several late-type dwarf stars which were previously known to be strong sources of X-rays and chromospheric emission (Gary & Linsky, 1981; Topka & Marsh, 1982; Pallavicini, Willson & Lang, 1985). This emission may emanate from plasma loops analogous to those associated with hot, non-flaring microwave loops in the solar corona (Section 3.5), although the high flux density of the stellar microwave emission calls this interpretation into question. Another interpretation is that the 'quiescent' emission is in fact the superposition of a large number of radio flare events.

Time-dependent microwave emission has been observed from several of the stars known to be strong X-ray sources, including several RS CVn systems (Mutel & Lestrade, 1985), pre-main sequence stars (Feigelson, 1987) and active main-sequence stars (Pallavicini *et al.*, 1985; Slee *et al.*, 1987; Bastian & Bookbinder, 1987). Although only a few observations have been reported to date, it does appear that a number of different emission mechanisms may be operative – they may be classified according to the temporal dependence of the flux density, the form of the dynamic spectrum and the polarization (Dulk, 1985). One form of emission involves highly polarized radiation which changes on time scales ranging from milliseconds to a few seconds and which may display burst structure with very narrow bandwidths (Bastian & Bookbinder, 1987). The rapid rate of change implies a very small source volume and hence brightness temperatures that may be higher than 10^{12} K. The emission process is almost certainly a coherent mechanism, perhaps an electron-cyclotron maser. Another form exhibits slower variations with only a moderate degree of polarization and inferred brightness temperatures below 10^{10} K. This emission may be analogous to the microwave bursts seen in solar flares (Sections 4.6, 4.8.5), which are known to be related to hard X-ray emission (Section 4.7.5). There are good prospects for further progress in the application of microwave observations to study the physics of stellar coronae.

The technique of very long baseline radio interferometry (VLBI) offers the chance to spatially resolve the stellar radio emission from a few stars. In particular, Mutel *et al.* (1985) have resolved the 5 GHz radiation from the RS CVn system UX Ari into a 'halo' occupying a volume on the scale of the binary separation, and a 'core' apparently associated with the cooler (more active) component. There is no evidence of loop-like structure in the radio source, although the reported resolution of about 0.001 arc seconds is far larger than the radius of either star in the system. The apparently large geometrical extent of the microwave emission from UX Ari has a possible parallel in the X-ray emission from the RS CVn system AR Lac. As discussed above, Walter *et al.* (1983) have reported *Einstein* observations consistent with the presence of X-ray emission from regions far removed from the surfaces of the two stars in this system.

6.3.2 Cool loops in stellar coronae

There have been few attempts to pursue the analogy between cool plasma loops in the solar corona (as described in Chapter 2) and similar loops in stellar coronae. This situation may reflect the circumstance that, although cool ($T_e < 1$ MK) coronal loops are readily observed in the solar atmosphere (cf. Chapter 2), most cool emission in fact comes from low-lying structures closely associated with the transition region between the chromosphere and the corona (Feldman, Doschek & Mariska, 1979) and not yet convincingly identified with loops or with portions of loops. Without spatially resolved observations, it is of course particularly difficult to confirm that cool stellar emission emanates from objects embedded in the stellar corona, rather than lying beneath it. There are, however, three tenuous lines of observational evidence regarding the presence of cool material in stellar coronae: (i) EUV data, obtained mainly from the *International Ultra-violet Explorer* (*IUE*) satellite, (ii) optical spectroscopy of eclipsing binary stars and (iii) evidence of the ejection of cool material from rapidly rotating dwarf stars.

IUE observations cover spectral lines formed at temperatures ranging up to about 0.2 MK, i.e. the domain of *cool* loops as defined in Chapter 2. There is little evidence that this stellar EUV emission occurs in loop-like structures, and the data are almost always interpreted in terms of a model corresponding to a compact, low-lying chromosphere and transition zone that is often assumed to be laterally homogeneous. This approach is not inconsistent with solar studies, since it appears that only a small part of the solar EUV emission emerges from clearly delineated cool coronal loops (Section 2.4.1).

A strong correlation between the X-ray (F_x) and EUV (F_{EUV}) surface fluxes extends over essentially all late-type stars (e.g. Ayres *et al.*, 1981; Schrijver, 1987). The correlation takes the form of a power-law ($F_x \sim F_{EUV}^\alpha$, where α is a constant of order 1.5 to 2.0), which is tightest when a *basal* component (i.e. an uncorrelated part) is removed from the EUV data. The tightness of the correlation between the (excess) fluxes implies that there is a single parameter controlling the level of emission, which varies from star to star in concert with the variation in emission level. This parameter is believed to be the total magnetic flux in the stellar atmosphere, but it is by no means clear that both the EUV and the X-ray emitting regions are linked by the *same* magnetic field lines. Indeed, the existence of *power-law correlations* between spectral features formed in different temperature regimes implies that the emission flux in at least one spectral line is not directly proportional to the amount of magnetic flux entering the atmosphere. This result perhaps supports the idea that cool and hot material lie on different field lines. It is important to note that very few attempts have been made to construct atmospheric models encompassing the (cool) EUV and (hot) X-ray emitting regions in stellar atmospheres, although it is clear that models based on hot coronal loops having cool footpoints are quite inadequate since they predict far too little EUV emission (Section 6.3.3).

The second line of evidence regarding possible cool loops in stellar coronae concerns the few eclipsing binary systems (consisting of a late-type giant and an early-type main sequence star) in which the optical spectra at ingress and egress display clear evidence of cool material extending far out in the atmosphere of the cool star. The systems 31 Cyg, 32 Cyg and ζ Aur are the best studied (Wilson, 1964; Reimers, 1989). Since it is unclear whether there is a hot corona associated with these systems, it is not presently possible to draw an analogy between these observations and properties of cool loops in the solar corona. In fact, the observed spectroscopic effects are generally interpreted in terms of analogues of solar prominences rather than loops, although a loop model could be attractive in view of the need to explain the confinement of a large amount of cool material at large distances from the star.

The third line of evidence regarding the presence of cool material in stellar coronae is the spectroscopic observation of 'clouds' being ejected from rapidly rotating, apparently young main-sequence stars (Cameron & Robinson, 1989). It is inferred from the systematic changes seen in the Hα line that the cool material co-rotates with the stellar photosphere as it flows outward, implying the presence of a substantial and widespread coronal magnetic field. However, the fact that the cool material appears to flow outward suggests that it is not located in coronal loops: rather, the ejection phenomenon may be analogous to erupting filaments or coronal mass ejections.

6.3.3   *Physical conditions in stellar coronal plasma loops*

Before stellar X-ray emission was detected, several authors had made predictions based on the hypothesis that stellar chromospheres and coronae are heated by the dissipation of a flux of acoustic energy generated in the stellar convection zone (e.g. Ulmschneider, 1979). In line with this model, an early observation by the ANS satellite of X-ray emission from Capella was interpreted by Mewe *et al.* (1975) in terms of a plane-parallel, isothermal model of the coronal source. The observed X-ray flux and known spectral response of the detector constrained the possible values of coronal base pressure (P), coronal temperature (T) and emission measure (EM) which would be consistent with the data. One acceptable model (assuming the source were Capella A) had $T = 1.5$ MK, $P = 0.005$ Pa and $N_e = 10^{14}$ m^{-3}, values not greatly different from those of the quiet solar corona. Mewe *et al.* (1975) noted that theoretical estimates of the convective acoustic flux from Capella A were only a few times larger than the estimated solar acoustic flux – the unstated implication being that this early observation was seen to provide support for the then-current shock heating model of stellar coronae.

However, weaknesses in the acoustic heating hypothesis were soon exposed when *HEAO 1* detected very high levels of X-ray emission from RS CVn stars, since the hypothesis offered no explanation of widely different heating rates in otherwise similar stars. This problem did not survive for long since the observations were made against a background of increasing awareness of the importance of *loops in the solar corona* (Section 3.4), and the emergence of new ideas regarding the importance of electrodynamic heating processes in the solar corona (Section 6.3). Thus it was natural to interpret the X-ray emission from RS CVn stars in terms of the newer picture of solar coronal structure and heating. For example, Walter *et al.* (1978) conjectured that in the active RS CVn stars the '. . . X-ray emission is due to a plasma contained by the dipolar magnetic fields expected to be associated with the star spots. The corona will consist of the hot plasma trapped above the active surface regions and a cooler stellar wind flowing out of the coronal holes . . .'. An important basis for this conjecture was the prior observation of stellar flares and photospheric starspots suggestive of extensive magnetic activity on the surface of these stars.

The magnetic plasma loop model for the X-ray coronae of RS CVn stars was developed quantitatively by Walter, Charles & Bowyer (1980). These authors assumed that the X-ray emission comes from plasma loops having a structure described by the constant-pressure loop model of Rosner *et al.*

Table 6.3. *Models of stellar coronal loops*

(a) Walter *et al.* (1980a)

Star	T (1)	EM (2)	P (3)	f (4)	L (5)	$N\alpha^2$ (6)
UX Ari (RS CVn)	10	4×10^{60}	1	16	300	2000
Capella (RS CVn)	10	1.2×10^{59}	0.15	0.12	200	9
ξ Boo A (active dwarf)	10	2.3×10^{57}	3	0.02	100	3
40 Eri A (active dwarf)	10	4×10^{56}	3	0.01	100	0.6
Sun (quiet)	3	10^{56}	0.1	0.11	300	20
Sun (covered by plage)	10	4×10^{58}	1	1.3	300	20
Subgiant	10	4×10^{59}	1	1.3	300	200

(1) Coronal temperature (MK) from HEAO-1 broad-band energy channels.
(2) Emission measure (m^{-5}) from HEAO-1 flux.
(3) Coronal pressure (Pa) from EUV or other data.
(4) Fractional coverage of star by loops.
(5) Loop length (Mm).
(6) Product of number N and loop radius-to-length ratio α.

(b) Golub *et al.* (1982)

Star	T (1)	EM (2)	F_x (3)	H_P (4)	f (5)	P_c (6)
α Cen A (dwarf)	2.1	2×10^{59}	13	140	0.05	0.034
α Cen B (dwarf)	2.1	5×10^{59}	61	90	0.15	0.056
Proxima Cen (flare star)			700	70	0.19	0.38
Sun (average)	2.2	3×10^{59}	33	100	0.09	0.046

(1) Coronal temperature (MK) from a one-temperature fit to *Einstein* IPC data.
(2) Emission measure (m^{-5}) from IPC and HRI data.
(3) X-ray surface flux (W m^{-2}).
(4) Coronal pressure scale height (Mm) from (1) and the stellar gravity.
(5) Filling factor, assuming loop length is equal to H_P.
(6) Coronal pressure (Pa), assuming loop length is equal to H_P.

(c) Mewe *et al.* (1982)

Star	T (1)	EM (2)	f (3)	H_P (4)	L (5)	P (6)
Capella (RS CVn system)						
component 1 (cooler)	5	6×10^{58}	0.9	5000	300	0.13
component 2 (hotter)	10	6×10^{58}	0.01	10000	60	6.0

(1) Coronal temperature (MK) from a two-temperature fit to *Einstein* HRI data.
(2) Emission measure (m^{-3}) from HRI data.
(3) Filling factors, assumed to correspond to the active and quiet Sun.
(4) Coronal pressure scale height (Mm) from (1) and surface gravity.
(5) Loop length (Mm), having assumed (3).
(6) Coronal pressure (Pa), having assumed (3).

Table 6.3. (*cont.*)

(d) Giampapa *et al.* (1985)

Star	T_{max} (1)	P (2)	f (3)	L (4)
Sun (min)	2.5	0.01	0.56	700
Sun (max)	3.0	0.15	0.03	69
α Cen B (KIV)	2.4	0.056	1.68	95
ι Per (G4V)	2.6	0.065–0.085	1.25–1.69	75–98
μ Her (G5IV)	2.6	0.035	0.27	180
σ Dra (K0V)	2.3	0.4–1.6	0.04–0.16	2.8–11
HR 3538 (G3V)	3.2	0.5–1.0	0.085–0.17	11–22
ε Eri (K2V)	3.5	0.5–2.0	0.54–0.22	8–30
HD 206860-1 (G0V)	4.4	0.6–0.8	0.35–0.47	38–50
HD 5303 (G2V+F)	24	2.5	3.6	(?)

(1) Maximum loop temperature (MK) based on a one-temperature fit to *IPC* data.
(2) Loop pressure (Pa) by fitting EUV data.
(3) Filling factor, derived by fitting X-ray and EUV data.
(4) Loop length (Mm) from loop model.

(e) Landini *et al.* (1985)

Star	T (1) I	II	P (2) I	II	f (3) I	II	L (4) I	II
α Cen A (G2V)	2.5	3.8	0.49	0.11	0.15	0.056	28	440
BD+15°640 (active dwarf)	5.0	7.0	1.3	0.22	0.13	0.8	76	1400
BD+14°693 (active dwarf)	20	25	27	1.5	0.06	0.11	250	1200
HD 165590 (active triple)	19	24	25	1.4	0.05	0.9	240	9800

(1) Temperature at loop apex (MK) deduced by matching to X-ray and EUV data.
(2) Loop base pressure (Pa) derived by fitting X-ray and EUV data.
(3) Filling factor, derived.
(4) Loop length (Mm), derived.

(f) Mewe & Schrijver (1986)

Star	T (1)	EM (2)	fP (3)	PL (4)
λ And (RS CVn system)				
hot component	20	4.3×10^{59}	0.073	1.1×10^6
cool component	8	4.4×10^{58}	0.019	70×10^3

(1) Mean loop temperature (MK) in hot and cool components from a two-temperature fit to *EXOSAT* data.
(2) Emission measure (m^{-3}) from the observed flux.
(3) Filling factor times pressure (Pa).
(4) Pressure (Pa) times length (Mm).

Table 6.3. (*cont.*)

(g) Stern *et al.* (1987)

Star	T (1)	L (2)	P (3)	f (4)
BD+14693	15	100	29	0.43
BD+14690	7	10	21	0.37
BD+15640	15	10	290	0.02
71 Tau	15	100	37	0.57

(1) Mean loop temperature (MK) from fit to IPC data.
(2) Loop length (Mm).
(3) Loop pressure (Pa).
(4) Filling factor.
Note: These are the best-fit models having $f < 1$.

(1978). As explained in Section 5.4, this model predicts the existence of a fixed relation between the summit temperature T, the pressure P and the length L of a loop, of the form

$$T = \left(\frac{4k\kappa}{\xi}\right)^{-\frac{1}{6}}(PL)^{\frac{1}{3}}, \qquad (6.17)$$

where κ is the coefficient of thermal conductivity and ξ is related to the radiation loss function $\mathscr{L}$ (Eqn. (5.39)) by

$$\mathscr{L} = \xi N_e^2 T^{\frac{1}{2}}. \qquad (6.18)$$

Assuming that the observed total emission measure is due to the superposition of N identical loops, Walter *et al.* (1980b) exhibited functional relations between f, the fraction of the surface covered by loops, $N\alpha$, where α is the ratio of the radius of the loop to its length, and L in the form

$$L = C_1 T^4 (EM)^{-1}(R_\odot/R_*)^2 f, \qquad (6.19)$$

$$f = C_2 T^{-1}(EM)(R_\odot/R_*)^2 P^{-1}, \qquad (6.20)$$

$$N\alpha^2 = C_3 T^{-7}(EM)P. \qquad (6.21)$$

Here, T is the temperature and EM the emission measure, while C_1, C_2 and C_3 are constants whose numerical values depend on the detailed loop models adopted. By themselves these relations are not completely defined by the available X-ray data, which give (at most) only the temperature and the X-ray luminosity. However, definite models may be constructed by specifying the length L (e.g. taking $L = R_*$, the stellar radius) or by adopting a value of P based on spectroscopic studies of the underlying chromosphere. Table 6.3 summarizes the results of this study by Walter *et al.* (1980b).

These values provide a remarkable picture of the postulated coronal plasma loops in RS CVn systems. For UX Ari, the implication is that the corona is extremely closely packed – even overfilled – with loops each about as long as a 'typical' solar loop. As noted by Walter *et al.* (1980b), this is perhaps consistent with the deduction from optical photometry that such stars possess large numbers of spots covering the stellar photosphere. It should be noted, however, that the large number and close proximity of loops deduced by this application of the loop model of Rosner *et al.* (1978) may vitiate several of the assumptions underlying the original model.

The corona of Capella appears to contain a far smaller number of loops than UX Ari, but each loop appears to be much longer in the giant star (Capella). Why this should be the case is unclear, although Capella rotates more slowly than UX Ari, and its stellar components have a smaller surface gravity. Walter *et al.* (1980b) suggested that the enhanced coronal activity of RS CVn stars could be due to an increased rate of generation of magnetic flux tubes – perhaps due to forced rapid rotation resulting from tidal coupling between the binary components. Were this the case, we would expect to detect a dependence of the level of coronal emission on the rotation rate; we might also expect to observe different kinds of X-ray variability on Capella and UX Ari.

There have been many subsequent attempts to interpret the observed X-ray emission from several types of star in terms of loop models, following the approach developed by Walter *et al.* (1980b) to study RS CVn systems but including a number of refinements. Landini & Monsignori Fossi (1981) included mass motions and gravity in their theory, and adopted a heating function determined by the properties of the convection zone and photosphere. They deduced that the emission measure (in effect, the X-ray luminosity) would vary from star to star as the quantity $qg_*^4R_*^5$, where q is the fraction of the stellar surface occupied by loops, and g_* and R_* are the stellar surface gravity and radius. Assuming q is fixed, these authors claimed 'rather satisfying' agreement with observation for dwarfs, giants and supergiants with spectral types ranging from O5 to M0. However, deviations from the predictions amount to more than two orders of magnitude and the claim seems unwarranted, particularly since other data imply that star-to-star variations in the quantity q are crucial in explaining different activity levels.

Golub *et al.* (1982), in a detailed analysis of the α Cen system, re-emphasized that the X-ray data alone are not sufficient to fully constrain models in which the coronal emission is supposed to come from a number of identical plasma loops, each one characterized by the physical processes discussed in Section 5.4 and embodied in Eqns. (6.17)–(6.21). The temperature may be inferred by fitting theoretical response curves to the measured X-ray spectrum, and then the emission measure may be deduced from the measured flux. However, a given X-ray flux and spectrum is consistent with a range of possible loop base pressures, and each pressure value leads to a different total number of loops. In essence, the observations are unable to distinguish between coronae that are dense and compact, or tenuous and extensive. In view of this ambiguity, other constraints must be introduced: Golub *et al.* (1982) assumed that the length of the coronal loops might be equal to the pressure scale height although they noted that there is no *a priori* reason why this should be so. As shown in Table 6.3, the resulting coronal and loop parameters inferred for α Cen A and B are quite similar to those of the Sun.

Attempts to infer the temperature of coronal material from stellar X-ray observations usually follow the procedures described in Section 3.6.4. Specifically, the theoretical response of the observing equipment to a range of isothermal sources is predicted by calculating the ionization and excitation balance, estimating the strength of line and continuum emission, and folding this with the instrumental spectral response function. For given data, the temperature may then be determined by a least-squares fit to the theoretical response function. This method yields an estimate of the temperature of the 'best' isothermal source, and also provides the spectral distribution of the deviation from this best fit. By studying such deviations, Holt *et al.* (1979) and Swank *et al.* (1981) showed that a single temperature fit to the *Einstein* X-ray spectrum of Capella was inadequate, and that there was strong evidence for the presence of a very hot component (T in excess of 25 MK).

In a more detailed study of Capella (see Table 6.3) using the objective grating spectrometer on board *Einstein*, Mewe *et al.* (1982) demonstrated that a two-temperature model (with components at 5 MK and 10 MK) could greatly improve the match to the spectral line data. These authors pointed out that two- or three-component temperature models are only convenient approximations, and that a continuous distribution of temperatures will in fact occur. They also made the important point that the EUV emission from Capella measured by *IUE* is far too bright to be ascribed to conduction from the loops inferred from X-ray observations, and suggested that '. . . . the chromosphere will be found not only at the base of hot coronal loops, but also in rather small loops which are expected to be at chromospheric temperatures along the entire loop length. In these loops the local heating is balanced by the local radiative losses, and thermal conduction is unimportant. This would increase the volume of the chromosphere by a large amount . . .'. As discussed below, this proposal has since been developed much further by Antiochos & Noci (1986).

EXOSAT observations of the X-ray spectrum of the RS CVn star λ And were analysed by Mewe & Schrijver (1986) in a manner similar to the studies of Capella (see Table 6.3). Again, a two-temperature model was derived, with components having $T = 8$ and 20 MK. The observed strong variability of the hot component points to occultation of the putative hot loops by the stellar photosphere, allowing Mewe & Schrijver (1986) to place an upper limit on the length (4×10^6 km) and fractional area coverage (3 %) and a lower limit on the base presure (2.5 Pa) of these structures.

In an attempt to avoid the approximations involved in models having temperature 'components', Giampapa *et al.* (1985) and Landini *et al.* (1985) computed in two separate studies the expected X-ray response of the *Einstein* satellite to emission from *self-consistent* models of loops (see also Schmitt *et al.*, 1987). In these studies, loop models describing the distribution of temperature, density and so forth are constructed using a variant of the theory outlined in Section 5.4, the spatial variation of the ionization and excitation balance is then calculated, and finally the emitted spectrum and the corresponding instrumental response is estimated (cf. Section 3.6.4). Loop properties are then deduced by systematically searching for models that minimize the differences between theory and observation (see Table 6.3). The methods can be readily extended to predict the corresponding EUV emission from the cooler parts at the footpoints of the loop, and this offers some chance of removing one unknown (the loop base pressure) from the analysis.

The results of the two studies are ambiguous: Giampapa *et al.* (1985) identified significant discrepancies between the X-ray models and the EUV (N V, C IV, Si IV) data for most objects in their sample of 9 stars, while Landini *et al.* (1985) appeared to have achieved agreement to within a factor of two for a sample of 4 stars including not only the lines listed above but also C II. Part of the ambiguity might lie in the fact that Giampapa *et al.* (1985) used the X-ray/EUV comparison to determine the fractional area covered by loops, while Landini *et al.* (1985) used it to calibrate the ratio of apex to base pressure in the loop. The latter procedure seems to involve one more degree of freedom than the former. There is, unfortunately, no overlap in the objects selected by these two groups of authors. The inferred coronal loop conditions range widely in the two studies, consistently with the selection of stars having markedly different X-ray and EUV properties: temperatures span the range 2–25 MK, base pressures 0.01–25 Pa, fractional area coverages by loops 0.03– > 1.0 and loop lengths 3×10^3–10^7 km. There are few clear trends in the inferred loop properties, beyond the generalization that the more active stars have higher temperatures and higher pressures in the loops.

A similar approach to the analysis of X-ray and EUV data has been adopted by Schmitt *et al.* (1985) in a close study of Procyon. This is an important object since the data are of good quality and there are extensive EUV data from *IUE* and *Copernicus*. A prior study of the EUV emission measure distribution of Procyon by Brown & Jordan (1981) did not provide strong evidence for the existence of coronal material in this star, but the analysis by Schmitt *et al.* (1985) showed clear evidence of material at temperatures of about 2.5 MK. Loop models consistent with the X-ray data can also account for the *IUE* data published by Brown & Jordan (1981) (i.e. C IV, Si IV and N V), but they cannot account simultaneously for the *Copernicus* observations of O V and O VI which was the reason why Brown & Jordan avoided the inclusion of a corona. To explain the discrepancy, Schmitt *et al.* (1985) pointed out the possibility of intrinsic variability or intercalibration problems.

Stern, Antiochos & Harnden (1986) have criticized the use of 'transition region' spectral data (i.e. *IUE* data) to constrain stellar loop models, on the grounds that solar studies suggest that the lower transition region and the corona are decoupled so that the '. . . application of simple static loop models to connect the corona and transition region may not be valid . . .'. Further, these authors criticize the use of two-temperature models on the grounds that they lack internal consistency and, it transpires, involve more free parameters than models based on an integration over loop models with non-uniform conditions. Stern *et al.* (1986) showed that a superposition of many such non-uniform but identical loop models is able to account for the *Einstein* observations of the bright X-ray dwarfs in the Hyades, provided the maximum loop temperature is about 25 MK, the pressure greater than 40 Pa, the length less than 10^5 km and the fractional area coverage less than 20%. These conditions correspond to loops that are hotter, more compact and at a higher pressure than most previous workers have deduced from this kind of analysis, perhaps suggesting that there are similarities between the apparently 'quiescent' emission from these very active stars and the extreme conditions encountered in large solar flares.

The question of the relation between X-ray and EUV emission from stars has been discussed extensively by Antiochos & Noci (1986) and Antiochos, Haisch & Stern (1986), who have noted that current coronal loop models do not account for the marked increase in emission from EUV lines formed at temperatures below 0.1 MK, as observed in both the Sun and stars. They also note that such explanations of the increase as have been advanced usually introduce a second mechanism (beyond coronal loops) to provide the cooler emission, despite the fact that observations show a strong correlation between the amount of hot material and the amount of cool material. They then argued that the hot and the cool emission can be explained by a unified loop model. The hot component corresponds to previous coronal loop models, while the cool component arises as a consequence of the thermal instability of dense, low-lying loops (Section 5.5). One important implication is that stellar surface gravity is predicted to have a marked influence on the temperature of the cool loops (since it determines the weight of cool material that can exist in a loop), and on the volume occupied by hot loops (since gravity determines the altitude below which thermal instability destroys hot loops). Antiochos *et al.* (1986) suggested that these gravitational effects could play an important role in the formation of the 'dividing line' discussed in Section 6.3.1.

As noted above, the high temperatures deduced by Stern *et al.* (1986) for the Hyades stars and by many workers for RS CVn and T Tauri stars are encountered in the solar corona only in flares. This fact, combined with time-series observations suggesting that there is a persistent low-level variability in the 'quiescent' X-ray emission of some stars (Ambruster, Sciortino & Golub, 1987; Pallavicini, 1988), supports the notion that at least part of the X-ray flux from stars may be produced by a host of overlapping 'flares', perhaps occurring in analogues of solar flare loops. Such a concept has been advanced to account for 'quiescent' solar X-ray emission (e.g. Tucker, 1973; Parker, 1988b), although there appears to be little *direct* evidence in its favour. Several authors have noted that there is a correlation between the quiescent X-ray emission and the rate of optical flares from late-type dwarf stars (e.g. Whitehouse, 1985). This implies a connection between the level of 'activity' in the quiescent state and the frequency and amplitude of detectible single flare events. However, it is not clear whether a single continuum of processes is necessary (or sufficient) to account for all significant modes of coronal heating in late-type stars (Ambruster *et al.*, 1987).

Table 6.4. *Properties of stellar X-ray flares*

Object	T_{max} (1)	L_x (2)	EM (3)	V (4)	L (5)
Sun	10–30	10^{23}–10^{24}	$<10^{56}$	10^{22}	45
Flare stars	12–40	10^{24}–10^{26}	10^{57}–10^{60}	10^{22}	45
HD27130	40–100	$>3 \times 10^{27}$	$>3 \times 10^{59}$	4×10^{24}	150
Pre-main sequence	10–20	$<10^{29}$	10^{59}–10^{61}	10^{24}–10^{26}	100–450
RS CVn	>30	10^{28}–10^{31}	10^{61}–10^{62}		
π^l UMa	30	10^{23}	7×10^{58}	1×10^{23}	230

(1) Maximum temperature in loop (MK).
(2) Flare energy (J).
(3) Emission measure (m^{-5}).
(4) Flare volume (m^3).
(5) Flare size (Mm) from volume.
Table after Stern, Underwood & Antiochos (1983) and Landini *et al* (1986).

A few authors have estimated the physical conditions in large, well-observed stellar flares by assuming that the X-ray emission arises in plasma loops. In particular, the initial decay phases of the flares have been interpreted by assuming that the observed decay time of the X-ray flux, τ_D, is determined by the radiative (τ_R) and conductive (τ_C) cooling times. These time scales are thought to be roughly equal to each other in many solar flares. Following the arguments advanced by Stern *et al.* (1983) we then have

$$\tau_R = 2N_e(\tfrac{3}{2}kT)/N_e^2 Q(T), \tag{6.22}$$

where $Q(T)$ is the radiation loss function defined in Eqn. (5.40), and

$$\tau_C = 2N_e(\tfrac{3}{2}kT)/(\kappa T/L^2). \tag{6.23}$$

Assuming these times are equal, adopting a simple functional dependence of $Q(T)$ on T, and supposing that the length:diameter ratio of the postulated flare loop is 10, Haisch (1983) derived expressions for the temperature T, the electron density N_e and the length L of the loop in terms of the observed emission measure and decay rate. The theory also offers an estimate of the gas pressure in the coronal flare, and if it is assumed that the magnetic pressure at the time of the flare exceeds this gas pressure, it is possible to estimate the strength of the coronal magnetic field.

The theory has been applied to observations of several stellar X-ray flares by Haisch (1983) with the results summarized in Table 6.4. An analysis of *EXOSAT* observations of a flare on π^1 UMa using a similar (but not identical) approach by Landini *et al.* (1986) gave the physical parameters also listed in the Table. It can be seen that the detected stellar X-ray flares are far larger than the largest solar flares. Little that can be said about the nature of any coronal plasma loops that might exist within such powerful flares, although it is clear that the solar analogy must be pressed hard to account for the observations, given the theoretical problems that are encountered even in the more powerful of solar flares.

6.3.4 *Coronal emission and other stellar properties*

Observations of stellar coronal emission provide, in essence, information only on the temperature and the total emission measure. These data may be interpreted in terms of coronal loop theory, and in a few cases time-resolved observations may be used to reinforce conclusions regarding the geometrical disposition of coronal material, but the models are weakly constrained. Given the coarse character of the available data and the consequent lack of constraints for model-building, there seem to be few opportunities to use the stellar data to reveal significant new information about the detailed physical processes occurring in coronal plasma loops. However, the stellar observations are useful in another direction, for they reveal a wider range of coronal states than is encountered on the Sun. The fact that this range of states is observed to be correlated with other stellar properties provides an opportunity to explore the processes *underlying* coronal activity in a variety of circumstances. This work promises new insight into the origin of magnetic activity, and the global electrodynamics of the Sun and stars.

Although it has long been thought that magnetic fields play the central role in the atmospheric electrodynamics of active stars, it is only in recent years that the first tentative direct measurements of magnetic fields have been made. Almost all positive detections of magnetic fields in late-type stars have been based on the study of *photospheric* fields using a method first described by Robinson (1980), in which the Zeeman signature of the intensity (not the polarization) profile of a magnetically sensitive spectral line is identified by comparing it with a similar line having a different magnetic sensitivity.

Applications of the method (e.g. Marcy, 1984; Saar, 1988) have provided estimates of the flux, field strength and surface area covered by magnetic fields in a number of selected active stars. Field strengths of 100–500 mT have been reported, with coverage in the range 20–80 % (smaller coverages cannot be detected). There are indications (Saar & Schrijver, 1987) that the field strengths vary from star to star in accord with the variation of the magnetostatic limit B_p (Eqn. (6.4)), so that stars with larger photospheric pressures, such as the cooler dwarf stars, have higher field strengths. Equally, stars with low photospheric pressures (i.e. giants) may have weak magnetic fields which would be hard to detect, even if the total flux were quite large. The inferred area coverage in the active dwarf stars is enormous when compared with that of the Sun, and it is the change in fractional area, f, rather than the change in field intensity, B, that accounts for the larger part of the star to star variation in the total magnetic flux, $\langle fB \rangle$.

Of considerable interest to the question of the origin of stellar coronae is the discovery by Saar & Schrijver (1987) that the total magnetic flux is correlated with the flux density of X-rays, F_x, over a sample of late-type stars that includes the Sun. The correlation takes the form of a power law,

$$F_x \sim \langle fB \rangle^{0.9 \pm 0.1}, \tag{6.24}$$

holding over almost 3 orders of magnitude in F_x. This observation provides very strong evidence that magnetic fields are implicated in the formation of stellar coronae and, moreover, places significant constraints on the classes of models that need to be considered in attempts to construct unified models of coronal structure.

In addition to coronal X-ray and radio emission, stars often exhibit other characteristics apparently analogous to solar activity (Zwaan, 1981; Noyes, 1983). Ground-based observations reveal evidence of starspots and flares in the optical continuum, and evidence of chromospheres in the Ca II resonance lines and in Hα. Observations made by the *IUE* satellite have provided data on the behaviour of EUV emission from regions believed to be analogous to the solar chromosphere and transition region. There are rather tight correlations between X-ray emission and the indicators of photospheric, chromospheric and transition zone activity. In particular, stellar X-ray fluxes from a broad sample of stars are related to Ca II emission fluxes and other radiative losses by power laws (Schrijver, 1987). Insofar as the X-ray emission arises in coronal plasma loops, this result points to a common origin of the activity indicators, presumably related to magnetic fields. Equally, the absence of a simple proportionality shows that the simplest geometrical interpretation of the correlations is unlikely to be correct.

The average, maximum and minimum strength of most indicators of stellar activity, including X-rays, varies systematically over large swathes of the Hertzsprung–Russell (HR) diagram. Phenomena similar to solar activity are most commonly detected in stars lying near the cooler end of the main sequence, leading some authors to suggest that solar-like activity is confined to stars later than about F0 (e.g. Böhm-Vitense & Dettmann, 1980). However, the claimed success of Landini & Monsignori Fossi (1981) and Mangeney & Praderie (1984) in unifying observations of X-ray fluxes across much of the HR diagram might indicate that magnetic activity is more widespread than hitherto recognized. Changes in the structure of the stellar atmosphere associated with changes in luminosity, surface gravity, effective temperature or metallicity may modify the observable behaviour of magnetic fields and, in some situations, make them particularly hard to detect.

Nevertheless, it is clear that there is not a unique relation between the position of a star in the HR diagram and its level of (magnetic) activity. This implies that some factor or factors other than the luminosity, effective temperature, mass, radius, surface gravity or metallicity must influence the degree of activity (cf. Rosner *et al.* (1978), Belvedere *et al.* (1981)). There is very strong evidence that the age and the rotation rate of a star are two factors that are particularly potent in this respect, almost certainly through the control they exert over the magnetic fields in the atmospheres of the stars.

Early indications that rotation and age may influence the level of stellar activity were obtained by Wilson (1963, 1964) and Kraft (1967), who noted that strong chromospheric Ca II emission was to be found in stars that are young and/or rotating rapidly. Skumanich (1972) suggested that the activity (as measured by Ca II emission strength) decayed as the inverse square root of the age. A correlation also exists between the strength of EUV and X-ray emission and the rotation rate (e.g. Pallavicini *et al.*, 1982; Walter, 1982). It has been suggested (e.g. Noyes, 1983) that the primary correlation is between activity level and rotation rate, whilst any age dependence arises because the rotation rate itself depends on age. One argument in favour of this proposition is that the highly active RS CVn stars are believed to be fairly old but nevertheless rotating rapidly as a consequence of the tidal coupling and high angular momentum density in the close binary system. On the other hand, the direct evidence for a connection between rotation rate and age among single stars is rather sparse (e.g. Soderblom, 1982). In particular, it is now clear that among active young stars near the lower main sequence there is a complex relation between age and rotation rate (Stauffer, 1987) and the detailed relation between activity and rotation in these objects is not well understood.

There has been considerable discussion regarding the precise measure of stellar rotation that is most closely associated with magnetic activity. For example, one may use either the period (angular velocity) or the surface equatorial speed to characterize the rotation, and use either the total luminosity or the surface flux (possibly normalized to the bolometric energy flow) of coronal or chromospheric emission to characterize the level of activity. Because the various quantities depend not only on activity but also on stellar radius and temperature, it is extremely difficult to isolate the separate factors that directly control activity.

In one fruitful approach to the problem, Noyes (1983) has shown that the chromospheric Ca II surface flux normalized to the stellar brightness, R_{HK}', depends on functions $f(P)$ of the period P and $g(B-V)$ of the colour $(B-V)$, according to

$$\log(R_{HK}') = f(P) + g(B-V), \qquad (6.25)$$

where $f(P) = -0.80 \log(P)$ and $g(B-V)$ can be expressed as a polynomial in $(B-V)$. This dependence can be transformed to an analogous result for coronal emission (and ultimately to loop properties, although this has not yet been done) using the power-law relations derived by Schrijver (1987). The separate dependence on P and $(B-V)$ can be unified by noting that the function $g(B-V)$ may be transformed always to another function $\tau_c(B-V)$ corresponding to the theoretical turnover time of convection at the base of the stellar convection zone, computed using a mixing-length model. Noyes *et al.* (1984) have shown that the normalized chromospheric emission R_{HK}' then depends on the single quantity $R_0 = P/\tau_c$, where R_0 is known as the Rossby number. As we note below, the Rossby number is believed to be a crucially important parameter in the theory of stellar dynamos.

The results discussed above refer to the *mean* level of activity of stars. Over many rotation periods, the activity levels of stars (as indicated by chromospheric Ca II data) vary systematically in a manner analogous to the solar activity cycle (see Baliunas & Vaughan, 1985). There is, however, little if any evidence available at this time for cyclic variations in coronal emission (X-ray or radio) from stars, although the instrumentation needed to make such observations has not yet been dedicated to synoptic observations. Nevertheless, both the solar analogy and the observed correlation between chromospheric and coronal activity observed in samples of stars at a given epoch implies that stellar coronal cycles probably exist, and future observations of the phenomenon should reveal important and informative changes in atmospheric structure or loop properties as a stellar cycle progresses.

6.3.5   *Theory of stellar magnetism*

We have seen that solar activity is always intimately connected to the behaviour of magnetic fields, and that plasma loops in the solar corona are fundamentally an electrodynamic phenomenon. It is believed that stellar activity is also connected with the behaviour of magnetized plasma, and there are strong indications that the major part of the radiation from stellar coronae is often due to electrodynamic processes. However, despite a great deal of productive work on the problem we are far from a satisfactory picture of the physical nature of stellar magnetism and its consequences such as plasma loops in stellar coronae. The aim of this section is to review the patterns of thought that currently guide studies of stellar magnetism and its connection with stellar coronal structure.

In Section 6.1.1 we outlined aspects of current views regarding the origin of solar magnetic fields. We saw that three key processes appear to be significant: regenerative dynamo action, presumably as a result of the interaction between turbulent convection and differential rotation; the mutual exclusion of magnetic fields and turbulent convective cells which is a feature of magnetoconvection at high magnetic Reynolds number; and the emergence of the dynamo-generated, convectively-structured magnetic flux into the atmosphere, perhaps as a consequence of magnetic buoyancy.

The first tentative steps towards a theory of stellar magnetic fields and activity have naturally attempted to include the important features of these processes, although the subject is by no means as advanced as the study of solar magnetism. In particular, a pioneering model developed by Durney, Mihalas & Robinson (1981), Robinson & Durney (1982), Durney & Robinson (1982) and Durney (1988) has proved to be useful in illuminating some of the more significant aspects of the observations described above. The model is based on a simplified description of the generation and regeneration processes in an elementary dynamo. Let

$$\mathbf{B}_P = \nabla \times (0, 0, A) \tag{6.26}$$

be the poloidal magnetic field determined by the poloidal vector potential $(0, 0, A)$ in spherical polar coordinates. Then the magnetic induction Eqn. (6.1) can be written in the form

$$\frac{\partial A}{\partial t} = \alpha B_\phi + \eta \nabla^2 A \tag{6.27}$$

$$\frac{\partial B_\phi}{\partial t} = \nabla \Omega \times \nabla (Ar \sin \theta) + \eta \nabla^2 B_\phi. \tag{6.28}$$

These equations describe the production of poloidal from toroidal field by the α-effect and of toroidal from poloidal field by the differential rotation $\Omega(r, \theta, \phi)$, as well as dissipation by ohmic losses (regarded as turbulent and not coulombic).

This system can be simplified without excluding the basic physical processes by approximating the field in terms of its value at the base of the convection zone, $r = r_c$, and adopting a simple latitudinal planiform for the field:

$$A = A_c \exp^{ikr_c\theta}; \quad B_\phi = B_c \exp^{ikr_c\theta}. \tag{6.29}$$

With these assumptions, the coupled dynamo equations become

$$\frac{\partial A}{\partial t} = \alpha B_\phi - \eta k^2 A \tag{6.30}$$

$$\frac{\partial B_\phi}{\partial t} = ikr_c \frac{\Delta_L \Omega}{L} - \eta k^2 B_\phi - \frac{B_\phi u}{L}. \tag{6.31}$$

Here, L is the scale height at the base of the convection zone and $\Delta_L \Omega$ is the shear in the rotational velocity profile over a radial scale of L. The term $-B_\phi u/L$ has been added to represent the possibility that magnetic buoyancy may carry toroidal magnetic flux away from the dynamo region.

The behaviour of the model described by these equations is controlled by four physical factors: the generation of field components by the α-effect (α) and differential rotation $(\Delta_L \Omega)$, and the removal of field components by ohmic dissipation (η) and by buoyancy (Bu/L). The influence of the first three factors may be characterized by the non-dimensional dynamo number, N_D, defined by

$$N_D^2 = (\alpha)\left(\frac{\Delta_L \Omega}{L}\right)\left(\frac{1}{\eta^2}\right)\left(\frac{r_c}{k^3}\right). \tag{6.32}$$

A larger dynamo number signifies a more potent dynamo and presumably a larger magnetic flux. For the Sun it is believed that $N_D \approx 1$. Durney & Robinson (1982) suggest that the terms α and $\Delta_L \Omega$ scale as

$$\alpha \propto L^2 \Omega/r_c, \tag{6.33}$$

$$\Delta_L \Omega \propto \Omega(L/r_c)^2, \tag{6.34}$$

so that

$$N_D^2 \propto \frac{\Omega^2}{\eta^2} f(r_c, L, k). \tag{6.35}$$

This model thus reveals a clear connection between the rate of rotation of the star and its dynamo-generated magnetic field. Calculations reported by

Durney & Robinson (1982) imply that for stars of given differential rotation (i.e. $\Omega\Delta\Omega$ = constant), the dynamo number and the predicted magnetic field strength in the stellar convection zone will increase markedly as one considers main-sequence stars of lower mass (i.e. later spectral types). A marked increase in the magnetic field strength is also predicted as one considers shorter rotation periods at fixed mass. While it is difficult to make contact in detail between the observational data and such a simple theory, Durney & Robinson (1982) claim that '...the predictions of the theory and the observations are clearly in good provisional agreement...'.

It is important to note that the dynamo number, N_D, and the Rossby number, R_O, are likely to be closely related. The former represents the interplay between α-effect, differential rotation and turbulent dissipation, while the latter relates the rotation rate to the convective turnover time. Although the connections between α-effect, turbulent diffusion and differential rotation are not fully understood at present, Noyes *et al.* (1984) suggest that the physical basis of the two dimensionless quantities will lead to the approximate relation

$$N_D R_O = 1 \tag{6.36}$$

(in our notation). Since Noyes *et al.* (1984) have demonstrated the importance of R_O as a unifying feature of *observations* of stellar activity, it is reasonable to conclude that further work along the lines suggested by the model of Durney and his co-workers is likely to be fruitful.

Nevertheless, it must also be stressed that we are far from a satisfactory theory of magnetic fields in stars. In addition to the problems already exposed in our discussions of the theory of solar magnetic fields, there are further complications. These include the question of the patterns taken up by magnetic fields that penetrate the stellar atmosphere (i.e. the spatial correlation function of the photospheric magnetic flux), the photospheric field strength maintained by magnetostatic and other forces, and the fascinating problems that must arise in the most active stars where observations indicate that a substantial part of the entire stellar surface ($>10\%$) is occupied by strong magnetic fields.

The dynamo theory outlined above represents one important part of a complete theory of stellar activity. The other major part concerns the question of the electrodynamic coupling between the interior of the star and the active phenomena that take place in the atmosphere. As we have seen, the main features of such coupling are (i) the production of a Poynting flux of electromagnetic energy as a consequence of the interaction between turbulent convection and the magnetic field in the stellar photosphere and subphotosphere, and (ii) the dissipation of this flux as heat and kinetic energy in the outer atmosphere.

The production of a Poynting flux is characterized by Eqns. (6.10) or (6.15) and (6.16). Although the photospheric twisting velocity v_ϕ and magnetic field B would be expected to vary from star to star as the luminosity, surface gravity and photospheric pressure and temperature vary, it seems that the major part of the variation is likely to come from changes in the fractional area of the stellar surface covered by magnetic field, f (often called the filling factor). In the same way that studies of the Sun have shown repeatedly that sufficient Poynting flux to power that corona can be generated by plausible values of v_ϕ, B and f (or the area A of active regions), it may also be shown that the energy losses from the outer layers of active stars can be met by plausible values of these parameters.

Although little work has yet been done on the subject of the dissipation of electrodynamic energy flows in stellar atmospheres, it appears that a major part of the star-to-star variation of activity may be ascribed simply to changes in f or A, i.e., in the area covered by magnetic field rather than the field strength in the photosphere. Changes in the efficiency of the electrodynamic coupling may also be important (e.g. Mullan, 1984). The observation that *power law* correlations exist between different indicators of activity represents an important clue that should assist the development of this aspect of the subject.

6.3.6 *Prospects*

Plasma loops in the solar corona are fascinating and important phenomena. We have seen that they exhibit remarkably complex and diverse characteristics, and that significant progress has been made towards a scientific understanding of their behaviour. A particularly important outcome of work on plasma loops has been the recognition of their crucial role in *unifying the observational picture* of the solar corona. Equally, the *utility of a global theory* of the behaviour of loops is now abundantly clear.

We have also seen that analogues of plasma loops in the solar corona exist in the coronae of stars. The stellar loops apparently bear many similarities to solar loops, but occur on a far more impressive scale. The development of a theory of stellar coronal loops has helped to unify our understanding of the connections between rotation, convection, atmospheric structure and internal evolution in the stars.

Magnetic fields occur widely in the cosmos: they are not restricted to stars. For example, our own galaxy is permeated by a large-scale magnetic field which, although weak by comparison with fields in stellar atmospheres, is certainly sufficiently strong to significantly influence the dynamical and physical state of the dust, gas and cosmic rays that comprise the interstellar medium. Other spiral galaxies also exhibit large-scale magnetic fields, and there is some evidence that the magnetic field lines 'balloon out' into the intergalactic medium in a possible analogue of coronal plasma loops (Ruzmaikin, Sokoloff & Shurkurov, 1988). The presence of polarized synchrotron emission from the so-called 'radio galaxies' indicates the existence of enormous magnetic structures, and there is growing evidence of a considerable level of fine structure in the radio lobes of these objects.

Near the centre of our galaxy, there exists an assembly of intriguing structures that may indicate the presence of close analogues of coronal loops. Among these structures, the so-called 'continuum arc' exhibits filamentary fine structure reminiscent of high-quality photographs of solar loops (Yusef-Zadeh, Morris & Chance, 1984), and several authors have speculated on possible parallels between the two kinds of structure. The remarkable morphological similarities between coronal loops and the galactic centre arc may be apparent also in some of the fine-scale alignments seen in the neighbourhood of certain galactic supernova remnants.

The past decade has witnessed a rapid improvement in our understanding of the observational aspects of solar and stellar coronal loops, accompanied by significant developments in our theoretical concepts. Of course, much more work remains to be done. The growth of a prolific and fruitful synergy between solar and stellar studies of coronal plasma loops may be the starting-point for a deeper and more complete understanding of the behaviour of cosmic magnetic fields.

References

Ambruster, C.W., Sciortino, S. & Golub, L. (1987). Rapid, low-level X-ray variability in active late-type dwarfs. *Astrophysical Journal Supplement Series*, **65**, 273–305.

Antiochos, S.K. & Noci, G. (1986). The structure of the static corona and transition region. *Astrophysical Journal*, **301**, 440–7.

Antiochos, S.K., Haisch, B.M. & Stern, R.A. (1986). On the dividing line for stellar coronae. *Astrophysical Journal* (*Letters*), **307**, L55–9.

Antiochos, S.K. (1987). The topology of force-free magnetic fields and its implications for coronal activity. *Astrophysical Journal*, **312**, 886–94.

Arter, W. (1983). Magnetic-flux transport by a convecting layer – topological, geometrical and compressible phenomena. *Journal of Fluid Mechanics*, **132**, 25–48.

Athay, R.G. (1982). Responses of transition region models to magnetic field geometry and downflow velocities. *Astrophysical Journal*, **263**, 982–6.

Ayres, T.R., Linsky, J.L., Vaiana, G.S., Golub, L. & Rosner, R. (1981). The cool half of the H–R diagram in soft X-rays. *Astrophysical Journal*, **250**, 293–9.

Babcock, H.W. (1961). The topology of the sun's magnetic field and the 22-year cycle. *Astrophysical Journal*, **133**, 572–82.

Baliunas, S.L. & Vaughan, A.H. (1985). Stellar activity cycles. *Annual Review of Astronomy and Astrophysics*, **23**, 379–412.

Bastian, T.S. & Bookbinder, J.A. (1987). First dynamic spectra of stellar microwave flares. *Nature*, **326**, 678–80.

Belvedere, G., Chiuderi, C. & Paternó, L. (1981). Stellar X-ray emission as a consequence of magnetic activity. *Astronomy and Astrophysics*, **96**, 369–72.

Böhm-Vitense, E. & Dettmann, T. (1980). The boundary line in the H–R diagram for stellar chromospheres and the theory of convection. *Astrophysical Journal*, **236**, 560–6.

Brants, J.J. & Steenbeek, J.C.M. (1985). Morphological evolution of an emerging flux region. *Solar Physics*, **96**, 229–52.

Bray, R.J. & Loughhead, R.E. (1964). *Sunspots*. London: Chapman and Hall.

Bray, R.J. & Loughhead, R.E. (1974). *The Solar Chromosphere*. London: Chapman and Hall.

Bray, R.J., Loughhead, R.E. & Durrant, C.J. (1984). *The Solar Granulation* (2nd ed.). Cambridge: Cambridge University Press.

Brown, A. & Jordan, C. (1981). The chromosphere and corona of Procyon (α CMi, F5 IV–V). *Monthly Notices of the Royal astronomical Society*, **196**, 757–79.

Browning, P.K., Sakurai, T. & Priest, E.R. (1986). Coronal heating in close-packed flux tubes: a Taylor–Heyvaerts relaxation theory. *Astronomy and Astrophysics*, **158**, 217–27.

Byrne, P.B. & Rodonò, M. (1983). Activity in red dwarf stars. *IAU Colloquium 71*. Dordrecht: D. Reidel Publishing Company.

Catura, R.C., Acton, L.W. & Johnson, H.M. (1975). Evidence for X-ray emission from Capella. *Astrophysical Journal (Letters)*, **196**, L47–9.

Cameron, A.C. & Robinson, R.D. (1989). Fast Hα variations on a rapidly rotating cool main sequence star – I. Circumstellar clouds. *Monthly Notices of the Royal astronomical Society*, **236**, 57–87.

Chiuderi, C. (1983). The role of magnetic fields in the heating of stellar atmospheres. In *Solar and Magnetic Fields: Origins and Coronal Effects*, IAU Symposium 102, ed. J.O. Stenflo, pp. 375–86. Dordrecht: D. Reidel Publishing Company.

Cram, L.E. & Kuhi, L.V. (1989). *FGK Stars and T Tauri Stars*. NASA SP-502. Washington: USA Government Printing Office.

Dere, K.P., Bartoe, J.-D.F. & Brueckner, G.E. (1983). Chromospheric jets: possible extreme-ultraviolet observation of spicules. *Astrophysical Journal (Letters)*, **267**, L65–8.

Dulk, G.A. (1985). Radio emission from the sun and stars. *Annual Review of Astronomy and Astrophysics*, **23**, 169–224.

Dunn, R.B. & Zirker, J.B. (1973). The solar filigree. *Solar Physics*, **69**, 281–304.

Durney, B.R. (1988). A simple dynamo model and the anisotropic α-effect. *Astronomy and Astrophysics*, **191**, 374–80.

Durney, B.R., Mihalas, D. & Robinson, R.D. (1981). A preliminary interpretation of stellar chromospheric Ca II emission variations within the framework of stellar dynamo theory. *Publications of the Astronomical Society of the Pacific*, **93**, 537–43.

Durney, B.R. & Robinson, R.D. (1982). On an estimate of the dynamo-generated magnetic fields in late-type stars. *Astrophysical Journal*, **253**, 290–7.

Durney, B.R. & Sofia, S. (1987). *The Internal Solar Angular Velocity*. Dordrecht: D. Reidel Publishing Company.

Feigelson, E.D. (1987). Microwave observations of non-thermal phenomena in pre-main sequence stars. In Linsky & Stencel (1987), pp. 455–65.

Feigelson, E.D. & DeCampli, W.M. (1981). Observations of X-ray emission from T Tauri stars. *Astrophysical Journal (Letters)*, **234**, L89–93.

Feigelson, E.D. & Kriss, G.A. (1989). Soft X-ray observations of pre-main-sequence stars in the Chamaeleon Dark Cloud. *Astrophysical Journal*, **338**, 262–76.

Feldman, U., Doschek, G.A. & Mariska, J.T. (1979). On the structure of the solar transition zone and lower corona. *Astrophysical Journal*, **229**, 369–74.

Gahm, G.F. (1980). X-ray observations of T Tauri stars. *Astrophysical Journal (Letters)*, **242**, L163–6.

Gaizauskas, V., Harvey, K.L., Harvey, J.W. & Zwaan, C. (1983). Large-scale patterns formed by solar active regions during the ascending phase of cycle 21. *Astrophysical Journal*, **265**, 1056–65.

Galloway, D.J. & Moore, D.R. (1979). Axisymmetric convection in the presence of a magnetic field. *Astrophysical and Geophysical Fluid Dynamics*, **12**, 73–106.

Galloway, D.J. & Weiss, N.O. (1981). Convection and magnetic fields in stars. *Astrophysical Journal*, **243**, 945–53.

Gary, D.E. & Linsky, J.L. (1981). First detection of non-flare microwave emission from the coronae of single late-type dwarf stars. *Astrophysical Journal*, **250**, 284–92.

Giacconi, R., Branduardi, G., Briel, Y., Epstein, A. & 27 others (1979). The *Einstein* (*HEAO 2*) X-ray Observatory. *Astrophysical Journal*, **230**, 540–50.

Giampapa, M.S., Golub, L., Peres, G., Serio, S. & Vaiana, G.S. (1985). Closed coronal structures. VI. Far-ultraviolet and X-ray emission from active late-type stars and the applicability of coronal loop models. *Astrophysical Journal*, **289**, 203–12.

Gibson, D.M. (1983). Quiescent and flaring radio emission from dMe stars. In Byrne & Rodonò (1983), pp. 273–86.

Giovanelli, R.G. (1980). An exploratory two-dimensional study of the coarse structure of network magnetic fields. *Solar Physics*, **68**, 49–69.

Giovanelli, R.G. & Jones, H.P. (1982). The three-dimensional structure of atmospheric magnetic fields in two active regions. *Solar Physics*, **79**, 267–78.

Gold, T. (1963). Magnetic energy shedding in the solar atmosphere. In *AAS-NASA Symposium on the Physics of Solar Flares*, ed. W.N. Hess, pp. 389–95, NASA SP-50. Washington: National Aeronautics and Space Administration.

Golub, L., Maxson, C., Rosner, R., Serio, S. & Vaiana, G.S. (1980). Magnetic fields and coronal heating. *Astrophysical Journal*, **238**, 343–8.

Golub, L., Harnden, Jr., F.R., Pallavicini, R., Rosner, R. & Vaiana, G.S. (1982). *Einstein* detection of X-rays from the Alpha Centauri system. *Astrophysical Journal*, **253**, 242–7.

Gondoin, P., Mangeney, A. & Praderie, F. (1987). Solar-type giants: new X-ray detections from *EXOSAT* observations. *Astronomy and Astrophysics*, **174**, 187–96.

Haisch, B.M. (1983). X-ray observations of stellar flares. In Byrne & Rodonò (1983), pp. 255–72.

Haisch, B.M. (1987). The coronal dividing line. In Linsky & Stencel (1987), pp. 269–82.

Haisch, B.M., Linsky, J.L., Bornmann, P.L., Stencel, R.E. & 3 others (1983). Coordinated *Einstein* and *IUE* observations of a *Disparitions Brusque* type flare event and quiescent emission from Proxima Centauri. *Astrophysical Journal*, **267**, 280–90.

Haisch, B.M., Linsky, J.L., Lampton, M., Paresce, F. & 2 others (1977). Extreme-ultraviolet observations of a flare on Proxima Centauri and implications concerning flare–star scaling theory. *Astrophysical Journal* (*Letters*), **213**, L119–23.

Hall, D.S. (1975). The RS CVn binaries and binaries with similar properties. In

Multiple Periodic Variable Stars: IAU Colloquium 29, ed. W.S. Fitch, pp. 287–348. Dordrecht: D. Reidel Publishing Co.

Hartmann, L., Dupree, A.K. & Raymond, J.C. (1982). Ultraviolet observations of stellar chromospheric activity. *Astrophysical Journal*, **252**, 214–29.

Heise, J., Brinkman, A.C., Schrijver, J., Mewe, R. & 3 others (1975). Evidence for X-ray emission from flare stars observed by *ANS*. *Astrophysical Journal (Letters)*, **202**, L73–6.

Hénoux, J.C. & Somov, B.V. (1987). Generation and structure of the electric currents in a flaring activity complex. *Astronomy and Astrophysics*, **185**, 306–14.

Heyvaerts, J. & Priest, E.R. (1984). Coronal heating by reconnection in dc current systems: a theory based on Taylor's hypothesis. *Astronomy and Astrophysics*, **137**, 63–78.

Holt, S.S., White, N.E., Becker, R.H., Boldt, E.A. & 3 others (1979). X-ray line emission from Capella. *Astrophysical Journal (Letters)*, **234**, L65–8.

Howard, R. & LaBonte, B.J. (1980). The Sun is observed to be a torsional oscillator with a period of 11 years. *Astrophysical Journal (Letters)*, **239**, L33–6.

Hurlburt, N.E. & Toomre, J. (1988). Magnetic fields interacting with non-linear compressible convection. *Astrophysical Journal*, **327**, 920–32.

Ionson, J.A. (1982). Resonant electrodynamic heating of stellar coronal loops: an *LRC* circuit analogy. *Astrophysical Journal*, **254**, 318–34.

Ionson, J.A. (1983). Electrodynamic coupling in magnetically confined X-ray plasmas of astrophysical origin. *Astrophysical Journal*, **271**, 778–92.

Ionson, J.A. (1984). A unified theory of electrodynamic coupling in coronal magnetic loops: the coronal heating problem. *Astrophysical Journal*, **276**, 357–68.

Ionson, J.A. (1985). Coronal heating by resonant (ac) and non-resonant (dc) mechanisms. *Astronomy and Astrophysics*, **146**, 199–203.

Kahn, S.M., Linsky, J.L., Mason, K.O., Haisch, B.M. & 3 others (1979). *HEAO* 1 observations of X-ray emission from flares on dMe stars. *Astrophysical Journal (Letters)*, **234**, L107–11.

Kan, J.R., Akasoufu, S.-I. & Lee, L.C. (1983). A dynamo theory of solar flares. *Solar Physics*, **84**, 153–67.

Kalkofen, W., Rosner, R., Ferrari, A. & Massaglia, S. (1986). The equilibrium structure of thin magnetic flux tubes. *Astrophysical Journal*, **304**, 519–25.

Kraft, R.P. (1967). Studies of stellar rotation. V. The dependence of rotation on age among solar-type stars. *Astrophysical Journal*, **150**, 551–70.

Krause, F. & Rädler, K.-H. (1980). *Mean-Field Magnetohydrodynamics and Dynamo Theory*. Oxford: Pergamon Press.

Kuperus, M. (1983). Electrodynamics of the outer solar atmosphere. *Space Science Reviews*, **34**, 47–54.

Landini, M. & Monsignori Fossi, B.C. (1981). Coronal loops in the sun and in the stars. *Astronomy and Astrophysics*, **102**, 391–400.

Landini, M., Monsignori Fossi, B.C., Paresce, F. & Stern, R.A. (1985). Extreme-ultraviolet emission from cool star outer atmospheres. *Astrophysical Journal*, **289**, 709–20.

Landini, M., Monsignori Fossi, B.C., Pallavicini, R. & Piro, L. (1986). *EXOSAT* detection of an X-ray flare from the solar type star π^1 UMa. *Astronomy and Astrophysics*, **157**, 217–22.

Leighton, R.B., Noyes, R.W. & Simon, G.W. (1962). Velocity fields in the solar atmosphere, I. Preliminary report. *Astrophysical Journal*, **135**, 474–99.

Linsky, J.L. & Haisch, B.M. (1979). Outer atmospheres of cool stars. I. The sharp division into solar-type and non-solar-type stars. *Astrophysical Journal (Letters)*, **229**, L27–32.

Linsky, J.L. & Stencel, R.E. (1987). Cool stars, stellar systems, and the Sun. *Lecture Notes in Physics*, **291**. Berlin: Springer-Verlag.

Livi, S.H.B., Wang, J. & Martin, S.F. (1985). The cancellation of magnetic flux. I. On the quiet Sun. *Australian Journal of Physics*, **38**, 855–73.

Low, B.C. (1981). Eruptive solar magnetic fields. *Astrophysical Journal*, **251**, 352–63.

Low, B.C. (1982). Magnetostatic atmospheres with variations in three dimensions. *Astrophysical Journal*, **263**, 952–69.

Low, B.C. (1987). Electric current sheet formation in a magnetic field induced by continuous magnetic footpoint displacements. *Astrophysical Journal*, **323**, 358–67.

Mangeney, A. & Praderie, F. (1984). The influence of convection and rotation on X-ray emission in main sequence stars. *Astronomy and Astrophysics*, **130**, 143–50.

Marcy, G. (1984). Observations of magnetic fields on solar-type stars. *Astrophysical Journal*, **276**, 286–304.

Mewe, R., Gronenschild, E.H.B.M., Westergaard, N.J., Heise, J. & 6 others (1982). X-ray spectrum of Capella and its relation to coronal structure and ultraviolet emission. *Astrophysical Journal*, **260**, 233–9.

Mewe, R., Heise, J., Gronenschild, E.H.B.M., Brinkman, A.C., Schrijver, J. & den Boggende, A.J.F. (1975). Detection of X-ray emission from stellar coronae with *ANS*. *Astrophysical Journal*, **202**, L67–71.

Mewe, R. & Schrijver, C.J. (1986). A two-component model for the corona of λ Andromedae. *Astronomy and Astrophysics*, **169**, 178–84.

Moffatt, H.K. (1978). *Magnetic Field Generation in Electrically Conducting Fluids*. Cambridge: Cambridge University Press.

Montmerle, T., Koch-Miramond, L., Falgarone, E. & Grindley, J.E. (1983). *Einstein* observations of the Rho Ophiuchi dark cloud: an X-ray Christmas tree. *Astrophysical Journal*, **269**, 182–201.

Moore, R. & Rabin, D. (1985). Sunspots. *Annual Review of Astronomy and Astrophysics*, **23**, 239–66.

Mullan, D.J. (1984). On the possibility of resonant electrodynamic coupling in the coronae of red dwarfs. *Astrophysical Journal*, **282**, 603–11.

Mutel, R.L., Lestrade, J.F., Preston, R.A. & Phillips, R.B. (1985). Dual polarization VLBI observations of stellar binary systems at 5 GHz. *Astrophysical Journal*, **289**, 262–8.

Mutel, R.L. & Lestrade, J.L. (1985). Radio emission from RS CVn binaries. I. VLA survey and period-radio luminosity relationship. *Astrophysical Journal*, **290**, 493–8.

Noyes, R.W. (1983). The relation between rotation and magnetic activity on lower main sequence stars as derived from chromospheric indicators. In *Solar and Magnetic Fields: Origins and Coronal Effects*, IAU Symposium 102, ed. J.O. Stenflo, pp. 133–47. Dordrecht: D. Reidel Publishing Co.

Noyes, R.W., Hartmann, L.W., Baliunas, S.L., Duncan, D.K. & Vaughan, A.H. (1984). Rotation, convection, and magnetic activity in lower main-sequence stars. *Astrophysical Journal*, **279**, 763–77.

Nugent, J. & Garmire, G. (1978). X-rays from α Centauri. *Astrophysical Journal (Letters)*, **226**, L38–85.

Pallavicini, R. (1988). *EXOSAT* observations of quiescent and flaring emission from M dwarf flare stars. In *Activity in Cool Star Envelopes*, eds. O. Haynes *et al.*, pp. 249–52. Dordrecht: Kluwer Academic Publishers.

Pallavicini, R.P., Golub, L., Rosner, R., Vaiana, G.S., Ayres, T. & Linsky, J.L. (1981). Relations among stellar X-ray emission observed from *Einstein*, stellar rotation and bolometric luminosity. *Astrophysical Journal*, **248**, 279–90.

Pallavicini, R., Willson, R.F. & Lang, K.R. (1985). Microwave observations of late-type stars with the Very Large Array. *Astronomy and Astrophysics*, **149**, 95–101.

Parker, E.N. (1955a). The formation of sunspots from the solar toroidal field. *Astrophysical Journal*, **121**, 491–507.

Parker, E.N. (1955b). Hydromagnetic dynamo models. *Astrophysical Journal*, **122**, 293–314.

Parker, E.N. (1975). The generation of magnetic fields in astrophysical bodies. X. Magnetic buoyancy and the solar dynamo. *Astrophysical Journal*, **198**, 205–9.

Parker, E.N. (1979). *Cosmical Magnetic Fields*. Oxford: Clarendon Press.

Parker, E.N. (1983a). Absence of equilibrium among close-packed twisted flux tubes. *Geophysical and Astrophysical Fluid Dynamics*, **23**, 85–102.

Parker, E.N. (1983b). Magnetic neutral sheets in evolving fields. I. General theory. *Astrophysical Journal*, **164**, 635–41.

Parker, E.N. (1984). Depth of origin of solar active regions. *Astrophysical Journal*, **280**, 423–7.

Parker, E.N. (1987a). Magnetic reorientation and the spontaneous formation of tangential discontinuities in deformed magnetic fields. *Astrophysical Journal*, **318**, 876–87.

Parker, E.N. (1987b). The dynamical oscillation and propulsion of magnetic fields in the convective zone of a star. I. General considerations. *Astrophysical Journal*, **312**, 868–79.

Parker, E.N. (1988a). The dynamical oscillation and propulsion of magnetic fields in the convective zone of a star. IV. Eruption to the surface. *Astrophysical Journal*, **325**, 880–90.

Parker, E.N. (1988b). Nanoflares and the solar X-ray corona. *Astrophysical Journal*, **330**, 474–9.

Pizzo, V.J. (1986). Numerical solution of the magnetostatic equations for thick flux tubes, with application to sunspots, pores, and related structures. *Astrophysical Journal*, **302**, 785–808.

Pneuman, G.W. & Kopp, R.A. (1977). Downflow of spicular material and transition region models. *Astronomy and Astrophysics*, **55**, 305–6.

Pneuman, G.W., Solanki, S.K. & Stenflo, J.O. (1986). Structure and merging of solar magnetic fluxtubes. *Astronomy and Astrophysics*, **154**, 231–42.

Pope, T. & Mosher, J. (1975). Evidence for non-radial fields in the Sun's photosphere and a possible explanation of the polar magnetic signal. *Solar Physics*, **44**, 3–12.

Proctor, M.R.E. & Weiss, N.O. (1982). Magnetoconvection. *Reports on Progress in Physics*, **45**, 1317–79.

Reimers, D. (1989). Observations of the chromospheres, coronae and winds of F, G and K stars. In Cram & Kuhi (1989).

Robinson, R.D. (1980). Magnetic field measurements on stellar sources: a new method. *Astrophysical Journal*, **239**, 961–7.

Robinson, R.D. & Durney, B.R. (1982). On the generation of magnetic fields in late-type stars: a local time-dependent dynamo model. *Astronomy and Astrophysics*, **108**, 322–5.

Rosner, R., Tucker, W.H. & Vaiana, G.S. (1978). Dynamics of the quiescent solar corona. *Astrophysical Journal*, **220**, 643–65.

Rosner, R., Golub, L. & Vaiana, G.S. (1985). On stellar X-ray emission. *Annual Review of Astronomy and Astrophysics*, **23**, 413–52.

Rosner, R., Low, B.C. & Holtzer, T.E. (1986). Physical processes in the solar corona. In *Physics of the Sun II. The Solar Atmosphere*, ed. P.A. Sturrock, pp. 135–80. Dordrecht: D. Reidel Publishing Co.

Ruzmaikin, A., Sokoloff, D. & Shurkurov, X. (1988). Magnetism of spiral galaxies. *Nature*, **336**, 341–7.

Saar, S.H. & Schrijver, C.J. (1987). Empirical relations between magnetic fluxes and atmospheric radiative losses for cool dwarf stars. In Linsky & Stencel (1987), pp. 38–40.

Saar, S.H. (1988). Improved methods for the measurement and analysis of stellar magnetic fields. *Astrophysical Journal*, **324**, 441–65.

Schmitt, J.H.M.M., Harnden, F.R., Peres, G., Rosner, R. & Serio, S. (1985). The X-ray corona of Procyon. *Astrophysical Journal*, **288**, 751–5.

Schmitt, J.H.M.M., Pallavicini, R., Monsignori Fossi, B.C. & Harnden, Jr., F.R. (1987). A comparison of coronal X-ray emission observed with the *Einstein* and *EXOSAT* observatories. *Astronomy and Astrophysics*, **179**, 193–201.

Schrijver, C.J. (1987). Magnetic structure in cool stars. XI. Relations between radiative fluxes measuring stellar activity, and evidence for two components in stellar chromospheres. *Astronomy and Astrophysics*, **172**, 111–23.

Sheeley, N.R. (1981). The overall structure and evolution of active regions. In *Solar Active Regions*, Skylab Solar Workshop III, ed. F.Q. Orrall, pp. 17–43. Boulder: Colorado Associated University Press.

Sheeley, N.R., Jr., Nash, A.G. & Wang, Y.-M. (1987). The origin of rigidly rotating magnetic field patterns on the Sun. *Astrophysical Journal*, **319**, 481–502.

Simon, G.W., Title, A.M., Topka, K.P. & 4 others (1988). On the relation between photospheric flow fields and the magnetic field distribution on the solar surface. *Astrophysical Journal*, **327**, 964–7.

Slee, O.B., Nelson, G.J., Stewart, R.T., Wright, A.E. & 3 others (1987). A microwave survey of southern active stars. *Monthly Notices of the Royal astronomical Society*, **229**, 659–77.

Soderblom, D.R. (1982). Rotational studies of late-type stars. I. Rotational velocities of solar-type stars. *Astrophysical Journal*, **263**, 239–51.

Skumanich, A. (1972). Time scales for Ca II emission decay, rotational braking and lithium depletion. *Astrophysical Journal*, **171**, 565–7.

Spicer, D.S. (1983). Coronal heating and photospheric boundary conditions. *Solar Physics*, **88**, 43–9.

Spicer, D.S., Mariska, J.T. & Boris, J.P. (1986). Magnetic energy storage and conversion in the solar atmosphere. In *Physics of the Sun II. The Solar Atmosphere*, ed. P.A. Sturrock, pp. 181–248. Dordrecht: D. Reidel Publishing Co.

Spruit, H.C. (1976). Pressure equilibrium and energy balance of small photospheric fluxtubes. *Solar Physics*, **50**, 269–95.

Spruit, H.C. (1981). Magnetic flux tubes. In *The Sun as a Star*, ed. S.D. Jordan, pp. 385–412. NASA SP-50. Washington: NASA Information Branch.

Spruit, H.C. & Roberts, B. (1983). Magnetic flux tubes on the Sun. *Nature*, **304**, 401–6.

Stauffer, J.R. (1987). Rotational velocity evolution on and prior to the main sequence. In Linsky & Stencel (1987), pp. 182–91.

Steiner, O., Pneuman, G.W. & Stenflo, J.O. (1986). Numerical models for solar magnetic fluxtubes. *Astronomy and Astrophysics*, **170**, 126–37.

Steinolfson, R.S. & Tajima, T. (1987). Energy buildup in coronal magnetic fluxtubes. *Astrophysical Journal*, **322**, 503–11.

Stenflo, J.O. (1984). Fine-scale structure of solar magnetic fields. *Advances in Space Research*, **4**(8), 5–16.

Stern, R.A., Underwood, J.H. & Antiochos, S.K. (1983). A giant X-ray flare in the Hyades. *Astrophysical Journal* (*Letters*), **264**, L55–9.

Stern, R.A., Antiochos, S.K. & Harnden, Jr., F.R. (1986). Modeling of coronal X-ray emission from active cool stars. I. Hyades cluster. *Astrophysical Journal*, **305**, 417–32.

Strauss, H.R. & Otani, N.F. (1988). Current sheets in the solar corona. *Astrophysical Journal*, **326**, 418–24.

Sturrock, P.A. & Uchida, Y. (1981). Coronal heating by stochastic magnetic pumping. *Astrophysical Journal*, **246**, 331–6.

Swank, J.H., White, N.E., Holt, S.S. & Becker, R.H. (1981). Two-component X-ray emission from RS Canum Venaticorum binaries. *Astrophysical Journal*, **246**, 208–14.

Topka, K., Fabricant, D., Harnden, Jr., F.R., Gorenstein, P. & Rosner, R. (1979). Detection of soft X-rays from α Lyrae and η Bootis with an imaging X-ray telescope. *Astrophysical Journal*, **229**, 661–8.

Topka, K. & Marsh, K.A. (1982). Detection of microwave emission from both components of the red dwarf binary EQ Pegasi. *Astrophysical Journal*, **254**, 641–5.

Tucker, W.H. (1973). Heating of solar active regions by magnetic energy dissipation: the steady-state case. *Astrophysical Journal*, **186**, 285–9.

Ulmschneider, P. (1979). Stellar chromospheres. *Space Science Reviews*, **24**, 71–100.

Underhill, A.B. & Michalitsianos, A.G. (eds) (1985). *The Origin of Non-radiative Energy and Momentum in Hot Stars*, NASA CP-2358. Washington: US Government Printing Office.

Vaiana, G.S., Cassinelli, J.P., Fabbiano, G., Giacconi, R. & 12 others (1981). Results from an extensive *Einstein* stellar survey. *Astrophysical Journal*, **244**, 163–82.

van Ballegooijen, A.A. (1985). Electric currents in the solar corona and the existence of magnetostatic equilibrium. *Astrophysical Journal*, **298**, 421–30.

Vrabec, D. (1973). Streaming magnetic features near sunspots. In *Chromospheric Fine Structure*, IAU Symposium 56, ed. R.G. Athay, pp. 201–31. Dordrecht: D. Reidel Publishing Co.

Walter, F.M. (1982). On the coronae of rapidly rotating stars III. An improved coronal rotation-activity relation in late-type dwarfs. *Astrophysical Journal*, **253**, 745–51.

Walter, F.M. (1986). X-ray sources in regions of star formation. I. The naked T Tauri stars. *Astrophysical Journal*, **306**, 573–86.

Walter, F., Charles, P. & Bowyer, S. (1978). X-ray emission from UX Arietis: RS Canum Venaticorum systems as a class of coronal X-ray sources. *Astrophysical Journal (Letters)*, **225**, L119–22.

Walter, F.M., Cash, W., Charles, P.A. & Bowyer, C.S. (1980). X-rays from RS Canum Venaticorum systems: *HEAO I* survey and the development of a coronal model. *Astrophysical Journal*, **236**, 212–18.

Walter, F.M. & Kuhi, L.V. (1981). The smothered coronae of T Tauri stars. *Astrophysical Journal*, **250**, 254–61.

Walter, F.M., Gibson, D.M. & Basri, G.S. (1983). First observations of stellar coronal structure: the coronae of AR Lacertae. *Astrophysical Journal*, **267**, 665–81.

Whitehouse, D.R. (1985). Coronal heating and stellar flares. *Astronomy and Astrophysics*, **145**, 449–50.

Wilson, O.C. (1963). A probable correlation between chromospheric activity and age in main-sequence stars. *Astrophysical Journal*, **138**, 832–48.

Wilson, O.C. (1964). Eclipses by extended atmospheres. In *Stellar Atmospheres*, ed. J.L. Greenstein, pp. 346–64. Chicago: University of Chicago Press.

Wilson, P.R. (1987). Solar cycle workshop. *Solar Physics*, **110**, 1–9.

Yusef-Zadeh, F., Morris, M. & Chance, D. (1984). Large, highly-organised radio structures near the galactic centre. *Nature*, **310**, 557–61.

Zel'dovich, Ya. B. & Ruzmaikin, A.A. (1983). Dynamo problems in astrophysics. *Astrophysics and Space Science Reviews*, **2**, 333–83.

Zirin, H. (1975). The helium chromosphere, coronal holes and stellar X-rays. *Astrophysical Journal (Letters)*, **199**, L63–6.

Zwaan, C. (1978). On the appearance of magnetic flux in the solar photosphere. *Solar Physics*, **60**, 213–40.

Zwaan, C. (1981). Stellar structure and activity. In *Solar Phenomena in Stars*

and Stellar Systems, eds. R.M. Bonnet & A.K. Dupree, pp. 463–86.
Dordrecht: D. Reidel Publishing Co.

Zwaan, C. (1987). Elements and patterns in the solar magnetic field. *Annual Review of Astronomy and Astrophysics*, **25**, 83–111.

Zweibel, E.G. & Li, He-Sheng (1987). The formation of current sheets in the solar corona. *Astrophysical Journal*, **312**, 423–30.

Additional notes

1. The geometrical reconstruction technique devised and applied by R. E. Loughhead and colleagues has been used by A. B. Delone *et al.* (*Soviet Astronomy Letters*, **15**, 110–11, 1989) to determine the geometrical and kinematical characteristics of an Hα active region loop, including shape, tilt, height, and velocities and accelerations. Axial flow speeds obtained by the Russian workers closely agree with the Australian results displayed in Fig. 2.8 (Sections 2.3.3 and 2.3.4).

2. Excellent photographs of hot active region loops taken in λ637.4, λ530.3 and the continuum at the 1980 Kenya eclipse have been published by Y. Hanaoka and H. Kurokawa (*Publications of the Astronomical Society of Japan*, **40**, 369–82, 1988). They analysed the observations to obtain electron densities and absolute intensities as functions of height (Sections 3.2 and 3.6.2). New observations in soft X-rays have been reported by B. M. Haisch *et al.* (*Astrophysical Journal Supplement Series*, **68**, 371–405, 1988); the loop properties studied include time variability, height, length, light curves, physical conditions, emission measure and total SXR luminosity (Sections 3.4 and 3.6.4).

3. A particularly clear example of a radio loop has been illustrated by D. E. Gary *et al.* (*Astrophysical Journal*, **355**, 321–8, 1990), observed near 8.5 GHz with the VLA (Sections 3.5 and 3.6.5).

4. A major advance in soft X-ray imaging technique has been reported by L. Golub *et al.* (*Nature*, **344**, 842–4, 1990), namely the use of normal incidence multi-layer-coated optics instead of the grazing incidence telescopes used previously. A resolution of better than 1″ of arc was achieved during a rocket flight in 1989. Loop structure *within* flare ribbons was detected in addition to the usual loops connecting the ribbons (Sections 3.4, 4.5.1, 4.5.2, and Frontispiece).

5. A detailed study of the relationship between Hα and SXR loops throughout the course of three flares has been carried out by Z. Mouradian *et al.* (*Astronomy and Astrophysics*, **224**, 267–83, 1989). The X-ray images were obtained with the Hard X-ray Imaging Spectrometer on SMM (Sections 4.3 and 4.5.2).

6. By comparing the theoretical ratio of the microwave and hard X-ray emission for a non-thermal thick target model with observational results from no less than 51 flares, E. T. Lu and V. Petrosian (*Astrophysical Journal*, **338**, 1122–30, 1989) have concluded that this model provides the best fit to the data (Sections 4.5.3, 4.6, 4.7.5, 4.8.4 and 4.8.5).

7. There are a number of fairly recent reviews of various aspects of flares not cited in the text. Firstly, E. L. Chupp and A. B. C. Walker have edited a volume of the journal

Solar Physics (**118**, 1–371, 1988) containing several extensive reviews of various high-energy aspects of solar flares, including HXR, γ-ray lines and continuum, microwave, neutron emission and future instrumentation (Sections 4.5.3, 4.5.4, 4.6, 4.7.5, 4.8.4 and 4.8.5). The same journal has an entire volume devoted to solar and stellar flares, edited by B. M. Haisch and M. Rodono (*Solar Physics*, **121**, 1–502, 1989; *IAU Colloquium 104*) (Chapters 4 and 6). A review of high-resolution observations of flares observed in Hα has been published by H. Kurokawa (*Space Science Reviews*, **51**, 49–84, 1989). This paper discusses pre-flare activity and temporal relationships between Hα and HXR and microwave emission (Sections 4.2, 4.3, 4.7.2 and 4.7.3). A review of SXR and HXR observations of flares, with particular emphasis on the results from Hinotori, has been prepared by S. Tsuneta (In *Activity in Cool Star Envelopes*, ed. O. Havnes *et al.*, pp. 325–40, 1988. Dordrecht: Kluwer). This paper also describes the Solar-A satellite, due for launch in 1991, with improved instruments compared to their counterparts on Hinotori, SMM and Skylab (Sections 4.5, 4.7.5, 4.8.4 and 4.9). Finally B. R. Dennis *et al.* (*NASA Technical Memorandum* No. 4036, 1988) has compiled an event listing of the HXR bursts detected with the HXRBS instrument on SMM from its launch in February 1980 to December 1987 – more than 8600 events (Section 4.5.3).

8. Skylab data (1973–4) continue to attract detailed analysis, an example of which is an excellent study of EUV flare loops by J. G. Doyle and K. G. Widing (*Astrophysical Journal*, **352**, 754–9, 1990). Four prominent loops in an arcade were photographed; properties studied include 'nesting', expansion, morphology, location of footpoints, altitude, aspect ratio and physical conditions (Sections 4.4 and 4.8.3).

9. The extrapolation of linear force-free fields from measurements of the vector magnetic field in the photosphere has been discussed by S. Cuperman, L. Ofman and M. Semel (*Astronomy and Astrophysics*, **216**, 265–77, 1989; **227**, 227–34, 1990). The theory of force-free fields has been extended in a discussion of the helicity and energy of non-linear fields by M. A. Berger (*Astronomy and Astrophysics*, **201**, 355–61, 1988), and in a generalization of the minimum energy principle by A. M. Dixon, M. A. Berger, P. K. Browning and E. R. Priest (*Astronomy and Astrophysics*, **225**, 156–66, 1989) (Section 5.2.2).

10. The radiative loss function has been re-evaluated using coronal abundances (which appear to differ from the photospheric abundances previously employed) by J. W. Cook, C.-C. Cheng, V. L. Jacobs and S. K. Antiochus (*Astrophysical Journal*, **338**, 1176–83, 1989). These give generally higher values. Of more significance, though, is the shape of the curve at lower temperatures where Lα is formed. R. G. Athay (*Astrophysical Journal*, **308**, 975–81, 1986) gives a series of estimates for different optical thicknesses (Sections 5.3.2, 5.4.2, 5.9).

11. In an effort to produce more cool plasma emission from hot loops D. T. Woods, T. E. Holzer and K. B. MacGregor (*Astrophysical Journal*, **355**, 295–320, 1990) have examined models in which local heating and not a conductive flux gradient balances the radiative losses. They emphasize the importance of the treatment of Lα losses but were unable to construct a model with mechanical heating which would produce the observed emission measure curve at low temperatures. P. S. Cally (*Astrophysical Journal*, **355**, 693–9, 1990), on the other hand, was able to obtain good agreement in models with a lowered temperature gradient achieved by replacing the classical thermal conductivity with a much higher turbulent conductivity based on the mixing-length model (Section 5.4.2).

12. More realistic models of magnetic flux loops have been analysed for stability by R. M. Lothian and A. W. Hood (*Solar Physics*, **122**, 227–44, 1989), who considered the shape of twisted tubes, and by T. Chiueh and E. G Zweibel (*Astrophysical Journal*, **338**, 1158–70, 1989), who allowed for interactions between different current loops. For simple twisted tubes with line-tying, both M. Velli, G. Einaudi and A. W. Hood (*Astrophysical Journal*, **350**, 428–36, 1990) and B. J. Foote and I. J. D. Craig (*Astrophysical Journal*, **350**, 437–51, 1990) find the linear growth rates of the kink mode to be small. The non-linear development was followed by I. J. D. Craig and A. D. Sneyd (*Astrophysical Journal*, **357**, 653–61, 1990) to a new kinked equilibrium. Little energy was lost in the process, though small-scale regions were produced, where resistive effects might lead to larger energy release (Sections 5.6.1 and 5.6.2).

13. The resonant absorption of Alfvén waves in an axisymmetric straight cylinder has been simulated numerically by S. Poedts, M. Goosens and W. Kerner (*Solar Physics*, **123**, 83–115, 1989) and by H. R. Strauss and W. S. Lawson (*Astrophysical Journal*, **346**, 1035–40, 1989). The former authors confirm that Alfvén wave absorption is efficient under solar coronal conditions, much more so than slow-mode absorption. The latter, however, argue that the time scale for the formation of a resonance layer is too long to account for coronal loop heating (Section 5.6.3).

14. An extensive discussion of flare loop evolution through evaporation, steady and condensation phases with reference to scaling laws has been provided by G. H. Fisher and S. L. Hawley (*Astrophysical Journal*, **357**, 243–58, 1990) (Section 5.8.1). Another simulation, using an electron beam as the heating source, is described by J. T. Mariska, A. G. Emslie and P. Li (*Astrophysical Journal*, **341**, 1067–74, 1989).

15. According to P. Amendt and G. Benford (*Astrophysical Journal*, **341**, 1082–7, 1989) the problem of spreading the heat released in a narrow layer of a loop to the surrounding region may be resolved by cross-field transport caused by turbulent ion-cyclotron waves (Section 5.9).

16. A new way of exploring stellar coronae has been developed by D. Mullan, E. Sion, F. Bruhweiler and K. Carpenter (*Astrophysical Journal Letters*, **339**, L33–L36, 1989), who found evidence of a cool coronal wind in the UV absorption spectrum of V471 Tau, a K2 star having a DA2 white dwarf companion. The conjecture that the inferred high mass loss rate ($10^{11} M_\odot$ yr^{-1}) is magnetically driven is related to the problem of the stability of coronal magnetic loops. D. Reimers, R. Baade and K.-P. Schroeder (*Astronomy and Astrophysics*, **227**, 133–40, 1990) have reported a similar discovery of a highly ionized wind and a corona in HR6902, a G9II star which eclipses its B-star companion. This system is particularly significant since it lies on the 'hot' side of the coronal dividing line, and has a warm corona (87000 K) and a compact chromosphere (Sections 6.3.1 and 6.3.3).

17. J. Lemen, R. Mewe, C. Schrijver and A. Fludra (*Astrophysical Journal*, **341**, 474–83 and 484–92, 1989) have described *Exosat* observations of X-ray emission from Capella, σ^2 CrB and Procyon. The observations are consistent with two-temperature coronal models (0.6 and 3 MK in Procyon, 5 and 25 MK in the other two stars). The authors suggest that these results might indicate the presence of 3 components in stellar coronae: a 'basal' cool component having a temperature < 1 MK, 'cool' (3–5 MK) and 'compact' loop systems in which the loop radius may expand by a factor of 25 between the footpoint and the apex, and 'hot' (25 MK) loop systems having smaller expansion. L. Pasquine, J. Schmitt and R. Pallavicini (*Astronomy and Astrophysics*,

226, 225–34, 1989) have described *Exosat* observations of 17 RS CVn stars which also show that multi-temperature coronal models are needed to fit the spectra. Moreover, the form of the emission measure distribution implies a different temperature stratification to that in the Sun, and might indicate that the coronae of RS CVn stars have structures different from solar coronal loops (Section 6.3.3).

18. *Exosat* observations of flares in stellar coronae have been reported by R. Pallavicini *et al.* (*Astronomy and Astrophysics*, **227**, 483–9, 1990) and R. Pallavicini, G. Tagliaferri and L. Stella (*Astronomy and Astrophysics*, **228**, 403–25, 1990). The data reveal a wide variety of light curves, little evidence for 'microvariability', and a tight correlation between the quiescent X-ray luminosity and the bolometric luminosity. This correlation suggests that for dwarf flare stars, there might be a correlation between the generation of magnetic fields and the general processes of energy generation and transfer in the star. Optical observations of a flare on AD Leo have been interpreted by G. Fisher and S. Hawley (*Astrophysical Journal*, **357**, 243–58, 1990) using a simple time-dependent model of a coronal loop. The results allow the authors to suggest a pattern for coronal evolution during the flare (Sections 6.3.3, 6.3.5).

Name index

Subject index